Achieving sustainable cultivation of temperate zone tree fruits and berries

Volume 1: Physiology, genetics and cultivation

It is widely recognised that agriculture is a significant contributor to global warming and climate change. Agriculture needs to reduce its environmental impact and adapt to current climate change whilst still feeding a growing population, i.e. become more 'climate-smart'. Burleigh Dodds Science Publishing is playing its part in achieving this by bringing together key research on making the production of the world's most important crops and livestock products more sustainable.

Based on extensive research, our publications specifically target the challenge of climate-smart agriculture. In this way we are using 'smart publishing' to help achieve climate-smart agriculture.

Burleigh Dodds Science Publishing is an independent and innovative publisher delivering high quality customer-focused agricultural science content in both print and online formats for the academic and research communities. Our aim is to build a foundation of knowledge on which researchers can build to meet the challenge of climate-smart agriculture.

For more information about Burleigh Dodds Science Publishing simply call us on +44 (0) 1223 839365, email info@bdspublishing.com or alternatively please visit our website at www.bdspublishing.com.

Related titles:

Achieving sustainable cultivation of temperate zone tree fruits and berries Volume 2: Case studies
Print (ISBN 978-1-78676-212-2) Online (ISBN 978-1-78676-214-6; 978-1-78676-215-3)

Achieving sustainable cultivation of apples
Print (ISBN 978-1-78676-032-6) Online (ISBN 978-1-78676-034-0; 978-1-78676-035-7)

Achieving sustainable cultivation of mangoes
Print (ISBN 978-1-78676-132-3) Online (ISBN 978-1-78676-134-7; 978-1-78676-135-4)

Achieving sustainable cultivation of tomatoes
Print (ISBN 978-1-78676-040-1) Online (ISBN 978-1-78676-042-5; 978-1-78676-043-2)

Achieving sustainable cultivation of tree nuts
Print (ISBN 978-1-78676-224-5) Online (ISBN 978-1-78676-226-9; 978-1-78676-227-6)

Chapters are available individually from our online bookshop:
https://shop.bdspublishing.com

climate•SMART•publishing
IN AGRICULTURAL SCIENCE

BURLEIGH DODDS SERIES IN AGRICULTURAL SCIENCE
NUMBER 53

Achieving sustainable cultivation of temperate zone tree fruits and berries

Volume 1: Physiology, genetics and cultivation

Edited by Professor Gregory A. Lang,
Michigan State University, USA

burleigh dodds
SCIENCE PUBLISHING

Published by Burleigh Dodds Science Publishing Limited
82 High Street, Sawston, Cambridge CB22 3HJ, UK
www.bdspublishing.com

Burleigh Dodds Science Publishing, 1518 Walnut Street, Suite 900, Philadelphia, PA 19102-3406, USA

First published 2019 by Burleigh Dodds Science Publishing Limited

Library of Congress Control Number: 2019941166

British Library Cataloguing in Publication Data
A catalogue record for this book is available from the British Library

ISBN 978-1-78676-208-5 (Print)
ISBN 978-1-78676-211-5 (PDF)
ISBN 978-1-78676-210-8 (ePub)
ISSN 2059-6936 (print)
ISSN 2059-6944 (online)

DOI 10.19103/AS.2018.0040.1

Typeset by Deanta Global Publishing Services, Chennai, India

Contents

Series list

Title	Series number
Achieving sustainable cultivation of maize - Vol 1 From improved varieties to local applications Edited by: Dr Dave Watson, CGIAR Maize Research Program Manager, CIMMYT, Mexico	001
Achieving sustainable cultivation of maize - Vol 2 Cultivation techniques, pest and disease control Edited by: Dr Dave Watson, CGIAR Maize Research Program Manager, CIMMYT, Mexico	002
Achieving sustainable cultivation of rice - Vol 1 Breeding for higher yield and quality Edited by: Prof. Takuji Sasaki, Tokyo University of Agriculture, Japan	003
Achieving sustainable cultivation of rice - Vol 2 Cultivation, pest and disease management Edited by: Prof. Takuji Sasaki, Tokyo University of Agriculture, Japan	004
Achieving sustainable cultivation of wheat - Vol 1 Breeding, quality traits, pests and diseases Edited by: Prof. Peter Langridge, The University of Adelaide, Australia	005
Achieving sustainable cultivation of wheat - Vol 2 Cultivation techniques Edited by: Prof. Peter Langridge, The University of Adelaide, Australia	006
Achieving sustainable cultivation of tomatoes Edited by: Dr Autar Mattoo, USDA-ARS, USA & Prof. Avtar Handa, Purdue University, USA	007
Achieving sustainable production of milk - Vol 1 Milk composition, genetics and breeding Edited by: Dr Nico van Belzen, International Dairy Federation (IDF), Belgium	008
Achieving sustainable production of milk - Vol 2 Safety, quality and sustainability Edited by: Dr Nico van Belzen, International Dairy Federation (IDF), Belgium	009
Achieving sustainable production of milk - Vol 3 Dairy herd management and welfare Edited by: Prof. John Webster, University of Bristol, UK	010
Ensuring safety and quality in the production of beef - Vol 1 Safety Edited by: Prof. Gary Acuff, Texas A&M University, USA & Prof. James Dickson, Iowa State University, USA	011
Ensuring safety and quality in the production of beef - Vol 2 Quality Edited by: Prof. Michael Dikeman, Kansas State University, USA	012
Achieving sustainable production of poultry meat - Vol 1 Safety, quality and sustainability Edited by: Prof. Steven C. Ricke, University of Arkansas, USA	013
Achieving sustainable production of poultry meat - Vol 2 Breeding and nutrition Edited by: Prof. Todd Applegate, University of Georgia, USA	014

Acknowledgements

We wish to acknowledge the following for their help in reviewing particular chapters:

- Chapter 8: Prof. Guglielmo Costa, University of Bologna, Italy
- Chapter 11: Prof. Dan Cooley, University of Massachusetts, USA; and Emeritus Prof. David Rosenberger, Cornell University, USA
- Chapter 12: Dr James Mattheis, USDA-ARS Wenatchee, USA

Introduction

Temperate fruits include stone/drupe fruits (such as peach), pome fruits (such as apple) and berries (such as strawberries). Like other crops, cultivation of these fruits faces a number of challenges. These include the need to optimize yields, sensory and nutritional quality; the dynamic threats from biotic and abiotic stresses in a changing climate; the increasing cost and decreasing availability of labour; and the need for more efficient use of resources to minimise environmental impact. Volume 1 of *Achieving sustainable cultivation of temperate zone tree fruits and berries* summarises the wealth of research addressing these challenges, focusing on key advances across the value chain, from breeding improved varieties to developing better techniques for cultivation and crop protection.

This volume reviews the latest research in tree fruit physiology, as well as discussing recent developments in genetics and their implications for improved breeding techniques. The volume also provides comprehensive coverage of key stages in cultivation, from nursery plants to water, nutrient and pest management.

Part 1 Physiology and genetics

The first part of the volume discusses current research on physiology and genetics. Chapter 1 is focussed on advances in understanding fruit tree root-rhizosphere relationships that enhance plant health. Host-microbe interactions in the rhizosphere influence numerous processes that determine plant productivity and health. The rhizo-microbiome influences functions ranging from protection of the plant from pathogen attack to enhanced nutrient availability and uptake. The chapter reviews the development of 'omics' technologies as a way to explore the effects of the rhizosphere microbiome on plant interactions with their environment. The chapter provides two case studies demonstrating ways to manipulate the rhizosphere microbiome to optimize crop production on orchard replant sites.

Moving from the soil to the root system, Chapter 2 discusses advances in development and utilization of rootstocks for fruit tree cultivation. In the previous century, apple rootstocks were chosen on the basis of suitability for a particular climate, required degree of vigour and plant availability, with only one or two choices available to growers in a particular location. However, with the proliferation of improved apple rootstocks, there is now a dizzying array of choices for apple growers. Advances in the development and utilization of fruit tree rootstocks are considered using apple as a case study. It starts by reviewing the use of clonal rootstocks, followed by sections on rootstock improvement, propagation and evaluation. Rootstock effects on the scion and rootstock tolerance to both abiotic and biotic stresses are then considered. The chapter concludes with a discussion on trends in apple rootstock use. Although this chapter focuses mostly on apple, it exemplifies the various inherent rootstock properties and rootstock-induced effects on grafted scions that can be found in other temperate fruit species.

Adopting a broader focus, Chapter 3 looks at advances in understanding fruit tree growth in general. Owing to the importance of photosynthesis for the efficient functioning of plants as solar energy collectors, scientists have been intensively studying the process of photosynthesis for more than a century, with the hope of increasing its efficiency. There is substantial evidence that fruit trees distribute their nutrient resources to maximize photosynthetic competency, and adjust the angles of leaves in different parts of their canopy to optimize use of resources and sunlight as it passes through the canopy. The chapter examines the factors that influence photosynthesis in fruit trees, as well as the key principles governing the distribution and utilization of photosynthates. Finally, the chapter looks at the features of fruit tree canopy architecture which affect photosynthesis efficiency.

Chapter 4 examines advances in understanding reproductive development in fruit-bearing plants. In recent years, knowledge of the reproductive physiology of fruit trees has increased through numerous molecular and genetic studies. Beginning with an introductory overview, the chapter includes sections on the physiology and genetic analysis of reproductive meristem development, bud dormancy and flowering, sexual determination and development of flower structures. The physiology of pollination, fertilization and gametogenesis is discussed. The chapter concludes with sections on fruit development and phytohormones and, lastly, fruit maturation and ripening.

The first section of the volume concludes with Chapter 5 on advances in fruit genetics. Changing weather patterns, innovative production systems, increased pest and pathogen pressure, and fluctuating consumer preferences are forcing fruit producers to adjust at a higher pace than ever before. The chapter provides an overview of the advances in fruit genetics as they pertain to discoveries in fruit quality and disease resistance, and their application to the improvement of fruit production and sustainability. Genomic mapping, marker-assisted breeding and biotechnology are also explored, as are the genetics of fruit quality and disease resistance.

Part 2 Optimizing sustainable cultivation and quality

The second part of the volume reviews advances in temperate zone fruit cultivation from orchard design to optimizing water and nutrient management and harvesting operations. The subject of Chapter 6 is optimizing the production of quality nursery plants for fruit tree cultivation. It is now possible to purchase specific nursery tree types that are more compatible with an intended training system, improving planting efficiency and encouraging early production. Although single axis, well-feathered scions suitable for high-density spindle training systems have dominated nursery production over the past two decades, some nurseries increasingly produce double axis scions for pear, apple, and even stone fruits, helping growers obtain uniform and early-producing fruiting walls. The chapter summarizes current knowledge of the physiology of grafting and recent innovations in fruit tree nursery techniques. New approaches to regulating nursery plant growth and development are discussed, including *in vitro* propagation, which is becoming a more powerful technique for producing rootstocks and self-rooted varieties of several fruit species. The chapter covers scion grafting and budding techniques as well as the biochemical and physiological aspects of graft incompatibility. The chapter concludes with a section on tree production in the nursery.

Chapter 7 moves to tree fruit nutrition, focussing on optimizing precision in orchard irrigation and nutrient management. Precise resource management in fruit production systems has become a key area of study over recent years, not only due to the changing climate but also due to increased demand for fruit from a growing population as part of a healthy diet. Irrigated fruit production has proven to be both reliable and more efficient than rain-fed production, and the trend is towards high efficiency irrigated systems. The chapter examines both precision management of irrigation and nutrients in orchards (mainly apple, pear, sweet cherry and peach). Precision management of irrigation is detailed, including the use of efficient irrigation systems, matching water supply to plant demand, reducing evaporation from the soil surface and responding to drought. A further section on nutrient management covers both ferti-irrigation or fertigation, in which nutrients are delivered in irrigation water, and foliar application, in which nutrients are applied directly to the fruit tree in dilute water sprays.

The subject of Chapter 8 is the optimization of plant growth, yield and fruit quality with plant bioregulators (PBRs). Both natural and synthetic plant bioregulators are used extensively in tree fruit production. They influence many processes in plants, including shoot growth, branch angle, bud break, flower bud formation, and fruit abscission, ripening, shape and finish. As a result, they can be used to optimize not only plant growth but also fruit yield and quality. The chapter starts with a classification of PBRs, followed by a detailed discussion on the application of PBRs and their use in the development and maintenance of tree structure, including the control of vegetative growth. The chapter reviews the use of PBRs in crop load management and their influence on flowering and fruit set, concluding with sections on both the preharvest application of PBRs and their use in improving fruit shape and appearance.

Moving from plant growth to fruit production, Chapter 9 examines the challenge of optimizing fruit production efficiencies via mechanization. Many technological advances in farming have been adopted over the last century, transforming various tasks. However, some processes in fruit and vegetable production such as harvesting have changed little, and crops such as for fresh market apples, cherries, peaches, blueberries and strawberries are harvested manually, an operation that is highly labour intensive, demanding a huge seasonal workforce in rural areas around the world. Starting with an introductory section on challenges faced in tree fruit harvesting, the chapter goes on to discuss harvest aid (or labour assist) machines (or platforms), which as the name suggests, are designed to assist human fruit pickers. The chapter also covers mass and robotic harvesting systems, which are designed to remove the need for manual picking, with specific examples included. The chapter also discusses fruit conveyance and handling, as well as mechanical pruning.

Chapter 10 shifts to the theme of fruit pests and the importance of optimizing pest management in fruit cultivation. Tree fruit growers in temperate regions have faced challenges in managing the complex of arthropod pests present in these systems because of increasing pesticide regulatory restrictions, public concerns about food safety and environmental quality, and resistance to older pesticides by key pests. Growers are attempting to adopt newer tactics such as reduced-risk (more selective) pesticides and increased reliance on sampling and monitoring techniques plus pheromone-based mating disruption, but these are more expensive and require more time and precise use patterns. In addition, most previous integrated pest management protocols were designed for control programs based on the use of older conventional pesticides. As this chapter shows, a more current and practical approach stresses a basic understanding of pest biology and development, which can be used in predictive models to support the effective use of more selective crop protectants and other sustainable tactics, such as mating disruption and biological control.

Chapter 11 shifts the focus from pest to disease management in fruit cultivation, and discusses optimizing plant protection from diseases. Sustainable management of plant diseases is a perennial challenge for pome, stone, and berry fruit producers in humid regions throughout the world. Delayed orchard profitability, decreased orchard longevity, reduction in fruit quality, and unsalable fruit caused by fungal and bacterial diseases result in devastating economic losses annually. Globally, tree fruit growers are increasingly relying on an integrated management approach through the incorporation of predictive pathogen life-cycle modeling and biological, chemical, and cultural control methods to mitigate disease-associated losses. The chapter provides an introduction to general tree fruit production management paradigms, followed by a discussion of the main pre-harvest fungal and bacterial diseases, methods of integrated disease management, and advances in technology for pathogen detection and disease management.

The challenges of prolonging the life of harvested fruit are reviewed in Chapter 12, including pre- and post-harvest strategies to optimize fruit quality and shelf-life. The adoption of new, non-destructive quality assessment technologies is important for quantifying, evaluating, and managing fruit quality in postharvest systems. The chapter examines the challenge of managing biological variance to improve the success of postharvest technologies, and describes ways of measuring harvest maturity and improving sorting operations, as well as postharvest technologies and temperature control. The chapter includes a detailed case study of post-harvest strategies in cherry cultivation in British Columbia.

The volume's final chapter, Chapter 13, addresses research on bioactive/nutraceutical compounds in fruit that optimize human health benefits. In the past decade, numerous reports demonstrate that high consumption of fruits and vegetables is beneficial for health, being associated with a reduced risk of degenerative diseases. The chapter provides a brief description of the chemistry of bioactive compounds (BCs), including polyphenols, carotenoids, and Vitamin C, their presence in temperate fruits, production practices that influence BC synthesis, and discusses recent advances in strategies toward improving sustainable crop production for nutraceuticals.

Part 1

Physiology and genetics

Chapter 1

Advances in understanding tree fruit-rhizosphere microbiome relationships for enhanced plant health

Mark Mazzola and Shashika S. Hewavitharana, USDA-ARS, USA

1 Introduction

The plant microbiome influences a spectrum of outcomes in terms of plant growth, development, resilience to abiotic stress and overall health and productivity. Contribution of the microbiome in directing plant responses to its environment and the instrumental role of host species or genotype in determining functional composition of the microbiome have been established (Berendsen et al., 2012; Mendes et al., 2013). The rhizosphere microbiome, in particular, may regulate important physiological and developmental processes that will directly determine plant productivity, some of which may be described as unanticipated. For instance, Panke-Buisse et al. (2015) demonstrated that the rhizosphere microbiome can modulate flowering time of a plant host and that flowering phenotype can be altered through manipulation of the rhizosphere microbiome. The rhizosphere microbiome clearly influences other qualities such as plant nutrient status through various processes such as nitrogen fixation, phosphorous acquisition via mycorrhizal fungi (Gianinazzi et al., 2010) and uptake of trace elements (Lemanceau et al., 2009).

There exists expansive study of the rhizosphere microbiome as it relates to plant health and soil-borne disease suppression. The microbial community

http://dx.doi.org/10.19103/AS.2018.0040.01
Published by Burleigh Dodds Science Publishing Limited, 2019.

inhabiting the rhizosphere can serve as a first line of plant defense against deleterious organisms, including plant pathogens, that inhabit the soil system. Historically, exploitation of microbial resources to address yield-limiting aspects of a crop production system has relied upon the introduction of specific microbes possessing a known functional trait (Raaijmakers and Mazzola, 2016). Interest in understanding how the indigenous soil microbiome can be managed to the benefit of plant production systems is of rather recent occurrence (Mazzola and Freilich, 2017). However, although there exists abundant information regarding a limited number of taxonomic groups that function in the rhizosphere to limit disease development, a vast portion of this microbial resource remains to be defined. There exists a need to not only identify composition of the microbiome in the rhizosphere environment, but also establish the role these microbes may have as determinants of plant health.

Various tactics may be implemented that strategically amplify functional components of the soil microbiome with the goal of subsequent crop plant rhizosphere colonization. Such strategies commonly involve the application of specific soil inputs that select for microbial consortia or traits that act to suppress soil-borne plant pathogens (Mazzola et al., 2015; Inderbitzin et al., 2017). Alternatively, management practices such as altering plant spatial patterns in an orchard system (Rumberger et al., 2007) or maintenance of a living tree row understory (Wardle et al., 2001; Hoagland et al., 2008) have been employed to expose the plant to a soil microbiome that will either directly improve plant production or minimize potential exposure to pathogenic microorganisms. Successful use of this approach will clearly rely upon intimate knowledge of microbial groups regulating the desired plant response and a highly directed protocol that amplifies the targeted functional group.

Significant interest also resides in potential to use the crop plant itself as a means to recruit a rhizosphere microbiome that will optimize crop performance. Composition of the rhizosphere microbial community may differ in a plant genotype-dependent manner (Micallef et al., 2009), and this selection may operate to the level of determining genotype of a functional microbial attribute. For instance, wheat (*Triticum aestivum*) cultivar influenced both the density and genotype of the 2,4-diacetylphloroglucinol (2,4-DAPG)-producing fluorescent *Pseudomonas* spp. population detected in the rhizosphere (Mazzola et al., 2004). Production of 2,4-DAPG by fluorescent pseudomonads contributes to the biological disease control potential of these bacteria towards a diversity of plant pathogens (Weller et al., 2007). Drivers determining plant recruitment of the rhizosphere microbiome are likely multifaceted including root architecture, metabolic composition of root exudates (Micallef et al., 2009) and multitrophic interactions that occur in this habitat. Composition of the apple (*Malus domestica*) rhizosphere microbiome also was shown to differ in a rootstock genotype-dependent manner (Reed and Mazzola, 2015). An understanding of the factors that influence differential composition of the apple rhizosphere microbiome and its effect on tree health could be utilized to develop improved rootstocks with enhanced capacity to withstand biotic and abiotic stresses.

2 Contribution of the microbiome in directing plant responses to their environment

2.1 Role of host species or genotype in determining rhizosphere microbiome functional composition

Plant fitness and crop productivity are a function of the plant as such and its attendant microbiota, which in total comprise the plant holobiont (Vandenkoornhuyse et al., 2015). All plant tissues act as host to microbial communities and support distinct consortia that interact to optimize plant function. For instance, the rhizosphere, phyllosphere and endosphere provide unique ecological niches to host different microbial communities that are adapted to environmental and nutritional conditions available. As these microbiomes are highly dependent on the plant genome, they form a second genome or pan-genome (Turner et al., 2013).

The most adjacent layer of soil influenced by plant roots via processes such as rhizodeposition of exudates, mucilage and sloughed cells is defined as the rhizosphere. The plant root may release 5–20% of plant photosynthate into the rhizosphere (Marschner, 1995) with the elevated substrate availability promoting increases in microbial biomass and activity. Interaction between the rhizosphere microbiome and different factors of the host plant is bidirectional. Root exudates are composed of chemically and functionally diverse compounds including organic acids, sugars, amino acids, fatty acids, vitamins, growth factors, hormones and antimicrobials (Bertin et al., 2003). Plant species, cultivars (Mark et al., 2005), age and developmental stage (Micallef et al., 2009) are all determinants of root exudate composition. Root exudates initiate and may modify the interaction between roots and soil microorganisms, and certain metabolites are known to function as signals to prompt symbiotic or pathogenic relationships (Bais et al., 2006; Bertin et al., 2003). The comprehensive root/rhizosphere metabolite profile (metabolome) will also be influenced by the microbiome through microbial transformations or altered exudation patterns. The observation of differential phenolic compound profiles between roots cultivated in field soil and roots cultivated in pasteurized soil indicated a role for the microbiome in altering the metabolic profile from apple roots (Emmett et al., 2014).

Tree fruit root exudates may function to directly promote or limit colonization of the rhizosphere and plant roots. Genotype-specific phenolic concentrations were detected in fine distal roots of apple (Emmett et al., 2014), and a number of compounds were associated with higher or lower root colonization by plant pathogenic fungi and oomycetes. A recent study employed an untargeted metabolomics approach to assess the potential role of apple rootstock genotype-dependent exudates contributing to apple replant disease field tolerance (Leisso et al., 2017). Findings of the study revealed that the four rootstock genotypes assayed (G.935, G.41, M.9Nic29 and M.26) differed significantly based on the water-soluble phenolic rhizodeposits collected from water percolated through rhizosphere planted in pasteurized quartz sand. The G.935 rootstock, which demonstrates higher field tolerance to soil-borne plant pathogens, generated higher levels of benzoic acid in root exudates. Abundance of this metabolite in the tolerant genotype was linked to a higher population of *Burkholderia* spp. relative to susceptible genotypes. Members of this bacterial genus are known for elevated abundance in the rhizosphere of tolerant apple rootstock genotypes and providing biological control against soil-borne plant pathogens (Mazzola et al., 2015; St Laurent et al., 2010). The preference of *Burkholderia* spp. towards benzoic acid as a substrate (Pumphrey and Madsen, 2008) supports the hypothesis of genotype-specific selection of the rhizosphere microbiome. Malic acid is a significant metabolite in apple root exudates and was shown to selectively signal and recruit the plant beneficial rhizobacterium *Bacillus subtilis* FB17 to the rhizosphere of *Arabidopsis thaliana* in a dose-dependent manner (Rudrappa et al., 2008). It could be speculated that differential metabolite production by tree fruit rootstocks may be exploited to selectively recruit or support specific microbial consortia to limit the activity of plant root pathogens.

2.2 Rhizosphere microbiome influence on nutrient acquisition and growth promotion

The capacity of any genotype to acquire essential nutrients from the soil relies essentially on two factors: (1) nutrient interception, which is dependent on root size and architecture and (2) the acquisition process as affected by efficient uptake of available nutrients and retention or loss of plant nutrients through various microbial transformations (Hinsinger et al., 2005; Ismail et al., 2007; Lambers et al., 2006). Several root-structure-related traits may also offer possibilities for improving nutrient acquisition. These include enhancing density or length of root hairs (Gahoonia and Nielsen, 2004), adjusting root to shoot ratio to balance acquisition of limiting nutrients and modifying root architecture to preferentially concentrate root biomass in soil strata containing the maximum amount of a limiting nutrient (Rubio et al., 2003). These plant attributes may be directed or modified by the associated rhizosphere microbiome. For instance, application of certain nitric-oxide-producing *Streptomyces* spp. recovered from orchard soils induced apple seedling lateral root development (Mazzola, unpublished data). Differential abundance in the microbiota functional in nitrogen transformations will yield varied levels of availability; in this regard, it is of interest to note that abundance of the denitrifying bacterium *Steroidobacter denitrificans* was significantly higher in the rhizosphere of apple rootstock M.26 than G.210 (Reed and Mazzola, 2015). Abundance of 16S rDNA sequences of the N-fixing free-living bacterium *Azospirillum brasilense* was 54% higher in the rhizosphere of M.9 than G.11 (Mazzola et al., 2015). Thus, it is plausible that N acquisition efficiency may involve complex interactions between soil/ rhizosphere microorganisms and apple rootstock genotype.

The community of mycorrhiza that colonize a plant has been studied for decades, but the diversity of fungi that form the relationship with a given plant species has only been recognized more recently (Vandenkoornhuyse et al., 2002). Of great importance is knowledge that the capacity of tree fruit roots to form functional mycorrhizal relationships differs with the fungal species or isolate. Infection of apple and effective transport of phosphorous from soil to plant was shown to diverge widely among *Glomus* spp., with some species exhibiting a complete inability to do so (Covey et al., 1981). Similarly, Forge et al. (2001) demonstrated that enhanced apple seedling or rootstock biomass in response to colonization by *Glomus* spp. occurred in a species-specific manner. In addition, apple genotype appears to influence the extent of mycorrhizal colonization. When the microbiome of two apple rootstocks cultivated in a single orchard soil was assessed, notable differences in fungal attributes were observed, including the fact that the glomeromycota comprised less than 1% of the fungal ITS sequences detected in the M.26 rhizosphere, but represented as great as 2–8% of the sequences detected in the G.210 rhizosphere (Reed and Mazzola, 2015).

Modification of the rhizosphere microbiome through targeted introduction of microbial inoculants has long been examined as a means to enhance crop productivity, including tree fruits. Response of tree growth and yield of apple by selected plant-growth-promoting rhizobacteria (PGPR) applied as root inoculants was assessed using two rootstocks (M.9 and MM.106) and two scion cultivars (Granny Smith and Stark Spur Golden) (Aslantas et al., 2007). The PGPR strains utilized exhibited indoleacetic acid (IAA) and cytokinin production as well as phosphate solubilization ability. Among those studied, the isolate BA-8 (*Pseudomonas* sp.) demonstrated the greatest capacity to enhance shoot proliferation. However, all PGPR strains examined in the study increased fruit yield in a manner dependent on rootstock, scion cultivar and PGPR strain. In a field trial, all four bacterial strains independently increased tree fruit yields; however, trees treated with *Pseudomonas* BA-8 produced yields significantly greater than all other treatments. The higher fruit yields in response to BA-8 were associated with significantly higher quantities of IAA and cytokinin produced *in vitro* by this strain (Aslantas et al., 2007), suggesting the possible contribution of exogenous hormone production to the observed growth response.

2.3 Managing the rhizosphere microbiome for enhanced tree health

During the extended lifespan of orchard systems, there exists ongoing transformations of the rhizosphere microbiome that will influence overall productivity of both the current and succeeding orchard. Changes in the soil microbiome and corresponding effects on plant health were demonstrated in studies that employed plant bioassays conducted in soils that had been cropped to apple for zero to five growing seasons. Planting a non-orchard soil to apple (Fuji/M.26) resulted in progressive changes in the soil microbiome over the five growing seasons that were characterized by decreased abundance of growth-promoting rhizobacteria, such as *Burkholderia cepacia* and *Pseudomonas putida*, and concomitant increases in prevalence of *Agrobacterium* and *Variovorax* spp. (Mazzola, 1999). *Paecilomyces* and *Ulocladium*, which have been reported as potential biological disease control agents, were significant components of the fungal community in year 1 and negligible in year 5. In contrast, the plant pathogens *Phytophthora*, *Pythium* and *Rhizoctonia* were not detected in year 1 but together represented approximately 50% of the oomycete/fungal community recovered from apple in year 5. Similar changes in the microbiome subsequently were documented using DNA-sequencing-based protocols (Jiang et al., 2017). This microbial transformation and proliferation of potential soil-borne pathogens has the capacity to reduce productivity of the current orchard (Mazzola et al., 2015). These plant-induced changes in the soil microbiome yield a complex of pathogens that serve as primary impediment to the successful replanting of tree fruits on sites previously planted to the same or closely related species; this phenomenon is termed 'replant disease'.

It has been implied that robust soil systems that are resilient to disturbance, including soil-borne disease, rely on the activity of a soil/rhizosphere microbial community possessing an elevated level of overall diversity (Brussard et al., 2007). As such, organic management practices that lead to soil microbial communities that are more heterogeneous and diverse than those found in conventional systems (Hartmann et al., 2015; Lupatini et al., 2016; Wang et al., 2017) are commonly viewed to possess the potential for resilience when exposed to a disturbance such as introduction of a soil-borne plant pathogen. In coffee (*Coffea arabica*), the rhizosphere microbiome was shown to be more diverse, relative to its bulk soil complement, in orchards under organic or transitional management relative to that cultivated using conventional practices (Caldwell et al., 2015). While correspondence of elevated diversity and enhanced system resilience seems intuitive, increased soil/rhizosphere microbial diversity may not be linked directly to improved productivity or resistance to disease development within a production system. In apple, the potential benefit of organic systems, and perceived elevated rhizosphere microbial diversity, to overall root health is not clear. In a replicated complete block field trial comparing organic and conventional management systems, infection of Fuji/M.9 roots by the oomycete *Pythium* spp. was significantly lower in the organic than conventional blocks, but *Rhizoctonia* spp. root infection was significantly higher in the organic blocks (Mazzola et al., 2002). Certain soil management strategies that reduced overall diversity of the apple rhizosphere microbiome have generated systems that demonstrate enhanced resistance to pathogen invasion and reduced disease incidence (Mazzola et al., 2015; Hewavitharana and Mazzola, 2017). In these orchards, the altered rhizosphere microbiome possessed unique bacterial and fungal profiles which included treatment-specific microbial taxa previously associated with the suppression of plant-pathogenic fungi, oomycetes and nematodes. Thus, in terms of predicting resilience of a soil system to pathogen invasion, it will be of greater value to assess the effect of a management strategy on specific functional attributes of the rhizosphere microbiome rather than merely investigating overall microbial diversity.

Management of replant diseases has traditionally involved the application of pre-plant soil fumigation prior to establishment of the new orchard. Although the method can be effective, the microbiome in fumigated soil systems does not provide long-term suppression of soil-borne plant pathogens and may not universally provide optimal fruit yields (Mazzola et al., 2015). Active management of the rhizosphere microbiome can diminish the activity of orchard replant pathogens and substantially enhance crop yields (Mazzola et al., 2015). Manipulation of the rhizosphere microbiome can be achieved merely by altering the tree row spatial arrangement of the new planting relative to the previous orchard site (Rumberger et al., 2007). Fungal and bacterial community composition in the rhizosphere varied significantly between apple rootstocks cultivated in the old tree row and the same rootstock genotype cultivated in the previous orchard aisle which had been cultivated to grass (Rumberger et al., 2007). These differences were maintained over a period of 2–3 years and were associated with differences in tree growth performance for replant disease-susceptible rootstocks M.7 and M.26. Groundcover management systems have also been explored as a measure to alter the rhizosphere microbiome for improved apple tree performance on replant sites (Hoagland et al., 2008; St Laurent et al., 2010). In both field (Hoagland et al., 2008) and pot trials (St Laurent et al., 2010), altered soil/rhizosphere microbial activity and composition were observed in response to management system. Although differences in tree performance were observed, plant productivity was associated with replant disease severity (St Laurent et al., 2010), but was not correlated with commonly reported indicators of soil quality including soil organic matter content and nutrient availability (Hoagland et al., 2008). Other orchard parameters, including weed competition, likely contributed to the differential tree growth observed in response to groundcover management system.

Host genotype affects composition of the rhizosphere microbiome and endophyte community, which in turn will alter root development (Molina-Favero et al., 2008), disease susceptibility (Mazzola and Gu, 2002) and nutrient uptake (Wissuwa et al., 2009). Differential selection of microbial species among closely related variants of the same plant species has been documented in a variety of systems (Dalmastri, 1999; Mazzola et al., 2004; Micallef et al., 2009). As noted earlier, this differential selection is likely driven by metabolic composition of root exudates and rhizodeposits, a primary source of substrate in soil systems, which differ in a genotype-dependent fashion (Micallef et al., 2009; Leisso et al., 2017). Interestingly, the selective nature of the plant-microbe interaction is a dynamic process, yet most studies concerning the relationship have employed a limited timepoint assessment. In terms of perennial systems such as tree fruits, this limited exploration may yield findings that cannot be extrapolated across season or orchard life in a manner that can effectively predict responses to management options.

Genotype-dependent microbiome composition recruited to the rhizosphere not only will influence production of the current orchard system, but may also determine relative performance of a subsequent planting when orchard renovation is required. Such an effect can be demonstrated in the following example: growth and disease severity observed for Gala apple seedlings planted in orchard soils from the same orchard site but previously cropped to Honeycrisp grafted to different apple rootstocks differed depending on the previous rootstock genotype (Fig. 1a). Differential seedling growth performance corresponded with the relative disease tolerance of the prior rootstock genotype cultivated in the orchard soil, with tolerance of G.935 > G.41 > M.9 rootstock.

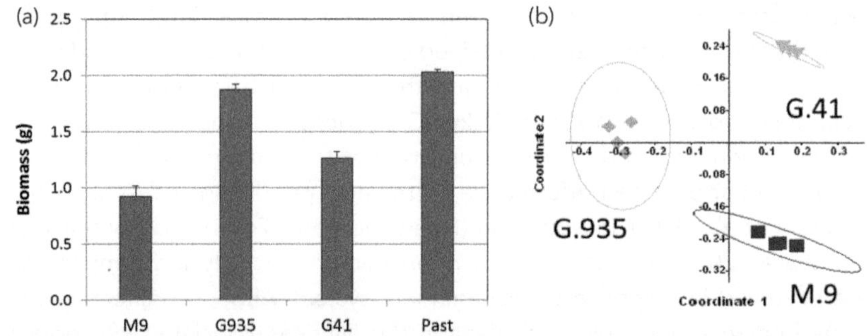

Figure 1 (a) Total biomass of Gala apple seedlings planted in soil previously cultivated to Honeycrisp apple grafted to M.9, G.935 or G.41 rootstock, relative to the same orchard soil pasteurized (Past) prior to seedling planting. (b) Relative similarity of the Gala seedling rhizosphere microbiome as assessed by nonmetric multidimensional scaling of T-RFLP derived data. Gala apple seedlings were planted into the same orchard soil which had previously been planted with Honeycrisp apple on the respective indicated rootstock. Symbols signify the following rootstock genotypes: ■ = M.9; ▼= G.41; ◆ = G.935.

Interestingly, composition of the rhizosphere microbiome was highly similar for all Gala seedlings cultivated in soil that was preceded by a given rootstock, but divergent for samples obtained from soils planted to different rootstocks in the prior orchard (Fig. 1b). This finding demonstrates that the rhizosphere microbiome conferred to a subsequent orchard planting is likely to be determined, in part, in a manner that is dependent upon the previous rootstock genotype (Fig. 1). This knowledge may be valuable in assessing strategies for long-term orchard sustainability and performance, particularly when considering the ever-diminishing time span between orchard renovation events, often driven by the rapid development and establishment of new tree fruit cultivars.

2.4 Induction of host-induced systemic resistance

Elements of the rhizosphere microbiome have been studied for their functional attributes other than plant growth promotion and nutrient acquisition of tree fruits. Various rhizosphere-inhabiting microbes, including certain PGPR, have the capacity to yield induced systemic resistance (ISR) towards both soil-borne and aerial pathogens (Pieterse et al., 1998, 2014). In apple, certain PGPR strains selected by different orchard soil management practices induced systemic resistance against the fungal pathogens *Colletotrichum gloeosporioides* and *Glomerella cingulata*, causal agents of leaf spot of apple (Dos Passos et al., 2014). Bacteria were isolated from roots and rhizosphere soil of 11-year-old Gala apple trees grafted on Marubakaido rootstock planted in conventional or organic orchards, and from undisturbed native field bulk soil. *Pseudomonas* spp., *Cedecea* spp. and *Leclercia* spp. were abundant in conventional orchard soil, whereas *Stenotrophomonas* and *Rahnella* spp. were overrepresented in the organic orchard soil. Organic practices stimulated siderophore producers and indole compound producers, phytohormones that stimulate root growth, resulting in greater access to soil nutrients through elevated production of root surface area. *Burkholderia* strains exhibited the greatest ability for siderophore production and demonstrated antagonistic action against *C. gloeosporioides*, and delayed the leaf spot symptoms in plant bioassays using Gala/Marubakaido trees. Inoculation of apple roots with *Burkholderia* sp. strain 89 prior to inoculation of leaves with *C. gloeosporioides* resulted in a significant suppression of disease development relative to the non-treated control (Dos Passos et al., 2014). This observation documented the capacity of certain rhizosphere-inhabiting bacteria to provide control of aerial plant pathogens.

Amplification of *Streptomyces* spp. in response to seed meal soil amendment also demonstrated capacity to promote ISR in apple. Although numerous strains produce metabolites that directly suppress root disease incited by the fungal pathogen *R. solani*, control was also conferred by *Streptomyces* sp. isolates that did not demonstrate antagonism towards this pathogen (Cohen et al., 2005). Rather, it was shown that various *Streptomyces* isolates were able to confer root-to-root systemic resistance to *R. solani* as demonstrated in split-root assays conducted using Gala apple seedlings as the host plant (Cohen and Mazzola, 2006).

2.5 Abiotic factors affecting root architecture, rhizosphere microbiome and plant nutrition

Abiotic factors, such as soil physical and chemical properties, irrigation, photoperiod and light quality, are well known to impact fruit tree growth and yield. Some of these factors are manipulated to obtain high yield of better-quality fruits in intensive tree fruit orchard systems. For instance, different coloured netting was employed to modulate soil temperature and moisture level using 3-year-old Honeycrisp apple on B.9 rootstock (Kalcsits et al., 2017). Red light (600–700 nm) and blue light (400–500 nm) were increased by using red and blue colour netting, respectively. Temperature at a soil depth of 20 cm and 40 cm was higher under red netting compared to blue and pearl colour netting, but not compared to the uncovered control. Root respiration was shown to increase as the soil temperature increases in hardwood forest trees (Boone et al., 1998) and increased soil temperature also has increased fine root turnover in forest trees (Pregitzer et al., 2000). Soil temperatures as high as 30°C may cause significant abiotic stress on apple trees (Kalcsits et al., 2017). Soil moisture was higher in netted treatments compared to the uncovered treatments. Based on these significant reductions of soil temperature and increased soil moisture under netting of specific colours, it was suggested that netting treatments may enhance tree growth through improved root health. As environmental factors such as soil temperature, light and soil moisture influence exudation processes (Badri and Vivanco, 2009), activity of the rhizosphere microbiome could be indirectly affected by these elements. There is evidence of this from non-tree fruit plant species in which tannin and phenolic root exudate secretion was affected by temperature (Bekkara et al., 1998). These studies demonstrate that the somewhat overlooked effects of horticultural practices on soil biological function should be considered when devising optimal management practices in tree fruit systems.

Similar studies were conducted using plastic mulch of different colours in tree rows using 5- to 6-year-old apricot trees in a cold and arid region in India (Singh, 2013). Plastic mulching contributed to preservation of soil moisture, stabilization of rhizosphere temperature, increased rhizosphere microbial activity and enhanced growth. Red colour mulch harboured higher fungal and bacterial soil densities compared to other mulch colours (blue, black and transparent). It was hypothesized that this response resulted from the effect of specific wavelengths filtered by the mulch on microbial growth or indirectly due to better root growth of the host.

In most semi-arid tree fruit production systems, irrigation is employed to fulfil soil moisture requirements necessary to maintain sufficient tree growth and bring the crop to harvest. Irrigation management can support nutrient availability through regulating specific plant–microbe associations. For instance, in peach (*Prunus persica*) orchards, the root infection rate by mycorrhizal fungi was significantly higher under irrigated than non-irrigated management conditions (Zhang et al., 2018). Alternatively, although ability exists to adequately control soil moisture conditions through irrigation, such irrigation events are sometimes predicated using a time-based rather than a need-based schedule, which can lead to suboptimal soil water status. Although not studied extensively, this may lead to conditions that impede the activity and function of the rhizosphere microbiome, particularly when instances of ponding or flooding arise due to excessive irrigation. Root aeration level, which will be influenced by irrigation events, may affect rhizosphere soil nutrient content, root activity, root architecture, photosynthetic capacity and growth of peach (Xiao et al., 2015). Peach tree root respiration rate is known to be high, and hence requires an elevated oxygen environment. Therefore, in events when rhizosphere aeration is poor, root respiration and energy metabolism are hindered. It was shown that enhanced rhizosphere ventilation resulted in increased rhizosphere soil-available N and P content. Increased nutrient availability corresponded with elevated rhizosphere soil enzyme activity, an important indicator of microbial activity, including urease and phosphatase activity with associated increases in nutrient availability and tree growth (Xiao et al., 2015).

3 Case studies: manipulation of the rhizosphere microbiome to optimize crop production on orchard replant sites

Multiple cultural management practices have been utilized to directly modify the soil/rhizosphere microbiome in a manner that reduces disease and improves crop production on orchard replant sites. Three methods have been examined extensively in apple, and although multiple mechanisms of action contribute to pathogen suppression, activity of a transformed rhizosphere microbiome contributes significantly to the overall plant growth response (Fig. 2). Evidence for the functional role of the transformed rhizosphere microbiome in disease suppression will be presented for the practice of Brassica seed meal soil amendment, anaerobic soil disinfestation (ASD) and wheat cover cropping during the process of apple orchard renovation.

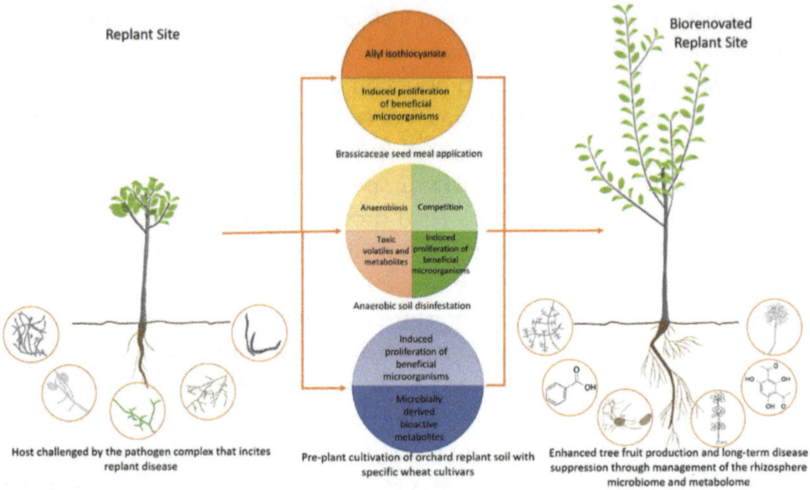

Figure 2 Management of apple replant disease by rhizosphere microbiome-mediated mechanisms that are generated through *Brassica* seed meal soil amendment, the application of anaerobic soil disinfestation or the use of wheat cover cropping prior to orchard renovation.

3.1 *Brassica-seed-meal-directed modification of the apple rhizosphere microbiome*

Pre-plant soil fumigation has long been utilized as an effective and economically viable practice for the control of orchard replant diseases (Mai and Abawi, 1981; Smith, 1994; Willett et al., 1994). Although regulatory actions have restricted or abolished use of certain fumigant chemistries, there continues to be significant study of this practice with emphasis on reducing product concentration or minimizing atmospheric emissions (Browne et al., 2013; Gao et al., 2016). Fumigation has been advocated as the optimal means to attain growth and yield performance through control of soil-borne disease when establishing new orchards on old orchard ground. However, there have been few comparative experiments designed that directly address this statement due to the short-term nature of research studies and the requirement of multi-year monitoring of the trial to adequately validate this assumption. As such, few studies have examined pathogen re-colonization of fumigated orchard soils and the subsequent post-plant effects on limiting apple yields. Reports from multiple field trials indicate that the fumigation effect, in terms of tree growth promotion, is a short-lived feature persisting only in the first growing season of orchard establishment (Auvil et al., 2011). Therefore, opportunity may exist to promote prolonged growth response of tree fruit on replant sites through selective engineering of the microbiome to limit re-infestation of the soil system by plant pathogens.

The deployment of new apple rootstocks from the USDA-Agricultural Research Service rootstock breeding program in Geneva, NY, has been viewed as an additional tool for the management of replant disease (Auvil et al., 2011). The site-specific variable growth performance response of these rootstocks indicates that the genotypes likely exhibit some level of tolerance to one or more pathogens contributing to overall disease expression. In controlled environment studies, when challenged with the root pathogen *Pythium ultimum*, distinct differences in survival rate and plant biomass were observed between the tolerant rootstock G.935 and susceptible rootstock B.9 (Zhu et al., 2016). Limited progression of the pathogen was observed in roots of G.935 relative to B.9, and differential expression of genes involved in host defense responses was observed between the two rootstock genotypes. Functional resistance within certain of these rootstocks at the field level remains unresolved based on recent findings. For instance, in a field trial, root colonization by *P. ultimum* and *Rhizoctonia solani* was lower or the same in the disease-susceptible rootstock M.9 than the tolerant rootstocks G.935 and G.41 (Fig. 3).

Certain cultural methods that result in transformation of the rhizosphere microbiome have elevated tree yields beyond those attained in response to the conventional pre-plant soil fumigation treatment. For instance, pre-plant soil incorporation of specific Brassica seed meal formulations significantly altered composition of the orchard soil microbiome, including amplification

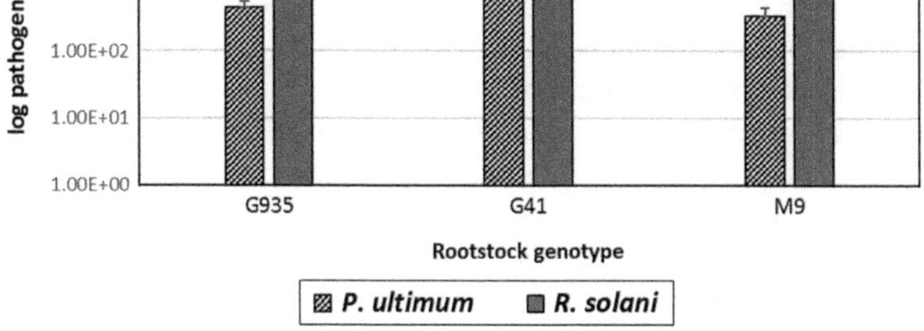

Figure 3 Effect of genotype on quantity of *Pythium ultimum* and *Rhizoctonia solani* DNA detected in roots of apple rootstocks planted in a replant orchard in Orondo, Washington. Apple rootstock genotypes included G.935 (tolerant to apple replant disease [ARD]), G.41 (moderately tolerant to ARD) and M.9 (susceptible to ARD). Values represent mean pathogen DNA (pg g^{-1} root) with n = 36. Means designated with the same letter are not significantly different ($p < 0.05$).

of fungal antagonists such as *Trichoderma* spp. and *Streptomyces* spp. (Cohen and Mazzola, 2006; Weerakoon et al., 2012). The rhizosphere microbiome also exhibited compositional shifts in response to a formulated seed meal soil amendment, and these changes in community composition were persistent for multiple years under field conditions. In controlled environment and field studies, the microbiome detected in the rhizosphere of apple planted in seed-meal-amended soil possessed unique or elevated microbial attributes (e.g. *Arthrobotrys* spp., *Dactylella oviparasitica*, *Oidiodendron* spp.) that limited infestation of the orchard system by known apple root pathogens including *Pythium* spp. and the root lesion nematode *Pratylenchus penetrans* (Weerakoon et al., 2012; Mazzola et al., 2015). In contrast, within 2 years of orchard establishment, the microbial profile obtained from the rhizosphere of trees grown in fumigated soil could not be distinguished from that of the no-treatment control soil. This lack of microbial profile difference corresponded with yields of Gala/M.9, Gala/G.11 and JonaGold/G.11 in fumigated soil that were significantly lower than those attained for trees planted in the seed-meal-amended soil system.

Disease control attained in response to Brassica seed meal soil amendment is mechanistically complex, with the functional process varying in a temporal manner and target pathogen. Disease control can initially operate exclusively through generation of active chemistries, but thereafter may rely upon compositional changes within the rhizosphere microbiome that lead to the dominance of specific microbial elements, as noted above, that directly or indirectly control the causal pathogen (Mazzola et al., 2007; Weerakoon et al., 2012). Seed-meal-induced amplification of specific microorganisms may result in disease control through direct suppression of the pathogen (Weerakoon et al., 2012) or function via the induction of host defense responses (Cohen et al., 2005). Given the previously demonstrated effect of host genotype on composition of the rhizosphere microbiome (Reed and Mazzola, 2015), it can be hypothesized that rootstock genotype could modulate disease control efficacy of these seed meal amendments through (1) differential recruitment of functional microbial elements and (2) capacity to express plant defense responses that are microbe induced.

The premise that rootstock-dependent rhizosphere microbiome composition influences disease control efficacy of Brassica seed meal amendments has not been experimentally addressed directly; however, ancillary studies support the hypothesis. As an example, transcriptome analysis was conducted on rootstocks that were planted in seed-meal-amended or non-treated orchard replant soil. Seed-meal-amended soil was incubated for 6 weeks prior to planting to allow a sufficient period for appropriate transformation of the soil microbiome (Wang and Mazzola, 2015) and degradation of herbicidal chemistries generated. Numerous differentially expressed genes were detected in roots of apple planted in the seed-meal-amended soils, a number of which are associated with host defense responses to pathogen challenge (Wang and Mazzola, unpublished). These included altered regulation of several genes involved in phytohormone signalling pathways that function in ISR in plants (Pieterse et al., 1998). However, the responses at the level of the transcriptome differed among rootstock genotypes and were more persistent in roots of the disease tolerant G.210 rootstock than the susceptible M.26 rootstock. Transformation of the rhizosphere microbiome in response to seed meal application transpired in a rootstock genotype-dependent manner. Relative to the control, changes in composition of the rhizosphere microbiome were detected at lower seed meal application rates and occurred more rapidly after planting of G.210 than MM.106 rootstock (Fig. 4) (Mazzola and Wang, 2015). These differences corresponded with lower incidence of root infection by the fungal pathogen *Rhizoctonia solani* for G.210 compared to MM.106 rootstock.

3.2 Modified rhizosphere microbiome and enhanced growth through anaerobic soil disinfestation

ASD has received significant attention as a soil-borne disease control alternative to pre-plant soil fumigation in a variety of high-value crop production systems (Rosskopf et al., 2015). Growth promotion has been attributed to the activity of specific chemistries derived through the ASD process that directly inhibit a diversity of plant pathogens. Alternatively, ASD may enhance plant growth through actions of an altered rhizosphere microbiome that results in enhanced nutrient availability, direct growth promotion or soil-borne disease control

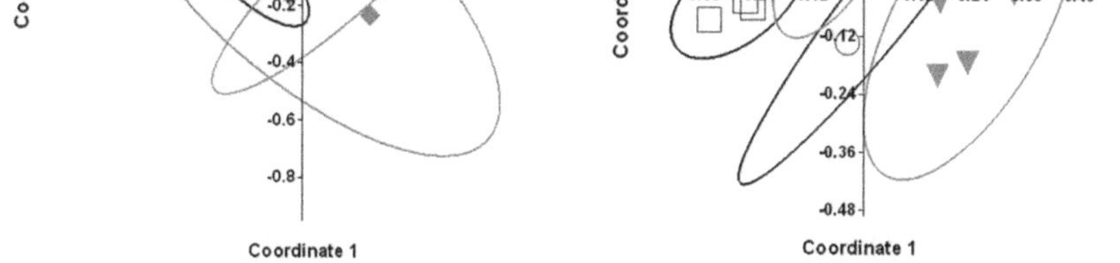

Figure 4 Similarity of fungal community composition in the rhizosphere of (a) MM.106 and (b) G.210 apple rootstocks 1 month after planting, as assessed by nonmetric multidimensional scaling of terminal restriction fragment length polymorphism-derived data using the Dice similarity coefficient. Symbols signify the following soil treatments: □ = no treatment control; ♦ = Brassica seed meal (SM) 2.2 t ha^{-1}; ○ = SM 4.4 t ha^{-1}; ▼ = 6.6 t ha^{-1}.

(Hewavitharana and Mazzola, 2016). Compositional and functional changes in the apple rhizosphere microbiome were observed in response to ASD conducted in both controlled environment (Hewavitharana and Mazzola, 2016) and long-term field trials (Fig. 5). In greenhouse trials, ASD-induced transformation of the apple rhizosphere microbiome and corresponding suppression of soil-borne disease were carbon input-dependent. In the field, ASD conducted using orchard grass as the carbon input generated changes in the apple rhizosphere microbiome that were distinguishable even after two growing seasons (Fig. 5). ASD conducted using orchard grass resulted in amplification of numerous fungal species, demonstrating production of cytotoxic or fungitoxic metabolites (Hewavitharana and Mazzola, 2016). The documented changes in the rhizosphere microbiome were linked to significantly improved tree growth for all rootstock genotypes examined (M.9, G.41, G.935), which represented a range of field-level tolerance to replant disease (Auvil et al., 2011).

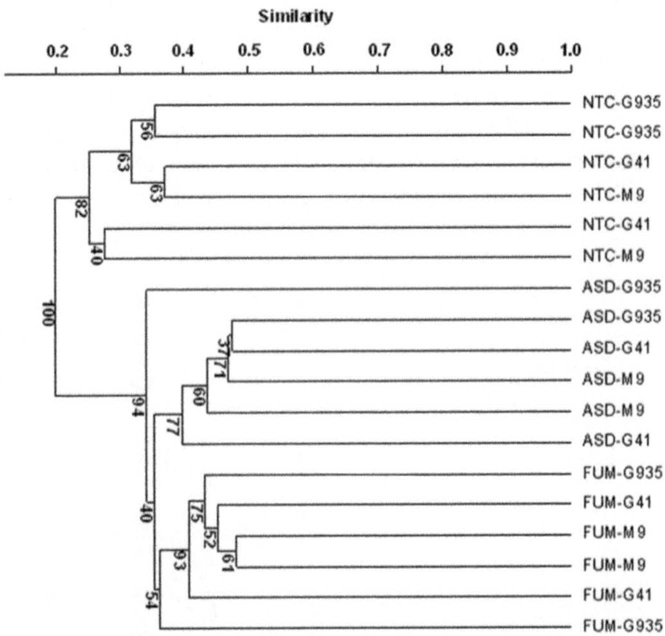

Figure 5 Effect of soil treatment on relative similarity of fungal communities detected after two growing seasons in the rhizosphere of apple rootstocks planted on a replant orchard site in Orondo, Washington, USA. Similarity of fungal community profiles was assessed by cluster analysis of terminal restriction fragment length polymorphism (T-RFLP)-derived data. Treatments: NTC = No treatment control; ASD = anaerobic soil disinfestation using 20 t ha^{-1} grass as the carbon input; FUM = pre-plant soil fumigation with 1,3-dichloropropene/chloropicrin (Telone® C35).

4 Conclusion

Study of the soil and rhizosphere microbiome, in terms of its effects on plant health and productivity, has attained significant interest across the majority of plant cropping systems. Recognition of the value of this resource in crop production has in large part arisen from the needs of organic producers who have more often relied upon an understanding of ecosystem function as a means to solve management problems. In turn, this grower interest, in concert with the advent of analytical technologies that allow for in-depth examination of the microbiome, has led to the proliferation of research concerning rhizosphere science. However, exploration of the rhizosphere microbiome in perennial production systems, including tree fruits, has been limited and has focused primarily on its relation to soil-borne disease development or suppression.

The perennial nature of tree fruits amplifies the complexity of interactions in the rhizosphere and creates additional challenges in our ability to effectively describe structural and functional attributes of the relationships between the rhizosphere and its microbiome. Plant age will significantly influence composition of the microbiome, as will management conditions and the ever-changing environment. Nonetheless, many different approaches have been undertaken in orchard systems to manipulate the microbiome in a manner that leads to enhanced plant health. These include attempts to establish synthetic microbiomes through inoculation of tree roots prior to planting, and the application of specific soil amendments to selectively amplify rhizosphere microbial communities that function in the suppression of soil-borne plant pathogens.

Effective measures for modification of the rhizosphere microbiome have been identified through controlled environment studies and subsequently validated in the field. In apple, microbial recruitment to the rhizosphere and ultimate composition of the rhizosphere microbiome is influenced by rootstock genotype. There is evidence that these rootstock genotype-dependent differences affect growth of apple subsequently planted into the previous orchard soil and may influence the health of the current orchard. This is likely to be the case for other tree fruit species and their diverse rootstock genotypes. Thus, breeding for rootstock rhizosphere-related traits has the potential to significantly improve crop productivity over the extended life of a tree fruit orchard.

A multitude of factors prevalent in the orchard environment will influence composition and function of the rhizosphere microbiome as it relates to tree health. Opportunities to modify the rhizosphere microbiome can be attained through the manipulation of this environment, such as by management of soil moisture or temperature. However, altering qualitative and quantitative attributes of substrate availability in orchard soils has yielded the greatest benefit with regard to sustaining a modified rhizosphere microbiome that promotes plant health. This includes the use of organic inputs with qualities that directly result in development of niches that are populated by dominant microbial taxa with the potential to suppress plant pathogenic organisms. A primary concern in perennial tree fruit production systems is a lack of knowledge concerning the successional events in the rhizosphere microbiome over successive growing seasons. However, in a multi-year study, although significant changes in the soil/rhizosphere microbiome were observed in response to Brassica seed meal amendment, the overall system retained its resistance to pathogen invasion after two growing seasons. This observation suggests such treatments can attain long-term benefits for a rhizosphere microbiome that will support tree fruit health.

5 Future trends

Based on the foundational research conducted thus far concerning the tree fruit rhizosphere microbiome, achieving progress on a range of production issues, including enhancing plant nutrient availability, resilience to environmental stress and improving overall plant health, would benefit from integrating findings reported in studies of non-tree fruit systems. Advancement in the realm of rhizosphere research in general has focused on the process of rhizosphere engineering, but limited effort has been made to analyse the system comprehensively. Even well-studied plant–microbial mutualist interactions (e.g. plant–mycorrhizal fungi) have been examined primarily from either the plant or microbe perspective. Rather, additional progress will require an approach of engineering not one but all three components of the rhizosphere: soil, plant and microbiome (Dessaux et al., 2016). Concomitant with this approach, the former practice of considering plants and associated microorganisms as individuals needs to be discarded and their interaction considered as an association, which has been termed the holobiont (Zilber-Rosenberg and Rosenberg, 2008). The holobiont is considered to be the unit of evolution with changes made to one component reflected in the other associated components.

While major advances have been made in our capacity to profile the rhizosphere microbiome and the influence of plant genotype on its composition, the need to equate these associations with desired phenotypic outcomes is extensive. Numerous questions concerning rhizosphere interactions remain to be considered. For instance, although seed-meal-induced manipulation of the apple rhizosphere microbiome exhibited potential to induce expression of genes involved in resistance responses (Wang and Mazzola, unpublished), activation of plant defense systems may come at a cost to overall plant fitness (Tian et al., 2003). Mobilization of the rhizosphere microbiome to enhance tree fruit productivity requires knowledge concerning the identity of members that contribute to the function of interest (e.g. nutrient acquisition or stress tolerance). In certain instances, information is readily available, such as the importance of plant-mycorrhizal fungi associations in phosphorous acquisition. However, a large fraction of the microbial community detected in the rhizosphere consists of unidentified organisms, and an even greater portion has yet to be described in terms of potential function in association with plants. Functional analysis of the rhizosphere microbiome will require additional experimentation prior to exploring the engineering of this resource in a manner that results in predictable outcomes for crop production.

Employing multi-omics data analysis has potential to fill the aforementioned gaps of knowledge in the field of rhizosphere research. White et al. (2017) extensively discussed the challenges associated with acquiring and analysing multi-omics data in an attempt to answer questions relative to the complex interactions that occur in the rhizosphere. By using the modern system biology tools of integrated multi-omics, interactions occurring in the rhizosphere can be explored using a holistic community approach (White et al., 2017). Metagenomics, metatranscriptomics, metaproteomics and metametabolomics technologies have been used most commonly in study of rhizosphere processes that influence plant development. While there exist a number of studies that have applied various -omics approaches for the study of structural and function aspects of the rhizosphere, there is a general absence of rhizosphere studies that have integrated multi-omics data. There exist even fewer studies which have been conducted to describe interactions or processes in the tree fruit rhizosphere.

One of these few examples included the systematic research conducted to discern management-induced changes in the apple tree rhizosphere microbiome, metatranscriptome and metametabolome in relation to subsequent effects on replant disease severity. Changes occurring in response to pre-plant treatments on replant orchard sites were identified using next-generation sequencing technology (Mazzola et al., 2015). Specific functional groups became abundant or scarce in the apple rhizosphere microbial community over time as a result of soil treatments such as pre-plant soil fumigation and the use of specific soil amendments. These specific changes in structural composition of the rhizosphere microbiome were linked to differences in relative growth of the host. The role of host genotype in directing these microbial community shifts was also detected using this technology. Metatranscriptome analysis was used to assess temporal changes in gene expression occurring in roots of disease-tolerant and -susceptible apple rootstocks in response to soil amendments (Wang and Mazzola, unpublished). Furthermore, characterization of the metametabolome from rhizodeposits of tolerant and susceptible rootstock genotypes revealed significant differences in specific root exudates that can be associated with the host phenotype (Leisso et al., 2017).

White et al. (2017) emphasized that experimental techniques and technologies, and data analysis software, need to be developed simultaneously to meet the major impediments to progress in the field of rhizosphere research. Most specifically, these challenges arise from the generation of complex and voluminous data which require the development of robust analytical methods and corresponding advancements in software. Application of each -omics analytical method for examining the rhizosphere has advantages and challenges. For instance, two of the major gaps identified in metametabolomics analysis are the lack of rhizosphere-specific metabolite reference libraries and limited availability of software for data analysis in stable isotopic-labelling experiments.

Integration of multi-omics data is commonly carried out by network-based and clustering-based approaches. There are a few examples of integrated studies using network-based approaches in rhizosphere research in non-tree fruit crop species (Ofaim et al., 2017; Poudel et al., 2016) and similar approaches could be employed in investigation of microbe-driven processes in the tree fruit rhizosphere. These reports are significant as they provide a framework for the application of integrated multi-omics studies. Metagenome analysis of microbial communities in the rhizosphere of wheat and cucumber (*Cucumis sativus*) was conducted to identify environment-specific gene catalogues and then processed into metabolic network representation (Ofaim et al., 2017). Based on the dominance level of specific groups, network edges which represented metabolic functions were assigned to taxonomic annotations. Treatment differences (bulk soil vs rhizosphere) were evident for environment-dependent effects such as metabolic activity of root exudates, which were linked to rhizosphere organisms but were not represented in taxonomic groups from the bulk soil.

Poudel et al. (2016) used microbiome networks to generate testable hypotheses on candidate microorganisms that influence plant health. Candidate taxa were identified if there were positive direct or indirect associations with desirable outcomes, or negative associations with undesirable outcomes. Four types of microbiome network analyses were described: general network analysis, host-focused network analysis, pathogen-focused network analysis and disease-focused network analysis that will be helpful to choose among based on the scope of the rhizosphere network analysis study. Another study used random matrix theory-based network in the analysis of wild oat (*Avena fatua*) rhizosphere and bulk soil bacterial 16S rRNA genes, taking the temporal factor of host age into account as well (Shi et al., 2016). As would be expected, the networks established in assessment of the rhizosphere bacterial community were highly complex compared to that established for the bulk soil community. Network complexity of the rhizosphere increased with increasing plant maturity, although diversity decreased, while networks in surrounding bulk soil remained relatively static. Changes in environmental properties, such as altered substrate availability, are known to alter the functional complexity of microbial networks in soil (Zhou et al., 2011). Elevated network size and complexity in the rhizosphere was interpreted as an indicator of increased microbial community organization which is driven, in part, by increased microbial interactions. The elevated interdependence of microbial entities in the rhizosphere argues for the need to study relationships among organisms in this niche, rather than independently, to more accurately assess the function of the rhizosphere microbiome in directing plant development.

Manipulation of the soil microbial communities as a means to enhance orchard ecosystem function and crop productivity requires additional knowledge concerning the functionality of the rhizosphere microbiome. For instance, the occurrence of quorum sensing among bacteria is well known and its operation is instrumental in the process of antibiotic production by certain bacteria that are antagonistic to plant pathogenic fungi (Loh et al., 2002). Quorum sensing also regulates functions related to the pathogenicity of certain plant pathogenic bacteria, including the transfer of the conjugative plasmid in *Rhizobium* (=*Agrobacterium*) *tumefaciens* (Uroz et al., 2003), causal agent of crown gall disease that affects tree fruits. Thus, manipulation of quorum sensing in the rhizosphere through quorum quenching strategies could limit the activity of certain biocontrol rhizobacteria, but alternatively could be utilized as a biocontrol mechanism to inhibit bacterial plant pathogens (Dong et al., 2001; Uroz et al., 2003). The interplay of these dynamic processes (of which we are aware) has not been addressed. The question remains as to the volume of rhizosphere interactions that we have yet to uncover or recognize as important determinants of plant health and productivity.

6 Where to look for further information

The study of the rhizosphere microbiome and its', influence on plant health has recently attained enormous interest not only in the research environment, but also among the community of crop producers. While these studies have the potential to be transformative in terms of crop management protocols, the practice of moving the science to application is in its infancy. Regardless, a number of professional organizations have led exploration in this scientific discipline, including the International Society for Microbial Ecology and more recently the American Phytopathological Society, which launched the plant-focused journal, *Phytobiomes*. Numerous mini-reviews are available that consider the value of the rhizosphere microbiome and its manipulation for enhanced plant productivity, including that by Philippot et al. (2013) and Dessaux et al. (2016). Several composite books exploring the topic of rhizosphere science are available, with an extensive overview in the book edited by Dessaux et al. (2010) entitled *Rhizosphere: Accomplishments and Challenges*. Due to the recent nature of the field of study, there continues to be a multitude of questions relative to the various plant processes (known and unknown) which the rhizobiome may influence. As such, the capacity to apply knowledge obtained from studies of the tree fruit rhizosphere microbiome to enhance production system efficiency continues to develop. Much of the available information on tree fruits relates to the engineering of the microbiome as a means to attain disease suppression (Mazzola and Manici, 2012; Watson et al., 2017) or enhanced orchard soil fertility.

The state-of-the-art in rhizosphere science, and particularly research relating to the rhizosphere microbiome, can be attained at noteworthy meetings that are held on a periodic basis. These include the biennial symposium hosted by the International Society for Microbial Ecology and the conference on the Rhizosphere every 5 years. When available, publications that emanate from these meetings often provide the most up-to-date and broad-based reviews on rhizosphere science and its role in the function of plant production systems.

7 References

Aslantas, R., Cakmakci, R. and Sahin, F. (2007), 'Effect of plant growth promoting rhizobacteria on young apple tree growth and fruit yield under orchard conditions', *Sci. Hort.*, 111(4), 371–7.

Auvil, T. D., Schmidt, T. R., Hanrahan, I., Castillo, F., McFerson, J. R. and Fazio, G. (2011), 'Evaluation of dwarfing rootstocks in Washington apple replant sites', *Acta Hort.*, 903(33), 265–71.

Badri, D. V. and Vivanco, J. M. (2009), 'Regulation and function of root exudates', *Plant Cell Environ.*, 32(6), 666–81.

Bais, H. P., Weir, T. L., Perry, L. G., Gilroy, S. and Vivanco, J. M. (2006), 'The role of root exudates in rhizosphere interactions with plants and other organisms', *Annu. Rev. Plant Biol.*, 57(1), 233–66.

Bekkara, F., Jay, M., Viricel, M. R. and Rome, S. (1998), 'Distribution of phenolic compounds within seed and seedlings of two *Vicia faba* cvs differing in their seed tannin content and study of their seed and root phenolic exudations', *Plant Soil*, 203(1), 27–36.

Berendsen, R. L., Pieterse, C. M. J. and Bakker, P. A. H. M. (2012), 'The rhizosphere microbiome and plant health', *Trends Plant Sci.*, 17(8), 1360–85.

Bertin, C., Yang, X. H. and Weston, L. A. (2003), 'The role of root exudates and allelochemicals in the rhizosphere', *Plant Soil*, 256(1), 67–83.

Boone, R. D., Nadelhoffer, K. J., Canary, J. D. and Kaye, J. P. (1998), 'Roots exert a strong influence on the temperature sensitivity of soil respiration', *Nature*, 396(6711), 570–2.

Browne, G. T., Lampinen, B. D., Holtz, B. A., Doll, D. A., Upadhyaya, S. K, Schmidt, L. S., Bhat, R. G., Udompetaikul, V., Coates, R. W., Hanson, B. D., Klonsky, K. M., Gao, S., Wang, D., Gillis, M., Gerik, J. and Johnson, R. S. (2013), 'Managing the almond and stone fruit replant disease complex with less soil fumigant', *Calif. Agric.*, 67(3), 128–38.

Brussaard, L., de Ruiter, P. C. and Brown, G. G. (2007), 'Soil biodiversity for agricultural sustainability', *Agric. Ecosyst. Environ.*, 121(3), 233–44.

Caldwell, A. C., Silva, L. C. F., da Silva C. C. and Ouverney, C. C. (2015), 'Prokaryotic diversity in the rhizosphere of organic, intensive and transitional coffee farms in Brazil', *PLoS One*, 10(6), e0106355.

Cohen, M. F. and Mazzola, M. (2006), 'Resident bacteria, nitric oxide emission and particle size modulate the effect of *Brassica napus* seed meal on disease incited by *Rhizoctonia solani* and *Pythium* spp.', *Plant Soil*, 286(1-2), 75-86.

Cohen, M. F., Yamasaki, H. and Mazzola, M. (2005), '*Brassica napus* seed meal soil amendment modifies microbial community structure, nitric oxide production and incidence of Rhizoctonia root rot', *Soil Biol. Biochem.*, 37(7), 1215-27.

Covey, R. P., Koch, B. L. and Larsen, H. J. (1981), 'Influence of vesicular arbuscular mycorrhizae on the growth of apple and corn in low-phosphorous soil', *Phytopathology*, 71(7), 712-15.

Dalmastri, C., Chiarini, L., Cantale, C., Bevivino, A. and Tabacchioni, S. (1999), 'Soil type and maize cultivar affect the genetic diversity of maize root-associated *Burkholderia cepacia* populations', *Microb. Ecol.*, 38(3), 273-84.

Dessaux, Y., Hinsinger, P. and Lemanceau, P. (2010), *Rhizosphere: Achievements and Challenges*, Springer Science Press, Berlin.

Dessaux, Y., Grandclément, C. and Faure, D. (2016), 'Engineering the rhizosphere', *Trends Plant Sci.*, 21(3), 266-78.

Dong, Y. H., Wang, L. H., Xu, J. L., Zhang, X. F. and Zhang, L. H. (2001), 'Quenching quorum-sensing-dependent bacterial infection by an N-acyl homoserine lactonase', *Nature*, 41(6839), 813-7.

Dos Passos, J. F. M., da Costa, P. B., Costa, M. D., Zaffari, G. R., Nava, G., Boneti, J. I., de Oliveira, A. M. R. and Passaglia, L. M. P. (2014), 'Cultivable bacteria isolated from apple trees cultivated under different crop systems: Diversity and antagonistic activity against *Colletotrichum gloeosporioides*', *Genet. Mol. Biol.*, 37(3), 560-72.

Emmett, B., Nelson, E. B., Kessler, A. and Bauerle, T. L. (2014), 'Fine-root system development and susceptibility to pathogen colonization', *Planta*, 239(2), 325-40.

Forge, T., Muehlchen, A., Hackenberg, C., Nielsen, G. and Vrain, T. (2001), 'Effects of preplant inoculation of apple (*Malus domestica* Borkh.) with arbuscular mycorrhizal fungi on population growth of the root-lesion nematode, *Pratylenchus penetrans*', *Plant Soil*, 236(3), 185-96.

Gahoonia, T. S. and Nielsen, N. E. (2004), 'Barley genotypes with long root hairs sustain high grain yields in low-P field', *Plant Soil*, 262(1-2), 55-62.

Gao, S., Sosnoskie, L. M., Cabrera, J. A., Qin, R., Hanson, B. D., Gerik, J. S., Wang, D., Browne, G. T. and Thomas, J. E. (2016), 'Fumigation efficacy and emission reduction using low-permeability film in orchard soil fumigation', *Plant Manag. Sci.*, 72(2), 306-14.

Gianinazzi, S., Gollotte, A., Binet, M. N., van Tuinen, D., Redecker, D. and Wipf, D. (2010), 'Agroecology: The key role of arbuscular mycorrhizas in ecosystem services', *Mycorrhiza*, 20(8), 519-30.

Hartmann, M., Frey, B., Mayer, J., Mader, P. and Widmer, F. (2015), 'Distinct soil microbial diversity under long-term organic and conventional farming', *ISME J.*, 9(5), 1177-94.

Hewavitharana, S. S. and Mazzola, M. (2016), 'Carbon source-dependent effects of anaerobic soil disinfestation on soil microbiome and suppression of *Rhizoctonia solani* AG-5 and *Pratylenchus penetrans*', *Phytopathology*, 106(9), 1015-28.

Hewavitharana, S. S. and Mazzola, M. (2017), 'Deciphering potential mechanisms of anaerobic soil disinfestation (ASD)-mediated control of *Pratylenchus penetrans*', *Phytopathology*, 107(5), 47.

Hinsinger, P., Gobran, G. R., Gregory, P. J. and Wenzel, W. W. (2005), 'Rhizosphere geometry and heterogeneity arising from root-mediated physical and chemical processes', *New Phytol.*, 168(2), 293–303.

Hoagland, L., Carpenter-Boggs, L., Granatstein, D., Mazzola, M., Smith, J., Peryea, F. and Reganold, J. P. (2008), 'Orchard floor management effects on nitrogen fertility and soil biological activity in a newly established organic apple orchard', *Biol. Fertil. Soils*, 45(11), 11–18. doi:10.1007/s00374-008-0304-4.

Inderbitzin, P., Ward, J., Barbella, A., Solares, N., Izyumin, D., Burman, P., Chellemi, D. and Subbarao, K. V. (2017), 'Soil microbiomes associated with Verticillium wilt suppressive broccoli and chitin amendments are enriched with potential biocontrol agents', *Phytopathology*, 108(1), 31–43. doi:10.1094/PHYTO-07-17-0242-R.

Ismail, A. M., Heuer, S., Thomson, J. T. and Wissuwa, M. (2007), 'Genetic and genomic approaches to develop rice germplasm for problem soils', *Plant Mol. Biol.*, 65(4), 547–70.

Jiang, J., Song, Z., Yang, X., Mao, Z., Nie, X., Guo, H. and Peng, X. (2017), 'Microbial community analysis of apple rhizosphere around Bohai Gulf', *Sci. Rept.*, 7(1), 8918.

Kalcits, L., Musacchi, S., Layne, D. R., Schmidt, T., Mupambi, G., Serra, S., Mendoza, M., Asteggiano, L., Jarolmasjed, S., Sankaran, S., Khot, L. R. and Espinoza, C. Z. (2017), 'Above and below-ground environmental changes associated with the use of photoselective protective netting to reduce sunburn in apple', *Agric. For. Meteorol.*, 237–238, 9–17. doi:10.1016/j.agrformet.2017.01.016.

Lambers, H., Shane, M. W., Cramer, M. D., Pearse, S. J. and Veneklaas, E. J. (2006), 'Root structure and functioning for efficient acquisition of Phosphorus: Matching morphological and physiological traits', *Ann. Bot.*, 98(4), 693–713.

Leisso, R., Rudell, D. and Mazzola, M. (2017), 'Metabolic composition of apple rootstock rhizodeposits differ in a genotype-specific manner and affects growth of subsequent plantings', *Soil Biol. Biochem.*, 113, 201–14.

Lemanceau, P., Expert, D., Gaymard, F., Bakker, P. and Briat, J. F. (2009), 'Role of iron in plant-microbe interactions', pp. 491–549, *in* Kader, J. and Delseny, M. (Eds), *Advances in Botanical Research*, vol. 51. Academic Press, San Diego, CA.

Loh, J., Pierson, E. A., Pierson, L. S., Stacey, G. and Chatterjee, A. (2002), 'Quorum sensing in plant-associated bacteria', *Curr. Opin. Plant Biol.*, 5(4), 285–90.

Lupatini, M., Korthals, G. W., de Hollander, M., Janssens, T. K. S. and Kuramae, E. E. (2016), 'Soil microbiome is more heterogeneous in organic than in conventional farming system', *Front. Microbiol.*, 7, 2064. doi:10.3389/fmicb.2016.02064.

Mai, W. F. and Abawi, G. S. (1981), 'Controlling replant disease of stone fruits in northeastern United States by preplant fumigation', *Plant Dis.*, 65, 859–64.

Mark, G. L., Dow, J. M., Kiely, P. D., Higgins, H., Hayes, J., Baysse, C., Abbas, A., Foley, T., Franks, A., Morrissey, J. and O'Gara, F. (2005), 'Transcriptome profiling of bacterial responses to root exudates identifies gene involved in microbe-plant interactions', *Proc. Natl. Acad. Sci. USA*, 102(48), 17454–9.

Marschner, H. (1995), *Mineral Nutrition in Higher Plants*, 2nd ed., Academic Press, London.

Mazzola, M. (1999), 'Transformation of soil microbial community structure and *Rhizoctonia*-suppressive potential in response to apple roots', *Phytopathology*, 89(10), 920–7.

Mazzola, M. and Freilich, S. (2017), 'Prospects for biological soilborne disease control: Application of indigenous versus synthetic microbiomes', *Phytopathology*, 107(3), 256–63.

Mazzola, M. and Gu, Y.-H. (2002). Wheat genotype-specific induction of soil microbial communities suppressive to disease incited by *Rhizoctonia solani* AG 5 and AG 8. *Phytopathology*, 92, 1300–7.

Mazzola, M. and Manici, L. M. (2012), 'Apple replant disease: Role of microbial ecology in cause and control', *Annu. Rev. Phytopathol.*, 50, 45–65.

Mazzola, M. and Wang, L. (2015), 'Reduced rate seed meal amendment efficacy is plant genotype-dependent', pp. 14.1–14.4, *in Proc. Annu. Int. Res. Conf. Methyl Bromide Altern. Emiss. Reduct.*, MBAO, San Diego, CA.

Mazzola, M., Andrews, P. K., Reganold, J. P. and Lévesque, C. A. (2002), 'Frequency, virulence, and metalaxyl sensitivity of *Pythium* spp. isolated from apple roots under conventional and organic production systems', *Plant Dis.*, 86(6), 669–75.

Mazzola, M., Funnell, D. L. and Raaijmakers, J. M. (2004), 'Wheat cultivar-specific selection of 2,4-diacetylphloroglucinol-producing fluorescent *Pseudomonas* species from resident soil populations', *Microbial. Ecol.*, 48(3), 338–48.

Mazzola, M., Brown, J., Izzo, A. D. and Cohen, M. F. (2007), 'Mechanism of action and efficacy of seed meal-induced pathogen suppression differ in a Brassicaceae species and time-dependent manner', *Phytopathology*, 97(4), 454–60.

Mazzola, M., Hewavitharana, S. S. and Strauss, S. L. (2015), '*Brassica* seed meal soil amendments transform the rhizosphere microbiome and improve apple production through resistance to pathogen reinfestation', *Phytopathology*, 105(4), 460–9.

Mendes, R., Garbeva, P. and Raaijmakers, J. M. (2013), 'The rhizosphere microbiome: Significance of plant beneficial, plant pathogenic and human pathogenic microorganisms', *FEMS Microbiol. Rev.* 37(5), 634–63.

Micallef, S. A., Shiaris, M. P. and Colón-Carmona, A. (2009), 'Influence of *Arabidopsis thaliana* accessions on rhizobacterial communities and natural variation in root exudates', *J. Exp. Bot.*, 60(6), 1729–42.

Molina-Favero, C., Creus, C. M., Simontacchi, M., Putarulo, S. and Lamattina, L. (2008), 'Aerobic nitric oxide production by *Azospirillum brasilense* Sp245 and its influence on root architecture in tomato', *Mol. Plant Microbe Interact.*, 21(7), 1001–9.

Ofaim, S., Ofek-Lalzar, M., Sela, N., Jinag, J., Kashi, Y., Minz, D. and Freilich, S. (2017), 'Analysis of microbial functions in the rhizosphere using a metabolic based framework for metagenomics interpretation', *Front. Microbiol.*, 8, 1606. doi:10.3389/fmicb.2017.01606

Panke-Buisse, K., Poole, A. C., Goodrich, J. K., Ley, R. E. and Kao-Kniffin, J. (2015), 'Selection on soil microbiomes reveals reproducible impacts on plant function', *ISME J.*, 9, 980–9. doi:10.1038/ismej.2014.196.

Philippot, L., Raaijmakers, J. M., Lemanceau, P. and van der Putten, W. H. (2013), 'Going back to the roots: The microbial ecology of the rhizosphere', *Nat. Rev. Microbiol.*, 11, 789–99.

Pieterse, C. M., van Wees, S. C., van Pelt, J. A., Knoester, M., Laan, R., Gerrits, H., Weisbeek, P. J. and van Loon, L. C. (1998), 'A novel signaling pathway controlling induced systemic resistance in Arabidopsis', *Plant Cell*, 10(9), 1571-80.

Pieterse, C. M. J., Zamioudis, C., Berendsen, R., Weller, D. M., van Wees, S. C. M. and Bakker, A. H. M. (2014), 'Induced systemic resistance by beneficial microbes', *Annu. Rev. Phytopathol.*, 52, 347–75.

Poudel, R., Jumpponen, A., Schlatter, D. C., Paulitz, T. C., Gardener, B. B. M., Kinkel, L. and Garrett, K. A. (2016), 'Microbiome networks: A systems framework for identifying candidate microbial assemblages for disease management', *Phytopathology*, 106(10), 1083–96.

Pregitzer, K. S., King, J. S., Burton, A. J. and Brown, S. E. (2000), 'Responses of tree fine roots to temperature', *New Phytol.*, 147(1), 105–15.

Pumphrey, G. M. and Madsen, E. L. (2008), 'Field-based stable isotope probing reveals the identities of benzoic acid-metabolizing microorganisms and their in situ growth in agricultural soil', *Appl. Environ. Microbiol.*, 74(13), 4111–18.

Raaijmakers, J. M. and Mazzola, M. (2016), 'Soil immune responses: Soil microbiomes may be harnessed for plant health', *Science*, 352(6292), 1392–3.

Reed, A. J. and Mazzola, M. (2015), 'Characterization of apple replant disease-associated microbial communities over multiple growth periods using next-generation sequencing', *Phytopathology*, 105(suppl. 4), 117.

Rosskopf, E. N., Serrano-Perez, P., Hong, J. C., Shrestha, U., Rodriguez-Molina, M., Martin, K., Burelle, N. K., Shennan, C., Muramoto, J. and Butler, D. (2015), 'Anaerobic soil disinfestation and soil borne pest Management', pp. 277–305, *in* M. K. Meghvansi and A. Varma (Eds), *Organic Amendments and Soil Suppressiveness in Plant Disease Management*, Soil Biology, vol. 46, Springer, Cham.

Rubio, G., Zhu, J. and Lynch, J. P. (2003), 'A critical test of the two prevailing theories of plant response to nutrient availability', *Amer. J. Bot.*, 90(10), 143–52.

Rudrappa, T., Czymmek, K. J., Paré, P. W. and Bais, H. P. (2008), 'Root-secreted malic acid recruits beneficial soil bacteria', *Plant Physiol.*, 148, 1547–56.

Rumberger, A., Merwin, I. A. and Thies, J. E. (2007), 'Microbial community development in the rhizosphere of apple trees at a replant disease site', *Soil Biol. Biochem.*, 39(7), 1645–54.

Shi, S., Nuccio, E. E., Shi, Z. J., He, Z., Zhou, J. and Firestone, M. K. (2016), 'The interconnected rhizosphere: High network complexity dominates rhizosphere assemblages', *Ecol. Lett.*, 19(8), 926–36.

Singh, H. V. (2013), 'Soil carbon sequestration and rhizosphere microbial population in apricot orchards following plastic film mulching under cold arid region', *Int. J. Hort.*, 3(8), 35–41.

Smith, T. J. (1994), 'Successful management of orchard replant disease in Washington', *Acta Hort.*, 394, 161–8.

St Laurent, A., Merwin, I. A. and Thies, J. E. (2010), 'Long-term orchard groundcover management systems affect soil microbial communities and apple replant disease severity', *Plant Soil* 304(1–2), 209–25.

Tian, D., Traw, M. B., Chen, J. Q., Kreitman, M. and Bergelson, J. (2003), 'Fitness costs of R-gene-mediated resistance in *Arabidopsis thaliana*', *Nature*, 423, 74–7.

Turner, T. R., James, E. K. and Poole, P. S. (2013), 'The plant microbiome', *Genome Biol.*, 14(209), 209. doi:10.1186/gb-2013-14-6-209.

Uroz, S., D'Angelo-Picard, C., Carlier, A., Elasri, M., Sicot, C., Petit, A., Oger, P., Faure, D. and Dessaux, Y. (2003), 'Novel bacteria degrading N-acylhomoserine lactones and their use as quenchers of quorum-sensing-regulated functions of plant-pathogenic bacteria', *Microbiology*, 149(8), 1981–9.

Vandenkoornhuyse, P., Baldauf, S. L., Leyval, C., Straczek, J. and Young J. P. W. (2002), 'Extensive fungal diversity in plant roots', *Science*, 295(5562), 2051.

Vandenkoornhuyse, P., Quaiser, A., Duhamel, M., Le Van, A. and Dufresne, A. (2015), 'The importance of the microbiome of the plant holobiont', *New Phytol.*, 206(4), 1196–206.

Wang, L. and Mazzola, M. (2015), 'Integration of apple rootstock genotype with reduced *Brassica* seed meal application rates for replant disease control', *Phytopathology*, 105(4), 145.

Wang, Y., Li, C., Tu, C., Hoyt, G. D., DeForest, J. L. and Hu, S. (2017), 'Long-term no-tillage and organic input management enhanced the diversity and stability of soil microbial community', *Sci. Total Environ.*, 609, 341–7. doi:10.1016/j.scitotenv.2017.07.053.

Wardle, D. A., Yeates, G. W., Bonner, K. I., Nicholson, K. S. and Watson, R. N. (2001), 'Impacts of ground vegetation management strategies in a kiwifruit orchard on the composition and functioning of the soil biota', *Soil Biol. Biochem.*, 33(7–8), 893–905.

Watson, T. T., Nelson, L. M., Neilsen, D., Neilsen, G. H. and Forge, T. A. (2017), 'Soil amendments influence *Pratylenchus penetrans* populations, beneficial rhizosphere microorganisms, and growth of newly planted sweet cherry', *Appl. Soil Ecol.*, 117, 212–20.

Weerakoon, D. M. N., Reardon, C. L., Paulitz, T. C., Izzo, A. D. and Mazzola, M. (2012), 'Long-term suppression of *Pythium abappressorium* induced by *Brassica juncea* seed meal amendment is biologically mediated', *Soil Biol. Biochem.*, 51, 44–52.

Weller, D. M., Landa, B. B., Mavrodi, O. V., Schroeder, K. L., de la Fuente, L., Bankhead, S. B., Allende Molar, R., Bonsall, R. F., Mavrodi, D. V. and Thomashow, L. S. (2007), 'Role of 2,4-diacetylphloroglucinol-producing fluorescent *Pseudomonas* spp. in the defense of plant roots', *Plant Biol.*, 9(1), 4–20.

White, R. A. III, Borkum, M. I., Rivas-Ubach, A., Bilbao, A., Wendler, J. P., Colby, S. M., Köberl, M. and Jansson, C. (2017), 'From data to knowledge: The future of multi-omics data analysis for the rhizosphere', *Rhizosphere*, 3, 222–9.

Willett, M., Smith, T. J., Peterson, A. B., Hinman, H., Stevens, R. G., Ley, T., Tvergyak, P., Williams, K. M., Maib, K. M. and Watson, J. W. (1994), 'Growing profitable apple orchards in replant sites: An interdisciplinary team approach in Washington state', *HortTechnology*, 4(2), 175–81.

Wissuwa, M., Mazzola, M. and Picard, C. (2009), 'Novel approaches in plant breeding for rhizosphere-related traits', *Plant Soil*, 321, 409–30.

Xiao, Y., Peng, F., Dang, Z., Jiang, X., Zhang, J., Zhang, Y. and Shu, H. (2015), 'Influence of rhizosphere ventilation on soil nutrient status, root architecture and the growth of young peach trees', *Soil Sci. Plant Nutr.*, 61(5), 775–87.

Zhang, Y., Wang, L., Yuan, Y., Xu, J., Tu, C., Fisk, C., Zhang, W., Chen, X., Ritchie, D. and Hu, S. (2018), 'Irrigation and weed control alter soil microbiology and nutrient availability in North Carolina Sandhill peach orchards', *Sci. Total Environ.*, 15, 517–25.

Zhou, J., Deng, Y., Luo, F., He, Z. and Yang, Y. (2011), 'Phylogenetic molecular ecological network of soil microbial communities in response to elevated CO_2', *mBio*, 2(4), e00122-11.

Zhu, Y., Shin, S. and Mazzola, M. (2016), 'Genotype responses of two apple rootstocks to infection by *Pythium ultimum* causing apple replant disease', *Can. J. Plant Pathol.*, 38(4), 483–91.

Ziber-Rosenberg, I. and Rosenberg, E. (2008), 'Role of microorganisms in the evolution of animals and plants: The hologenome theory of evolutions', *FEMS Microbiol. Rev.* 32(5), 723–35.

Chapter 2

Advances in the development and utilization of fruit tree rootstocks: a case study for apple

Gennaro Fazio, USDA-ARS Plant Genetics Resources Unit, USA; and Terence Robinson, Cornell University, USA

1 What are rootstocks?

All commercial temperate zone fruit trees are composed of an aerial 'scion' cultivar grafted or budded on another cultivar which serves as the support root system referred to as the 'rootstock.' The practice of budding or grafting desirable scion cultivars on rootstocks has been practiced for centuries due to the highly heterozygous nature of tree fruits (Tukey, 1978), most of which do not reproduce 'true to type' by seed. Thus, in general, the seed from a desirable fruit variety will not result in a tree which produces the same fruit characteristics as the parent. To overcome this problem, fruit growers learned many centuries ago that a desirable genotype could be propagated asexually by budding (a single bud) or grafting (a small section of shoot with several buds) onto other plants with roots (usually the same or a closely related species) and then allowing only the bud of the desirable cultivar to grow and develop into each tree's canopy, thus creating multiple trees of the desirable cultivar (Cummins, 1973; Larsen, 1976). For example, all of the 'Red Delicious' apple (*Malus domestica*) trees in the world originated from a single tree discovered

http://dx.doi.org/10.19103/AS.2018.0040.02
Published by Burleigh Dodds Science Publishing Limited, 2019.

in Peru, Iowa, in the 1800s. Thereafter, buds from the original tree were budded onto other apple seedlings, and later interspecific hybrid clones, that serve as rootstocks to produce the millions of 'Delicious' trees that have been grown around the world from until the present day. Similarly, 'Montmorency' sour cherry (*Prunus cerasus*) is a 400+ year-old cultivar that originated in France, but comprises the majority of sour cherry production in the United States, where it is grown primarily on *Prunus mahaleb* seedling rootstocks. The general history of all major fruit cultivars, including subtropicals such as *Citrus* as well as temperate zone fruit and nut trees, is similarly based on the propagation of superior fruiting genotypes on different rootstock genotypes. Not only does this provide a way to reliably reproduce the superior traits of the scion, but also a way to adapt its production to different localized soils, climates, and production systems.

Historically, seedlings were used as rootstocks for deciduous fruit tree species such as apple, pear (*Pyrus communis*), peach (*Prunus persica*), tart and sweet (*Prunus avium*) cherry, apricot (*Prunus armeniaca*), and plum (*Prunus domestica* and *Prunus salicina*) (Sax, 1949). The classic way to produce a rootstock is to plant seeds and when the young seedling is 30–50 cm tall, bud or graft onto that seedling the desirable scion cultivar. However, with the exception of some peach and almond rootstocks, since each seedling rootstock is a unique genotype, there can be considerable variability in its own growth characteristics as well as the characteristics it may impart to the scion due to the heterozygosity of each fruit species, resulting in variability in tree performance in the orchard. Potential variations in seedling rootstocks include vegetative vigor, tree shape, and size, yield, precocity of fruit bearing, fruit size, and susceptibility to root diseases and abiotic stresses. Nevertheless, almost all commercial orchards used seedling rootstocks until the twentieth century. While this chapter focuses mostly on apple, it exemplifies the various inherent rootstock properties and rootstock-induced qualities on grafted scions that can be found in other temperate fruit rootstocks.

2 History and modern use of clonal rootstocks in apple

Several millennia ago (possibly by the fourth century AD), rootstocks that had unique desirable characteristics began to be selected and propagated asexually by rooted cuttings or via stoolbed or layerbed techniques (Tukey, 1964). One such apple rootstock named 'Paradise' was dwarfing. It was propagated vegetatively in Europe and used in home gardens for several centuries before the modern era of rootstock improvement (Lindley, 1828; Loudon, 1822). A number of other dwarfing apple rootstocks were in use and propagated clonally by the late nineteenth century, but duplication of names and confusion of rootstock identity hindered adaptation for commercial production. To solve this problem, Hatton at East Malling Research Station (EMRS) in the United Kingdom (UK) began collecting and categorizing apple rootstocks from all over Europe. These were named using the letters EM (East Malling) and roman numerals I-IX (Hatton, 1917, 1919). Later, the EM designation was changed to simply M and the numbers from Roman to Arabic numbers. The series was expanded to 16 genotypes in 1914, and later additional rootstocks 17–24 were listed in 1924.

The most dwarfing rootstocks of the Malling series (M.8 and M.9) initially were considered too dwarfing for commercial orchards and more suitable for home gardens. However, from the semi-dwarfing Malling rootstock series (M.2, M.4, M.7, M.13 etc.), several were adopted by commercial apple growers in England and other European countries (Hatton, 1920). By the late 1950s, clonal rootstocks began to replace seedling rootstocks in most of Europe, North America, Australia, and New Zealand. This facilitated the development of planting systems at double or triple the planting density of trees on seedling rootstocks, which typically were planted at 150 trees/ha. At roughly the same time in the 1960s, Don Heinicke in the United States and Don McKenzie in New Zealand independently developed the central leader tree training system for use with semi-dwarfing clonal rootstocks (Heinicke, 1975; McKenzie, 1964, 1985). This planting system revolutionized apple growing and was adopted worldwide. Its primary advantage was earlier production because of greater precocity of the semi-dwarfing Malling rootstocks and greater cumulative production due to the higher planting density (Palmer et al., 1989; Sansavini et al., 1981; Sansavini and Musacchi, 2000).

The more dwarfing rootstocks of the Malling series, particularly M.9, began to be used in some German and Dutch orchards in the 1960s, leading to the development of the slender spindle tree form by Bob Wertheim in the late 1960s (Wertheim, 1978). These dwarf slender spindle trees were planted at 1500–2000 trees/ha. Initially, this concept was only accepted in Northern Europe where land for orchards was limited (Wertheim, 1981; Wertheim and Callesen, 2000). In areas of the world where land was more plentiful, most growers preferred to plant semi-dwarfing rootstocks on large land areas.

In France, a different tree form that could be used with semi-dwarfing or dwarfing clonal rootstocks, named the Vertical Axis, was developed by Jean Marie Lespinasse in the mid-1970s. Trees were planted at a density of 1000–1500 trees/ha. Although in France this tree form was developed mostly with M.9, growers in many other parts of the world mostly used semi-dwarfing clonal Malling rootstocks in the 1980s and 1990s (Barden, 1995; Crassweller and Smith, 2001).

In the late 1980s, many research and extension personnel around the world began to evaluate and promote higher tree densities on M.9 rootstock trained to various versions of the slender spindle tree form. However, growers in most regions were hesitant to adopt M.9 rootstock. A notable exception was Northern Italy where Herman Oberhofer, an extension specialist, began to take groups of growers to Holland to observe high-density orchards on dwarfing rootstocks, and within 10 years the vast majority of Northern Italy apple farms had converted to slender spindle on M.9 rootstocks (Comai and Widmann, 1972).

Through the 1990s and 2000s, most other apple-growing areas of the world switched from clonal semi-dwarfing rootstocks to clonal dwarfing rootstocks (Hampson et al., 2002; Robinson et al., 1991a). This happened more rapidly in some countries than others. In some countries, an intermediate step was taken by using a semi-dwarfing rootstock with the scion grafted on M.9 as an interstock. These interstem trees were more dwarfing than those on semi-dwarfing rootstocks, but not as dwarfing as those directly on M.9 rootstock (Domoto, 1982; Ferree et al., 1982; Koike and Tsukahara, 1988; Lord, 1983). In almost all regions of the world that used interstem trees, these have now been replaced by fully dwarfing rootstocks. Currently, most areas of Europe, North and South America, and Australia, New Zealand, and South Korea, use dwarfing stocks and planting densities greater than 2000 trees/ha and some as high as 6000 trees/ha. However, in some areas of the world, particularly China, Japan, and India, adoption of dwarfing clonal rootstocks has been slow and growers continue to use seedling, or semi-dwarfing, rootstocks with limited use of interstem trees (Ma et al., 2013; Tamai et al., 2002, 2003).

3 Rootstock improvement efforts

After the initial effort to name and categorize European rootstocks in the early 1900s, Preston at EMRS conducted controlled crosses of rootstocks which resulted in the release of M.26 in 1959 and later M.27 in 1975 (Preston, 1967; Preston and Belcher, 1982). M.26 was a cross of M.16 and M.9 and found widespread acceptance around the world since it was slightly more vigorous than M.9, but less vigorous than M.7. M.27 has found only limited use because it is even more dwarfing than M.9 and often with reduced fruit size (Wertheim and Scholtens, 1994).

The introduction of Malling rootstocks to Australia and South Africa revealed an important weakness, their susceptibility to woolly apple aphid (*Eriosoma lanigerum*) (Dozier et al., 1974). These aphids colonize the tops of trees, but in areas with cold winters, they are killed and then re-colonize slowly the next year. However, in areas with mild or warm winters, they colonize the root system and then re-infest the aerial parts of the tree rapidly the next year. The need for woolly apple aphid-resistant rootstocks led to a joint breeding program between EMRS and the Merton Research Station in the UK. Crosses of Malling rootstocks with Northern Spy resulted in a new series of rootstocks, the

Malling-Merton (MM) series numbered from 101 to 114. All are semi-dwarfing. Of these, the most important were MM.104, MM.106 and MM.111 (Preston, 1966). They were adopted by growers in many apple regions in the world and were utilized in the Central Leader system at densities from 500 to 800 trees/ha. MM.106 is highly productive, but also highly susceptible to *Phytophthora* root rot, which limited its use in wet soils (Browne and Mircetich, 1993). MM.111 is less productive and slightly more vigorous than MM.106, but is very durable and tolerant to drought stress (Atkinson et al., 1997; Tworkoski et al., 2016).

The need for better rootstocks has prompted many institutions around the world to make crosses for breeding objectives that have varied by institution, and have been as simple as improved rooting in the propagation bed or as complex such as multiple resistances to rootstock biotic and abiotic stresses. Rootstock breeding programs have been conducted in Sweden (Alnarp 2), Poland (P series) (Czynczyk and Omiecinska, 1989), Germany (Supporter® and Pillnitz series) (Fisher, 1994), Czech Republic (JTE series) (Dvorak, 1983; Webster and Tobutt, 1994), Romania (Voinesti series) (Mazilu et al., 1999), Russia (Budagovsky series) (Kuldoshin and Sadowski, 1999; Webster and Tobutt, 1994), China (SM series) (Gao et al., 2011; Rong et al., 2011; Wan et al., 2011), Japan (Morioka series) (Bessho and Soejima, 1992; Tsuchiya, 1988), Canada (Ottawa, KSC, SJM and Vineland series) (Elfving et al., 1993; Embree, 1985; Khanizadeh et al., 2005; Spangelo et al., 1974), New Zealand (IFO series) (Bus et al., 2008), Michigan, USA (MAC series) (Carlson and Perry, 1986), and Geneva, New York, USA (Geneva® series) (Cummins and Aldwinckle, 1974; Fazio et al., 2015b). There are now more than 100 named rootstocks in the world (Table 1).

One of the more impactful breeding programs started in 1937 at the Michurinsk Research Station in Russia. The primary objective of this program was increased winter hardiness. They used Russian red leaf rootstock as their source of cold hardiness and M.8 as their source of dwarfing. They released Budagovsky 9 (B.9) in 1975 as an M.9-sized stock with greater cold hardiness than M.9 (Czynczyk, 1979). It has had a worldwide impact and has been planted widely in the United States and Northern Europe. Researchers in the United States (LoGiudice et al., 2006; Russo et al., 2008b) showed B.9 is also resistant to fire blight (caused by *Erwinia amylovora*). The nature of the resistance is unusual since the young plant is sensitive to fire blight, but with age the grafted tree shows field-level resistance. Other rootstocks which have had limited acceptance are B.491 and B.118. A fourth and more recent rootstock, B.10 (B.62-396) is rapidly gaining acceptance in the United States (Autio et al., 2017a,b). It is slightly more vigorous than M.9, but is highly productive like M.9 and shows fire blight tolerance similar to B.9. Several other selections from the Budagovsky breeding program have been evaluated in North America, but none has shown high productivity and dwarfing (Autio et al., 2017a,b).

Table 1 A partial listing of apple rootstocks worldwide, with origin, characteristic vigor, parentage, and tree size relative to industry standards

Roostock	Type	Origin	Parentage	Tree size	References
B.54-118	Semi-dwarf	Michurinsk College, Russia	Unknown	–	Hulko et al. (1999), Kuldoshin (1999)
B.9	Dwarf	Michurinsk College, Russia	Unknown	M.9	Hulko et al. (1999), Kuldoshin (1999)
B.10	Dwarf	Michurinsk College, Russia	Unknown	M.9	Hulko et al. (1999), Kuldoshin (1999)
B.118	Semi-Dwarf	Michurinsk College, Russia	Unknown	M.9	Hulko et al. (1999), Kuldoshin (1999)
B.490	Dwarf	Michurinsk College, Russia	Unknown	M.9	Hulko et al. (1999), Kuldoshin (1999)
B.491	Dwarf	Michurinsk College, Russia	Unknown	M.9	Hulko et al. (1999), Kuldoshin (1999)
CG.2022	Dwarf	Geneva, USA	Malling 9 × Ottawa 11	M.27	Russo et al. (2007)
CG.2034	Dwarf	Geneva, USA	Dolgo crab × Malling 27	M.27	G. Fazio, pers. comm.
CG.4003	Dwarf	Geneva, USA	(Antonovka Kamienaja × Ottawa 3) × Robusta 5	M.26	Norelli et al. (2003)
CG.4004	Dwarf	Geneva, USA	722506-004 × OP	M.26	G. Fazio, pers. comm.
CG.5257	Dwarf	Geneva, USA	Ottawa 3 × Robusta 5	M.26	G. Fazio, pers. comm.
CG.6006	Semi-dwarf	Geneva, USA	PK-14 × Robusta 5	M.26 to M.7	G. Fazio, pers. comm.
CG.8189	Semi-dwarf	Geneva, USA	Ottawa 3 × Robusta 5	M.7-MM.106	G. Fazio, pers. comm.
G.202	Dwarf	Geneva, USA	M.27 × Robusta 5	M.26	Norelli et al. (2003)
G.210	Semi-dwarf	Geneva, USA	Ottawa 3 × Robusta 5	M.7	Norelli et al. (2003)
G.214	Dwarf	Geneva, USA	Ottawa 3 × Robusta 5	M.26	Norelli et al. (2003)
G.222	Dwarf	Geneva, USA	M.27 × Robusta 5	M.26	G. Fazio, pers. comm.
G.814	Dwarf	Geneva, USA	Ottawa 3 × Robusta 5	M.7	Norelli et al. (2003)
G.890	Semi-Dwarf	Geneva, USA	Ottawa 3 × Robusta 5	M.26 to M.7	Norelli et al. (2003)
G.935	Dwarf	Geneva, USA	Ottawa 3 × Robusta 5	M.26 to M.7	Norelli et al. (2003)
G.969	Semi-dwarf	Geneva, USA	Ottawa 3 × Robusta 5	M.7	Norelli et al. (2003)
JM.1	Dwarf	Morioka, Japan	Unknown	M.9	G. Fazio, pers. comm.

JM.2	Dwarf	Morioka, Japan	Unknown	M.7	Norelli et al. (2003)
JM.4	Dwarf	Morioka, Japan	*M. prunifolia* Seishi × M.9	M.26	Norelli et al. (2003)
JM.10	Dwarf	Morioka, Japan	*M. prunifolia* Seishi × M.9	M.9	Norelli et al. (2003)
JTE-B	Dwarf	Czech Republic	Unknown	M.7	Norelli et al. (2003)
JTE-C	Dwarf	Czech Republic	Unknown	M.7	Norelli et al. (2003)
Mark	Dwarf	Michigan, USA	M.9 open pollinated	M.9	Carlson and Perry (1986)
M.26	Dwarf	HRI-East Malling, UK	M.16 × M.9	M.26	Preston (1974), Rogers and Beakbane (1957), Proctor et al. (1974)
M.27	Dwarf	HRI-East Malling, UK	M.13 × M.9	M.27	Preston (1974), Barritt et al. (1995)
M.7	Semi-dwarf	East Malling, UK	Unknown	M.7	Hatton (1917), Preston (1967)
M.9	Dwarf	Reselected at HRI-East Malling, UK	Unknown	M.9	Hatton (1917), Van Oosten (1977), Van Oosten and Groene (1984), Webster and Hollands (1999)
MM.106	Semi-dwarf	East Malling, Uk	Northern Spy × M.1	MM.106	Preston (1955, 1966)
P.1	Semi-dwarf	Skierniewice, Poland	M.4 × Antonovka	M.7	Czynczyk and Omiecinska (1989)
P.2	Dwarf	Skierniewice, Poland	M.9 × Antonovka	M.9	Czynczyk and Omiecinska (1989)
P.16	Dwarf	Skierniewice, Poland	Longfield × M.11	M.9	Czynczyk and Omiecinska (1989)
P.22	Dwarf	Skierniewice, Poland	M.9 × Antonovka	M.27	Czynczyk and Omiecinska (1989)
PiAu.51-11	Dwarf	Pillnitz, Germany	M 4 open pollinated	M.26	Norelli et al. (2003)
PiAu.51-4	Semi-dwarf	Pillnitz, Germany	M 4 open pollinated	M.26	Norelli et al. (2003)
PiAu.56-83	Semi-dwarf	Pillnitz, Germany	M 11 open pollinated	M.9	Norelli et al. (2003)
V.1	Dwarf	Vineland, Ontario, Canada	"Kerr" open pollinated	M.9-M.26	Elfving et al. (1993), Hampson et al. (2012)
V.2	Semi-dwarf	Vineland, Ontario, Canada	"Kerr" open pollinated	M.7	Elfving et al. (1993), Hampson et al. (2012)
V.3	Dwarf	Vineland, Ontario, Canada	"Kerr" open pollinated	M.27-M.9	Elfving et al. (1993), Hampson et al. (2012)
V.4	Semi-dwarf	Vineland, Ontario, Canada	"Kerr" open pollinated	M.7-MM.106	Elfving et al. (1993), Hampson et al. (2012)
V.7	Semi-dwarf	Vineland, Ontario, Canada	"Kerr" open pollinated	M.7	Elfving et al. (1993), Hampson et al. (2012)

Another breeding program that has had significant impact has been the German program at Pillnitz. Fisher has released four rootstocks, the Supporter® series 1-4. Supporter® 4 is similar to M.26, being highly productive, and has been planted to some extent in several European countries (Fischer et al., 1997). It has had little importance outside of Europe because it is susceptible to fire blight and replant disease (Autio et al., 2013; Auvil et al., 2011).

The breeding program in Poland has released a number of cold-hardy rootstocks which have been used in Poland and to a limited extent in other European countries. The most important have been P.22 (very dwarfing, similar to M.27) and P.16 (similar to M.9, but more winter hardy). While these have been tested worldwide, their implementation has been somewhat slow because they did not offer significant improvements over the current standard rootstocks (Marini et al., 2006).

Mark rootstock, bred by Robert Carlson and released by Michigan State University in the early 1980s, had its most significant impact in the United States from 1985 to 1995. Mark is slightly more vigorous than M.9 and is winter hardy and very productive. However, as the trees aged, a proliferation of non-organized tissue developed just below the soil line, leading to weak tree growth and small fruit size (Travis and Rytter, 1995; Travis et al., 1999; Warmund et al., 1993).

Of the four Canadian breeding programs, Ottawa 3 has had the most impact. It is very winter hardy and very productive, with vigor similar to M.9. It was planted to a limited extent in the late 1980s and early 1990s, but problems with transplant losses, virus sensitivity, and difficulty in propagation limited its use (Ferree, 1992; Rioux et al., 1984; Spangelo et al., 1974). Currently, five rootstocks from Vineland, Ontario, are under development and may have importance in the future (Hampson, 2012; Hampson et al., 2012). The KSC and SJM rootstocks have not had commercial importance.

Another impactful rootstock breeding program has been at the Geneva campus of Cornell University. The program was started by James Cummins in 1969 and was joined by Herb Aldwinckle in 1971 to develop a series of rootstocks that not only conferred high productivity and dwarfing, but also resistances to the most important rootstock diseases and biotic stresses (Aldwinckle et al., 1972; Cummins and Aldwinckle, 1974). They extensively used 'Robusta 5' rootstock as a parent since it is resistant to fire blight and other diseases (Gardner et al., 1980). Other parents were either M.9, M.26, M.27, or Ottawa 3. They screened all progeny for resistance to fire blight and tolerance/resistance to crown and root rot caused by *Phytophthora cactorum*. They also screened for woolly apple aphid resistance and selected for low number of root suckers or burr-knots (Cummins et al., 1983). These were then selected for high productivity, dwarfing, and cold hardiness. Interestingly, some of the Geneva® rootstocks also have shown tolerance/resistance to apple replant disease (ARD), although that was not a breeding objective. The program was converted to a joint breeding venture between Cornell University (Geneva) and the USDA-ARS in 1998, and it continues to make crosses and release new rootstocks under the leadership of Gennaro Fazio. The breeding objectives have evolved through time. While these include previous objectives of fire blight resistance, crown rot resistance, cold hardiness, low root suckers, and low burr-knots, newer objectives include replant disease tolerance, specific nutrient uptake (especially Ca), low chill induction requirement, drought tolerance, water-use efficiency (WUE), tolerance to sodic soils, tolerance to particular soil pH levels (high and low), and graft union strength (Fazio et al., 2015b). As of 2019, 14 rootstocks have been released by the Geneva® program and several have achieved importance in the United States and some other parts of the world. Those that are being produced in large volumes (>500,000 plants per year) include, in order of importance, G.41, followed by G.11, G.935, G.969, G.890, and G.213. The total worldwide sales of these rootstocks was 8.8 million in 2017.

Currently in the world, there are only five to seven rootstock breeding programs. In addition to the Geneva® program, there are three programs in China, one in New Zealand, and one in Russia. A unique objective of one of the Chinese programs is apomixis in rootstocks, which would allow propagation of rootstocks by seed. This would drastically change the propagation industry in the world. Primary objectives of these programs include cold hardiness (Russia, China), fire blight resistance (New Zealand), and drought tolerance (China) (Gao et al., 2011; Ma et al., 2012; Sha et al., 2011; Wan et al., 2011; Zhang et al., 2011).

4 Rootstock propagation

Apple rootstocks are propagated either by seed, cuttings, layering, stooling, or by tissue culture. When propagated by seed, the grafted trees are usually vigorous, but also variable in tree size and productivity due to the variability inherent in seeds (Visser and Schaap, 1967). Thus, almost all apple rootstocks in the world are propagated asexually by cuttings, layerbed, or stoolbed. Only since about 2008 have apple rootstocks been propagated commercially by tissue culture (Castillo et al., 2015).

Propagation in all areas of the world, except some Asian countries including China, is mostly done by layerbed or stoolbed. However, in China most rootstocks are propagated by rooting of hardwood cuttings (Kwon et al., 1999; Yoshida and Muramatsu, 1998). This is due to the difficulty of propagating Malling stocks (which are used in the west) by cuttings (Sun and Bassuk, 1991), while rootstocks used in China have the genetic makeup to root well from cuttings.

Typically, propagation by stoolbed (plants planted vertically) or by layerbed (plants planted on an incline and then laid flat at the end of the first season) is done by planting rootstock plants in a row in a shallow trench (Hartmann et al., 1997). After the first year's growth, the plants are cut back to three buds on each shoot for a stoolbed or the shoots are laid horizontal along the ground in the bottom of the trench for a layerbed. Later in the spring of the second year, when shoots from the 'mother' plants reach about 30 cm tall, sawdust, peat, or soil is mounded up to cover the lower parts of the new shoot. The sawdust is kept moist by irrigation and additional layers of sawdust or soil are mounded up through the season to a height of 30 cm. In the late summer and fall and during the winter in climates with mild winters, the shoots develop roots from one to five nodes along the lower stem of the 1-year-old shoot. These shoots are harvested from the mother plants in late fall, winter, or early spring by cutting the stem below the new roots, but leaving intact the mother plant. The process is repeated each year by sweeping away the sawdust or soil from the mother plants in the spring, exposing the horizontal shoots in the layerbed or the upright plant in the stoolbed, and then adding more sawdust or soil again as new shoots reach 30 cm in height.

The ability to produce roots in a stoolbed or as a rooted cutting differs among rootstocks (Villeneuve, 1986). A related rootstock trait is the tendency to produce burr-knots which are aboveground masses of root initials. Burr-knots are considered a defect and create a risk of rootstock infection by the bacteria that causes fire blight (Marini et al., 2003). Good rooting in a stoolbed generally is associated with a tendency to produce burr-knots. Many of the Malling stocks produce burr-knots, but also can be propagated easily in a stoolbed, while many of the Geneva® rootstocks root poorly in a stoolbed and do not produce burr-knots.

The stoolbed/layerbed method has been used for several centuries to propagate apple rootstocks, with only small improvements in technique. This method is an extension of what happens naturally with some apple trees where root-derived suckers come up from the ground season after season (Costante et al., 1983). Rootstock clones that root well with this system have been successful commercially, while those that root poorly in this system usually have been discarded (Robinson et al., 1997). However, the 2005 introduction of G.41 (which does not root well in a layerbed) stimulated the development of improved new techniques. Adams (2010) found that applications of the gibberellin biosynthesis inhibitor, prohexadione-calcium, to the shoots of rootstock layerbeds when the shoots were 90–100 cm tall, resulted in a reduction in shoot growth, but better rooting at the base of the shoot. Fazio (unpublished data) also observed in commercial nursery settings that planting G.41 in a vertical position in a stoolbed at double or triple the normal density resulted in better rooting of each shoot. This was due to the more limited number of shoots produced by each stooling mother plant (3–5 on a high-density stool plant vs. 10–15 from a layerbed plant) and the competition for resources between shoots. Lastly, Adams (2010) showed that if the stoolbed was established using tissue culture mother plants, rooting was improved significantly, and the increased rooting lasted for several years. These three improvements in stool/layerbed technology have allowed the successful stoolbed propagation of difficult-to-root rootstocks such as G.41.

A second important method of apple rootstock propagation has been the use of hardwood cuttings. This is the main method of propagation of rootstocks in China, but it is uncommon in other parts of the world. The most common rootstock in China is *Malus prunifolia*, which roots readily from cuttings (Yao et al., 2001). Typically, dormant (hardwood cuttings) are dipped in a synthetic auxin (indolebutyric acid, IBA) and then planted in a rooting bed of sand/soil in a plastic-covered high tunnel greenhouse and kept in high humidity until cuttings have rooted. They are then transplanted to a nursery where they are budded with a scion variety (Rong et al., 2011; Wen et al., 2018). This is seldom successful with Malling or Budagovsky rootstocks due to their low rooting percentage (Bassuk and Howard, 1980). Recently in the United States, several of the new Geneva® rootstocks have been propagated by softwood cuttings. Typically, green cuttings consisting of the tops of tissue-cultured plants are removed and dipped in rooting hormone (IBA), planted in a rooting bed of artificial media (vermiculite and peat moss), and kept under a plastic tunnel with misting until rooted. When the source of the green wood is micro-propagated material, these cuttings are more successful (Quamme and Hogue, 1994). These rooted plants are transplanted to a nursery and then budded in the late summer with a scion variety (Fleming, pers. comm.).

The newest large-scale commercial method of propagation is via tissue (tip) culture. Tissue culture consists of harvesting a shoot apex (an explant) and growing it on an artificial medium with a complete set of nutrients (Castillo et al., 2015; Geng et al., 2015). The plant hormones (or synthetic versions of plant hormones) are placed in the medium and their relative concentration is modified to obtain specific growth characteristics. By varying the balance of auxins and cytokinins, the explant is induced first to multiply by producing callus and shoots. These plants are divided and subdivided multiple times in an iterative process that produces thousands of new explants from an original plant. Later, the hormone balance is modified by increasing auxins to induce rooting. These small, sterile, rooted plantlets are then transplanted into a soil-less medium and grown in a mist tunnel for several weeks for the first phase of acclimation ('hardening off'). Then they are moved to larger pots in a regular greenhouse to acclimate to higher light levels, and finally are moved to the open air. These rooted plants can then be planted in a nursery and budded with a scion variety. This method was tried in the 1980s for propagating M.9

and Mark, but problems arose when the plants developed differently in the field, with many more burr-knots and vigorous growth (James and Thurbon, 1979; Webster and Jones, 1989). This may have been due to epigenetic effects of the hormones used in the tissue culture process, or it may have been simply a mix up of plant material and the propagation of a seedling instead of M.9 or Mark. This occurred before the era of DNA fingerprinting, thus the problem was never resolved. Nevertheless, because of those bad experiences in Europe, the use of tissue culture to propagate apple rootstocks was banned and fell into disfavor. In the mid-2000s, Gennaro Fazio began working with various tissue culture labs in the United States in an effort to propagate G.41 via tissue culture since it is difficult to propagate by stoolbed or cuttings. Field trials of trees from these tissue-cultured rootstocks performed similarly to stoolbed propagated rootstocks (Autio et al., 2005, b); by 2010, commercial quantities of G.41 and other Geneva® rootstocks were being propagated by tissue culture. This success stimulated others in the world to accept tissue-cultured Geneva® rootstocks. By 2017, there were more than three million Geneva® rootstocks being propagated by tissue culture each year.

Important advances in both the techniques of apple rootstock tissue culture and the improvements in tree performance from tissue-cultured plants have been achieved. Each tissue culture company has developed proprietary methods to achieve commercial success. These individual trade secrets are not widely shared. However, the results in the nursery and in the orchard have been published. Adams (2010) showed that liners from tissue-cultured plants have more roots and a more fibrous root system than stoolbed plants. The improvement in rooting carried over to the stoolbed where liners from a stoolbed that had tissue-cultured mother plants had more roots than liners from a stoolbed started with conventional plants. Because of the more fibrous root system, tissue culture plants establish better in the orchard. In addition to propagation benefits, micro-propagated plants feature a more vigorous root system with many more primary roots than conventional liners. Some nurseries are offering these well-developed root systems in a potted tree nursery production system that can be transplanted with few losses even during mid-summer.

Currently, the price of a stoolbed-produced rooted rootstock liner or a rooted cutting is less than a tissue-cultured, rooted rootstock liner. Nevertheless, all three methods are used commercially to propagate apple rootstocks worldwide. However, the stoolbed/layerbed method predominates.

5 Rootstock evaluation

Systematic rootstock evaluation began with the work of Hatton at EMRS in the early 1900s (Hatton, 1917). Their published work was a guide for growers and researchers alike. Individual researchers in Europe and North America continued individual comparative trials of rootstocks through the 1960s (Carlson, 1974; Nelson and Tukey, 1955). Later researchers in other countries also began comparative rootstock trials, including New Zealand, Australia, South Africa, Japan, and more recently China (Bergh, 1992; George and Nissen, 1986; Racsko et al., 2011; Tustin and Cashmore, 1994; Tustin et al., 1993). However, the trial results in one climate and soil type often differed from results of other trials in other climates. This led to substantial confusion and differing opinions among researchers and growers. In 1976, a group of researchers in eastern North America launched a coordinated rootstock testing program named the NC-140 project (Ferree, 1991, 1992; Ferree and Perry, 1989). This group began conducting uniform multilocation orchard comparisons of rootstocks and met annually to compare results. The project was later expanded across the United States and now includes participants from Canada and Mexico as well. The group also conducts comparative research trials with peach, cherry and pear rootstocks (Cowgill et al., 2017). Over the 45-year existence of this project, it has conducted 18 trials of apple rootstocks (approximately one every 3 years). With each new trial, the latest rootstocks from around the world have been included. The primary rootstock characteristics evaluated in the coordinated NC-140 trials include tree survival in various climates, level of dwarfing, precocity, yield, yield efficiency, fruit size, number of root suckers, and burr-knots. A similar group of researchers from Europe was organized in 2003 and is conducting coordinated uniform multilocation rootstock trials in several European countries (Kviklys, 2011). Another group led by Leo Rufato began multilocation coordinated trials in Brazil in 2014.

In addition to coordinated trials, individual researchers have focused on evaluations of cold hardiness, graft union strength, virus susceptibility, tolerance to replant disease, nematode tolerance or resistance, and mineral nutrient profiles.

6 Rootstock effects on scion traits and mechanisms

Rootstock genotype has numerous effects on the scion, including vigor, precocity, yield efficiency, partitioning of carbon, mineral nutrient profile, branch angle, and graft union strength (Fig. 1). In addition, rootstock tolerance or susceptibility to soil characteristics, climate stress, and biotic stresses determine if the tree survives, is stunted, or grows well.

6.1 Dwarfing

The most desirable rootstock characteristic is dwarfing. In most cases, seedling rootstocks confer the most vigor. The current list of clonal rootstocks range in

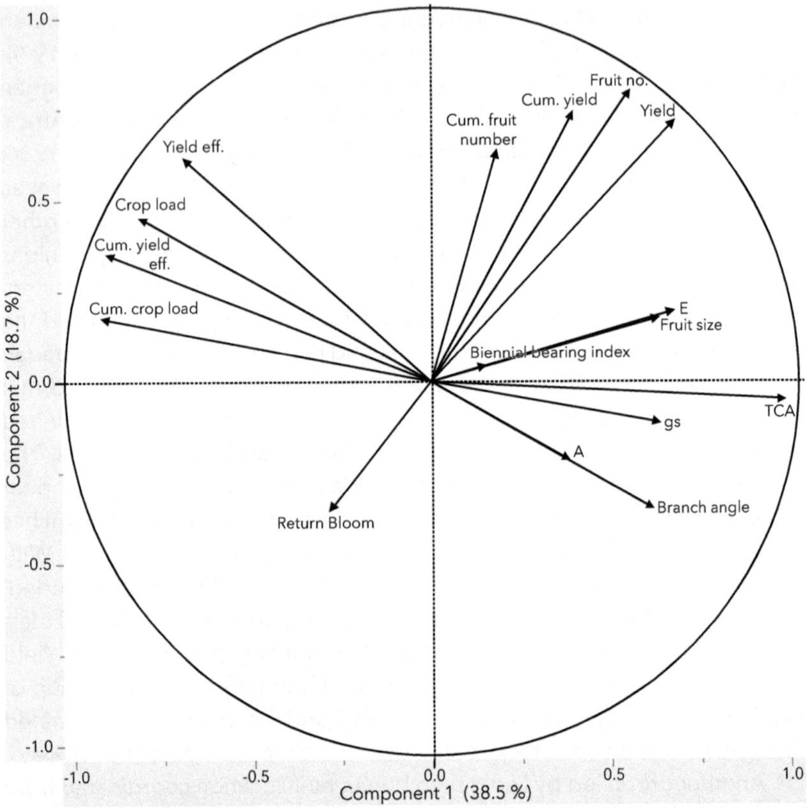

Figure 1 Principal components (PCA analysis) of variables affected by rootstocks: cumulative fruit number, cumulative yield, fruit number, yield, mean biennial bearing index, fruit size, transpiration (E), trunk cross-sectional area (TCA), stomatal conductance (*gs*), carbon assimilation (A), branch angle, return bloom, cumulative crop load, cumulative yield efficiency, crop load, and yield efficiency. The arrows indicate the effects that are positively correlated (in the same quadrant) and those that are negatively correlated (opposite quadrants).

vigor from 100% to only 10% of seedling (Table 1). This vast range of dwarfing has allowed a 10- to 20-fold increase in planting density in modern orchards. Although the mechanism of dwarfing has been studied intensively for more than 50 years, the complete explanation of apple dwarfing is still not clear. In the last 10 years, the genetic basis of dwarfing has been linked to several genes, *Dw1* and *Dw2*, and possibly a third (Fazio et al., 2014; Foster et al., 2015; Harrison et al., 2016). The physiological expression of these genes is less clear as they have not been characterized yet at the molecular level. Physiologists have measured effects of dwarfing rootstocks on root-supplied hormones and distinct carbon partitioning, with dwarfing rootstocks inducing a much greater use of carbon for fruit production compared to root system development. This is probably why trees on dwarfing rootstocks have much smaller root systems. It has been shown that dwarfing in apple is not caused by restricted water supply to the scion (Olien and Lakso, 1986). While the rootstock may cause some changes in vessel size, xylem element and vascular resistance to water is similar in vigorous and dwarfing rootstocks (Tworkoski and Fazio, 2011). In contrast, dwarfing in peach has been linked to water stress induced by dwarfing rootstocks (Johnson et al., 2011). Currently, the physiological basis for dwarfing appears to be a combination of root-supplied signals (maybe hormones) to the scion, inducing increased flowering, and reduced partitioning of carbon to the root system and early termination of vegetative growth (Foster et al., 2017; Van Hooijdonk et al., 2010).

6.2 Precocity

Precocity or induction of early bearing is defined as the earliness of an apple tree to flower and begin fruiting (Fazio et al., 2014). The physiological trait of juvenility describes young trees that are grown from seeds which do not produce any flowers or fruits for several (5–8) years, a phenomenon that seems to be linked to changes in methylation of the apple genome (Hafiz et al., 2008). As the trees age, they transition from juvenility to a reproductive stage of flowering and fruiting. The basis of juvenility may be linked to a plethora of root signals, including hormones. Gene-altering approaches that modulate genes involved in flower induction have been shown to reduce this juvenile phase for breeding purposes (Kotoda et al., 2006; Schlatholter et al., 2018; Schouten et al., 2009); however, thus far they have not proven successful in graft transmissible alteration of the juvenile period. When a mature (non-juvenile) apple scion is grafted on a seedling rootstock (which is juvenile), the scion may revert to a juvenile-like phase and flowering may be delayed for 5–8 years after grafting. However, when grafted onto a precocious clonal rootstock, there is no reversion to a juvenile-like phase, with the potential for flowering to occur in the first or second year in the orchard (Schmidt, 1986). Rootstocks differ in their effect on scion precocity (Fallahi and Mohan, 2000). Most of the semi-dwarfing rootstocks from the Malling series, and the Budagovsky series, are more precocious than seedling rootstocks, but flowering is still delayed 3–5 years after grafting. However, the dwarfing Malling, Budagovsky, and Geneva® rootstocks are much more precocious, with flowering in the first year in the orchard or even in the nursery. G.11 rootstock is highly precocious and can have flowers itself in the nursery. Interestingly, many of the semi-dwarf Geneva® rootstocks are also highly precocious such as the more dwarfing rootstocks. Precocity was a selection factor for semi-dwarf Geneva® rootstocks, since the lack of precocity is a serious flaw in Malling and Budgovsky semi-dwarfing rootstocks. Precocity induced by the rootstock has a large effect on orchard economics, since early production in years 1–5 is important for repaying the capital investment in a new orchard. The increase in precocity of dwarfing rootstocks has allowed the planting of much higher tree densities, with associated higher orchard establishment costs, since the investment can be paid off rapidly (Lordan et al., 2018a; Reig et al., 2019).

6.3 Yield efficiency and harvest index

A primary criterion to compare rootstocks has been the calculation of yield efficiency, which is defined as the weight of fruit produced (kg) per unit of trunk cross-sectional area (TCA in cm^2) measured at a set distance above the graft union (Robinson et al., 1991b). This is a rootstock-scion performance efficiency measurement since it relates the tree's fruit output relative to the size of tree as measured by TCA, which tends to be proportional to canopy size up to a point. This facilitates the comparison of trees on rootstocks of vastly different vigor and tree size on an orchard-area basis. However, this measure was developed for comparing trees on rootstocks of varying vigor that were minimally constrained by allotted orchard space in evaluation trials for which the filling of orchard space and canopy maturation typically took 6–10 years. It has recently been recognized that, as long as the primary determinant of canopy size is rootstock vigor, yield efficiency allows comparative evaluation of differences in rootstock-induced productivity relative to vegetative growth. However, as orchards have become more dense, with a primary training/production system focus on filling allotted orchard space and reaching full production rapidly, the point at which the canopy completes the filling of its allotted space occurs much earlier, and therefore pruning intervention to maintain the canopy in that space becomes a confounding factor to yield efficiency calculations, since the canopy's productive area is no longer expanding but the TCA continues to increase for the life of the tree. Therefore, yield efficiency comparisons are primarily of value only until the allotted orchard space is filled. Furthermore, as modern orchards are trained and pruned to more two-dimensional 'fruiting wall' canopy architectures, the inherent rootstock effect on yield efficiency is further confounded due to non-rootstock-based horticultural interventions.

When fruit production over several years (traditionally 10 in rootstock comparison trials, perhaps less for higher density training systems) is summed and the cumulative production is divided by the tree TCA (at the point at which allotted orchard space is filled), an estimate of harvest index is obtained (Palmer, 2011; Strong and Azarenko, 2000). Harvest index is the measure of fruit production compared to vegetative production (branches, leaves, trunk, and roots) of the tree. However, harvest index is difficult and expensive to measure, thus the measure of cumulative yield efficiency is used since final TCA is an estimate of the cumulative vegetative growth of the tree. In the few cases where harvest index has been measured directly, dwarfing rootstocks (M.9) were found to partition 70–80% of annual carbon into fruit and only 20–30% to vegetative growth, while vigorous rootstocks induce much more partitioning of carbon into vegetative growth (>50%) (Strong and Miller Azarenko, 1991).

An annual estimate of partitioning of carbon into fruit vs. vegetative growth can be obtained from the ratio of fruit production (kg) to the incremental increase in TCA (cm^2 increase); however, this estimate of annual harvest index is less commonly used by rootstock researchers. If this annual estimate was used more, it could show how the partitioning of carbon between fruit production and vegetative production changes over time, and would allow the identification of rootstocks that will be problematic over time as the orchard matures since they continue to have a high fraction of carbon partitioned into vegetative growth even when the tree is mature.

Rootstock vigor is linked loosely to yield efficiency, with most vigorous, semi-vigorous rootstocks, and semi-dwarfing having lower yield efficiency than dwarfing stocks. Interestingly, many of the semi-dwarfing Geneva® rootstocks have yield efficiencies that are similar to dwarfing rootstocks (Reig et al., 2018; Russo et al., 2007). This is because high yield efficiency was a selection criterion for semi-dwarfing Geneva® stocks. The increase in yield efficiency has resulted in much higher yields per ha for dwarf trees if the dwarf trees are planted at their optimum tree density, as predicted from their inherent tree size (Lordan et al., 2018b). The indiscriminate use of high yield efficiency values to plan orchard designs has resulted in serious errors for the apple industry when scion vigor is low, and growers choose the most dwarfing rootstocks available because of their high yield efficiency. This has resulted in many orchards where the trees do not fill the space allocated to each tree, resulting in moderate yields per ha even though the yield efficiency of the rootstocks is high. Nevertheless, if rootstock vigor is sufficient to fill the space quickly, then the high yield efficiency of a dwarfing rootstock will result in higher yields than a less-efficient rootstock. This improved mature production is an important factor for sustainable long-term profitability of high-density orchards on dwarfing rootstocks, compared to medium and low-density orchards on semi-dwarfing or vigorous rootstocks (Lordan et al., 2018a).

It seems clear that the impact of rootstocks on carbon partitioning and flowering is intertwined with the dwarfing effect of the rootstock on the scion. Since a dwarfing rootstock induces the partitioning of 70–80% of annual fixed carbon into the fruit, the amount of carbon left for vegetative growth is a small fraction of that available in a tree on a vigorous rootstock. The mechanism of increased yield efficiency of dwarfing rootstocks is not completely clear. Evidence to date indicates that increased root-supplied hormones differ among rootstocks (Adams et al., 2018; Lordan et al., 2017) and may cause increased flower initiation and fruit set, which leads to early production in the first or second years accompanied by higher partitioning of carbon to fruits. However, that explanation does not fully explain the dwarfing effect of rootstocks, since shoots of the scion on a dwarfing rootstock also stop growth earlier in the season that those on a vigorous rootstock. Nor does it explain why interstems of varying length decrease vigor and increase early bearing in apple trees (Carlson and Oh, 1975).

6.4 Influence on leaf and fruit nutrient concentrations

Recently, there has been great interest in rootstock effects on fruit mineral nutrient profiles. This has been driven by the widespread planting of the variety 'Honeycrisp,' which suffers from several fruit disorders including the calcium-related disorder bitter pit (Baugher et al., 2017; Biggs and Peck, 2015; Rosenberger et al., 2004). While the modulation of nutrients in the scion by rootstocks had been described in the past (Lockard, 1976; Rom and Rom, 1991; Tukey et al., 1962), the types of rootstocks used in such research were genetically very similar (Duan et al., 2017; Gharghani et al., 2010; Jin et al., 2012). Wide crosses performed in the Geneva® breeding program have revealed significantly different nutrient profiles induced by rootstocks (Fazio et al., 2012, 2013). Some rootstocks cause greater levels of K in the leaves or in the fruit, while others induce higher levels of Ca in the fruit or the leaves. Such rootstock-induced differences have been shown for several other nutrients, including N, P, S, Mg, and B (Reig et al., 2018). Changes in soil pH, for example, caused differences in the expected absorption curves for metal ions such as manganese and iron, indicating that some rootstocks perform better at certain pH values than others (Fazio et al., 2012). Soil pH is one of the most important predictors of soil fertility, and developing a set of rootstocks well adapted to specific pH profiles may improve orchard performance and open marginal land to apple cultivation. The genetic inheritance of nutrient absorption and translocation to different parts of the scion is quite complex, as there are many mechanisms that contribute to a rootstock's differential efficiency for a particular nutrient (differential evapotranspiration, crop load, root morphology, water availability and use efficiency, interaction with soil biota, active and passive transport, vessel composition, and size etc.) and the genetic landscape described by Fazio et al. (2013) shows a very dynamic multi-locus model intertwined with between-nutrient to nutrient positive and negative correlations. It is possible to identify rootstocks with high calcium effects on the scion; however, given the complex genetic nature of each nutrient profile, the combinatorial probability of developing a rootstock that features multiple desirable nutrient profiles decreases with the addition of more nutrient requirements (Fazio et al., 2015a; Reig et al., 2018). Therefore, nutrient-based selection of new apple rootstocks may have to be limited to a few nutrients at a time.

6.5 Branch angle and hormones

Within rootstock breeding populations, breeders have noticed the induction of differences in scion branch angle as well as level of sylleptic branching. Fazio and Robinson (2008a,b) reported certain Geneva® rootstocks induced flatter scion branch angles than other Malling stocks. G.935 is particularly adept at this in a nursery environment; it consistently promotes more feathers (sylleptic branches) than other traditional rootstocks. More recently, Lordan et al. (2017) showed that certain Geneva® rootstocks (e.g. G.11 and G.41) had higher levels of root-supplied cytokinins and abscisic acid than other Malling stocks (Lordan et al., 2017). This hormone profile was somewhat associated with flatter branch angles. Flatter branch angle also is associated with the potential for more flowers (Lauri and Lespinasse, 2001). This trait is potentially quite valuable in high-density orchard production systems, since trees with flat branch angles require less branch manipulation to control tree vigor.

7 Rootstock tolerance to abiotic and biotic stresses

7.1 Cold hardiness and lack of winter chilling

In northern climates, fall, mid-winter, and early spring cold temperatures are a serious risk and limitation to apple tree survival (Moran et al., 2018). In warmer climates, the lack of winter cold (sufficient to complete the endodormancy to ecodormancy transition) is also a limitation to uniform bud burst in the spring. Rootstocks can affect both cold hardiness and the chilling requirement for bud burst in the spring.

When fully cold-acclimated, apple flower buds can withstand temperatures of −30°C; however, roots can only withstand temperatures of −10°C. Soil buffering capacity for cold winter temperatures and snow cover usually protect roots from the temperatures below −10°C. However, if there is little snow cover and there are prolonged air temperatures below −30°C, then damaging soil temperatures below −10°C can occur in the root zone. A second type of rootstock damage during winter can affect the rootstock shank (the trunk-like portion of the root system below the graft union). If cold winter temperatures occur with no snow cover, the part of the rootstock that is exposed above ground and the part just below the soil surface can be damaged. If the entire cambium is killed in this zone just above and below the soil line, the tree will die in the spring about blossom time or during mid-summer when temperatures get hot (Embree and McRae, 1991; Prive et al., 2001).

The hardiness of the rootstock shank varies considerably among rootstocks. Differential tolerance to cold temperatures has been studied and certain Malling stocks such as M.7 are quite sensitive to winter damage at moderately cold temperatures of −20°C. In 1990, Quamme classified the rootstocks available at the time for winter hardiness. He classified M.7 as very tender, M.2, M.4, M.9, MM.106, and P.16 as tender, M.26, MM.111, MM.104, P.1, and J.9 as moderately hardy and Antonovka seedling, A.2, Beautiful Arcade, O.3, O.8, B.9, P.2, P.22, and P.18 as hardy (Quamme, 1990; Quamme et al., 1997). Under severe climatic conditions in Poland, tree mortality was greater on M.9 than on M.26 or B.9 (Czynczyk and Zagaja, 1984). Following the mid-winter cold event of 2004, Robinson et al. (2006) found tree survival with 'Honeycrisp' and 'McIntosh' as the scions was greatest (~90%) for O.3, V.1, V.3, G.16, G.30, and Mark. B.118, M.9T337, B.9, M.9 Nic 29, Supporter 4, M.26, and MM.111 had only 50% survival, while M.7 and MM.106 had very poor survival. Moran et al. (2011a) froze non-grafted rootstocks and found that G.41, G.11, G.30, B.9, P.2, and M.26 had similar hardiness, whereas G.935 had greater root hardiness than M.26 (Moran et al., 2011a,b). More recently, Moran et al. (2018) found that the Geneva® and Vineland series rootstocks exhibited a high degree of winter hardiness in January, but that some were more tender in the fall (October) or in the spring (April). G.30 was not hardy below −15°C in October or in April, while CG.4013 was not hardy below −15°C in the fall and CG.5257 was not hardy below −15°C in April. However, in November or in March, they had hardiness similar to mid-winter levels.

On the other extreme of low winter temperatures is the situation in some apple-growing regions of the world of too little winter cold to satisfy the endodormancy chilling requirement. Without adequate chilling, bud burst of the scion in the spring is delayed and variable, with a high percentage of buds failing to grow and subsequently dying (Midgley and Lotze, 2011; Rufato et al., 2010). Rootstocks can affect the percentage of buds on the scion that grow in spring following insufficient chilling. Recently, researchers in Brazil found that G.213 rootstock had a greater percentage of buds that grew in the spring after mild winters compared to M.9 or Marubakaido rootstocks. The same research revealed that other Geneva® rootstocks, such as G.210 or G.814, also may have a positive effect of bud development in climates with too little winter chilling (Macedo et al., 2018). Preliminary studies indicate that these rootstocks have higher levels of root-supplied cytokinins which may stimulate bud growth of the scion. Experiments conducted in Geneva, New York, that subjected 'Gala' budded on similar rootstocks found that budbreak occurred after accumulation of 550 chilling hours, about 150 less than the standard requirement for 'Gala.'

7.2 Drought tolerance and WUE

Apple tree WUE is a complex trait defined by the amount of photosynthesized carbon per unit of water transpired, and is commonly measured seasonally (units of seasonal dry-matter growth/units of water) or by measuring CO_2, O_2, and H_2O flux of tree canopies during short periods (Glenn, 2014). WUE in combination with phytohormones and root morphology are thought to be associated with drought tolerance in apples (Tworkoski et al., 2016; Zhang et al., 2014). Phenotypic diversity for WUE physiological and morphological components were found in domesticated apple and related wild species (Bassett et al., 2011), and several genes responding to water deficit have been described in apple roots (Bassett et al., 2014). While WUE may be related to tolerance to drought stress, the effective measure of tolerance to drought stress should be the maintenance of productivity and marketable fruit quality after the occurrence of stress (Atkinson et al., 1997). This is a difficult parameter to measure, because beyond the effect of apple rootstocks, soil conditions, scion variety, crop load, and other physiological variables all contribute to that parameter (Atkinson et al., 2000; Ebel et al., 2001; Lo Bianco et al., 2012).

7.3 Fire blight

One of the most serious risks to orchards on susceptible rootstocks is the bacterial disease fire blight. Although fire blight infects blossoms through the nectary in open flowers, it can travel in the plant through the xylem and then infect the cambium of the rootstock. If the rootstock is sensitive to fire blight, the cambium connecting the top of the tree and the root system is killed and the tree collapses a few months later or the next year (Norelli et al., 2001). Some of the Malling semi-dwarfing rootstocks, such as M.7 and MM.111, are partially resistant and thus there was little tree death in the era when they were the predominant rootstocks. However, M.9 and M.26 are extremely susceptible, and with the increased use of M.9 since the late 1990s, there have been numerous fire blight epidemics that have killed millions of trees and have cost apple growers millions of dollars in losses (Aldwinckle et al., 2004; Russo et al., 2007). A recent (2018) epidemic in Washington (USA) caused the death of an estimated 10% of the trees. Because of this risk, the primary objective of the Geneva rootstock breeding program was to develop fire blight-resistant rootstocks. If the rootstock is resistant, some flowers and then branches in the scion may become infected, but they can be removed by pruning and the tree will survive. The Geneva rootstock program has now released 14 fire blight-resistant rootstocks. The new rootstocks from New Zealand (IFO series) also are reported to be resistant.

The basis of the fire blight resistance used in the Geneva breeding program was the rootstock 'Robusta 5,' which is a descendant of an Asiatic crab apple species *Malus X robusta*. The Geneva progeny of 'Robusta 5' have broad resistance to fire blight strains, but there exists some variability in resistance to all known strains (Fazio et al., 2008). G.41 has some of the strongest resistance, while G.935 has shown some susceptibility to one strain. Nevertheless, they all have provided growers with a level of protection to the risk of tree death due to fire blight. Genetic inheritance of the 'Robusta 5'-type of resistance has been described as having a strain-specific component on chromosome 3 identified as a gene belonging to the NBS-LRR class of resistance genes (Broggini et al., 2014a,b; Fahrentrapp et al., 2013; Kost et al., 2015). Other minor QTLs on linkage groups 5, 7, 11, and 14, which do not seem to be strain-specific, were detected in a non-rootstock population ('Idared' × 'Robusta 5') (Wohner et al., 2014). Another non-strain-specific locus was discovered on linkage group 7 in a rootstock population derived from a cross between 'Ottawa 3' and 'Robusta 5' (Gardiner et al., 2012). Cis-genic approaches with the *LG03* gene proved only partially successful, suggesting a more complex pathway of resistance than just one gene recognition of the pathogen (Kost et al., 2015). There is some evidence of graft transmissible benefits conferred to the grafted scion from fire blight-resistant rootstocks (Jensen et al., 2003, 2011, 2012), including reports by large-scale apple growers that they see less mortality and incidence of strikes when a resistant rootstock is used.

A unique situation was discovered with B.9 rootstocks. It was shown to be susceptible when inoculated with the bacteria, but when used as a rootstock and the scion was inoculated, it exhibited good field-level resistance (LoGiudice et al., 2006; Russo et al., 2008a,b).

7.4 Apple replant disease (ARD)

When apple trees are replanted in the same orchard in which apples and pears were planted previously, the new trees often are stunted and do not grow well. This problem has been named apple replant disease (ARD) and is caused by a complex of several microorganisms which thrive on the roots of the previous apple trees in the soil where the trees grew. Mazzola (1998) has reported that the most important pathogens associated with the disease include *Phythium, Phytophthora* spp., *Rhizoctonia solani,* and *Cylindrocarpon* spp., as well as the root lesion nematode (*Pratylenchus penetrans*) and bacteria (Mazzola, 1998). Research trials and grower observation indicate that M.26 is very sensitive to ARD (Robinson, 2011). In virgin soils, M.26 produces a larger tree than M.9, but in replant soils it often is similar in size or smaller than M.9. All rootstocks exhibit less vigorous growth on replant soils than virgin soils. This has resulted in the use of soil fumigation to kill pathogenic microorganisms prior to planting (Peryea and Covey, 1989; Yao et al., 2006), which sometimes (depending on soil type) is not effective.

Beginning in 2000, Merwin and students evaluated Geneva® rootstocks for replant disease tolerance and found that G.65, CG.6210, and G.30 show greater tolerance to the disease than Malling stocks (Isutsa and Merwin, 2000). Rumberger et al. (2004) found that trees on M.7, M.26, and G.16 remained smaller when growing in the previous tree rows compared with previous grass lanes, whereas the growth of trees on G.210 and G.30 planted in the two locations was similar. Leinfelder and Merwin (2006) suggested that using G.30 and G.210 rootstocks and planting in the previous grass lanes instead of the old rows may be an effective strategy against ARD. Based on field trials, Robinson et al. (2006) found that G.935 and G.202 had good tolerance to ARD. In a replant study in Washington, Mazzola et al. (2009b) found that G.11 and G.30 were more tolerant to lesion nematode than M.7, M.9, M.26, MM.106, and MM.111 (Mazzola et al., 2009a,b). Trees on M.26, MM.106, and MM.111 were more susceptible to *Pythium* spp. than trees on B.9 and rootstocks in the Geneva® series. Auvil et al. (2011) also reported that trees on several Geneva® rootstocks in several Washington locations outperformed the industry standards (B.9, M.9, and M.26) on replant sites (Auvil et al., 2011). In replant trials in North Carolina (USA), trees on G.30 and G.210 performed better in replant soils than trees on M.26 and M.7 (Parker et al., 2014).

The mechanism of ARD tolerance of Geneva® rootstocks is not clear. It is possible that the early screening for resistance to *Phytophthora* root rot fungi also co-screened for tolerance to other soil microorganisms. It is also possible that the root systems of the tolerant Geneva® rootstocks simply have a faster turnover rate and can essentially outgrow the pathogens (Atucha et al., 2013). What is remarkable is that the microbial community in the rhizosphere of these new rootstocks is drastically changed compared to the Malling stocks (Rumberger et al., 2007), possibly by deposition of specific exudates (Leisso et al., 2017, 2018). Regardless of the mechanism, this tolerance to ARD for replanting apple orchards on previous orchard land is becoming more important as soil fumigation options become more limited. In some parts of the United States and the world, soil fumigation is no longer an available option.

7.5 Viruses

Some plant viruses are lethal to many apple varieties, but other viruses are not lethal to most apple varieties and rootstocks. Viruses that can exist in the plant and cause few symptoms are termed latent viruses, which can be spread to new trees by grafting infected wood on clean rootstocks or by grafting clean wood on infected rootstocks. There are five main latent viruses: apple stem pitting virus (ASPV), apple stem grooving virus (ASGV), chlorotic leaf spot virus (ACLSV), apple mosaic virus (ApMV) and tomato ringspot virus (ToRSV) (Fuchs et al., 2018). Over time, all common commercial rootstocks from the Malling series became infected with one or more latent viruses. During the 1950s and 1960s, the Malling stocks were heat-treated to eliminate known viruses, and were given the designation East Malling-Long Ashton (EMLA). The clean versions of the Malling stocks were slightly more vigorous than the infected versions. In addition to the effort at the East Malling and Long Ashton research stations, the Dutch organization NAKB and the French organization CTIFL produced their own versions of clean M.9. The Dutch clean M.9 is referred as M.9T337 and the French version is referred as M.9Pajam1 and Pajam2.

A specific case of lethality was with MM.106 rootstock. When 'Delicious' scions were grafted on MM.106 and the trees were later infected with ToRSV through nematode vectors, the trees developed a brown (dead) line of cells at the graft union and the trees died (Tuttle and Gotlieb, 1985). More recently, several of the Geneva® rootstocks have shown susceptibility to one or more of the latent viruses. G.16 was very sensitive to the three most common latent viruses and required the use of virus-free budwood; otherwise, the trees die in the nursery or in the first year in the orchard. G.814 and G.935 have shown lesser susceptibility to latent viruses. The case of G.935 is still unclear, since it seems tolerant of individual viruses, but possibly combinations of viruses result in poor growth although the trees do not die. Nevertheless, the solution to these sensitivities is the use of virus-free bud wood since the common latent viruses are only transmitted by grafting.

Some apple-growing regions, such as the European Union, have very good virus elimination programs and require both rootstocks and scion wood to be virus-free. However, other areas, such as the United States, have allowed virus elimination programs to lapse due to limited government funding, and currently there are widespread latent virus infections in New York orchards (Fuchs et al., 2018). It is imperative that apple regions worldwide strengthen their virus elimination programs.

7.6 Tree anchorage and graft union strength

In the era when trees were expected to be freestanding (i.e., before trees were supported by posts or trellises), the anchorage of a rootstock was an important characteristic. Most of the semi-dwarfing rootstocks from around the world are freestanding, but some semi-dwarfing and the dwarfing rootstocks are not. M.7 is a semi-dwarfing rootstock that is freestanding in most cases, but with heavy rains and winds it can lean. In many orchards with M.7, about 30% of the trees exhibit significant leaning. In the previous era of semi-dwarfing rootstocks, many orchards with M.7 required tree support. However, with the adoption of the dwarfing rootstocks M.9 and M.26, tree support is required because the rootstock does not provide sufficient anchorage to support the tree with a heavy crop load. Support is also required because dwarfing rootstocks are much more precocious, resulting in a heavy crop on a young tree with small diameter limbs and trunk which can break down the canopy without support. Thus, in the dwarf tree era and high planting densities, rootstock anchorage has become an unimportant rootstock characteristic.

Another tree structural issue is rootstock-scion graft union strength, which has become more important with the adoption of dwarfing rootstocks. Both M.9 and M.26 have weak graft unions with some scion cultivars (e.g. 'Gala'), which requires good trellis support. Geneva® 30 also has exhibited poor graft union strength with 'Gala,' as high winds from a 1990 hurricane caused the breakage of mature trees at the graft union. Recent nursery and field observations indicate that unions of some newer cultivars, such as 'Cripps Pink' and 'Scilate' on G.41, are also brittle and trees break in wind storms, as well as when digging trees in the nursery or planting trees in the orchard. Since tree breakage can have significant economic consequences for nurserymen and orchardists, researchers have evaluated methods for determining union strength and flexibility (Adams et al., 2017). The amount of force required to break graft unions currently is used in the Geneva breeding program to classify graft union strength of rootstocks. Rehkugler et al. (1979) found that 18-year-old 'Golden Delicious' on M.9 could withstand only one-third of the force required to cause breakage on vigorous rootstocks (Rehkugler et al., 1979). Robinson et al. (2003) found the graft union of 'Gala'/G.30 was more brittle than 'Gala'/M.26, but the strength of the 'Gala'/G.30 graft union increased with tree age (Robinson et al., 2003). Adams et al. (2017) found that 'Scilate'/G.41 graft unions were weaker than 'Scilate'/M.9 graft unions, and that grafting method did not improve the graft union strength. Application of plant growth regulators to graft unions in the nursery did improve graft union strength: foliar applications of prohexadione-calcium and benzyladenine applied to the union in latex paint increased the flexural strength per scion cross-sectional area and the flexibility of the union. To avoid tree breakage problems, support should be provided in the nursery, as well as in the orchard so branches can be tied to multiple wires to prevent the twisting of trees in the wind.

8 Trends in apple (and other tree fruit) rootstock use

Over the last 100 years, total worldwide apple rootstock production has increased many fold. This is due to increased acreage worldwide, but also due to the tenfold increase in planting density as the world's apple growers moved from low-density orchards to high-density orchards. Our estimate of worldwide production of apple rootstocks is two million plants in 1950, but in 2019 we estimate there are more than 120 million rootstocks produced worldwide.

The apple-producing world before 1950 primarily used seedling rootstocks. Excluding China, starting in the mid-twentieth century a slow transition from seedling rootstocks to semi-dwarfing Malling rootstocks (MM.106, MM.111, and M.7) began and accelerated in the 1960s and 1970s so that by 1980, semi-dwarfing rootstocks accounted for an estimated 80% of all apple rootstocks in the world. However, starting in the 1970s, the use of M.9 increased from less than 10% of all rootstocks to almost 70% of all rootstocks by 2000, while the use of semi-dwarfing rootstocks declined rapidly during the 1990s. As M.9 gained in popularity, M.26 also grew to account for about 20% of rootstocks produced by the year 2000. Since 2000, M.9 has dominated the worldwide production of rootstocks, but in some countries, B.9 has increased rapidly and accounts for 20% of all rootstocks produced today. Since 2006, the production of the Geneva® rootstock series has increased rapidly in the United States and, in 2019, accounts for about 40% of all rootstocks produced. On a worldwide scale, the Geneva® rootstocks account for about 8% of all rootstocks. It is expected that their production will rise to account for 50% of worldwide apple rootstock production (excluding China) in the next 10 years.

Apple rootstock usage in China has followed a different path. Apple production in China before 1980 was small; however, due to government promotion of growing apples, production exploded in the late 1980s and 1990s. This resulted in China becoming the largest apple-producing country in the world by 2000, with currently five times the production of the second largest producer, the United States. Almost all Chinese apple production is based on vigorous and semi-vigorous rootstocks of *Malus prunifolia*, *M. baccata*, *M. hupehensis*, *M. micromalus,* and *M. seversei*. Although China has a vast orchard area, the planting densities are moderate, thus resulting in an estimated annual rootstock production of 30 million plants, which is similar to the European apple rootstock production.

This historical evolution of apple production tied to rootstock innovations presents a template for the likely evolution of other temperate zone tree fruit production systems as well. Clearly, sweet cherry production has undergone a similar, even more rapid evolution and expansion since the first development of vigor-limiting, precocious rootstocks (primarily the Gisela® series) in the 1980s and 1990s. Significant advances in genetic development and selection of vigor-limiting peach and pear rootstocks also have been made since the turn of the century, with the subsequent research into their utilization for production system innovations now in their early stages.

9 Future trends in apple rootstocks

In the previous century, the primary criteria in choice of apple rootstock has been "will it survive in my climate, is it the right vigor and is it available?" Cold damage and fire blight have been the two primary and economically important causes of tree death in North America. In addition, *Phytophthora* root rots and waterlogging have also caused tree death. Thus, the rootstock decision in the past was usually quite simple, with only one or two choices available to growers. However, with the proliferation of improved apple rootstocks available around the world, there is now a dizzying array of choices for apple growers. The Geneva breeding program alone has released 14 apple rootstocks and is poised to release four more in the next few years. Even so, availability of new rootstocks is still a problem for some growing regions.

Researchers are attempting to evaluate new rootstocks in different locations and provide advice to apple growers regarding which rootstocks will perform best in each region. With so many rootstock choices, Fazio has suggested the term 'designer rootstocks' to indicate the possibility of choosing a rootstock suited for the specific climate, soil, cultivar, and planting system a grower chooses. Robinson further defined the four variables that need to be determined specifically for each orchard before choosing a rootstock: (1) vigor of the variety, (2) vigor imparted by the climate, (3) vigor imparted by the soil, and (4) the space allocated to each tree. Each of these should be considered as pieces of a puzzle specific to each orchard or areas in an orchard for selecting the rootstock. Robinson has further suggested that a rootstock in a modern orchard should be able to grow well enough to fill the space allocated to the tree in 2 years and begin production in either the first or second year, depending on the quality of available nursery trees. If rootstock vigor combined with scion, climate, and soil vigor do not result in sufficient growth to fill the space in 2 years, then substantial economic penalties in lost yield accrue to the grower. When rootstock choices were relatively limited, growers often planted an available rootstock that was not well matched with all of the vigor factors (including tree spacing), resulting in trees that took 5–8 years to fill their allotted space or that grew too vigorously for the allotted space and then were difficult to manage in later years. Robinson has estimated that with high-priced varieties, the lost yield when trees fail to fill their space by the end of the second or third year can cost up to $250 000/ha in lost yield over the first 8 years of orchard life. This economic reality often is not appreciated by growers who never see the un-realized income from lower-than-potential yields due to the wrong rootstock choice.

The introduction of the 'Honeycrisp' apple in the United States in the mid-1990s brought new challenges to growers for rootstock selection. It is a weak-growing cultivar that often fails to fill the orchard space allocated to the tree in 2 years when grafted to dwarfing rootstocks. However, due to its high market price, 'Honeycrisp' has been very profitable for growers even though it often fails to achieve this goal. In addition, its susceptibility to the Ca-related disorder, bitter pit, has resulted in the quest for rootstocks that not only have the appropriate vigor level but also have a genetically programmed specific mineral nutrient profile for higher Ca uptake and a better translocated K/Ca ratio in the fruit to reduce bitter pit. A national project involving a group of US researchers from the NC140 rootstock evaluation group has begun a

5-year project to speed the discovery of such rootstocks that are more ideally matched to specific varieties in locations where apples are produced. Similarly, this effort is also examining potential rootstocks for new areas of production that historically have been limited by high soil pH and/or high salt tolerance and drought tolerance. The project intends to develop an online decision aid tool to help growers choose the right rootstock for their specific soil, climate, variety, and spacing.

The future of rootstock improvement will assuredly lead to greater combinations of resistance/tolerance to biotic and abiotic stresses and positive horticultural traits. In the short term, efforts to characterize existing rootstocks or existing breeding populations for all of the desirable rootstock traits can be done relatively rapidly (10 years). However, to breed new 'designer' rootstocks for specific combinations of important traits is a much longer process (30 years). To speed up progress, marker-assisted breeding will improve the efficiency of selecting rootstocks with desirable traits at an early stage, potentially reducing the time to develop a new rootstock to only 13 years (2–3 years to select for one or more desirable traits and 10 years for propagation and field evaluation). Another possible way to accelerate rootstock development is to use genetic engineering through cis-gene transfer of specific apple genes through CRISPR-Cas9 technology. By this method, a specific gene could be inserted to an elite rootstock that already has many positive attributes. This may reduce the time to develop a new rootstock to only 11 years (1 year for gene transfer and 10 years for propagation and field testing).

Limitations to rapidly introducing a new rootstock worldwide include virus-certification, bulk propagation, and the need by growers and researchers in each production region worldwide to establish objective rootstock trials to confirm the performance of a rootstock in a given climate with the target varieties. Most growers are hesitant to plant a new rootstock that has not been proven in their area. For growers, a new orchard is at least a 20-year investment and if the choice of rootstock (or variety) does not result in a productive orchard of marketable fruit for 20 years, there can be large economic penalties.

There are few tree fruit rootstock breeding programs in the world. Some focus on only a few horticultural traits, such as improved rooting or cold hardiness. However, the world apple community needs rootstock breeding programs to focus on multiple resistances to biotic and abiotic stresses, in addition to superior horticultural characteristics including high yield of premium-quality fruit. Such broad breeding objectives require a large team and many cooperators who will evaluate the rootstocks in different climates and soils. We predict that rootstock breeding programs with the vision of developing rootstocks with multiple resistances and superior horticultural performance will produce an increasing array of new valuable rootstocks over the next 30 years.

10 References

Adams, R. R. 2010. *Increasing the Rooting in Apple Rootstock Stoolbeds, Horticulture.* Cornell University, Ithaca, NY.

Adams, S., Black, B. L., Fazio, G. and Roberts, N. A. 2017. The effect of plant growth regulators on apple graft union flexural strength and flexibility. *Journal of the American Pomological Society* 71, 8–18.

Adams, S., Lordan, J., Fazio, G., Bugbee, B., Francescatto, P., Robinson, T. L. and Black, B. 2018. Effect of scion and graft type on transpiration, hydraulic resistance and xylem hormone profile of apples grafted on Geneva (R) 41 and M.9-NIC (TM) 29 rootstocks. *Scientia Horticulturae* 227, 213–22. doi:10.1016/j.scienta.2017.09.052.

Aldwinckle, H. S., Cummins, J. N. and Gustafson, H. I. 1972. Resistance to *Phytophthora cactorum* in apple seedlings from controlled crosses. *Phytopathology* 62, 743-.

Aldwinckle, H. S., LoGiudice, N., Robinson, T. L., Holleran, H. T., Fazio, G., Johnson, W. C. and Norelli, J. L. 2004. Resistance of apple rootstocks to fire blight infection caused by internal movement of Erwinia amylovora from scion infections. *Acta Horticulturae* 663(663), 229–34. doi:10.17660/ActaHortic.2004.663.36.

Atkinson, C. J., Webster, A. D., Policarpo, M., Kuden, A., Ziraat Fakültesi, C. Ü. and Kappel, F. 1997. Drought sensitivity of apple rootstocks. *Acta Horticulturae* 451(451), 171–8. doi:10.17660/ActaHortic.1997.451.17.

Atkinson, C. J., Policarpo, M., Webster, A. D. and Kingswell, G. 2000. Drought tolerance of clonal *Malus* determined from measurements of stomatal conductance and leaf water potential. *Tree Physiology* 20(8), 557–63. doi:10.1093/treephys/20.8.557.

Atucha, A., Emmett, B. and Bauerle, T. L. 2013. Growth rate of fine root systems influences rootstock tolerance to replant disease. *Plant and Soil* 376(1–2), 337–46. doi:10.1007/s11104-013-1977-5.

Autio, W. R., Robinson, T. L., Barritt, B. H., Cline, J. A., Crassweller, R. M., Embree, C. G., Ferree, D. C., Garcia, M. E., Greene, G. M., Hoover, E. E., Johnson, R. S., Kosola, K., Masabni, J., Parker, M. L., Perry, R. L., Reighard, G. L., Seeley, S. D. and Warmund, M. 2005. Performance of 'Fuji' and 'McIntosh' apple trees after 5 years as affected by several dwarf rootstocks in the 1999 NC-140 apple rootstock trial. *Journal of the American Pomological Society* 59, 202–14.

Autio, W., Robinson, T., Archbold, D., Cowgill, W., Hampson, C., Quezada, R. P. and Wolfe, D. 2013. 'Gala' apple trees on supporter 4, P.14, and different strains of B.9, M.9 and M.26 rootstocks: final 10-year report on the 2002 NC-140 apple rootstock trial. *Journal American Pomological Society* 67, 62–71.

Autio, W., Robinson, T., Black, B., Blatt, S., Cochran, D., Cowgill, W., Hampson, C., Hoover, E., Lang, G., Miller, D., Minas, I., Quezada, R. P. and Stasiak, M. 2017a. Budagovsky, Geneva, Pillnitz, and malling apple rootstocks affect 'Honeycrisp' performance over the first five years of the 2010 NC-140 'Honeycrisp' apple rootstock trial. *Journal of the American Pomological Society* 71, 149–66.

Autio, W., Robinson, T., Black, B., Crassweller, R., Fallahi, E., Parker, M., Quezada, R. P. and Wolfe, D. 2017b. Budagovsky, Geneva, Pillnitz, and malling apple rootstocks affect 'Fuji' performance over the first five years of the 2010 NC-140 'Fuji' apple rootstock trial. *Journal of the American Pomological Society* 71, 167–82.

Auvil, T. D., Schmidt, T. R., Hanrahan, I., Castillo, F., McFerson, J. R. and Fazio, G. 2011. Evaluation of dwarfing rootstocks in Washington apple replant sites. *Acta Horticulturae* 903(903), 265–71. doi:10.17660/ActaHortic.2011.903.33.

Barden, J. A. 1995. Experience in Virginia with slender spindle, vertical axe, and central leader on several rootstocks with Delicious and Empire. *Compact Fruit Tree* 28, 9–11.
Barritt, B. H., Konishi, B. S. and Dilley, M. A. 1995. Performance of three apple cultivars with 23 dwarfing rootstocks during 8 seasons in Washington. *Fruit Varieties Journal* 49, 158–70.
Bassett, C. L., Glenn, D. M., Forsline, P. L., Wisniewski, M. E. and Farrell, R. E.. 2011. Characterizing water use efficiency and water deficit responses in apple (*Malus × domestica* Borkh. and *Malus sieversii* Ledeb.) M. Roem. *HortScience* 46(8), 1079–84. doi:10.21273/HORTSCI.46.8.1079.
Bassett, C. L., Baldo, A. M., Moore, J. T., Jenkins, R. M., Soffe, D. S., Wisniewski, M. E., Norelli, J. L. and Farrell, R. E. 2014. Genes responding to water deficit in apple (*Malus × domestica* Borkh.) roots. *BMC Plant Biology* 14, 182. doi:10.1186/1471-2229-14-182.
Bassuk, N. L. and Howard, B. H. 1980. Seasonal rooting changes in apple hardwood cuttings and their implications to nurserymen. *Combined Proceedings, International Plant Propagators' Society* 30, 289–93.
Baugher, T. A., Marini, R., Schupp, J. R. and Watkins, C. B. 2017. Prediction of bitter pit in 'Honeycrisp' apples and best management implications. *Hortscience* 52(10), 1368–74. doi:10.21273/HORTSCI12266-17.
Bergh, O. 1992. Apple rootstocks for high density plantings under South African conditions. *Deciduous Fruit Grower* 42, 369–73.
Bessho, H. and Soejima, J. 1992. Apple rootstock breeding for disease resistance. *Compact Fruit Tree* 25, 65–72.
Biggs, A. R. and Peck, G. M. 2015. Managing bitter pit in 'Honeycrisp' apples grown in the mid-Atlantic United States with foliar-applied calcium chloride and some alternatives. *HortTechnology* 25, 385–91. doi:10.21273/HORTTECH.25.3.385.
Broggini, G. A. L., Kost, T., Fahrentrapp, J., Patocchi, A., Wohner, T., Flachowsky, H., Peil, A., Hanke, M. V. and Gessler, C. 2014a. FB-MR5 is an apple gene providing resistance to fire blight. *Acta Horticulturae*, 273–6. doi:10.17660/ActaHortic.2014.1056.46.
Broggini, G. A. L., Wohner, T., Fahrentrapp, J., Kost, T. D., Flachowsky, H., Peil, A., Hanke, M. V., Richter, K., Patocchi, A. and Gessler, C. 2014b. Engineering fire blight resistance into the apple cultivar 'Gala' using the FB_MR5 CC-NBS-LRR resistance gene of *Malus * robusta* 5. *Plant Biotechnology Journal* 12(6), 728–33. doi:10.1111/pbi.12177.
Browne, G. T. and Mircetich, S. M. 1993. Relative resistance of thirteen apple rootstocks to three species of *Phytophthora*. *Phytopathology* 83(9), 744–9. doi:10.1094/Phyto-83-744.
Bus, V. G. M., Chagne, D., Bassett, H. C. M., Bowatte, D., Calenge, F., Celton, J. -M., Durel, C. -E., Malone, M. T., Patocchi, A., Ranatunga, A. C., Rikkerink, E. H. A., Tustin, D. S., Zhou, J. and Gardiner, S. E. 2008. Genome mapping of three major resistance genes to woolly apple aphid (*Eriosoma lanigerum* Hausm.). *Tree Genetics and Genomes* 4(2), 223–36. doi:10.1007/s11295-007-0103-3.
Carlson, R. F. 1974. Status and performance of 'M.26' rootstock. *Fruit Varieties Journal* 28, 71–2.
Carlson, R. F. and Oh, S. D. 1975. Influence of interstem lengths of M.8 clone *Malus sylvestris* Mill. on growth, precocity, yield, and spacing of 2 apple cultivars. *Journal of the American Society for Horticultural Science* 100, 450–2.
Carlson, R. F. and Perry, R. L. 1986. Mark, apple rootstock. *HortScience* 21(I), 165.

Castillo, A., Cabrera, D., Rodriguez, P., Zoppolo, R. and Robinson, T. 2015. *In vitro* micropropagation of CG41 apple rootstock. *Acta Horticulturae* 1083(1083), 569–74. doi:10.17660/ActaHortic.2015.1083.76.

Comai, M. and Widmann, L. 1972. The slender spindle method of growing golden delicious, and the time involved in pruning. *Esperienze e Ricerche, Stazione sperimentale Agraria Forestale di S. Michele all'Adige* 3, 37–56.

Costante, J. F., Lord, W. J., Howard, D. and Connington, L. 1983. Influences of planting depth on growth, root suckering, and yield on interstem apple trees. *HortScience* 18, 913-5.

Cowgill Jr., W. P., Autio, W. R., Hoover, E. E., Marini, R. P. and Domoto, P. A. 2017. NC-140 multi-state research project: improving economic and environmental sustainability in tree-fruit production through changes in rootstock use. *Journal of the American Pomological Society* 71, 34–46.

Crassweller, R. M. and Smith, D. E. 2001. Early growth and yield of three apple cultivars grown on four training systems. *HortScience* 36, 461.

Cummins, J. N. 1973. Systems for producing multiple-stock fruit trees in the nursery. *Plant Propagator* 19, 7–11.

Cummins, J. N. and Aldwinckle, H. S. 1974. Breeding apple rootstocks. *HortScience* 9, 367–72.

Cummins, J. N., Aldwinckle, H. S. and Janick, J. 1983. Breeding apple rootstocks. *Plant Breeding Reviews* 1, 294–394.

Czynczyk, A. 1979. Effect of M 9, B 9 and M 26 rootstocks on growth, fruiting and frost resistance of apple trees. *Fruit Science Reports* 6, 143–52.

Czynczyk, A. and Omiecinska, B. 1989. Effect of new rootstocks of Polish, Russian and Czechoslovakian breeds and two depths of planting of trees with interstems on growth and cropping of 3 apple cultivars. *Acta Horticulturae* 243(243), 71–8. doi:10.17660/ActaHortic.1989.243.7.

Czynczyk, A. and Zagaja, S. W. 1984. Evaluation of growth and cropping of apple trees grafted on dwarf rootstocks and interstems. *Compact Fruit Tree* 17, 37–49.

Domoto, P. A. 1982. The survival and performance of four cultivars on six dwarfing interstems. *Compact Fruit Tree* 15, 19–24.

Dozier Jr., W. A., Latham, A. J., Kouskolekas, C. A. and Mayton, E. L. 1974. Susceptibility of certain apple rootstocks to black root rot and woolly apple aphids. *HortScience* 9, 35–6.

Duan, N., Bai, Y., Sun, H., Wang, N., Ma, Y., Li, M., Wang, X., Jiao, C., Legall, N., Mao, L., Wan, S., Wang, K., He, T., Feng, S., Zhang, Z., Mao, Z., Shen, X., Chen, X., Jiang, Y., Wu, S., Yin, C., Ge, S., Yang, L., Jiang, S., Xu, H., Liu, J., Wang, D., Qu, C., Wang, Y., Zuo, W., Xiang, L., Liu, C., Zhang, D., Gao, Y., Xu, Y., Xu, K., Chao, T., Fazio, G., Shu, H., Zhong, G. Y., Cheng, L., Fei, Z. and Chen, X. 2017. Genome re-sequencing reveals the history of apple and supports a two-stage model for fruit enlargement. *Nature Communications* 8(1), 249. doi:10.1038/s41467-017-00336-7.

Dvorak, A. 1983. Cultural characteristics of the new J-TE series of apple rootstocks. In: *Progresivni Smery v Ovocnarske Vyrobe*. Vyzkumny a Slechtitelsky Ustav Ovocnarsky, Holovousy, Czechoslovakia, 117–33.

Ebel, R. C., Proebsting, E. L. and Evans, R. G. 2001. Apple tree and fruit responses to early termination of irrigation in a semi-arid environment. *HortScience* 36(7), 1197–201. doi:10.21273/HORTSCI.36.7.1197.

Elfving, D. C., Schecter, I. and Hutchinson, A. 1993. The history of the Vineland (V.) apple rootstocks. *Fruit Varieties Journal* 47, 52–8.

Embree, C. G. 1985. The new KSC (Kentville stock clones) apple rootstock series - their current and future use. *Compact Fruit Tree* 18, 34–8.

Embree, C. G. and McRae, K. B. 1991. An exploratory-study of reciprocal apple rootstock and scion hardiness with 2 methods of assessment. *HortScience* 26(12), 1523–5. doi:10.21273/HORTSCI.26.12.1523.

Fahrentrapp, J., Broggini, G. A. L., Gessler, C., Kellerhals, M., Peil, A., Malnoy, M. and Richter, K. 2013. Fine mapping of the fire blight resistance locus in *Malus * robusta* 5 on linkage group 3. *Acta Horticulturae* 976(976), 499–500. doi:10.17660/ActaHortic.2013.976.70.

Fallahi, E. and Mohan, S. K. 2000. Influence of nitrogen and rootstock on tree growth, precocity, fruit quality, leaf mineral nutrients, and fire blight in 'Scarlet Gala' apple. *HortTechnology* 10, 589–92.

Fazio, G. and Robinson, T. L. 2008a. Modification of nursery tree architecture by apple rootstocks. *HortScience* 43, 1271–.

Fazio, G. and Robinson, T. L. 2008b. Modification of nursery tree architecture with apple rootstocks: a breeding perspective. *New York Fruit Quarterly* 16, 13–6.

Fazio, G., Wan, Y., Russo, N. L. and Aldwinckle, H. S. 2008. Investigation on the inheritance of strain specific resistance to *Erwinia amylovora* in an apple rootstock segregating population. *Acta Horticulturae* 793(793), 331–5. doi:10.17660/ActaHortic.2008.793.49.

Fazio, G., Kviklys, D., Grusak, M. A. and Robinson, T. L. 2012. Soil pH, soil type and replant disease affect growth and nutrient absorption in apple rootstocks. *New York Fruit Quarterly* 20, 22–8.

Fazio, G., Kviklys, A., Grusak, M. A. and Robinson, T. L. 2013. Phenotypic diversity and QTL mapping of absorption and translocation of nutrients by apple rootstocks. *Aspects of Applied Biology* 119, 37–50.

Fazio, G., Wan, Y., Kviklys, D., Romero, L., Adams, R. R., Strickland, D. and Robinson, T. L. 2014. *Dw2*, a new dwarfing locus in apple rootstocks and its relationship to induction of early bearing in apple scions. *Journal of the American Society for Horticultural Science* 139(2), 87–98. doi:10.21273/JASHS.139.2.87.

Fazio, G., Cheng, L., Grusak, M. A. and Robinson, T. L. 2015a. Apple rootstocks influence mineral nutrient concentration of leaves and fruit. *New York Fruit Quarterly* 25, 11–5.

Fazio, G., Robinson, T. L. and Aldwinckle, H. S. 2015b. The Geneva apple rootstock breeding program. *Plant Breeding Reviews* 39, 379–424.

Ferree, D. C. 1991. Results over ten years: NC-140 apple rootstock trial. *Compact Fruit Tree* 24, 5–6.

Ferree, D. C. 1992. Ten-year summary of the performance of 9 rootstocks in the NC-140 trials. *Compact Fruit Tree* 25, 5–11.

Ferree, D. C. and Perry, R. L. 1989. NC-140 rootstock trials in North America. *Acta Horticulturae* 243, 51–8. doi:10.17660/ActaHortic.1989.243.5.

Ferree, D. C., Schmid, J. C. and Morrison, C. A. 1982. An evaluation over 16 years of Delicious strains and other cultivars on several rootstocks and hardy interstems. *Fruit Varieties Journal* 36, 37–45.

Fischer, M., Barritt, B. H. and Kappel, F. 1997. Pillnitzer supporter 4 (Pi 80) – a semi-dwarf apple rootstock from Dresden-Pillnitz. *Acta Horticulturae* 451(451), 99–104. doi:10.17660/ActaHortic.1997.451.7.

Fischer, M. 1994. Results of the apple rootstock breeding programme obtained at Pillnitz. *Rivista di Frutticoltura e di Ortofloricoltura* 56, 17-21.

Foster, T. M., Celton, J. M., Chagne, D., Tustin, D. S. and Gardiner, S. E. 2015. Two quantitative trait loci, *Dw1* and *Dw2*, are primarily responsible for rootstock-induced dwarfing in apple. *Horticulture Research* 2, 15001. doi:10.1038/hortres.2015.1.

Foster, T. M., McAtee, P. A., Waite, C. N., Boldingh, H. L. and McGhie, T. K. 2017. Apple dwarfing rootstocks exhibit an imbalance in carbohydrate allocation and reduced cell growth and metabolism. *Horticulture Research* 4, 17009. doi:10.1038/hortres.2017.9.

Fuchs, M., Kahke, C., Donahue, D., Wallis, A. and Basedow, M. 2018. Distribution of viruses in New York apple orchards. *Fruit Quarterly* 26, 5-9.

Gao, H., Dang, Z., Li, D., Mei, L., Wan, Y., Wang, L., Zhao, Z. and Fazio, G. 2011. The history of apple breeding in People's Republic of China. *Acta Horticulturae* 903(903), 199-205. doi:10.17660/ActaHortic.2011.903.24.

Gardiner, S. E., Norelli, J. L., de Silva, N., Fazio, G., Peil, A., Malnoy, M., Horner, M., Bowatte, D., Carlisle, C., Wiedow, C., Wan, Y. Z., Bassett, C. L., Baldo, A. M., Celton, J. M., Richter, K., Aldwinckle, H. S. and Bus, V. G. M. 2012. Putative resistance gene markers associated with quantitative trait loci for fire blight resistance in *Malus* 'robusta 5' accessions. *BMC Genetics* 13, 25. doi:10.1186/1471-2156-13-25.

Gardner, R. G., Cummins, J. N. and Aldwinckle, H. S. 1980. Inheritance of fire blight resistance in *Malus* in relation to rootstock breeding. *Journal of the American Society for Horticultural Science* 105, 912-6.

Geng, F., Moran, R., Day, M., Halteman, W. and Zhang, D. L. 2015. *In vitro* shoot proliferation of apple rootstocks 'B 9', 'G 30', and 'G 41' grown under red and blue light. *HortScience* 50(3), 430-3. doi:10.21273/HORTSCI.50.3.430.

George, A. P. and Nissen, R. J. 1986. Growth, yield and scion-rootstock interaction effects of the custard apple in south-east Queensland. *Queensland Journal of Agricultural and Animal Sciences* 43, 73-81.

Gharghani, A., Zamani, Z., Talaie, A., Fattahi, R., Hajnajari, H., Oraguzie, N. C., Wiedow, C. and Gardiner, S. E. 2010. The role of Iran (Persia) in apple (*Malus * domestica* Borkh) domestication, evolution and migration via the Silk Trade Route. *Acta Horticulturae*, 229-36. doi:10.17660/ActaHortic.2010.859.26.

Glenn, D. M. 2014. An analysis of ash and isotopic carbon discrimination (Δ13C) methods to evaluate water use efficency in apple. *Scientia Horticulturae* 171, 32-6. doi:10.1016/j.scienta.2014.03.031.

Hafiz, I. A., Abbasi, N. A., Ahmad, T. and Hussain, A. 2008. DNA methylation profiles differ between juvenile and adult phase leaves of crab apple (*Malus micromalus*) seedling tree. *Pakistan Journal of Botany* 40, 1025-32.

Hampson, C. R. 2012. The performance of four Vineland apple rootstocks in British Columbia, Canada. *Journal of the American Pomological Society* 66, 23-7.

Hampson, C. R., Quamme, H. A. and Brownlee, R. T. 2002. Canopy growth, yield, and fruit quality of 'Royal Gala' apple trees grown for eight years in five tree training systems. *HortScience* 37(4), 627-31. doi:10.21273/HORTSCI.37.4.627.

Hampson, C. R., Randall, P. and Sholberg, P. 2012. Tolerance of Vineland apple rootstocks to waterlogging and *Phytophthora* infestation. *Canadian Journal of Plant Science* 92(2), 267-9. doi:10.4141/cjps2011-132.

Harrison, N., Harrison, R. J., Barber-Perez, N., Cascant-Lopez, E., Cobo-Medina, M., Lipska, M., Conde-Ruiz, R., Brain, P., Gregory, P. J. and Fernandez-Fernandez, F. 2016. A new three-locus model for rootstock-induced dwarfing in apple revealed by genetic mapping of root bark percentage. *Journal of Experimental Botany* 67(6), 1871–81. doi:10.1093/jxb/erw001.

Hartmann, H. T., Kester, D. E., Davies, F. T. and Geneve, R. L. 1997. *Plant Propagation: Principles and Practices*. Prentice-Hall Inc., Upper Saddle River.

Hatton, R. G. 1917. Paradise apple stocks. *Journal of the Royal Horticultural Society* 42, 361–99.

Hatton, R. G. 1919. Paradise apple stocks their fruit and blossom described. *Journal of the Royal Horticultural Society* 44, 89–94.

Hatton, R. G. 1920. Suggestion for the right selection of apple stocks. *Journal of the Royal Horticultural Society* 45, 257–68.

Heinicke, D. R. 1975. High density apple orchards planning training and pruning. *US Department of Agriculture Agriculture Handbook*, pp. 1–34.

Hulko, I. P., Hulko, B. I. and Sadowski, A. 1999. Results of study of clonal apple rootstocks in stoolbeds and in nursery, Apple rootstocks for intensive orchards. In: *Proceedings of the International Seminar*, Warsaw Ursynow, Poland, 18–21 August 1999. Warsaw Agricultural University Department of Pomology, Warsaw, Poland, pp. 41–2.

Isutsa, D. K. and Merwin, I. A. 2000. *Malus* germplasm varies in resistance or tolerance to apple replant disease in a mixture of New York orchard soils. *HortScience* 35(2), 262–8. doi:10.21273/HORTSCI.35.2.262.

James, D. J. and Thurbon, I. J. 1979. Rapid *in vitro* rooting of the apple rootstock M.9. *Journal of Horticultural Science* 54(4), 309–11. doi:10.1080/00221589.1979.11514887.

Jensen, P. J., Halbrendt, N., Fazio, G., Makalowska, I., Altman, N., Praul, C., Maximova, S. N., Ngugi, H. K., Crassweller, R. M., Travis, J. W. and McNellis, T. W. 2012. Rootstock-regulated gene expression patterns associated with fire blight resistance in apple. *BMC Genomics* 13, 9. doi:10.1186/1471-2164-13-9.

Jensen, P. J., Rytter, J., Detwiler, E. A., Travis, J. W. and McNellis, T. W. 2003. Rootstock effects on gene expression and fire blight resistance in apple. *Phytopathology* 93, S40.

Jensen, P. J., McNellis, T. W., Halbrendt, N., Travis, J. W., Altman, N., Praul, C. A., Maximova, S. N., Crassweller, R. M. and Makalowska, I. 2011. Rootstock-regulated gene expression profiling in apple trees reveals genes whose expression levels are associated with fire blight resistance. *Acta Horticulturae* 903(903), 87–93. doi:10.17660/ActaHortic.2011.903.7.

Jin, W., Zhang, Q., Liu, S., Wei, Q., Jin, W., Cheng, Z., Xue, X. and Yang, T. 2012. Genetic diversity of 41 apple rootstocks based on simple sequence repeat markers. *Journal of the American Society for Horticultural Science* 137(1), 51–6. doi:10.21273/JASHS.137.1.51.

Johnson, S., Andersen, R., Autio, W., Beckman, T., Black, B., Byers, P., Cline, J., Chavez-Gonzalez, C., Cowgill, W., Godin, R., Greene, G., Kaps, M., Kamas, J., Larsen, H., Lindstrom, T., Miller, D., Newell, M., Ophardt, D., Ouellette, D., Parra-Quezada, R., Pokharel, R., Reighard, G., Robinson, T., Schupp, J., Stein, L., Taylor, K., Walsh, C., Ward, D., Warmund, M. and Whiting, M. 2011. Performance of the 2002 NC-140 cooperative peach rootstock planting. *Journal of the American Pomological Society* 65, 17–25.

Khanizadeh, S., Groleau, Y., Levasseur, A., Granger, R., Rousselle, G. L. and Davidson, C. 2005. Development and evaluation of St Jean-Morden apple rootstock series. *HortScience* 40, 521-2.

Koike, H. and Tsukahara, K. 1988. Various interstem effects in combination with 'Marubakaido N-1' rootstock on 'Fuji' apple growth. *HortScience* 23, 580-1.

Kost, T. D., Gessler, C., Jansch, M., Flachowsky, H., Patocchi, A. and Broggini, G. A. 2015. Development of the first cisgenic apple with increased resistance to fire blight. *PLoS ONE* 10(12), e0143980. doi:10.1371/journal.pone.0143980.

Kotoda, N., Iwanami, H., Takahashi, S. and Abe, K. 2006. Antisense expression of MdTFL1, a TFL1-like gene, reduces the juvenile phase in apple. *Journal of the American Society for Horticultural Science* 131(1), 74-81. doi:10.21273/JASHS.131.1.74.

Kuldoshin, I. A. and Sadowski, A. 1999. Perspective clonal rootstocks for intensive orchards in the central areas of Russia, apple rootstocks for intensive orchards. In: *Proceedings of the International Seminar*, Ursynow, Warsaw, Poland, 18-21 August 1999. Warsaw Agricultural University Department of Pomology, Warsaw, Poland, pp. 63-4.

Kviklys, D. 2011. Fruit rootstock research in Europe performed by EUFRIN Rootstock group. *Acta Horticulturae* 903(903), 349-53. doi:10.17660/ActaHortic.2011.903.45.

Kwon, S., Kim, K., Kim, H., Kim, M., Kwon, S. I., Kim, K. R., Kim, H. Y. and Kim, M. J. 1999. Effects of various combined-treatment by some growth regulators, sugars and inorganic materials with IBA on rooting in hardwood cuttings of apple rootstock M.26 (*Malus domestica* Bork.). *Journal of the Korean Society for Horticultural Science* 40, 447-50.

Larsen, F. E. 1976. Budding and grafting with polyethylene strips. *Plant Propagator* 22, 10-1.

Lauri, P. E. and Lespinasse, J. M. 2001. Genotype of apple trees affects growth and fruiting responses to shoot bending at various times of year. *Journal of the American Society for Horticultural Science* 126(2), 169-74. doi:10.21273/JASHS.126.2.169.

Leinfelder, M. M. and Merwin, I. A. 2006. Rootstock selection, preplant soil treatments, and tree planting positions as factors in managing apple replant disease. *HortScience* 41(2), 394-401. doi:10.21273/HORTSCI.41.2.394.

Leisso, R., Rudell, D. and Mazzola, M. 2017. Metabolic composition of apple rootstock rhizodeposits differs in a genotype-specific manner and affects growth of subsequent plantings. *Soil Biology and Biochemistry* 113, 201-14. doi:10.1016/j.soilbio.2017.06.011.

Leisso, R., Rudell, D. and Mazzola, M. 2018. Targeted metabolic profiling indicates apple rootstock genotype-specific differences in primary and secondary metabolite production and validate quantitative contribution from vegetative growth. *Frontiers in Plant Science* 9, 1336. doi:10.3389/fpls.2018.01336.

Lindley, J. (Ed.). 1828. *The Pomological Magazine; Or, Figures and Descriptions of the Most Important Varieties of Fruit Cultivated in Great Britain*. J. Ridgway, London.

Lo Bianco, R., Talluto, G. and Farina, V. 2012. Effects of partial rootzone drying and rootstock vigour on dry matter partitioning of apple trees (*Malus domestica* cvar Pink Lady). *The Journal of Agricultural Science* 150(1), 75-86. doi:10.1017/S0021859611000463.

Lockard, R. G. 1976. Effect of apple rootstocks and length and type of interstock on leaf nutrient levels. *Journal of Horticultural Science* 51(2), 289-96. doi:10.1080/00221589.1976.11514692.

LoGiudice, N., Aldwinckle, H. S., Robinson, T. L. and Fazio, G. 2006. The nature of resistance of the 'B.9' apple rootstock to fire blight. *Acta Horticulturae (Wageningen)* 704, 515–9.

Lord, W. J. 1983. Scion/rootstock and interstem effects on growth, nutrition, and fruiting of apple trees and on fruit quality. *Compact Fruit Tree* 16, 65–9.

Lordan, J., Fazio, G., Francescatto, P. and Robinson, T. 2017. Effects of apple (*Males × domestica*) rootstocks on scion performance and hormone concentration. *Scientia Horticulturae (Amsterdam)* 225, 96–105.

Lordan, J., Francescatto, P., Dominguez, L. I. and Robinson, T. L. 2018a. Long-term effects of tree density and tree shape on apple orchard performance, a 20 year study-Part 1, agronomic analysis. *Scientia Horticulturae* 238, 303–17. doi:10.1016/j.scienta.2018.04.033.

Lordan, J., Wallis, A., Francescatto, P. and Robinson, T. L. 2018b. Long-term effects of training systems and rootstocks on 'McIntosh' and 'Honeycrisp' performance, a 20-year study in a northern cold climate-part 2: economic analysis. *HortScience* 53(7), 978–92. doi:10.21273/HORTSCI13117-18.

Loudon, J. C. 1822. *An Encyclopœdia of Gardening*.

Ma, R., Huang, Y., Sha, G., Shi, Y., Li, J., Gong, X., Li, M., Ma, R. Q., Huang, Y., Sha, G. L., Shi, Y. J., Li, J., Gong, X. H. and Li, M. 2012. Identifying hybrids of apomictic apple rootstocks with SSR and flow cytometry. *Journal of Fruit Science* 29, 461–5.

Ma, L., Hou, C. W., Zhang, X. Z., Li, H. L., Han, D. G., Wang, Y., Han, Z. H. 2013. Seasonal growth and spatial distribution of apple tree roots on different rootstocks or interstems. *Journal of the American Society for Horticultural Science* 138(2), 79–87. doi:10.21273/JASHS.138.2.79.

Macedo, T. A., Sander, G. F., Michelon, M. F., Carminatti, J. F., Rufato, A. R., Rufato, L. and Robinson, T. L. 2018. Chilling requirement and budburst uniformity of cultivar 'Maxi Gala' grafted on different rootstocks. *Acta Horticulturae* 1228(1228), 241–6. doi:10.17660/ActaHortic.2018.1228.36.

Marini, R. P., Parker, M. L., Barden, J. A. and Unrath, C. R. 2003. The effect of eight dwarf rootstocks on burrknot development on 'Gala' apple trees at two locations. *Journal of the American Pomological Society* 57, 93–6.

Marini, R. P., Anderson, J. L., Autio, W. R., Barritt, B. H., Cline, J., Cowgill, W. P., Jr., Crassweller, R. C., Garner, R. M., Gauss, A., Godin, R., Greene, G. M., Hampson, C., Hirst, P., Kushad, M. M., Masabni, J., Mielke, E., Moran, R., Mullins, C. A., Parker, M., Perry, R. L., Prive, J. P., Reighard, G. L., Robinson, T., Rom, C. R., Roper, T. and Schupp, J. R. 2006. Performance of 'Gala' apple trees on 18 dwarfing rootstocks: ten-year summary of the 1994 NC-140 rootstock trial. *Journal of the American Pomological Society* 60, 69–83.

Mazilu, C., Viscol, I. I. and Sadowski, A. 1999. A Romanian apple rootstock for intensive orchards, apple rootstocks for intensive orchards. In: *Proceedings of the International Seminar*, Ursynow, Warsaw, Poland, 18–21 August 1999. Warsaw Agricultural University Department of Pomology, Warsaw, Poland, pp. 73–4.

Mazzola, M. 1998. Elucidation of the microbial complex having a causal role in the development of apple replant disease in Washington. *Phytopathology* 88(9), 930–8. doi:10.1094/PHYTO.1998.88.9.930.

Mazzola, M., Brown, J. and Fazio, G. 2009a. Interaction of Brassicaceous seed meal and rootstock on recovery of *Pratylenchus penetrans* from roots of apple grown in replant orchard soils. *Journal of Nematology* 41, 354.

Mazzola, M., Brown, J., Zhao, X., Izzo, A. D. and Fazio, G. 2009b. Interaction of Brassicaceous seed meal and apple rootstock on recovery of *Pythium* spp. and *Pratylenchus penetrans* from roots grown in replant soils. *Plant Disease* 93(1), 51–7. doi:10.1094/PDIS-93-1-0051.

McKenzie, D. W. 1964. Apple rootstock trials Jonathan on East Mailing, Merton and Malling-Merton rootstocks. *Journal of Horticultural Science* 39(2), 69–77. doi:10.1080/00221589.1964.11514092.

McKenzie, D. W. 1985. Apple rootstock trials cultivar golden-delicious on 8 semi-dwarfing rootstocks in Hawkes Bay New-Zealand. *Journal of Horticultural Science* 60(1), 7–11. doi:10.1080/14620316.1985.11515593.

Midgley, S. J. E. and Lotze, E. 2011. Climate change in the Western Cape of South Africa: trends, projections and implications for chill unit accumulation. *Acta Horticulturae* 903(903), 1127–34. doi:10.17660/ActaHortic.2011.903.157.

Moran, R. E., Sun, Y. P., Geng, F., Zhang, D. L. and Fazio, G. 2011a. Cold temperature tolerance of trunk and root tissues in one- or two-year-old apple rootstocks. *HortScience* 46(11), 1460–4. doi:10.21273/HORTSCI.46.11.1460.

Moran, R. E., Zhang, D. and Sun, Y. 2011b. Cold temperature tolerance of G.16 and G.935 apple roots. *Acta Horticulturae* 903, 289–93.

Moran, R. E., Peterson, B. J., Fazio, G. and Cline, J. 2018. Genotypic variation in apple rootstock low temperature tolerance during spring and fall. *Journal of the American Society for Horticultural Science* 143(5), 319–32. doi:10.21273/JASHS04470-18.

Nelson, S. H. and Tukey, H. B. 1955. Root temperature affects the performance of East Malling apple rootstocks. *Quarterly Bulletin. Michigan State University Agricultural Experiment Station* 38, 46–51.

Norelli, J., Aldwinckle, H., Momol, T., Johnson, B., DeMarree, A. and Reddy, M. V. B. 2001. Fire blight of apple rootstocks. *Compact Fruit Tree* 34, 12–5.

Norelli, J. L., Holleran, H. T., Johnson, W. C., Robinson, T. L. and Aldwinckle, H. S. 2003. Resistance of Geneva and other apple rootstocks to *Erwinia amylovora*. *Plant Dis.* 87(1), 26–32. doi:10.1094/PDIS.2003.87.1.26.

Olien, W. C. and Lakso, A. N. 1986. Effect of rootstock on apple (*Malus domestica*) tree water relations. *Physiologia Plantarum* 67(3), 421–30. doi:10.1111/j.1399-3054.1986.tb05757.x.

Palmer, J. W. 2011. Changing concepts of efficiency in orchard systems. *Acta Horticulturae* 903(903), 41–9. doi:10.17660/ActaHortic.2011.903.1.

Palmer, J. W., Sansavini, S., Winter, F., Bünemann, C. and Wagenmakers, P. S. 1989. The international planting systems trial. *Acta Horticulturae* 243(243), 231–42. doi:10.17660/ActaHortic.1989.243.29.

Parker, M. L., Hoyt, T. and Clark, B. 2014. Evaluating apple replant strategies in the southeastern United States. *Acta Horticulturae* 1058(1058), 645–50. doi:10.17660/ActaHortic.2014.1058.84.

Peryea, F. J. and Covey, R. P. 1989. Replant management strategies influence early growth of apple trees in a sand soil. *HortScience* 24, 947–9.

Preston, A. P. 1953. Five new apple rootstocks. Ann. Rept. East Malling Res. Sta. for 1952, pp. 169–70.

Preston, A. P. 1955. Apple rootstock studies: malling-Merton rootstocks. *Journal of Horticultural Science* 30(1), 25–33. doi:10.1080/00221589.1955.11513825.

Preston, A. P. 1956. The control of fruitfulness behaviour by the use of rootstocks. *Annals of Applied Biology* 44(3), 511–7. doi:10.1111/j.1744-7348.1956.tb02146.x.

Preston, A. P. 1966. Apple rootstock studies: fifteen years' results with Malling-Merton clones. *Journal of Horticultural Science* 41(4), 349–60. doi:10.1080/00221589.1966.11514181.

Preston, A. P. 1967. Apple rootstock studies: fifteen years' results with some M.IX crosses. *Journal of Horticultural Science* 42(1), 41–50. doi:10.1080/00221589.1967.11514191.

Preston, A. P., 1974. Apple rootstock studies: some rootstock and interstock comparisons. *Horticultural Research* 14:47–53.

Preston, A. P. and Belcher, D. E. 1982. Apple rootstock studies: Bramley's Seedling on dwarfing clones. *Experimental Horticulture* 32, 18–24.

Prive, J. P., Zhang, M. I. N., Embree, C. G., Hebb, D., Palmer, J. W. and Wunsche, J. N. 2001. The influence of freeze-thaw cycling in cold hardiness studies on apple rootstocks. *Acta Horticulturae* 557(557), 123–30. doi:10.17660/ActaHortic.2001.557.15.

Proctor, J. T. A., A. Hutchinson, and W. F. Pierce, 1974. A 10-year trial of seven apple cultivars on Malling 26 rootstock. *Canadian Journal of Plant Science* 54:661–5.

Quamme, H. A. 1990. Cold hardiness of apple rootstocks. *Compact Fruit Tree* 23, 11–6.

Quamme, H. A. and Hogue, E. J. 1994. Improved rooting of Ottawa 3 apple rootstock by soft wood cuttings using micropropagated plants as a cutting source. *Fruit Varieties Journal* 48, 170–3.

Quamme, H. A., Brownlee, R. T., Barritt, B. H. and Kappel, F. 1997. Cold hardiness evaluation of apple rootstocks. *Acta Horticulturae* 451(451), 187–94. doi:10.17660/ActaHortic.1997.451.19.

Racsko, J., Miller, D. D., Midgley, S. J. E., Lakatos, L., Soltesz, M., Szabo, Z., Nyeki, J. and Costa, C. 2011. Influence of cultivars and rootstocks on the incidence of sunburn damage on apple fruit in the Northern and Southern Hemispheres. *Acta Horticulturae* 903(903), 1041–8. doi:10.17660/ActaHortic.2011.903.145.

Rehkugler, G. E., Cummins, J. N. and Markwardt, E. D. 1979. Rupture strength of unions of 'Golden Delicious' apple with Malling 8, Malling 9, and vigorous rootstocks. *Journal of the American Society for Horticultural Science* 104, 226–9.

Reig, G., Lordan, J., Fazio, G., Grusak, M. A., Hoying, S., Cheng, L. L., Francescatto, P. and Robinson, T. 2018. Horticultural performance and elemental nutrient concentrations on 'Fuji' grafted on apple rootstocks under New York State climatic conditions. *Scientia Horticulturae* 227, 22–37. doi:10.1016/j.scienta.2017.07.002.

Reig, G., Lordan, J., Sazo, M. M., Hoying, S., Fargione, M., Reginato, G., Donahue, D. J., Francescatto, P., Fazio, G. and Robinson, T. 2019. Long-term performance of 'Gala', Fuji' and 'Honeycrisp' apple trees grafted on Geneva (R) rootstocks and trained to four production systems under New York State climatic conditions. *Scientia Horticulturae (Amsterdam)* 244, 277–93.

Rioux, J. A., Lemattre, P., Parcevaux, S. D. and De Parcevaux, S. 1984. *In vitro* propagation of apple rootstock Ottawa 3. *Quebec Vert.* 6, 22–7.

Robinson, T. 2011. Advances in apple culture worldwide. *Revista Brasileira De Fruticultura* 33(spe1), 37–47. doi:10.1590/S0100-29452011000500006.

Robinson, T. L., Lakso, A. N. and Carpenter, S. G. 1991a. Canopy development yield and fruit quality of empire and delicious apple trees grown in four orchard production systems for ten years. *Journal of the American Society for Horticultural Science* 116(2), 179–87. doi:10.21273/JASHS.116.2.179.

Robinson, T. L., Lakso, A. N. and Ren, Z. B. 1991b. Modifying apple tree canopies for improved production efficiency. *HortScience* 26(8), 1005–12. doi:10.21273/HORTSCI.26.8.1005.

Robinson, T. L., Cummins, J. N., Hoying, S. A., Smith, W. H., Barritt, B. H. and Kappel, F. 1997. Commercial orchard evaluation of the new Cornell-Geneva apple rootstocks. *Acta Horticulturae* 451(451), 113–20. doi:10.17660/ActaHortic.1997.451.9.

Robinson, T., Aldwinckle, H., Fazio, G., Holleran, T. and Janick, J. 2003. The Geneva series of apple rootstocks from Cornell: performance, disease resistance, and commercialization. *Acta Horticulturae* 622(622), 513–20. doi:10.17660/ActaHortic.2003.622.56.

Robinson, T. L., Fazio, G., Aldwinckle, H. S., Hoying, S. A. and Russo, N. 2006. Field performance of Geneva apple rootstocks in the Eastern USA. *Sodininkyste-ir-Darzininkyste* 25, 181–91.

Rom, C. R. and Rom, R. C. 1991. Rootstock effects on foliar nutrient content of 'Starkspur Supreme' in the NC-140 rootstock trials. *Compact Fruit Tree* 24, 11–4.

Rogers, W. S. and Beakbane, A. B. 1957. Stock and scion relations. *Annual Review of Plant Biology* 8(1), 217–36. doi:10.1146/annurev.pp.08.060157.001245.

Rong, Z., Yi, K., Yang, F., Liu, Z., Li, X., Wan, D. and Wan, Y. 2011. A dwarfing apple rootstock: 'Liaozhen 2'. *Acta Horticulturae* 903(903), 169–76. doi:10.17660/ActaHortic.2011.903.17.

Rosenberger, D. A., Schupp, J. R., Hoying, S. A., Cheng, L. and Watkins, C. B. 2004. Controlling bitter pit in 'Honeycrisp' apples. *HortTechnology* 14, 342–9. doi:10.21273/HORTTECH.14.3.0342.

Rufato, L., Kretzschmar, A. A., Brighenti, A. F., Macedo, T. A., Mendes, M. and Silva, L. C. 2010. Bud break in different cultivars of apple trees in two regions of Santa Catarina State, Brazil. *Acta Horticulturae* 884(884), 643–6. doi:10.17660/ActaHortic.2010.884.85.

Rumberger, A., Merwin, I. A. and Thies, J. E. 2007. Microbial community development in the rhizosphere of apple trees at a replant disease site. *Soil Biology and Biochemistry* 39(7), 1645–54. doi:10.1016/j.soilbio.2007.01.023.

Rumberger, A., Yao, S. R., Merwin, I. A., Nelson, E. B. and Thies, J. E. 2004. Rootstock genotype and orchard replant position rather than soil fumigation or compost amendment determine tree growth and rhizosphere bacterial community composition in an apple replant soil. *Plant and Soil* 264(1/2), 247–60. doi:10.1023/B:PLSO.0000047760.13004.94.

Russo, N. L., Robinson, T. L., Fazio, G. and Aldwinckle, H. S. 2007. Field evaluation of 64 apple rootstocks for orchard performance and fire blight resistance. *Hortscience* 42(7), 1517–25. doi:10.21273/HORTSCI.42.7.1517.

Russo, N. L., Aldwinckle, H. S., Robinson, T. L. and Fazio, G. 2008a. Budagovsky 9 rootstock: uncovering a novel resistance to fire blight. *Acta Horticulturae*, 321–4. doi:10.17660/ActaHortic.2008.793.47.

Russo, N. L., Robinson, T. L., Fazio, G. and Aldwinckle, H. S. 2008b. Fire blight resistance of Budagovsky 9 apple rootstock. *Plant Disease* 92(3), 385-91. doi:10.1094/PDIS-92-3-0385.

Sansavini, S. and Musacchi, S. 2000. New pear orchards: densities, rootstocks, training systems. *Rivista di Frutticoltura e di Ortofloricoltura* 62(84-94), 12.

Sansavini, S., Bassi, D. and Giunchi, L. 1981. Tree efficiency and fruit quality in high-density apple orchards. *Acta Horticulturae* 114(114), 114–36. doi:10.17660/ActaHortic.1981.114.13.

Sax, K. 1949. The use of *Malus* species for apple rootstocks. *Proceedings of the American Society for Horticultural Science* 53, 219–20.

Schlatholter, I., Jansch, M., Flachowsky, H., Broggini, G. A. L., Hanke, M. V. and Patocchi, A. 2018. Generation of advanced fire blight-resistant apple (*Malus* × *domestica*) selections of the fifth generation within 7 years of applying the early flowering approach. *Planta* 247(6), 1475–88. doi:10.1007/s00425-018-2876-z.

Schmidt, H. 1986. Ways of shortening the juvenile period of apples. *Erwerbsobstbau* 28, 6–7.

Schouten, H. J., Soriano, J. M., Joshi, S. G., Kortstee, A. J., Krens, F. A., Schaart, J. G., van der Linden, K., Allan, A. C., Hellens, R. P., Espley, R. V. and Jacobsen, E. 2009. Cisgenesis is a promising approach for fast, acceptable and safe breeding of pip fruit. *Acta Horticulturae* 814(814), 199-204. doi:10.17660/ActaHortic.2009.814.26.

Sha, G. L., Gong, X. H., Huang, Y., Shao, Y. C., Yin, T., Hao, Y. J. and Shu, H. R. 2011. 'Qingzhen 1' and 'Qingzhen 2', two apple rootstocks with high apomitic fruit setting ability. *Acta Horticulturae* 903, 159–62.

Spangelo, L. P. S., Fejer, S. O., Leuty, S. J. and Granger, R. L. 1974. Ottawa 3 clonal apple rootstock. *Canadian Journal of Plant Science* 54(3), 601–3. doi:10.4141/cjps74-107.

Strong, D. and Azarenko, A. N. 2000. Relationship between trunk cross-sectional area, harvest index, total tree dry weight and yield components of 'Starkspur Supreme Delicious' apple trees. *Journal of American Pomological Society* 54, 22–7.

Strong, D. and Miller Azarenko, A. 1991. Dry matter partitioning in 'Starkspur Supreme Delicious' on nine rootstocks. *Fruit Varieties Journal* 45, 238–41.

Sun, W. Q. and Bassuk, N. L. 1991. Stem banding enhances rooting and subsequent growth of M.9 and MM.106 apple rootstock cuttings. *HortScience* 26(11), 1368–70. doi:10.21273/HORTSCI.26.11.1368.

Tamai, H., Ono, T., Koike, H. and Shigehara, I. 2002. Comparison of growth, yield efficiency and fruit quality between 'Fuji' on M.9Nagano rootstock and 'Fuji' on M.9Nagano (interstock)/Marubakaido (*Malus prunifolia* Bork. var. ringo Asami) under high density planting. *Journal of the Japanese Society for Horticultural Science* 71, 670–4.

Tamai, H., Komatsu, H., Koike, H., Ono, T. and Shigehara, I. 2003. Performance of 'Rakuraku Fuji' apple trees on JM.7 and M.9Nagano rootstocks, and M.9 Nagano/Marubakaido interstem combination in Japan. *Journal of American Pomological Society* 57, 157–60.

Travis, J. W. and Rytter, J. L. 1995. The development of fire blight and black rot on MARK and EMLA 7 rootstocks. *Pennsylvania Fruit News* 75, 35–6.

Travis, J. W., Rytter, J. L., Hickey, K. D., Momol, M. T. and Saygili, H. 1999. The susceptibility of apple rootstocks and cultivars to *Erwinia amylovora*. *Acta Horticulturae* 489(489), 235–42. doi:10.17660/ActaHortic.1999.489.37.

Tsuchiya, S. 1988. Breeding of dwarfing rootstocks for fruit trees. *Research Journal of Food and Agriculture* 11, 3–11.

Tukey, H. B. 1964. *Dwarfed Fruit Trees, for Orchard, Garden and Home*. Macmillan, New York.

Tukey, H. B. 1978. *Dwarfed Fruit Trees for Orchard, Garden, and Home: With Special Reference to the Control of Tree Size and Fruiting in Commercial Fruit Production*. Cornell University Press, Ithaca, NY.

Tukey, R. B., Langston, R. and Cline, R. A. 1962. Influence of rootstock, bodystock and interstock on the nutrient content of apple foliage. *Proceedings of the American Society for Horticultural Science* 80, 73–8.

Tustin, D. S. and Cashmore, W. M. 1994. Rootstock and spacing effects on precocity, yield and fruit quality of 'Fuji' apple using slender pyramid tree management. *Compact Fruit Tree* 27, 60–8.

Tustin, D. S., Hirst, P. M., Cashmore, W. M., Warrington, I. J., Stanley, C. J., Erez, A. and Jackson, J. E. 1993. Spacing and rootstock studies with central leader apple canopies in a high vigour environment. *Acta Horticulturae*, 169–78. doi:10.17660/ActaHortic.1993.349.26.

Tuttle, M. A. and Gotlieb, A. R. 1985. Graft Union histology and distribution of tomato ringspot virus in infected McIntosh/Malling Merton 106 apple trees. *Phytopathology* 75(3), 347–51. doi:10.1094/Phyto-75-347.

Tworkoski, T. and Fazio, G. 2011. Physiological and morphological effects of size-controlling rootstocks on 'Fuji' apple scions. *Acta Horticulturae* 903(903), 865–72. doi:10.17660/ActaHortic.2011.903.120.

Tworkoski, T., Fazio, G. and Glenn, D. M. 2016. Apple rootstock resistance to drought. *Scientia Horticulturae* 204, 70–8. doi:10.1016/j.scienta.2016.01.047.

Van Hooijdonk, B. M., Woolley, D. J., Warrington, I. J. and Tustin, D. S. 2010. Initial alteration of scion architecture by dwarfing apple rootstocks may involve shoot-root-shoot signalling by auxin, gibberellin, and cytokinin. *The Journal of Horticultural Science and Biotechnology* 85(1), 59–65. doi:10.1080/14620316.2010.11512631.

Van Oosten, H. J. V. 1977. The new apple rootstock M.27. (2) M.27 in the nursery. *Fruitteelt* 67, 402–5.

Van Oosten, H. J. V. and Groene, J. M. D. 1984. Differences between sources of M.9. 3. The fruit tree nursery. *Fruitteelt* 74, 968–9.

Villeneuve, F. 1986. Burr-knots, their development and methods of control. *Arboriculture Fruitiere* 33, 44–8.

Visser, T. and Schaap, A. A. 1967. Preselection for juvenile period, flowering and picking time in apple seedlings. *Euphytica* 16(1), 109–21. doi:10.1007/BF00034104.

Wan, Y., Li, D., Zhao, Z., Mei, L., Han, M., Schwaninger, H. and Fazio, G. 2011. The distribution of wild apple germplasm in Northwest China and its potential application for apple rootstock breeding. *Acta Horticulturae* 903(903), 123–41. doi:10.17660/ActaHortic.2011.903.12.

Warmund, M. R., Barritt, B. H., Brown, J. M., Schaffer, K. L. and Jeong, B. R. 1993. Detection of vascular discontinuity in bud unions of 'Jonagold' apple on Mark rootstock with magnetic resonance imaging. *Journal of the American Society for Horticultural Science* 118(1), 92–6. doi:10.21273/JASHS.118.1.92.

Webster, A. D. and Hollands, M. 1999. Orchard comparisons of 'Cox's Orange Pippin' grown on selections of the apple rootstock M.9. *Journal of Horticultural Science and Biotechnology* 74(4), 513–21. doi:10.1080/14620316.1999.11511145.

Webster, C. A. and Jones, O. P. 1989. Micropropagation of the apple rootstock M.9: effect of sustained subculture on apparent rejuvenation *in vitro*. *Journal of Horticultural Science* 64(4), 421–8. doi:10.1080/14620316.1989.11515973.

Webster, T. and Tobutt, K. 1994. European and East Malling rootstock research – new and promising apple rootstocks. *Compact Fruit Tree* 27, 17–24.

Wen, T., Dong, L., Wang, L., Ma, F., Zou, Y. and Li, C. 2018. Changes in root architecture and endogenous hormone levels in two *Malus* rootstocks under alkali stress. *Scientia Horticulturae* 235, 198–204. doi:10.1016/j.scienta.2017.09.015.

Wertheim, S. J. 1978. Pruning of slender spindle type trees. *Acta Horticulturae* 65(65), 173–80. doi:10.17660/ActaHortic.1978.65.26.

Wertheim, S. J. 1981. High density planting development and current achievements in the Netherlands Belgium and West Germany. *Acta Horticulturae* 114(114), 318–30. doi:10.17660/ActaHortic.1981.114.44.

Wertheim, S. J. and Callesen, O. 2000. Results of multi-site interstem trials with apple trees. *Gartenbauwissenschaft* 65, 251–9.

Wertheim, S. J. and Scholtens, A. 1994. Trial results with M.27 as rootstock and interstock. M.27 can be used as an interstock. *Fruitteelt Den Haag* 84, 11-3.

Wohner, T., Hanke, M. V., Flachowsky, H., Peil, A., Richter, K. and Broggini, G. A. L. 2014. Investigation on fire blight resistance in the cross population 'Idared' * *Malus * robusta* 5 with different Erwinia amylovora strains. *Acta Horticulturae* 1056, 277–80.

Yao, Z., Bi, X., Gao, L., Wang, W., Yao, Z. Z., Bi, X. M., Gao, L. X. and Wang, W. S. 2001. The performance of the apple dwarf rootstock GM-256 in Mengluan area, Hebei Province. *China Fruits* 3, 12–5.

Yao, S. R., Merwin, I. A., Abawi, G. S. and Thies, J. E. 2006. Soil fumigation and compost amendment alter soil microbial community composition but do not improve tree growth or yield in an apple replant site. *Soil Biology and Biochemistry* 38(3), 587–99. doi:10.1016/j.soilbio.2005.06.026.

Yoshida, M. and Muramatsu, H. 1998. Performance and hardwood cutting propagation of new apple JM rootstocks in Hokkaido. *Bulletin of Hokkaido Prefectural Agricultural Experiment Stations* (75), 11–4.

Zhang, M. J., Ding, L. H., Wang, Q., Li, Y. B., Yan, X. K. and Xing, G. J. 2011. Development of cold resistant apple rootstocks in China. *Acta Horticulturae* 903(903), 183–6. doi:10.17660/ActaHortic.2011.903.19.

Zhang, L. S., Li, X. W., Zhang, L. X., Li, B. Z., Han, M. Y., Liu, F. T., Zheng, P. and Alva, A. K. 2014. Role of abscisic acid (aba) in modulating the responses of two apple rootstocks to drought stress. *Pakistan Journal of Botany* 46, 117–26.

Chapter 3

Advances in understanding fruit tree growth

Ted DeJong, University of California-Davis, USA

1 Introduction

In a sense, a fruit tree can be viewed as a massive network of solar energy collectors. The individual solar collection plates (grana stacks) are located in chloroplasts, green microscopic structures within biological cells of the leaves. Each leaf contains thousands of chloroplasts that function as solar energy cells and the tree, in turn, has thousands of leaves. The solar energy cells (chloroplasts) only function if they are in aqueous solution, so leaves are specially designed to maintain the solar cells in a hydrated state inside the biological cells even though leaves are usually exposed to dry ambient conditions (Taiz et al., 2015). In this analogy, the woody framework of the tree can be viewed as providing the structure by which the tree is capable of exposing optimal numbers of solar cells to light energy. In addition to providing the structural framework for optimum light exposure, the wood and bark provide a vascular tissue for transporting water and nutrients to the leaves and chemical energy (photosynthates) from the solar cells (chloroplasts) in the leaves to other parts of the plant. The efficiency of a fruit tree as a solar energy collector network depends on its efficiency in capture and conversion of light energy into chemical energy (photosynthesis) and the subsequent transport, storage and utilization of that chemical energy for fruit production. This concept is supported by the fact that maximum yields of fruit and nut orchards, when optimally managed, have been shown to be directly related to the percentage of daily solar radiation intercepted (Lampinen et al., 2012; Palmer et al., 2002; Wünsche and Lakso, 2000).

http://dx.doi.org/10.19103/AS.2018.0040.03

Looking at the functioning of fruit trees from this perspective is useful for both scientific and practical horticultural reasons. Because of the importance of photosynthesis to the efficient functioning of plants as solar energy collectors, scientists have been intensively studying the process of photosynthesis for more than 100 years with the hope of increasing its efficiency. However, there is little evidence that scientists have been or will be able to increase this efficiency in crop plants in the near future (Horton, 2000). On the other hand, there is substantial evidence that fruit trees distribute their nutrient resources and photosynthetic competency, and adjust the angles of leaves in different parts of their canopy to optimize use of resources and sunlight as it passes through a tree canopy (Auzmendi et al., 2013; DeJong and Doyle, 1985; DeJong et al., 1989; Niinemets, 1995, 1997; Rosati et al., 1999, 2000, 2002).

So why should horticulturists be concerned about studying and understanding photosynthesis and the distribution of photosynthates in fruit trees? Trees have evolved to optimize these processes in the context of survival and reproduction in diverse natural environments. The horticulturist's goal is to optimize orchard conditions for enabling fruit trees to carry out photosynthesis and the efficient distribution and use of photosynthates towards obtaining economic yields in managed, agricultural systems. An orchard manager's concern is to optimize inputs that influence these processes to produce an economically valuable crop. This requires a basic understanding of the plant's fundamental processes and the factors that influence them.

1.1 Photosynthesis

Simply summarized, photosynthesis is the process by which energy from the sun is trapped in green pigments (chlorophyll), converted into chemical energy and used to convert carbon dioxide (CO_2) and water into simple carbohydrates that eventually become sugars ([CH_2O]n) that can be distributed throughout the plant. Oxygen is given off in the process. The reverse process, whereby all plants and animals recover energy from these simple carbohydrates, is called respiration (Taiz et al., 2015).

The actual photosynthetic process is a complex set of reactions involving many of the nutrients green plants require. For example, nitrogen (N) is a constituent of photosynthetic enzymes and chlorophyll; phosphorus is important in the energy transfer process; magnesium is an essential part of the chlorophyll molecule; potassium, iron, manganese and other nutrients play important roles in specific photosynthetic reactions (Taiz et al., 2015). The carbohydrate products of photosynthesis are collectively called photosynthates. A principal product is glucose, a six-carbon sugar. It is transformed into other simple sugars, that is fructose, sucrose and sorbitol, a sugar alcohol. In most plants, sucrose is the predominant carbohydrate that is transported from the leaves to other parts of the plant; however, in most Rosaceous fruit trees, sorbitol is the primary transported carbohydrate (Bieleski, 1982).

The CO_2 for photosynthesis must come from the air surrounding the leaf, and the water comes up from the soil through the plant's vascular system. CO_2, which makes up only about 0.040% of the earth's atmosphere (~400 ppm in air), diffuses through the stomata located in the lower epidermis of most fruit tree leaves. The stomata not only allow entry of CO_2 into the leaf, but also allow water vapour to escape from the leaf. Thus, to minimize water loss from the leaf (transpiration), the stomata have an active mechanism for controlling their opening to permit just enough CO_2 into the leaf to allow photosynthesis to continue without excessive loss of water (Taiz et al., 2015).

1.2 Interactions between photosynthesis and water use

During daylight hours when photosynthesis occurs, the tree loses (transpires) as much as 400 molecules of water for every molecule of CO_2 that is absorbed depending on the ambient temperature and humidity (Taiz et al., 2015). The water that is lost from the leaf is replenished by transport of water through the tree from the soil. If the soil around the tree is moist, water also evaporates into the air directly from the soil. The total quantity of water lost by the tree and the soil is called orchard evapotranspiration or crop evapotranspiration (ETc). Tree transpiration usually accounts for most crop water use. In recent years, because of increasing water scarcity in many agricultural growing areas, there has been increased research emphasis placed on developing plants with increased photosynthetic water use efficiency. However, again most of this research has had limited success because photosynthesis can only be carried out in aqueous solution within cells of leaves, and the laws of physics (diffusion) dictate the amount of water loss when stomata open to allow CO_2 to be taken up to supply the photosynthetic process (Taiz et al., 2015; Blum, 2009). The primary successes in increasing orchard water use efficiencies have been achieved by enhancing efficiencies of delivery systems used for orchard irrigation, changing to localized irrigation systems that more precisely deliver water to trees where they need it and scheduling irrigation so that it is delivered when they most need it (Blum, 2009; Auzmendi et al., 2011; Lopez et al., 2016; Marsal et al., 2016; Behboudian et al., 2011).

2 Factors that influence photosynthesis in fruit trees

2.1 Light

Since a fruit tree's primary function is to be a solar energy collector, light is the most important driver of photosynthesis. However, there is seemingly a flaw in this system since photosynthesis of an individual fruit tree leaf, as well as leaves of most other crop plants, is light saturated at approximately one-third to one-half full sunlight if a leaf is exposed perpendicular to the sun's rays (Taiz et al., 2015; DeJong, 1983; Li and Lakso, 2004). However, only leaves on the outer surface of a tree canopy are ever exposed to direct sunlight for long periods and even those leaves are usually oriented vertically and often folded at the midrib. Thereby they only receive direct exposures for very short periods of the day as the orientation of the sun to the tree changes from east to west (DeJong and Doyle, 1985; Rosati et al., 1999). Each leaf, located in its zone of the tree canopy, has its own ever-changing light environment (DeJong and Doyle, 1985). Thus, light is shared among leaves in deciduous fruit tree canopies so that most leaves in a mature tree function most of the day on the steep, rather than the light saturated portion of the photosynthetic light response curve (Rosati et al., 2002). Light becomes limiting for photosynthesis along a gradient from the outer, exposed edge to the centre of the foliar canopy, and often this gradient is depicted as a continuous reduction of light intensity towards the interior of tree canopies (Robinson et al., 1991). However, much of the light intercepted by all leaves is in the form of sun flecks, and the reduction in light exposure of interior leaves is a function of the amount of time leaves are exposed to sun flecks as opposed to being in shade (DeJong and Doyle, 1985). Thus, interior leaves contribute less photosynthates to the local fruit bearing shoots to which they are attached, and those shoots develop less leaf area and are less productive than more exposed shoots. These shoots may eventually die if total light interception is below the threshold for shoot survival (Lampinen et al., 2011), reinforcing the importance of solar energy collection in a tree's economy.

2.2 Sink strength

In horticulture, much attention has been paid to the question of whether photosynthesis of fruit trees is strongly controlled by a tree's demands for carbohydrates, as opposed to environmental drivers of photosynthesis such as light (Neales and Incoll, 1968). Researchers have reported that photosynthesis can be substantially increased in the presence of high demand for carbohydrates by fruits in several fruit crops (Avery, 1975; Hansen, 1970; Maggs, 1963), and this led some researchers to assert that fruit demand for carbohydrates is a major factor controlling photosynthesis in fruit trees. However, other research has indicated that the effect of fruit on photosynthesis can be relatively minor (DeJong, 1986). Close analysis of much of the literature reporting strong effects of crop load on photosynthesis indicates that those effects are mainly present when there are factors such as dwarfing rootstocks (Palmer et al., 2005; Wünsche et al., 2000) or girdling (Ben Mimoun et al., 1996) that limit the flow of photosynthates to alternative sinks and cause a feed-back-mediated reduction in carbohydrate movement from the leaves. While there may be a tendency for stomata to function less conservatively in controlling the ratio of CO_2 uptake to water loss in the presence of fruit (DeJong, 1986), there does not appear to be strong evidence for crop load being a primary regulator of photosynthesis in the absence of some 'artificial' mechanism that limits the capacity of alternative sinks to utilize photosynthates for growth. This corresponds with the concept that a tree species' success and survival in nature is expected to be associated with garnering as much carbohydrate resource as possible to grow and compete for space in addition to reproducing (Stephenson, 1981).

2.3 Temperature

Photosynthesis functions optimally at leaf temperatures between about 20°C and 30°C in many temperate deciduous fruit trees (Ro et al., 2001). While the temperature-based limits for temperate deciduous fruit trees are dictated most often by winter and spring cold events or lack of winter chill, rather than photosynthetic temperature optima, growing season temperatures can influence fruit quality and yield through effects on photosynthesis. Crops that have fruits with high sugar contents, such as many stone fruits, tend to be sweeter in climates where daytime maximum temperatures are greater than 30°C. Whereas many starch-accumulating fruit species do better in areas where temperatures rarely exceed 30°C. More research is needed to explore whether this is related to the photosynthetic process or downstream carbohydrate metabolism in these species. It is important to note that even in regions where maximum temperatures often exceed 30°C, these temperatures usually only occur for a relatively small part of a day (afternoon) in most regions where temperate deciduous crops are commercially grown.

All commercial fruit bearing species use what is known as C_3 photosynthesis (the first carbon compound assimilated in the photosynthetic process has three carbon atoms) (Taiz et al., 2015). Some plant scientists have suggested that the productivity of temperate deciduous fruit trees could be enhanced if they could be converted to the C_4 photosynthetic pathway found in some other plants, such as corn (*Zea mays*). This is highly unlikely, since C_4 photosynthesis would not be as efficient as C_3 photosynthesis in early spring when temperatures are relatively low (Taiz et al., 2015) and there is no competitive advantage of C_4 photosynthesis under the shady conditions (Pearcy and Ehleringer, 1984) that are common within the canopies of most fruit tree species.

3 Distribution and use of photosynthates

Over the past couple of decades, the concept that carbohydrate partitioning at the whole plant level is primarily driven by growth and development of individual organs has become widely accepted (Gifford and Evans, 1981; Ho, 1988; Lacointe, 2000; Marcelis, 1994; Watson and Casper, 1984; Weinstein and Yanai, 1994). Grossman and DeJong (1995b) used this concept in the development of the PEACH model and later DeJong (1999) outlined the following four principles for applying this concept to logically understand carbon partitioning in fruit trees.

3.1 The tree as a collection of semiautonomous organs

The first principle is that a tree is a collection of semiautonomous organs and each organ has a genetically determined, organ-specific development pattern and growth potential. Although much emphasis is often placed on considering plants as highly integrated organisms, the concept of semi-autonomy among organs is widely recognized (Harper, 1980; Sprugel et al., 1991; Watson and Casper, 1984; White, 1979). Indeed, the primary morphological features that distinguish one species or cultivar from another are at the organ or sub-organ level (i.e. fruit or leaf shape and size, floral characteristics, etc.), not at the whole plant level. Furthermore, although variation exists, the developmental patterns and growth rates of individual organs under specific environmental conditions are generally predictable. Models have been developed for the growth of fruit (DeJong and Goudriaan, 1989; Genard and Huguet, 1996; Genard and Souty, 1996; Grossman and DeJong, 1995b; Pavel and DeJong, 1993b; Lakso et al., 1995), shoots and branches (Costes et al., 1993, 2014; Costes and Guédon, 1996; Genard et al., 1998; Grossman and DeJong, 1995c; Lescourret et al., 1998; Johnson and Lakso, 1986) and roots (Bidel et al., 2000). Although tree pruning and training can drastically alter the shape of trees, they generally have very little effect on individual organ characteristics other than those explained by changes in the local micro-environment of the organs or changes in the availability of carbohydrates due to the proximity of other sinks.

The fact that there appears to be some level of branch autonomy (Sprugel et al., 1991; Heerema et al., 2008) in fruit trees further reinforces this first principle. Branch autonomy tends to functionally isolate some sinks from sources of carbohydrates. When sinks are manipulated through pruning or fruit thinning to create an apparent abundance of photosynthate in one part of the tree and an under-supply somewhere else, the carbon does not freely move to the location of greatest demand. When one scaffold of Y-shaped peach trees were de-fruited, the remaining fruit on the fruited scaffold benefited very little from the carbon that should have been available for fruit growth from the de-fruited scaffold (Marsal et al., 2003). Interestingly, scaffold diameter growth appeared to be one of the sinks that benefited most from the removal of fruit, while root growth was only marginally affected. There is much to be learned about the movement of carbohydrates within the context of the whole tree. The role of branch autonomy in early spring, when much of the carbon used for growth is mobilized from storage in the root, trunk and major branches and is presumably transported in the xylem, is still being elucidated (Zwieniecki et al., 2015).

Carbon partitioning at the branch level has been studied explicitly with radioactive tracer studies (Corelli-Grappadelli et al., 1996) and by manipulating leaf number and fruit load in isolated branches (Genard et al., 1998). Implicit conclusions about carbon partitioning within shoots have also been drawn from fruit thinning studies to determine optimal fruit positioning for fruit size (Marini and Sowers, 1994; Spencer and Couvillon, 1975). These studies support the idea that fruit are strong sinks for carbon within shoots, but their influence on where recently fixed carbon goes varies substantially within the local context of the stem unit.

3.2 Activation of an organ's genetic potential

The second principle is that the genetic potential of an organ is activated or deactivated by organ-specific, endogenous and/or environmental signals. The semiautonomous nature of individual organs is further demonstrated by the fact that individual organs on a tree can be experimentally activated by manipulating factors that stimulate the growth of specific organs independently from processes occurring in organs elsewhere on the tree. For instance, exposing individual buds on a branch to rest-breaking treatments can induce bud break in those buds, while similar buds on other parts of the tree remain inactive (Chandler, 1942). Similarly, grafting multiple cultivars with differing chilling requirements onto one trunk will not influence the inherent chilling exposure required for activation by the branch of each specific cultivar. Also, removing the apical meristem on a shoot will promote the activation of growth of lateral buds on the remaining part of the shoot, while buds on other shoots are unaffected (Harris, 1983). Although the exact mechanism of the environment and/or endogenous signals that activate growth is not fully understood, the primary site of activation is clearly at the organ or sub-organ level. This is certainly one area where hormones play key roles in influencing carbon partitioning at the whole tree level, as suggested by data on hormone concentration in xylem sap (Sorce et al., 2002).

3.3 Interaction of current environmental conditions and genetic growth potential

The third principle is that after an organ is activated, current environmental conditions and genetic growth potential interact to determine conditional organ growth capacity. Although often overlooked, ambient temperature is probably the single most important environmental factor influencing organ growth. This importance is related to the strong dependence of respiration on temperature. All irreversible plant organ growth is dependent on metabolic activity and enzyme function, and these processes are linked to respiration. Plant respiration generally has a temperature response quotient (Q_{10}) of about 2 (respiration doubles for every 10°C increase in temperature between 5°C and 35°C, Amthor, 1989). Therefore, conditional growth capacity of any organ is highly dependent on ambient temperature. The conditional growth capacity of fruits growing under near-optimal field conditions has been modelled for several peach and apple cultivars using mathematical functions responsive to heat accumulation (Berman et al., 1998; DeJong and Goudriaan, 1989; Grossman and DeJong, 1995a; Lakso et al., 1999; Pavel and DeJong, 1993a; Reyes et al., 2016). That other environmental factors such as water status can also have a substantial effect on organ growth is well documented (Bradford and Hsiao, 1982). Extension growth of shoots has been successfully modelled by considering temperature and dynamic changes in shoot water status (Basile et al., 2003; Berman and DeJong, 1997a). Although fruit growth is generally quite sensitive to water stress, it is important to distinguish between growth in fresh vs. dry matter since the former is much more sensitive than the latter (Berman and DeJong, 1997b; Girona et al., 1993). Nutrient availability also can strongly influence conditional organ growth capacity because certain nutrients are required as constituents for growing organs. Accordingly, Saenz et al. (1997) have demonstrated that limited N availability can increase developmental rates of peach fruit.

3.4 The nature of organ growth

The fourth principle is that actual organ growth is a consequence of conditional organ growth capacity, resource availability (assimilate and nutrient supply) and inter-organ competition for those resources. Inter-organ competition for carbohydrates is a function of location relative to sources and sinks of carbohydrates, transport resistances, organ sink efficiency and organ microenvironment. When conditional growth capacity of an organ is set, organ growth should proceed at a rate equal to the conditional growth capacity as long as transport is not limited and enough resources (carbohydrates) are available to support that organ's growth and the growth of all other competing organs. However, if the tree does not have enough carbohydrate to support the conditional growth capacity of all organs, or carbohydrate transport within the tree is limited, then the growth of an individual organ will be a function of its ability to compete for available carbohydrates with other organs. When flowering and pollination occur under favourable conditions, many fruit tree cultivars set very heavy fruit loads. Therefore, lack of available assimilates and inter-fruit competition for carbohydrates are generally the primary factors that limit realized fruit growth in these situations, and fruit thinning is essential to manage this competition (Cain and Mehlenbacher, 1956; Costa and Vizzoto, 2000; DeJong and Grossman, 1995; Dorsey and McMunn, 1928; Grossman and DeJong, 1995b; Goffinet et al., 1995). Certainly, there are some limitations to carbohydrate transport within trees (DeJong and Grossman, 1995; Marsal et al., 2003), but these are difficult to specifically quantify. There is substantial evidence that fruit growth of many species can compete effectively for carbohydrates with shoot, trunk and root growth when the crop loads are high and all fruit are considered as a collective sink (Grossman and DeJong, 1995a; Marsal et al., 2003; Proebsting, 1958). Yet, there is some evidence to the contrary when pruning stimulates excessive vegetative shoot growth (Grossman and DeJong, 1998). There is also clear documentation of the capacity of individual fruit organs to compete with each other and/or vegetative sinks at the local branch level (Genard et al., 1998). A further complication is that the ability of fruit to compete for carbohydrates appears to vary with the stage of fruit development (DeJong and Grossman, 1995) and location within a tree (Basile et al., 2007).

Upon examining these principles for carbon partitioning in fruit trees, it becomes apparent that phenological patterns of organ growth are the main determinants of carbon partitioning. When experiments are conducted involving different crop load treatments or some other treatment that dramatically favours the growth of one type of organ over others, biomass data collected at the end of the season appear to indicate that some organs are in direct competition with others (Chalmers and Van den Ende, 1975; Proebsting, 1958). However, when seasonal patterns of growth are analysed, it is apparent that direct competition between different organ types is often limited by temporal separation of growth activities (Berman and DeJong, 2003; DeJong et al., 1987; Miller and Walsh, 1988; Rufat and DeJong, 2001). Generally, in late-maturing fruit cultivars, shoot and root growth is the dominant sink shortly after bud break in the spring. This period is followed by a peak of fruit growth and then there is a resurgence of root growth (Pace, unpublished data) and shoot diameter growth after harvest (Berman and DeJong, 2003; Grossman and DeJong, 1995a). It is interesting that breeding efforts to create cultivars with early fruit ripening times have apparently interfered with the natural temporal separation of dominant sink activities in fruit trees. The dominant period of fruit growth of early-maturing peach cultivars often coincides directly with the early peak of shoot growth. This increased competition between fruit and shoot growth results in decreased yield potential (DeJong et al., 1987; Grossman and DeJong, 1995a). There is also some evidence that selection for early-maturing cultivars has involved coincidental selection for decreases in the total fruit growth potential and dry matter content, and these factors account for some of the differences in yield potential between early- and late-maturing cultivars (Berman et al., 1998). Selection for early-maturing fruit has also increased the competition for carbohydrates between sub-organs within the fruit such that seed and endocarp development corresponds with the period of flesh enlargement (Pavel and DeJong, 1993a) as well as increasing the individual fruit relative growth rates so that the tree cannot support as many fruits at one time (Grossman and DeJong, 1995a,b).

3.5 Carbohydrate storage

Where does 'allocation to storage' fit into this scheme of carbon partitioning? Long-term carbohydrate storage is essential for tree survival during adverse conditions (particularly winter for temperate deciduous crops) and subsequent productivity. However, there has been confusion about factors controlling storage reserves in trees (Epron et al., 2012). The prevailing view has been that trees store carbohydrate reserves during times of 'excess' photosynthate production (when current supply exceeds demands for growth and tissue metabolism), and deplete reserves when the potential rate of carbohydrate utilization exceeds the rate of current photosynthate production (Oliveira and Priestley, 1988; Kozlowski et al., 1991; Dickson, 1991). This has created the notion that carbohydrate storage occurs only when photosynthates are in excess of demand.

Some researchers have challenged this passive concept of carbohydrate storage arguing that storage reserves are extremely important and storage sinks should not be conceived of as passive reservoirs (Cannell and Dewar, 1994). They cited examples of control mechanisms for the use of carbohydrate reserves, and that storage sinks are refilled at the same time as the growth of other carbohydrate sinks (Weinstein et al., 1991). Indeed, careful evaluation of seasonal dynamics of reserve mobilization and accumulation that correspond to periods of shoot and fruit growth indicates that, although rates of reserve accumulation are generally lower when fruit growth rates are high, reserve accumulation still occurs during this period even though potential fruit growth rates are likely not at a maximum (Ryugo and Davis, 1959; Priestley, 1970). Similarly, although autumn appears to be the main period for accumulation of carbohydrate reserves in temperate deciduous trees, some reserves are accumulated while growth is still occurring during summer (Barbaroux and Bréda, 2002; Landhäusser and Lieffers, 2003; Wong et al., 2003). Wargo (1979) reported that substantial storage of carbohydrates preceded radial growth of *Acer saccharum* roots and even speculated that root storage of that species may have priority for transported carbohydrates over growth.

Da Silva et al. (2014) pointed out that long-term carbohydrate storage in trees is a function of the volume of xylem and phloem parenchyma tissue in the tree. Furthermore, the volume of xylem parenchyma greatly exceeds the volume of phloem parenchyma. Thus, the collective storage 'organ' of a tree can be thought of as the woody parenchyma tissues of the tree. The storage capacity of that 'collective organ' is mainly comprised of the overall mass of xylem parenchyma, the maximum potential concentration of carbohydrates in the xylem parenchyma, the minimum amount of carbohydrates remaining in the xylem parenchyma after maximum mobilization and the relative change in storage activity with xylem ageing.

If tree carbohydrate storage capacity is determined primarily as wood is formed, and only current-year sapwood growth can be affected by environmental conditions in a given year, the overall ability for a tree to adjust its storage capacity in response to environmental conditions is very limited. However, this also opens up important questions for future research into the effects of growing conditions on development of carbohydrate storage capacity in trees, the dynamics of storage and mobilization over time and the transport of substances from tree's bottom to the top at different periods during the season (DeJong, 2016). While most carbohydrate transport is usually thought of as occurring in the phloem, it is clear that much of the upward transport of carbohydrate mobilized from xylem parenchyma in the spring occurs in the xylem (Bonhomme et al., 2010; Ameglio et al., 2002). Tixier et al. (2017) have proposed a novel concept for how carbohydrates stored in lower parts of a tree can be delivered to growing shoot tips in the xylem before there is little or no transpiration to facilitate xylem flow. In addition, while seasonal changes in stored carbohydrates have been known to occur for a long time (Kozlowski et al., 1991), recently it has been shown that diurnal and seasonal changes in temperature patterns cause dynamic changes in starch storage throughout dormancy and facilitate re-distribution of storage carbohydrates in response to changes in temperature (Zwieniecki et al., 2015; Sperling et al., 2017). It appears likely that the dynamics of carbohydrate storage in trees may influence tree phenology to a much greater extent than previously recognized.

4 Fruit tree canopy architecture

4.1 Tree architecture

In recent years, there have been major advances in the understanding of fruit tree architecture. While the growth characteristics of specific fruit tree species and cultivars have been generally recognized for many years, recent advances in statistically based analyses and descriptions of bud fate distributions and shoot and branch growth patterns of multiple species have revealed previously under-appreciated similarities and differences among growth characteristics of numerous fruit tree species (Durand et al., 2005; Costes et al., 2006). This has led to a greater understanding that trees are composed of repeating growth units with similarities in patterning of lateral vegetative and floral buds along their axes and that patterns at the shoot level lead to differences in fruit bearing at the shoot level and overall tree architecture at the whole tree level (Costes et al., 2014).

These types of statistical analyses of tree architecture have been used to describe differences in growth and architectural development of different apple (Costes and Guedon, 2002; Costes et al., 2003) and almond (*Prunus dulcis*) cultivars (Negron et al., 2013), the influence of dwarfing apple rootstocks on scion growth and flowering (Costes and García-Villanueva, 2007; Seleznyova et al., 2003) and similarities and differences in bud fate structures among Rosaceous species (Costes et al., 2014). It also has been used in developing simulation models to demonstrate canopy growth dynamics in apple (Renton et al., 2006), sweet cherry (*Prunus avium*) (Lang et al., 2004), peach (Lopez et al., 2008; Lescourret et al., 2011) and almond (DeJong et al., 2017).

4.2 Architecture-informed pruning

In fruit crops, it is well recognized that there are two objectives with regard to optimizing the capture of solar energy to achieve maximum economic yields: (1) optimizing the total light interception by the canopy and (2) distributing the light within the canopy to obtain as many high-quality fruit as possible while nurturing high-quality fruiting spurs/shoots for the following year's crop. The increased understanding derived from shoot growth and tree architecture models has been valuable in developing canopy management strategies that optimize pruning procedures that work with the natural growth characteristics of trees to achieve these goals (Costes et al., 2006; Lauri, 2002). This has led to the development of 'centrifugal' pruning techniques involving 'spur extinction' in apples (Lauri et al., 2004, 2009; Tustin et al., 2011), and less intrusive training systems in stone fruits that adapt pruning practices to the natural growth characteristics of trees (Day et al., 2013; Lang, 2001). Recent research in nut crops, for which concerns about distribution of light within tree canopies to maintain quality are less than for fruit crops but have been traditionally pruned similarly to fruit crops, has led to the realization that the growth habits and architectures of some nut tree species naturally lend themselves to efficient capture of solar energy. Thus, high yields can be achieved without extensive pruning and training (Tombesi et al., 2011). In California, it is now recommended that young almond and walnut (*Juglans regia*) orchards be managed without traditional, annual pruning (Duncan, 2010; Lampinen et al., 2015).

5 Conclusion

While it is important to continually increase knowledge and understanding of fundamental aspects of the physiology of fruit trees and there is still much to be learned, it also is important to recognize that researchers have had very little success in improving upon what aeons of natural selection has provided in terms of fundamental tree functioning. This should not be a surprise, since trees are solar energy collection systems that have been evolving these systems for millions of years. Most horticultural progress has been achieved through empirical research or using physiological knowledge and understanding to improve tree and orchard management practices, rather than improving physiological processes.

6 Where to look for further information

Research reports and articles on many of the subjects discussed in this chapter authored by the author of this chapter can be found at http://dejong.ucdavis.edu/research-overview-publications/.

Updates on fruit tree physiology can be obtained by following the publications from the International Society for Horticultural Science (https://www.ishs.org/) and the American Society for Horticultural Science (https://www.ashs.org/).

An interesting and very readable book focused on the structural growth of trees is *The Growing Tree* by Wilson (1970). Further information on detailed studies of fruit tree architecture has been reviewed by Costes et al. (2006).

Further information of the details of photosynthesis, physiology and growth of plants can be found in Taiz et al. (2015).

7 References

Améglio, T., Bodet, C., Lacointe, A. and Cochard, H. (2002). Winter embolism, mechanisms of xylem hydraulic conductivity recovery and springtime growth patterns in walnut and peach trees. *Tree Physiology* 22:1211–20.

Amthor, J. S. (1989). *Respiration and Crop Productivity*. Springer-Verlag, New York, NY.

Auzmendi, I., Mata, M., Lopez, G., Girona, J. and Marsal, J. (2011). Intercepted radiation by apple canopy can be used as a basis for irrigation scheduling. *Agricultural Water Management* 98:886–92.

Auzmendi, I., Marsal, J., Girona, J. and Lopez, G. (2013). Daily photosynthetic radiation use efficiency for apple and pear leaves: Seasonal changes and estimation of canopy net carbon exchange rate. *European Journal of Agronomy* 51:1–8.

Avery, D. J. (1975). Effects of fruits on photosynthetic efficiency. In: *Climate and the Orchard*. H. C. Pereira, ed. Res. Rev. Common W. Bur. Hort. Planta. Crops. No. 5, pp. 110–12.

Barbaroux, C. and Breda, N. 2002. Contrasting distribution and seasonal dynamics of carbohydrate reserves in stem wood of adult ring-porous sessile oak and diffuse-porous beech trees. *Tree Physiology* 22:1201–10.

Basile, B., Marsal, J. and DeJong, T. M. (2003). Daily shoot extension growth of peach trees growing on rootstocks that reduce scion growth is related to daily dynamics of stem water potential. *Tree Physiology* 23:695–704.

Basile, B., Solari, L. I. and DeJong, T. M. (2007). Intra-canopy variability of fruit growth rate in peach trees grafted on rootstocks with different vigour-control capacity. *Journal of Horticultural Science and Biotechnology* 82:243-56.

Behboudian, M. H., Marsal, J., Girona, J. and Lopez, G. (2011). Quality and yield responses of deciduous fruits to reduced irrigation. *Horticultural Reviews* 38:149-89.

Ben Mimoun, M., Longuenesse, J.-J. and Genard, M. (1996). Pmax as related to leaf: Fruit ratio and fruit assimilate demand in peach. *Journal of Horticultural Science* 71:767-75.

Berman, M. E. and DeJong, T. M. (1997a). Crop load and water stress effects on daily stem growth in peach (*Prunus persica*). *Tree Physiology* 17:467-72.

Berman, M. E. and DeJong, T. M. (1997b). Diurnal patterns of stem extension growth in peach (*Prunus persica*): Temperature and fluctuations in water status determine growth rate. *Physiologia Plantarum* 100:361-70.

Berman, M. E. and DeJong, T. M. (2003). Seasonal patterns of vegetative growth and competition with reproductive sinks in peach (*Prunus persica*). *Journal of Horticultural Science and Biotechnology* 78:303-9.

Berman, M. E., Rosati, A., Pace, L., Grossman, Y. and DeJong, T. M. (1998). Using simulation modeling to estimate the relationship between data of fruit maturity and yield potential in peach. *Fruit Varieties Journal* 52:229-35.

Bidel, L. P. R., Pages, L., Rivière, L. M., Pelloux, G. and Lorendeau, J. Y. (2000). MassFlowDyn I: A carbon transport and partitioning model for root system architecture. *Annals of Botany* 85:869-86.

Bieleski, R. L. (1982). Sugar alcohols. In: *Encyclopedia of Plant Physiology*, Volume 13A, Plant Carbohydrates I. Intracellular Carbohydrates. F. A. Loewus and W. Tanner, eds. Springer-Verlag, Berlin, pp. 158-92.

Blum, A. (2009). Effective use of water (EUW) and not water-use efficiency (WUE) is the target of crop yield improvement under drought stress. *Field Crops Research* 112:119-23.

Bonhomme, M., Peuch, M., Améglio, T., Rageau, R., Guilliot, A., Decourteix, M., Alves, G., Sakr, S. and Lacointe, A. (2010). Carbohydrate uptake from xylem vessels and its distribution among stem tissues and buds in walnut (*Juglans regia* L.). *Tree Physiology* 30:89-102.

Bradford, K. J. and Hsiao, T. C. (1982). Physiological responses to moderate water stress. In: *Physiological Plant Ecology II. Water Relations and Carbon Assimilation.* O. L. Lange, P. S. Nobel, C. B. Osmond and H. Ziegler, eds. Springer-Verlag, Berlin, pp. 263-324.

Cain, J. C. and Mehlenbacher, R. J. (1956). Effects of nitrogen and pruning on trunk growth in peaches. *Proceedings of American Society for Horticultural Science* 67:139-43.

Cannell, M. G. R. and Dewar, R. C. (1994). Carbon allocation in trees: A review of concepts for modeling. In: *Advances in Ecological Research*, volume 25. M. Begon and A. H. Fitter, eds. Academic Press, London, pp. 59-104.

Chalmers, D. J. and Van den Ende, B. (1975). Productivity of peach trees: Factors affecting dry-weight distribution during tree growth. *Annals of Botany* 39:423-32.

Chandler, W. H. (1942). *Deciduous Orchards*. Lea and Febiger, Philadelphia.

Corelli-Grappadelli, L., Ravaglia, G. and Asirelli, A. (1996). Shoot type and light exposure influence carbon partitioning in peach cv. Elegant Lady. *Journal of Horticultural Science* 71:533-43.

Costa, G. and Vizzoto, G. (2000). Fruit thinning of peach trees. *Plant Growth Regulation* 31:113–19.

Costes, E. and García-Villanueva, E. (2007). Clarifying the effects of dwarfing rootstock on vegetative and reproductive growth during tree development: A study on apple trees. *Annals of Botany* 100:347–57.

Costes, E. and Guédon, Y. (1996). Modeling the annual shoot structure of the apricot tree (cv Lambertin). in terms of axillary flowering and vegetative growth. *Acta Horticulturae* 416:21–8.

Costes E. and Guédon, Y. (2002). Modelling branching patterns on 1-year-old trunks of six apple cultivars. *Annals of Botany* 89:513–24.

Costes, E., Lanri, P. E., Guédon, Y. and de Reffye, P. (1993). Modelling growth of peach trees using the renewal theory. *Acta Horticulturae* 349:253–8.

Costes, E., Sinoquet, H., Kelner, J. J. and Godin C. (2003). Exploring within-tree architectural development of two apple tree cultivars over 6 years. *Annals of Botany* 91:91–104.

Costes, E., Lauri, P. E. and Regnard, J. L. (2006). Analyzing fruit tree architecture: Implications for tree management and fruit production. *Horticultural Reviews* 32:1–60.

Costes, E., Crespel, L., Denoyes, B., Morel, P., Demene, M.-N., Lauri, P.-E. and Wenden, B. (2014). Budstructure, position and fate generate various branching patterns along shoots of closely related Rosaceae species: A review. *Frontiers in Plant Science* 5:Article 666, 11pp.

Da Silva, D., Qin, L., DeBuse, C. and DeJong, T. M. (2014). Measuring and modelling seasonal patterns of carbohydrate storage and mobilization in the trunks and root crowns of peach trees. *Annals of Botany* 114:643–52.

Day, K. R., Johnson R. S. and DeJong, T. M. (2013). Developing a pedestrian plum orchard: the role of tree form, density and height. *Acta Horticulturae* 985:175–80.

DeJong, T. M. (1983). CO_2 assimilation characteristics of five *Prunus* tree fruit species. *Journal of the American Society for Horticultural Science* 108:303–7.

DeJong, T. M. (1986). Fruit effects on photosynthesis in *Prunus persica*. *Physiologia Plantarum* 66:149–53.

DeJong, T. M. (1999). Developmental and environmental control of dry-matter partitioning in peach. *HortScience* 34:1037–40.

DeJong, T. M. (2016). Demystifying carbohydrate allocation to storage in fruit trees. *Acta Horticulturae* 1130:329–34.

DeJong, T. M. and Doyle, J. F. (1985). Seasonal relationships between leaf nitrogen content (photosynthetic capacity). and leaf canopy light exposure in peach *Prunus persica*. *Plant, Cell and Environment* 8:701–6.

DeJong, T. M. and Goudriaan, J. (1989). Modeling peach fruit growth and carbohydrate requirements: reevaluation of the doublesigmoid growth pattern. *Journal of the American Society for Horticultural Science* 114:800–4.

DeJong, T. M. and Grossman, Y. L. (1995). Quantifying sink and source limitations on dry matter partitioning to fruit growth in peach trees. *Physiologia Plantarum* 95:437–43.

DeJong, T. M., Doyle, J. F., and Day, K. R. (1987). Seasonal patterns of reproductive and vegetative sink activity in early and late maturing peach (*Prunus persica*) cultivars. *Physiology Plantarum* 71:83–8.

DeJong, T. M., Day, K. R. and Johnson, R. S. (1989). Partitioning of leaf nitrogen with respect to within canopy light exposure and nitrogen availability in peach (*Prunus persica*). *Trees* 3:89–95.

DeJong, T. M., Da Silva, D., Negron, C., Cieslak, M. and Prusinkiewicz, P. (2017). The L-ALMOND model: a functional-structural virtual tree model of almond tree architectural growth, carbohydrate dynamics over multiple years. *Acta Horticulturae* 1160:43–9

Dickson, R. E. (1991). Assimilate distribution and storage. In: *Physiology of Trees*. A. S. Raghavendra, ed. John Wiley & Sons, Inc., New York, pp. 51–85.

Dorsey, M. J. and McMunn, R. L. (1928). The third report on the Illinois thinning investigations. *Proceedings of American Society for Horticultural Science* 25:269–76.

Duncan, R. (2010). Pruning almond trees does not pay. http://cestanislaus.ucanr.edu/files/111772.pdf (Cooperative Extension web post).

Durand, J.-B., Guédon, Y., Caraglio, Y and Costes, E. (2005). Analysis of the plant architecture via tree-structured statistical models: the hidden Markov tree models. *New Phytologist* 166:813–25.

Epron, D., Nouvellon, Y. and Ryan, M. G. (2012). Introduction to the invited issue on carbon allocation of trees and forests. *Tree Physiology* 32:639–43.

Genard, M. and Huguet, J. G. (1996). Modeling the response of peach fruit growth to water stress. *Tree Physiology* 16:407–15.

Genard, M. and Souty, M. (1996). Modeling the peach sugar contents in relation to fruit growth. *Journal of American Society for Horticultural Science* 121:1122–31.

Genard, M., Lescourret, F., Ben Mimoun, M., Besset, M. J. and Bussi, C. (1998). A simulation model of growth at the shoot-bearing fruit level. II. Test and effect of source and sink factors in the case of peach. *European Journal of Agronomy* 9:189–202.

Gifford, R. M. and Evans, L. T. (1981). Photosynthesis, carbon partitioning and yield. *Annual Review of Plant Physiology* 32:485–509.

Girona, J., Mata, M., Goldhamer, D. A., Johnson, R. S. and DeJong, T. M. (1993). Patterns of soil and tree water status and leaf functioning during regulated deficit irrigation scheduling in peach. *Journal of the American Society for Horticultural Science* 118:580–6.

Goffinet, M. C., Robinson T. L. and Lakso A. N. (1995). A comparison of 'Empire' apple fruit size and anatomy in unthinned and hand-thinned trees. *Journal of Horticultural Science* 70:375–87.

Grossman, Y. L. and DeJong, T. M. (1995a). Maximum fruit growth potential and seasonal patterns of resource dynamics during peach growth. *Annals of Botany* 75:553–60.

Grossman, Y. L. and DeJong, T. M. (1995b). Maximum fruit growth potential following resource limitation during peach growth. *Annals of Botany* 75:561–7.

Grossman, Y. L. and DeJong, T. M. (1995c). Maximum vegetative growth potential and seasonal patterns of resource dynamics during peach growth. *Annals of Botany* 76:473–82.

Grossman, Y. L. and DeJong, T. M. (1998). Training and pruning system effects on vegetative growth potential, light interception, and cropping efficiency in peach trees. *Journal of American Society for Horticultural Science* 123:1058–64.

Hansen, P. (1970). 14C studies on apple trees. VI. Influence of the fruit on the photosynthesis of the leaves, and the relative photosynthetic yields of fruits and leaves. *Physiologia Plantarum* 23:805–10.

Harper, J. L. (1980). Plant demography and ecological theory. *Oikos* 35:244–53.

Harris, R. W. (1983). *Arboriculture*. Prentice-Hall. Englewood Cliffs, NJ.

Heerema, R. J., Weinbaum, S. A., Pernice, F. and DeJong, T. M. (2008). Spur survival and return bloom in almond [*Prunus dulcis* (Mill.). D.A. Webb] varied with spur fruit load, specific leaf weight, and leaf area. *Journal of Horticultural Science and Biotechnology* 83:274–81.

Ho, L. C. (1988). Metabolism and compartmentation of imported sugars in sink organs in relation to sink strength. *Annual Review of Plant Physiology* 39:355–78.

Horton, P. (2000). Prospects for crop improvement through the genetic manipulation of photosynthesis: Morphological and biochemical aspects of light capture. *Journal of Experimental Botany* 51:475–85.

Johnson, R. S. and Lakso, A. N. (1986). Carbon model of a growing apple shoot: 1. Development of the model. *Journal of the American Society for Horticultural Science* 111:160–4.

Kozlowski, T. T., Kramer P. J. and Pallardy, S. G. 1991. *The Physiological Ecology of Woody Plants*. Academic Press, San Diego, CA, 657p.

Lacointe, A. (2000). Carbon allocation among tree organs: A review of basic processes and representation in functional-structural models. *Annual Forestry of Science* 57:521–34.

Lakso, A. N., Corelli-Grappadelli, L. and Goffinet, M. C. (1995). An expolinear model of the growth pattern of the apple fruit. *Journal of Horticultural Science* 70:389–94.

Lakso, A. N., Wünsche, J. N. and Palmer, J. W. (1999). Measurement and modeling of carbon balance of the apple tree. *HortScience* 34:1040–7.

Lampinen, B. D., Tombesi, S., Metcalf, S. G. and DeJong, T. M. (2011). Spur behaviour in almond trees: Relationships between previous year leaf area, fruit bearing and mortality. *Tree Physiology* 31:700–6.

Lampinen, B. D., Udompetaikul V., Browne, G. T., Metcalf, S. G., Stewart, W. L., Contador, L., Negron, C. and Upadhyaya, S. K. (2012). A mobile platform for measuring canopy photosynthetically active radiation interception in orchard systems. *HortTechnology* 22:237–44.

Lampinen, B. D., Edstrom, J. P., Metcalf, S. G., Stewart, W. L., Negron, C. M. and Contador, M. L. (2015). Howard walnut trees can be brought into bearing without annual pruning. *California Agriculture* 69:123–8.

Landhäusser, S. M. and Lieffers, V. J. 2003. Seasonal changes in carbohydrate reserves in mature northern *Populus tremuloides* clones. *Trees* 17:471–6.

Lang, G. A. (2001). Critical concepts for sweet cherry training systems. *The Compact Fruit Tree* 34:70–3.

Lang, G. A., Olmstead, J. W. and Whiting, M. D. (2004). Sweet cherry fruit distribution and leaf populations: Modeling canopy dynamics and management strategies. *Acta Horticulturae* 636:591–9.

Lauri, P.-É. (2002). From tree architecture to tree training – An overview of recent concepts developed in apple in France. *Journal of the Korean Society for Horticultural Science* 43:782–8.

Lauri P.-É., Willaume, M., Larrive, G. and Lespinasse, J.-M. (2004). The concept of centrifugal training in apple aimed at optimizing the relationship between growth and fruiting. *Acta Horticulturae* 636:35–42.

Lescourret, F., Ben Mimoun, M. and Genard, M. (1998). A simulation model of growth at the shoot-bearing fruit level I. Description and parameterization for peach. *European Journal of Agronomy* 9:173–88.

Lescourret, F., Moitrier, N., Valsesia, P. and Génard, M. (2011). QualiTree, a virtual fruit tree to study the management of fruit quality. I. Model development. *Trees* 25:519–30.

Li, K-T. and Lakso, A. N. (2004). Photosynthetic characteristics of apple spur leaves after summer pruning to improve exposure to light. *HortScience* 39:969–72.

Lopez, G., Favreau, R. R., Smith, C., Costes, E., Prusinkiewicz, P. and DeJong, T. M. (2008). Integrating simulation of architectural development and source-sink behavior of peach trees by incorporating Markov models and physiological organ function sub-models into L-Peach. *Functional Plant Biology* 35:761–71.

Lopez, G., Echeverria, G., Bellvert, J., Mata, M., Behboudian, M. H., Girona, J. and Marsal, J. (2016). Water stress for a short period before harvest in nectarine: Yield, fruit composition, sensory quality, and consumer acceptance of fruit. *Scientia Horticulturae* 211:1–7.

Maggs, D. H. (1963). The reduction in growth of apple trees brought about by fruiting. *Journal of Horticultural Science* 38:119–28.

Marcelis, L. F. M. (1994). A simulation model for dry matter partitioning in cucumber. *Annals of Botany* 74:43–52.

Marini, R. P. and Sowers, D. (1994). Peach fruit weight is influenced by crop density and fruiting shoot length but not position on the shoot. *Journal of American Society of Horticultural Science* 119:180–4.

Marsal, J., Basile, B., Solari, L. and DeJong, T. M. (2003). Influence of branch autonomy on fruit, scaffold, trunk and root growth during Stage III of peach fruit development. *Tree Physiology* 23:313–23.

Marsal, J., Casadesus, J., Lopez, G., Mata, M., Bellvert, J. and Girona, J. (2016). Sustainability of regulated deficit irrigation in a mid-maturing peach cultivar. *Irrigation Science* 34:201–8.

Miller, A. N. and Walsh, C. S. (1988). Growth and seasonal partitioning of dry matter in eight-year-old 'Loring peach trees'. *Journal of American Society for Horticultural Science* 113:309–14.

Neales, T. F. and Incoll, L. D. (1968). The control of leaf photosynthesis rate by the level of assimilate concentration in the leaf: A review of the hypothesis. *Botanical Review* 34:107–25.

Negron, C., L. Contador, B. D. Lampinen, S. G. Metcalf, Y. Guedon, E. Costes and DeJong, T. M. (2013). Systematic analysis of branching patterns of three almond cultivars with different tree architectures. *Journal of the American Society for Horticultural Science* 138:407–15.

Niinemets, Ü. (1995). Distribution of foliar carbon and nitrogen across the canopy of Fagus sylvatica: Adaptation to a vertical light gradient. *Acta Oecologia* 16:525–41.

Oliveira, C. M. and Priestley, C. A. 1988. Carbohydrate reserves in deciduous fruit trees. *Horticultural Reviews* 10:403–30.

Palmer, J. W., Wünsche, J. N., Meland, M. and Hann, A. (2002). Annual dry-matter production by three apple cultivars at four within-row spacings in New Zealand. *Journal of Horticultural Science and Biotechnology* 77:712–17.

Palmer, J. W., Wünsche, J. N., Greer, D. H., Laings, W. A. and Palmer, J. W. (2005). Physiological and biochemical leaf and tree responses to crop load in apple. *Tree Physiology* 25:1253–63.

Pavel, E. W. and DeJong, T. M. (1993a). Relative growth rate and its relationship to compositional changes of nonstructural carbohydrates in the mesocarp of developing peach fruits. *Journal of American Society for Horticultural Science* 118:503–8.

Pavel, E. W. and DeJong, T. M. (1993b). Seasonal CO_2 exchange patterns of developing peach (Prunus persica) fruits in response to temperature, light and CO_2 concentration. *Physiologia Plantarum* 88:322-30.

Pearcy, R. W. and Ehleringer, J. (1984). Comparative ecophysiology of C3 and C4 plants. *Plant, Cell and Environment* 7:1-13.

Priestley, C. A. (1970). Carbohydrate storage and utilization. In: *Physiology of Tree Crops*. L. C. Luckwill and C. V. Cutting, eds. Academic Press, London, pp. 113-27.

Proebsting, E. L. (1958). A quantitative evaluation of the effect of fruiting on growth of Elberta peach trees. *Proceedings of the American Society of Horticultural Science* 71:103-9.

Renton, M., Guédon, Y., Godin, C. and Costes, E. (2006). Integrating Markov chain models and L-systems to simulate the architectural development of apple trees. *Acta Horticulturae* 707:63-70.

Reyes, F. DeJong, T. Franceschi, P. Tagliavini M. and Gianelle D. (2016). Maximum growth potential and periods of resource limitation in apple trees. *Frontiers of Plant Science* 7:233.

Robinson, T. L., Lakso, A. N. and Ren Z. (1991). Modifying apple tree canopies for improved production efficiency. *HortScience* 26:1005-12.

Ro, H.-M., Kim, P.-G., Lee, I.-B., Yiem, M.-S. and Woo, S.-Y. (2001). Photosynthetic characteristics and growth responses of dwarf apple (*Malus domestica* Borkh. cv. Fuji). saplings after 3 years of exposure to elevated atmospheric carbon dioxide concentration and temperature. *Trees* 15:195-203.

Rosati, A., Esparza, G., DeJong, T. M. and Pearcy, R. W. (1999). Influence of canopy light environment and nitrogen availability on leaf photosynthetic characteristics and photosynthetic nitrogen-use efficiency of field-grown nectarine trees. *Tree Physiology* 19:173-80.

Rosati, A., Day, K. R. and DeJong, T. M. (2000). Distribution of leaf mass per unit area and leaf nitrogen concentration determine partitioning of leaf nitrogen within tree canopies. *Tree Physiology* 20:271-6.

Rosati. A., DeJong, T. M. and Esparza, G. (2002). Physiological basis for light use efficiency models. *Acta Horticulturae* 584:89-94.

Rufat, J. and DeJong, T. M. (2001). Estimating seasonal nitrogen dynamics in peach trees in response to nitrogen availability. *Tree Physiology* 21:1133-40.

Ryugo, K. and Davis, L. D. 1959. The effect of the time of ripening on the starch content of bearing peach branches. *Proceedings of the American Society for Horticultural Science* 74:130-3.

Saenz, J. L., DeJong, T. M. and Weinbaum, S. A. (1997). Nitrogen stimulated increases in peach yields are associated with extended fruit development period and increased fruit sink capacity. *Journal of the American Society for Horticultural Science* 122:772-7.

Seleznyova, A., Thorp, G., White, M., Tustin, S. and Costes E. (2003). Structural development of branches of 'Royal Gala' apple grafted on different rootstock/interstock combinations. *Annals of Botany* 91:1-8.

Sorce, C., Massai, R., Picciarelli, P. and Lorenzi, R. (2002). Hormonal relationships in xylem sap of grafted and ungrafted *Prunus* rootstocks. *Scientia Horticulturae* 93:333-42.

Spencer, S. and Couvillon, G. A. (1975). The relationship of node position to bloom date, fruit size and endosperm development of the peach, *Prunus persica* (L.). Batsch cv. Sullivan's Elberta. *Journal of American Society of Horticultural Science* 100:242-4.

Sperling, O., Silva, L. C. R., Tixier, A., Théroux-Rancourt, G. and Zwieniecki, M. A. (2017). Temperature gradients assist carbohydrate allocation within trees. *Science Reports* 7:3265.

Sprugel, D. G., Hinkley, T. M. and Schaap, W. (1991). The theory and practice of branch autonomy. *Annual Review of the Ecology System* 22:309–34.

Stephenson, A. G. (1981). Flower and fruit abortion: Proximate causes and ultimate functions. *Annual Review of Ecology and Systematics* 12:253–79.

Taiz, L., Zaiger, E., Moller I. M. and Murphy, A. (2015). *Plant Physiology and Development*. Sinauer Associates Inc., Sunderland, MA, 761pp.

Tixier, A., Sperling, O., Orozco, J., Lampinen, B., Roxas, A. A., Saa, S., Earles, J. M. and Zwieniecki, M. A. (2017). Spring bud growth depends on sugar delivery by xylem and water recirculation by phloem Münch flow in *Juglans regia*. *Planta* 246:495–508.

Tombesi, S., Lampinen, B. D., Metcalf, S. and DeJong, T. M. (2011). Relationships between spur- and orchard-level fruit bearing in almond (*Prunus dulcis*). *Tree Physiology* 31:1413–21.

Tustin, D. S., Dayatilake, G. A., Henriod, R. E., Breen, K. C. and Oliver, M. (2011). Changes in fruiting behaviour and vegetative development of 'Scifresh' apple in response to artificial spur extinction using centrifugal training. *Acta Horticulturae* 903:603–10.

Wargo, P. M. 1979. Starch storage and radial growth in woody roots of sugar maple. *Canadian Journal of Forest Research* 9:49–56.

Watson, M. A. and Casper, B. B. (1984). Morphogenetic constraints on patterns of carbon distribution in plants. *Annual Review of Ecology and Systematics* 15:233–50.

Weinstein, D. A. and Yanai, R. D. (1994). Integrating the effects of simultaneous multiple stresses on plants using the simulation model TREGRO. *Journal of Environmental Quality* 23:418–28.

Weinstein, D. A., Beloin, R. M. and Yanai, R. D. (1991). Modelling changes in red spruce carbon balance and allocation in response to interacting ozone and nutrient stresses. *Tree Physiology* 9:127–46.

White, J. (1979). The plant as a metapopulation. *Annual Review of Ecology and Systematics* 10:109–45.

Wilson, B. F. (1970). *The Growing Tree*. University of Massachusetts Press, Amherst, MA, 152pp.

Wong, B. L., Baggett, K. L. and Rye, A. H. 2003. Seasonal patterns of reserve and soluble carbohydrates in mature sugar maple (*Acer saccharum*). *Canadian Journal of Botany* 81:780–8.

Wünsche, J. N. and Lakso A. N. (2000). The relationship between leaf area and light interception by spur and extension shoot leaves and apple orchard productivity. *HortScience* 35:1202–6.

Wünsche, J. N., Palmer, J. W. and Greer, D. H. (2000). Effects of crop load on fruiting and gas-exchange characteristics of 'Braeburn'/M.26 apple trees at full canopy. *Journal of the American Society for Horticultural Science* 125:93–9.

Zwieniecki, M. A., Tixier, A. and Sperling, O. (2015). Temperature-assisted redistribution of carbohydrates in trees. *American Journal of Botany* 102:1216–18.

Chapter 4

Advances in understanding reproductive development in fruit-bearing plants

Tomoya Esumi, Shimane University, Japan; and Ryutaro Tao, Kyoto University, Japan

1 Overview of reproductive development in fruit trees

Flowering is an essential step in tree fruit production, because the fruit typically develops from specific parts of the flower. The process of floral bud and primordial flower development can be strictly divided into two steps: initiation and differentiation. In deciduous fruit trees, floral primordia are initiated more than 6 months before flowering. Flower initiation corresponds to the transition in plant development from the vegetative to the reproductive phase, in which active meristem tissue in the bud changes its developmental fate to produce flowers (Fig. 1). Following initiation, flower differentiation occurs in the bud to produce floral organs, including sepals, petals, stamens and pistils. Successful initiation and differentiation of flowers are considered important processes that affect fruit production in the following season. In deciduous trees, developing buds enter a dormancy period during the winter, and then resume development through flowering in early spring. Once the seasonal climate and environmental conditions become suitable for development in the spring, dramatic flowering episodes can occur.

To ensure successful fruiting, many important developmental and physiological steps must occur during the production of floral organs, especially the successful maturation of male and female organs during gametophyte

http://dx.doi.org/10.19103/AS.2018.0040.04

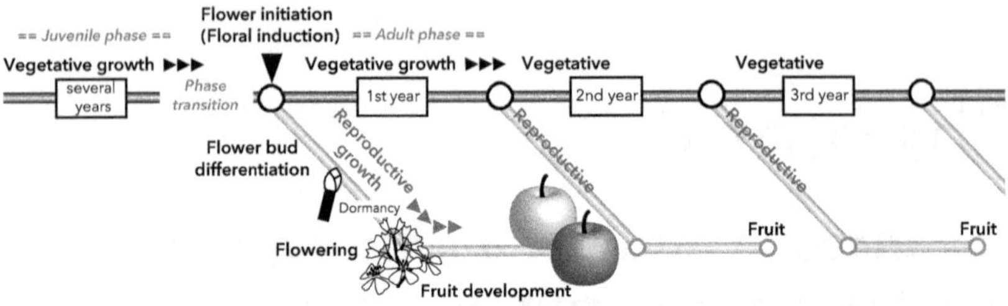

Figure 1 Reproductive development and life cycle of deciduous fruit trees. Reproductive development begins at floral induction and proceeds to flower and fruit development.

development. The sequential occurrence of sexual events, including pollen release (anther dehiscence), pollination, pollen tube elongation, fertilization and embryo development, are essential to plant reproduction. Many intriguing subjects remain to be researched in plant reproductive science. Following fertilization, fruit enlargement and maturation progress as the seed and embryo develop. In some fruit tree species or mutant cultivars, fruit development occurs autonomously, without seed or embryo development. This process is called parthenocarpy. Parthenocarpic fruits are prized by consumers for their lack of seeds; thus, parthenocarpy is frequently studied in fruit developmental research.

A fruit develops by increasing its volume through cell division and cell enlargement; many factors can influence its growth rate and ultimate size. Phytohormones are important internal messengers that stimulate and regulate fruit growth. Seeds in fruit can produce phytohormones that affect fruit development. Fruit maturation and ripening are the final steps of reproductive development in fruit trees, after which the fruit is ready for consumption. Fruit maturation involves physical, biochemical and physiological changes, including decreased firmness; altered texture; changes in fruit skin colour; degradation of skin chlorophyll; decreased respiratory activity; decreased acidity; and increased sugars, soluble solids and soluble pectin. The biosynthesis of secondary metabolites such as carotenoids, flavonoids and volatile compounds is enhanced during this step. Seeds within the fruit also mature during the fruit maturation stage.

In recent years, molecular biological and genetic studies have increased our physiological understanding of the reproductive development process of fruit trees. As a research trend, key scientific findings are frequently derived from model plant species such as *Arabidopsis thaliana*, rice (*Oryza sativa*) and tomato (*Solanum lycopersicum*). Such findings are generally immediately examined in fruit tree species; thus, the body of scientific knowledge on fruit tree reproduction is enriched and expanded. Conversely, interesting and notable findings in fruit tree reproduction research sometimes inspire new scientific ideas that can be directly applied to improve crop production. Although the physiological traits of perennial or woody fruit trees often hinder studies of the mechanisms driving their reproductive physiology, such studies can provide important new findings in plant science. This chapter introduces recent accomplishments and discoveries in research on the physiology of reproductive development in fruit trees.

2 The physiology of reproductive meristem development in fruit trees

In the 1980s, a fruit tree physiology textbook (Faust, 1989) stated that the nature of the signal for flower initiation was unknown, despite many past physiological experiments and observations on flower meristem induction. The advent of genetic analysis using various mutations in model plants, especially *Arabidopsis*, opened a new avenue for research, particularly with the floral meristem identity gene *LEAFY* (*LFY*), a plant-specific transcription factor first reported in *Arabidopsis* (Weigel et al., 1992). *LFY* can accelerate flowering in fruit tree species, as demonstrated in transgenic citrange (*Citrus sinensis* × *Poncirus trifoliata*) (Pena et al., 2001). However, *FT* (*FLOWERING LOCUS T*) in *Arabidopsis* was found to be a more suitable flowering signal candidate gene (Kobayashi et al., 1999), and *FT* was predicted to function as a more favourable floral inducer, encouraging the expansion of research to a broad range of plant species, including fruit trees. *FT* and *FT-like* homologous genes were isolated and characterized in many different fruit tree species, including apple (*Malus* × *domestica*) (Hattasch et al., 2008; Kotoda et al., 2010; Mimida et al., 2011a; Trankner et al., 2010), satsuma mandarin (*Citrus unshiu*) (Endo et al., 2005; Nishikawa et al., 2007, 2010, 2012) and grape (*Vitis vinifera*) (Carmona et al., 2007; Sreekantan and Thomas, 2006). *FT*, a small PEBP-like protein, is expressed in leaves and transferred to the apical meristem to induce floral development. Transgenic plants overexpressing *FT* and its homologous genes have exhibited precocious and accelerated flowering even in fruit tree species. In some species, *in vitro* flowering has been induced by *FT* overexpression (Endo et al., 2005; Kotoda et al., 2010; Matsuda et al., 2009; Song et al., 2013; Trankner et al., 2010). In citrus trees, *FT* (*CiFT*) expression in stems can be analysed to predict fruit production in the following season, because the amount of *CiFT* mRNA in stems is correlated with flower production in the following spring (Nishikawa et al., 2017).

Genetic transformation using Agrobacterium as a vector is now a conventional technique used to control *FT* gene expression in plants. The breeding of fast-track fruit trees, which exhibit a shortened breeding cycle, is currently being tested in several countries. Because *FT* is a mobile protein, breeding new rootstocks with high *FT* gene expression may reduce vegetative growth and shorten the time to flowering in scions. Other tools include plant virus vector systems using *Apple latent spherical virus (ALSV)* harbouring the *Arabidopsis thaliana FT* (*AtFT*) gene, which can induce early flowering after infection in apple seedlings (Yamagishi et al., 2011). A similar vector system using *Citrus leaf blotch virus* has been developed for citrus trees (Velazquez et al., 2016). Interestingly, the effect of an *FT* gene may differ according to its DNA sequence. In apple, precocious flowering was more stable following gene transformation using the *A. thaliana FT* (*AtFT*) gene than that using the apple *FT* gene (*MdFT*) (Tanaka et al., 2014), indicating that interaction between the *FT* protein and other flowering factors may be important. *FT* promotes flowering; however, *TERMINAL FLOWER 1* (*TFL1*), a member of a family of homologous genes encoding PEBP-like proteins in plants, exhibits the counter-effect of repressing flowering. Expression analyses of these genes have shown that *TFL1* is highly expressed below the apical meristem prior to floral induction, and that expression drastically decreases during flower differentiation (Esumi et al., 2007; Mimida et al., 2011b). *TFL1* represses flowering in fruit trees. Transgenic experiments with the *TFL1* gene, in which the endogenous expression of *TFL1* was suppressed by antisense or RNA interference (RNAi) methods, resulted in early flowering (Freiman et al., 2012; Kotoda et al., 2006). *TFL1* is also an antagonistic protein to *FT* in fruit trees. Recent studies have elucidated the key factors influencing floral induction or flower initiation in fruit trees. Controlling the expression of florigen (*FT*) and anti-florigen (*TFL1*) genes is among newly available tools with the potential to improve breeding efficiency and cultivation techniques.

The first visible sign of flower differentiation in buds is doming (swelling) of the apical meristem. Following doming, regular and rhythmic initiation of flower primordia can be observed. Recently developed stereoscopic imaging techniques using microscopes allow fine-scale observation of meristem development. *In situ* hybridization using sections is a technique applied to observe gene expression in buds (Fig. 2). The expression of flower meristem identity genes such as *LFY* and *APETALA 1* (*AP1*) and anti-florigen (*TFL1*) genes in fruit tree species can be observed using this technique (Carmona et al., 2002; Esumi et al., 2007; Kotoda et al., 2000; Mimida et al., 2011b; Sung et al., 1999, 2000).

3 Bud dormancy and flowering

Knowledge of bud dormancy in deciduous fruit trees has also advanced in recent years. The mechanisms of bud dormancy were previously attributed to environmental responses, hormonal changes and metabolic changes. Genetic approaches were recently applied to examine dormancy using quantitative trait locus (QTL) identification to determine chilling requirements in peach (*Prunus*

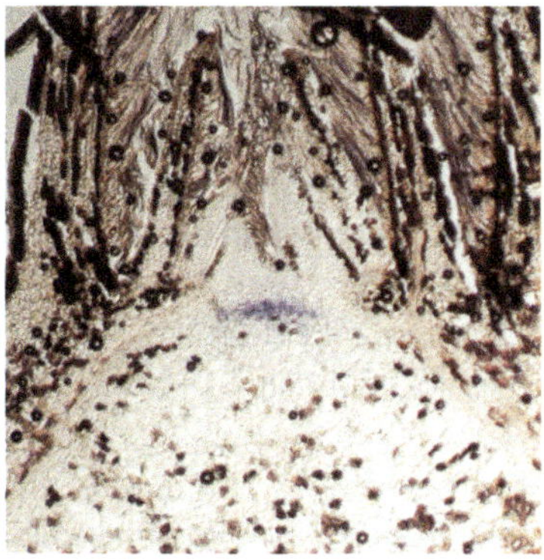

Figure 2 Expression of the *TFL1* gene, a floral induction repressor, detected by *in situ* hybridization in a Japanese pear (*Pyrus pyrifolia*) bud. *PpTFL1* expression (blue) was observed beneath the apical meristem prior to flower differentiation.

persica) (Castede et al., 2015; Fan et al., 2010). The involvement of transcriptional regulator genes such as MADS-box genes has also been explored. Dormancy-associated MADS-box (*DAM*) that belongs to *SHORT VEGETATIVE PHASE* (*SVP*) genes in *Prunus* and other fruit species were discovered using mutant varieties and transcriptome analyses (Yamane, 2014). *DAM* gene expression was found to increase by short photoperiod and low temperature during autumn, then decreased at the release of endodormancy, when trees were fully exposed to a chilling environment. *DAM* gene expression is a new marker used in the monitoring of bud dormancy status and modulation of transcriptional regulation at endodormancy release in buds. Other genes are involved in bud dormancy; overexpression of the *CBF* gene in a transgenic apple rootstock has been shown to indicate changes in the dormancy cycle by altering *DAM* gene expression and increasing freezing tolerance (Wisniewski et al., 2015). In a recent genome-wide analysis, *Dof* genes, which are also transcriptional factors, were discovered to be related to bud dormancy in peach (Chen et al., 2017a). These transcription factor genes are thought to be key regulators responding to low temperature and regulating bud development in winter.

In poplar (*Populus* spp.), a model woody plant, the genetic and molecular regulation of dormancy has recently been elucidated. The gene family that encodes PEBP-like proteins including FT and CEN/TFL1, which have important roles in flower initiation, was found to also be involved in the bud dormancy cycle and its regulation (Rinne et al., 2011; Mohamed et al., 2010). Interestingly, transgenic plum (*Prunus domestica*) with overexpressed poplar *FT* (*PtFT1*) did not enter a state of endodormancy upon cold treatment, or alternatively had no chilling requirement after dormancy was established (Srinivasan et al., 2012). The known functions of flowering-related genes such as *FT*, *TFL1* and MADS-box genes have broadened to include bud development during dormancy.

The involvement of phytohormones in promoting bud dormancy has long been discussed (Cooke et al., 2012). Gibberellic acids (GAs) and abscisic acid (ABA) are well-studied phytohormones that play roles in dormancy. Recent studies have analysed the genes involved in the biosynthesis and signalling of these phytohormones in Japanese apricot (*Prunus mume*) (Wen et al., 2016), apple (Wisniewski et al., 2015) and grape (Zheng et al., 2015), and the involvement of GA and ABA in bud dormancy has been confirmed. Additionally, perhaps through crosstalk with ABA, ethylene may be involved in bud dormancy. The *EBB1* gene, an APETALA2/ethylene-responsive transcription factor originally found in poplar (Yordanov et al., 2014), has been shown to regulate the timing of bud break in apple (Wisniewski et al., 2015). The importance of epigenetic changes during bud dormancy in poplar and apple has recently been recognized (Kumar et al., 2016; Rios et al., 2014). By reporting molecular-level physiological changes during dormancy, these studies have laid the foundation for a thorough understanding of the mechanisms controlling bud dormancy.

4 Sexual determination and development of flower structures

After dormancy, flower buds resume development. Differentiated floral organs then complete their structural formation and rapidly increase in volume until flowering. The post-bud burst stage is an important period for the development of male (pollen) and female (ovule) organs. In nature, inadequate climatic/weather conditions or nutrient/water deficiency during this period critically affect floral development. The successful development of male and female functions is essential for pollination, fertilization and fruit development (Fig. 3).

In several fruit species exhibiting dioecious or monoecious flower phenotypes, the mechanisms of sex determination were recently elucidated using genome sequencing and transcriptomic approaches. In melon (*Cucumis melo*), ethylene is a known suppressor of stamen development. Recent studies have elucidated the involvement of the *CmWIP1*, *CmACS11* and *CmACS-7* genes. *CmACS11* and *CmACS-7* encode 1-aminocyclopropane-1-carboxylic acid (ACC) synthase, which is involved in ethylene biosynthesis, and *CmWIP1* is a transcription factor. The interaction of these three genes controls sex determination in melon (Boualem et al., 2008, 2015, 2016; Martin et al., 2009). In the Caucasian persimmon (*Diospyros lotus*), *OGI* and *MeGI* genes encoding class I homeodomain transcription factors have been identified

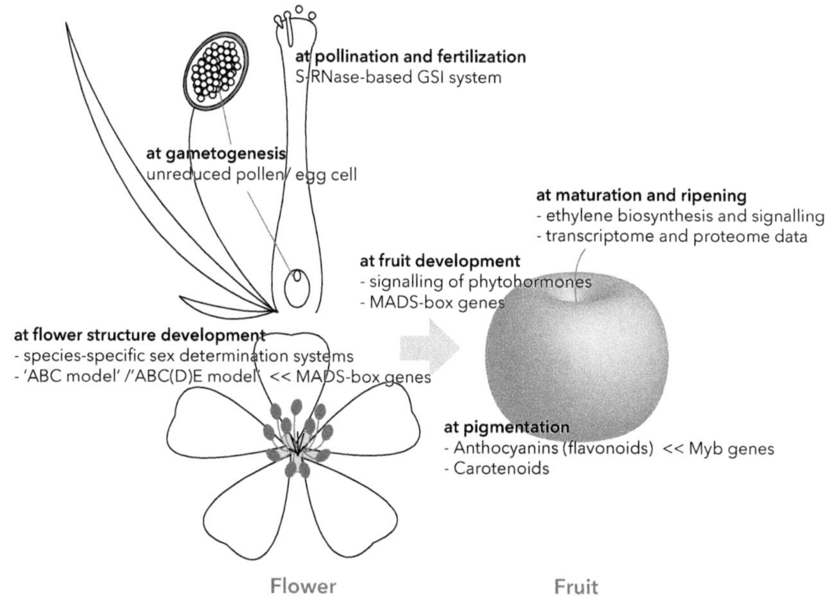

Figure 3 Recent advances and active research on the reproductive physiology of flower and fruit developmental stages.

(Akagi et al., 2014). The *MeGI* gene is a growth inhibitor for androecia, and small RNAs from the *OGI* gene on the Y sex chromosome suppress autosomal *MeGI* gene expression when the male flower is formed. In kiwifruit (*Actinidia* spp.), a type-C cytokinin response regulator has been identified as a potential sex determinant gene by genome-wide transcriptome analysis (Akagi et al., 2018). Synthetic cytokinin treatment has been shown to restore incomplete stigma development in the male flower. In fig/caprifig (*Ficus carica*), the *RAN1* gene, which encodes copper-transporting ATPase, has been identified as a candidate gene for sex determination by genome analysis (Mori et al., 2017); however, the role and function of *RAN1* remains unclear. These findings indicate that different plant species have developed different genetic systems for sex determination.

The 'ABC model' or 'ABC(D)E model' concept with minor modifications has been broadly accepted to explain flower structure development. In fruit tree species, mutant flower and fruit structures have been studied based on these models in annual herbaceous plant species. For example, apetalous flower mutants in apple were found to have transposon insertion mutations in the *MdPI* gene, which is a transcription factor gene homologous to *PISTILLATA* (*PI*) in *Arabidopsis* Class B flower homeotic genes. Similarly, *PI* mutant *Arabidopsis* and apple lack petals and stamens in their flowers. These apple mutants also exhibit parthenocarpic fruit development (Yao et al., 2001). In grape, the *FLESHLESS BERRY* somatic variant, which has abnormal berry development, was shown to be a *PI-like* gene (*VvPI*) mutant (Fernandez et al., 2013). Hence, the importance of the Class B homeotic gene *PI* in fruit development has been revealed. Other flower homeotic genes homologous to *AGAMOUS* (*AG*) in Class C, *APETALA 1* (*AP1*) in Class A and *SEPALLATA* (*SEP*) in Class E in *Arabidopsis* have been characterized in fruit tree species in numerous publications by many research groups (Calonje et al., 2004; Diaz-Riquelme et al., 2009; Grimplet et al., 2016; Klocko et al., 2016; Kotoda et al., 2000, 2002; Li et al., 2012a; Liu et al., 2010, 2013, 2017; Pillitteri et al., 2004; Sun et al., 2014; Sung et al., 1999, 2000; Tanaka et al., 2016; Wang et al., 2015; Wells et al., 2015; Yao et al., 1999; Zhou et al., 2017). These genes belong to the MADS-box gene family and are similar in DNA or protein sequences to conserved domains or motifs. MADS-box genes function as transcription factors to regulate the expression of other genes involved in organ development. MADS-domain proteins bind to DNA sequences termed 'CArG-boxes' (Theissen et al., 2016); however, because there are thousands of DNA-sequence elements similar to CArG-boxes within plant genomes, determining which target genes are regulated by MADS-box genes is a current challenge.

Inflorescence architecture is another intriguing aspect of floral morphological development. Branching and elongation of the inflorescence axis and peduncles form various structures in flower clusters. Inflorescence architecture is thought to be influenced by both genetic and environmental factors. The most well-studied gene influencing inflorescence development is *TERMINAL FLOWER1* (*TFL1*), a PEBP-like protein, in *Arabidopsis* (Bradley et al., 1997; Hanano and Goto, 2011; Serrano-Mislata et al., 2016). *TFL1* is involved in the determination of flower induction timing via anti-florigen, but also plays roles in maintaining undifferentiated meristem in the shoot apex. *TFL1* homologous genes *SELF PRUNING* (*SP*) in tomato (Pnueli et al., 1998, 2001; Thouet et al., 2008), *CENTRORADIALIS* (*CEN*) in *Antirrhinum* (Bradley et al., 1996) and *DETERMINATE* (*DET*) in pea (*Pisum sativum*) (Benlloch et al., 2015; Foucher et al., 2003) have been studied using their mutants and were found to be involved in inflorescence development and branching. In grape, based on the analysis of the reiterated reproductive meristem (RRM) somatic variant, which exhibits altered berry clustering (Fernandez et al., 2010), *VvTFL1A* was found to be associated with inflorescence meristem development and flowering time (Fernandez et al., 2014). However, inflorescence morphogenesis differs greatly among plant species; thus, it remains difficult to determine the mechanism for this process. Further investigation of the other factors (genes) involved in inflorescence development is necessary for each species.

5 Physiology during pollination, fertilization and gametogenesis

Successful pollination, fertilization and seed formation are essential in many fruit species. Fertile pollen grains are developed through meiosis and subsequent mitotic divisions, then mature prior to flowering. Pollen grains are released from the anther by dehiscence. Pollen is attached to the stigma, at the top of the pistil, and growth begins with the elongation of the pollen tube, which passes through transmitting tissue in the style (pistil) to reach the ovule, after which sperm cells are released into the embryo sac. One sperm fuses with an egg cell and the other fuses with the central cell. In this manner, double fertilization occurs in angiosperms. The fertilized egg cell and central cell develop into the embryo and endosperm, respectively.

Several reproductive barriers prevent self-pollination or self-fertilization in fruit tree species. Self-incompatibility (SI) is a rejection system for self-pollination in the pistil. This system was thought to be genetically regulated by two mechanisms: gametophytic SI (GSI) and sporophytic SI (SSI). In the past two decades, intensive molecular biological and genetic research have revealed these mechanisms in great detail. In Rosaceae fruit tree species, including apples, pears (*Pyrus* spp.), cherries (*Prunus avium*), apricots (*Prunus armeniaca*), Japanese apricot and plums, the GSI system functions during pollination; hence, outcrossing is necessary to obtain a sufficient fruit set. GSI is controlled by a single polymorphic *S* locus harbouring pistil *S* and pollen *S* determinant genes. A defect mutation in either of the *S* genes results in a self-compatible (SC) phenotype in *Prunus*. In the Rosaceae GSI system, S-RNase gene expression in the pistil and F-box gene expression in pollen have been shown to control self- or non-self-recognition reactions between pistil and pollen (Sassa, 2016). The S-RNase-based GSI system has also been observed in Solanaceae (Wang and Kao, 2012) and Plantaginaceae (Xue et al., 1996). Using further biochemical analyses such as protein-protein interaction and phylogenetic analyses of S-RNase-based GSI in Rosaceae, different self-recognition mechanisms were shown to act in *Prunus* (e.g. cherries and apricots) and Pyreae (e.g. apples and pears), despite their close phylogenetic relationship (Matsumoto and Tao, 2016). These findings have enabled the development of DNA markers that can be used in S-genotyping for compatible trees and marker-assisted selection of SC trees (Okada, 2015).

In gamete (i.e. pollen and egg cell) development, some fruit tree species exhibit abnormalities during meiosis, or in mitotic divisions following meiosis. Unreduced gametes can sometimes arise through meiotic defects. An unreduced gamete contains the same number of chromosomes as a somatic cell, and is termed a *2n* gamete. This irregular development is often observed in male gametes during the formation of giant pollen grains. Because it is very difficult to observe female gametogenesis, the irregular development of female gametes is indirectly detected by counting the chromosomes of the progeny plant or by analytically determining the ploidy level of formed seeds or progeny. Unreduced gametes have been observed in peach (Dermen, 1937), apple (Einset, 1952), citrus (Esen and Soost, 1971), *Vaccinium* spp. (Cockerham and Galletta, 1976; Megalos and Ballington, 1988; Ortiz et al., 1992), cherry (Iezzoni and Hancock, 1984), *Vitis* (Martens et al., 1989; Zhang et al., 1998), *Musa* (Ortiz, 1997) and kiwifruit (Yan et al., 1997). In recent years, following the development of easy and rapid flow cytometry-based analytical methods for the determination of ploidy levels, unreduced gametes have also been detected in Japanese persimmon (*Diospyros kaki*) (Sugiura et al., 2000; Yamada et al., 2005; Yamada and Tao, 2007), and current research is attempting their detection in citrus species (Aleza et al., 2010, 2016; Honsho et al., 2012, 2016; Rouiss et al., 2017a,b). Unreduced gametes are thought to be deeply related to the evolution of polyploidy in plant species. Higher polyploid plants can naturally arise through the hybridization of $2n \times n$, $n \times 2n$, or possibly $2n \times 2n$, derived from unreduced gametes. Moreover, unreduced gametes can be used to manipulate ploidy during plant breeding. The association of some gene- and cell cycle-related proteins with meiotic defects has been suggested in *Arabidopsis* (Brownfield and Kohler, 2011); however, the detailed mechanisms of these defects in fruit tree species remain obscure.

6 Fruit development and phytohormones

Fruit development begins during seed (embryo) development after fertilization. The seed is considered to be the *hormone centre*, where relatively high concentrations of phytohormones are produced. Phytohormones from the seed are thought to control cell division, cell expansion and component accumulation in fruit tissues. However, in parthenocarpy, the fruit can develop without fertilization or seed formation. Parthenocarpy can be induced in several ways; however, the threshold concentration of growth substances in the ovary is crucial. Therefore, applications of phytohormonal treatments and analyses of phytohormone biosynthesis and signalling pathways are frequently performed to contribute to the elucidation of this mechanism.

The key phytohormones regulating parthenocarpic development are GAs and auxins. Global gene expression analyses for their biosynthesis and signalling pathways have been applied to determine the genes and physiological mechanisms involved in parthenocarpy (Joldersma and Liu, 2018). In particular, studies of tomato, which is used as a model plant for fleshy fruit development, have contributed to elucidating the regulatory gene network for fruit set since transgenic and novel genome editing systems and mutants have become available (Ficcadenti et al., 1999; Ruiu et al., 2015; Shinozaki et al., 2018; Tang et al., 2015; Ueta et al., 2017). *YUCCA*, *ARF7/8*, *IAA*, *AUCSIA*, *PIN* and *TIR* genes in the auxin biosynthesis and signalling pathways, and *GA20ox* and *DELLA* genes in the GA biosynthesis and signalling pathways have been shown to be involved in parthenocarpic development (Pandolfini, 2009; Sotelo-Silveira et al., 2014). Similarly, GA and auxin interactions in the parthenocarpic regulatory network have been studied in *Citrus* spp. (Bermejo et al., 2018; Mesejo et al., 2016), loquat (*Eriobotrya japonica*) (Mesejo et al., 2010) and pear (Liu et al., 2018). The involvement of cytokinin (Ding et al., 2013; Matsuo et al., 2012) and transcription factor genes such as *SEPALLATA1/TM29* in tomato (Ampomah-Dwamena et al., 2002) have also been suggested. Apple *MdPI*, a MADS-box gene among the Class B flower homeotic genes, has been reported to be involved in parthenocarpic development (Yao et al., 2001); however, according to a recent study in transgenic apple, the depression of *MdPI* cannot explain fruit formation in parthenocarpic cultivars. Therefore, the involvement of some other unidentified genes is suggested for parthenocarpic development of apple (Tanaka et al., 2016). In agricultural practice, hormonal control is applied at flowering to produce seedless fruits and ensure fruit set. Treatment with GA or auxin-like substances is often applied to induce parthenocarpic fruit development in diverse species. Alternatively, gamma- or soft X-ray-irradiated pollen is used for parthenocarpic fruit production in watermelon (*Citrullus lanatus*) (Moussa and Salem, 2010; Sugiyama and Morishita, 2000; Sugiyama et al., 2002).

Fruit development consists of cell division and cell expansion stages. Fruit size is determined synergistically by cell division and cell expansion. Large differences in fruit size between domesticated cultivars and undomesticated wild species have been observed. The fruit of wild species are much smaller than those of cultivated species. Fruit size is genetically regulated by QTLs. Domestication and breeding processes have likely selected and accumulated advantageous alleles for the production of larger fruit.

Bud sport mutants for fruit size have been found in the 'Gala' apple (Malladi and Hirst, 2010), 'La France' pear (Isuzugawa et al., 2014) and 'Hiratanenashi' Japanese persimmon (Yamane et al., 2008; Habu et al., 2016), and are used in studies of fruit development. Analyses of these fruit size mutants have elucidated mechanisms underlying fruit cell division and expansion, such as genes involved in cell cycles, hormonal control and chimeric polyploidization or endoreduplication.

Fruit cell expansion is thought to be highly dependent on water uptake and assimilation into the cell. The large vacuole, which eventually occupies almost 90% of the cell volume, is developed through cell expansion, and sugars, organic acids, minerals and secondary metabolites are accumulated via water inflow. Many integral transmembrane proteins, such as pump, channel, transporter and aquaporin proteins in the plasma membrane or tonoplast, transport substances and water into the cell. Recent studies have characterized many genes that encode membrane transport proteins in fruit cells. The mechanism of cell expansion has been described in great detail due to recent work in molecular biology and biochemistry (Rogiers et al., 2017).

7 Physiology of fruit maturation and ripening

Fruit maturation and ripening involves major metabolic changes in fruit characteristics such as colour, texture, sugar and acid content, and aroma, which are regulated by phytohormone signals and genetic factors. Ethylene is a well-known phytohormone that plays a key role in fruit maturation and ripening in diverse plant species. Fruits can be classified into two physiological groups: climacteric and non-climacteric. The climacteric fruit group, which includes apples, bananas (*Musa accuminata*), kiwifruits and peaches, exhibits a rapid rise in respiration and a burst of ethylene production during ripening. In contrast, the non-climacteric group includes citrus, grapes and cherries, which do not produce ethylene during ripening (Cherian et al., 2014). The molecular mechanisms of ethylene biosynthesis, perception and signalling have been studied intensively, and the genes involved in this process are being elucidated. ACC synthase (ACS) and ACC oxidase (ACO) are two major enzymes for ethylene biosynthesis, and the ACS and ACO genes are well studied. Ethylene is perceived by ETR/ERS receptors localized on the endoplasmic reticulum membrane in association with CTR-interacting protein, and EIN3/EIL and ERF transcription factors are expressed during the signalling process (Cherian et al., 2014; Liu et al., 2015b).

Acting separately from ethylene signalling, MADS-box transcription factors play important roles in maturation and ripening induction. The tomato fruit-ripening mutant *rin* was analysed to discover the *SlMADS-RIN* gene, a member of the *SEPALLATA 4* subfamily of the MADS-box gene family; it regulates fruit maturation. *MdMADS8* and *MdMADS9* (in the *SEPALLATA 1/2-like* subfamily) were isolated from apple, and transgenic experiments demonstrated that fruits did not begin maturation following application of the antisense method to suppress the expression of these genes (Ireland et al., 2013). In tomato, *Sl*MADS-RIN forms a multimeric complex with other MADS-box proteins; thus, other MADS-box genes are also involved in fruit maturation. Other than MADS-box genes, NAC and HB1 transcription factors are also involved in fruit maturation and ripening (Karlova et al., 2014). The effects of epigenetic controls, DNA methylation changes and chromatin remodelling, on fruit ripening have also been examined (Farinati et al., 2017; Giovannoni et al., 2017).

Pigmentation by accumulation of anthocyanins or carotenoids in the fruit pericarp occurs during fruit maturation and ripening. Anthocyanins are responsible for red/purple/blue colours, and carotenoids for orange/yellow colours. Green fruit skin colour is generally decreased through the degradation of chlorophylls during fruit maturation and ripening. The colour change of fruit skin is thus an indicator of maturation and ripening. Among anthocyanins, *MYB* transcription factor genes play key roles in the regulation of biosynthesis genes in the fruit of many plant species (Azuma, 2018; Lin-Wang et al., 2010; Peng and Moriguchi, 2013; Tohge et al., 2017). The involvement of *MdMYB1* and *MdMYB10* in fruit colour development in apple and that of *VvMYBA1* in grape have been intensively analysed (Espley et al., 2007, 2009; Kobayashi et al., 2004; Takos et al., 2006; Yakushiji et al., 2006). Biosynthetic/metabolic pathway genes (*PSY, PDS, Z-ISO, ZDS, CRTISO, LCYE, LCYB, ZEP, CYP97A, CYP97C* etc.) have been identified in carotenoids, and their regulatory mechanisms elucidated (Liu et al., 2015a; Llorente et al., 2017; Sun et al., 2018; Yuan et al., 2015). The pathways and genes involved in chlorophyll degradation have also been clarified (Zhu et al., 2017a). In particular, ethylene signalling has been shown to activate chlorophyll degradation in sweet orange (*Citrus sinensis*) (Yin et al., 2016). The control of fruit pigmentation and colour change significantly affects the appearance of fresh fruit and increases their functional components for human health.

Fruit softening and texture changes also progress during maturation and ripening. Fruit firmness and flesh texture are related to the structure and composition of the cell wall, which contains several types of polysaccharides (cellulose, pectin and hemicellulose), non-enzymic proteins and lignin. During fruit softening, matrix polysaccharides are quantitatively and qualitatively altered by the action of many cell wall-degrading enzymes and cell wall-remodelling proteins including polygalacturonase (PG), endo-1,4-β-glucanase (EGase), pectate lyase (PL), pectin methylesterase (PME), β-galactosidase (β-Gal), α-arabinofuranosidase (α-Ara), xyloglucan endotransglucosylase/hydrolase (XTH) and expansins (Exp). Recent studies have attempted to elucidate the entire process of physiological changes in gene and protein expression and metabolite levels during fruit maturation and ripening using transcriptome, proteome and metabolome analysis techniques (Table 1). Genes and proteins expressed during fruit development and ripening (including those involved in cell wall remodelling and degradation) have been extensively analysed by these -omics approaches, which can provide a comprehensive understanding of fruit development and ripening, as well as new discoveries related to unexpected regulatory pathways and genes. New research paradigms may be proposed on the basis of these technologies.

Table 1 Physiological studies of fruit development and related properties by transcriptomic and proteomic approaches

Fruit species	Study subjects and features	Including proteomic approach, or not	References
Tomato	Fruit development		Alba et al. (2005)
	Ripening	Proteome	Rocco et al. (2006)
	Fruit development and ripening		Faurobert et al. (2007)
	Ripening		Kok et al. (2007)
	Polyamine accumulation		Srivastava et al. (2007)
	Ripening		Kok et al. (2008)
	Chromoplast	Proteome	Barsan et al. (2010)
	Protective proteins	Proteome	Page et al. (2010)
	Cuticle biogenesis		Yeats et al. (2010)
	Ripening		Karlova et al. (2011)
	Chloroplast-to-chromoplast transition	Proteome	Barsan et al. (2012)
	Ripening		Bemer et al. (2012)
	Ripening		Kumar et al. (2012)
	Ripening, carotenoid biosynthesis		Lee et al. (2012)
	Ripening		Fujisawa et al. (2013)
	Carotenoid biosynthesis	Proteome	Kilambi et al. (2013)
	Cell wall	Proteome	Konozy et al. (2013)
	Cell wall	Proteome	Ruiz-May and Rose (2013)
	Carotenoid biosynthesis		Smita et al. (2013)
	Fruit development		Tang et al. (2013)
	Ripening, MADS-box genes		Fujisawa et al. (2014)
	Fruit development and ripening		Nguyen et al. (2014)
	Ripening	Proteome	Pan et al. (2014)
	Wounding		Baldassarre et al. (2015)
	Fruit development		Ye et al. (2015)
	Ripening		Li et al. (2016a)
	Ripening		Wang et al. (2016)
	Environmental response		D'Esposito et al. (2017)
	Ripening, ethylene-auxin		Li et al. (2017a)
	Ripening, MADS-box RIN		Li et al. (2017b)
	Fruit set		Ru et al. (2017)
	Fruit development	Proteome	Szymanski et al. (2017)
	Ripening		da Costa et al. (2018)
	Fruit development and ripening		Shinozaki et al. (2018)

Fruit species	Study subjects and features	Including proteomic approach, or not	References
Grape	Fruit quality	Proteome	Sarry et al. (2004)
	Berry development		Terrier et al. (2005)
	Ripening		Waters et al. (2006)
	Berry development		Deluc et al. (2007)
	Ripening	Proteome	Deytieux et al. (2007)
	Berry development		Fernandez et al. (2007)
	Ripening	Proteome	Negri et al. (2008a)
	Cell wall	Proteome	Negri et al. (2008b)
	Ripening, plasma membrane	Proteome	Zhang et al. (2008)
	Ripening initiation	Proteome	Lucker et al. (2009)
	β-1,3-glucanase	Proteome	Wang et al. (2009)
	Ripening	Proteome	Giribaldi et al. (2010)
	Ripening		Koyama et al. (2010)
	Berry quality, viral infection	Proteome	Giribaldi et al. (2011)
	Ripening		Guillaumie et al. (2011)
	Ripening	Proteome	Negri et al. (2011)
	Ripening		Pastore et al. (2011)
	Ripening, viral infection		Vega et al. (2011)
	Ripening		Lijavetzky et al. (2012)
	Berry development		Sweetman et al. (2012)
	Ripening		Carbonell-Bejerano et al. (2013)
	Berry development and ripening	Proteome	Martinez-Esteso et al. (2013)
	Ripening, flavonoid biosynthesis		Pastore et al. (2013)
	Ripening		Cramer et al. (2014)
	Berry development, ripening (muscadine grape)	Proteome	Kambiranda et al. (2014)
	Ripening, heat stress		Rienth et al. (2014)
	Berry development	Proteome	Wang et al. (2014)
	Anthocyanin biosynthesis		Wu et al. (2014a)
	Ripening	Proteome	Fraige et al. (2015)
	Ripening, early-ripening mutants		Wu et al. (2015)
	Ripening (muscadine grape)	Proteome	Kambiranda et al. (2016)
	Berry development, root restriction		Leng et al. (2016)
	Phenylpropanoid and terpenoid biosynthesis		Savoi et al. (2016)
	Ripening		Rienth et al. (2016)

(*Continued*)

Table 1 (*Continued*)

Fruit species	Study subjects and features	Including proteomic approach, or not	References
	Ripening, berry size		Wong et al. (2016)
	Berry development		Lecourieux et al. (2017)
	Ripening, anthocyanin biosynthesis		Massonnet et al. (2017)
	Berry development, water deficit		Savoi et al. (2017)
	Berry development		Serrano et al. (2017)
	Berry development		Shangguan et al. (2017)
	Flavonoid biosynthesis, temperature		Pastore et al. (2017)
	Fruit quality	Proteome	Riebel et al. (2017)
	Aroma		Wang et al. (2017c)
	Berry size		Wang et al. (2017d)
	Ripening		Balic et al. (2018)
Apple	Soluble components in flesh	Proteome	Guarino et al. (2007)
	Fruit senescence	Proteome	Qin et al. (2009b)
	Maturation and ripening		Costa et al. (2010)
	Fruit firmness	Proteome	Marondedze and Thomas (2012)
	Ripening	Proteome	Zheng et al. (2013)
	Plastid-targeted proteins in fruit		Schaeffer et al. (2014)
	Maturation		Segonne et al. (2014)
	Skin colour mutation		El-Sharkawy et al. (2015)
	Fruit senescence	Proteome	Liu et al. (2016)
	Cell wall		Dheilly et al. (2016)
	Maturation		Qi et al. (2017)
Pear	Ripening		Nishitani et al. (2010)
	Fruit-specific DNA reduplication		Nashima et al. (2013)
	Fruit development and maturation		Xie et al. (2013)
	Ripening		Huang et al. (2014)
	Maturation	Proteome	Liu et al. (2014)
	Fruit development and quality	Proteome	Li et al. (2015b)
	Fruit development and ripening		Nham et al. (2015)
	Ripening	Proteome	Gao et al. (2016)
	Maturation		Zhang et al. (2016b)
	Anthocyanin biosynthesis		Bai et al. (2017)
	Skin colour fading		Wang et al. (2017e)

Fruit species	Study subjects and features	Including proteomic approach, or not	References
Loquat	Susceptibility to sunburn	Proteome	Jiang et al. (2015)
	Fruit development and ripening		Song et al. (2016)
Peach/ nectarine	Ripening		Trainotti et al. (2006)
	Ripening		Ziliotto et al. (2008)
	Ripening, fruit quality	Proteome	Lara et al. (2009)
	Fruit senescence, redox proteomics	Proteome	Qin et al. (2009a)
	Ripening		Vecchietti et al. (2009)
	Ripening, extracellular proteome	Proteome	Bustamante et al. (2012)
	Fruit softening	Proteome	Nilo et al. (2012)
	Aroma		Pirona et al. (2013)
	Ripening	Proteome	Kang et al. (2016)
	Fruit skin quality	Proteome	Karagiannis et al. (2016)
	Ripening	Proteome	Almeida et al. (2016)
	Ripening	Proteome	Wu et al. (2016)
Plum/ apricot	Ripening		Grimplet et al. (2005)
	Fruit development and ripening		Manganaris et al. (2011)
	Anthocyanin biosynthesis		Fang et al. (2016)
	Sugar metabolism, non-climacteric fruit mutant		Farcuh et al. (2017)
Cherry	Anthocyanin biosynthesis		Wei et al. (2015)
	Ripening		Prinsi et al. (2016)
Strawberry	Ripening	Proteome	Bianco et al. (2009)
	Fruit development		Bombarely et al. (2010)
	Ripening	Proteome	Li et al. (2013)
	Anthocyanin biosynthesis, lignin biosynthesis		Ring et al. (2013)
	Ripening		Li et al. (2015a)
	Flavonoid biosynthesis		Pillet et al. (2015)
	Anthocyanin biosynthesis		Li et al. (2016c)
	Fruit development and ripening		Medina-Puche et al. (2016)
	Ripening		Estrada-Johnson et al. (2017)
	Anthocyanin biosynthesis		Hartl et al. (2017)
	Ripening		Sanchez-Sevilla et al. (2017)
	Fruit shape development		Wang et al. (2017b)

(Continued)

Table 1 (*Continued*)

Fruit species	Study subjects and features	Including proteomic approach, or not	References
Citrus	Fruit quality	Proteome	Katz et al. (2007)
	Ethylene response		Fujii et al. (2007)
	Fruit quality		Quecini et al. (2007)
	Ripening	Proteome	Muccilli et al. (2009)
	Carotenoid biosynthesis	Proteome	Pan et al. (2009)
	Fruit development	Proteome	Katz et al. (2010)
	Fruit epidermis tissue, microdissection		Matas et al. (2010)
	Fruit quality	Proteome	Pignataro et al. (2010)
	Chromoplast	Proteome	Zeng et al. (2011)
	Carotenoid biosynthesis	Proteome	Pan et al. (2012)
	Fruit development and ripening		Yu et al. (2012)
	Ripening		Zeng et al. (2012)
	Colour development		Suh et al. (2013)
	Fruit development and ripening		Patel et al. (2014)
	Ripening	Proteome	Wu et al. (2014c)
	Chromoplast		Zeng et al. (2014)
	Ripening		Zhang et al. (2014)
	Fruit development, citrate accumulation		Lu et al. (2016)
	Fruit development, peel roughing disorder		Lu et al. (2017)
	Fruit development and ripening	Proteome	Wang et al. (2017a)
Banana	Ripening	Proteome	Toledo et al. (2012)
	Fruit quality	Proteome	Esteve et al. (2013)
	Ripening		Asif et al. (2014)
	Peel tissue, ethylene/high-temperature response	Proteome	Du et al. (2016)
	Fruit development and ripening		Hu et al. (2016)
	Ripening, ethylene		Xia et al. (2016)
	Ripening		Kuang et al. (2017)
Pineapple	Fruit development		Moyle et al. (2005)
Papaya	Ripening		Fabi et al. (2012)
	Ripening	Proteome	Nogueira et al. (2012)
	Ripening		Shen et al. (2017)
Water-melon	Fruit development		Guo et al. (2011)
	Ripening		Grassi et al. (2013)

Fruit species	Study subjects and features	Including proteomic approach, or not	References
	Fruit development		Guo et al. (2015)
	Fruit development and ripening		Zhu et al. (2017b)
	Soluble sugar and organic acid accumulation		Gao et al. (2018)
Melon	Fruit development and ripening		Zhang et al. (2016a)
	Fruit development		Shin et al. (2017)
	Fruit development		Yano et al. (2018)
Mango	Fruit development	Proteome	Wu et al. (2014)
	Ripening		Dautt-Castro et al. (2015)
	Ripening		Srivastava et al. (2016)
	Fruit development and ripening		Deshpande et al. (2017)
	Cuticle biogenesis		Tafolla-Arellano et al. (2017)
Mango-steen	Ripening		Abdul-Rahman et al. (2017)
Litchee	Maturation		Lai et al. (2015)
Kiwifruit	Fruit development, anthocyanin biosynthesis		Li et al. (2015c)
	Fruit development and ripening		Tang et al. (2016)
	Fruit development and ripening	Proteome	Ainalidou et al. (2016)
Persimmon (kaki)	Fruit development		Nakagawa et al. (2008)
	Fruit quality	Proteome	Chen et al. (2017b)
	Fruit softening		Jung et al. (2017)
	Flavonoid biosynthesis		Nishiyama et al. (2018)
Blueberry	Antioxidants		Li et al. (2012b)
	Ripening		Gupta et al. (2015)
	Fruit development		Li et al. (2016b)
Fig	Ripening		Freiman et al. (2014)
	Fruit development and ripening		Rosianski et al. (2016)
	Flavonoid biosynthesis		Wang et al. (2017f)
Pepper/ chili pepper	Ripening		Lee et al. (2010)
	Fruit development		Martinez-Lopez et al. (2014)
Olive	Fruit development		Alagna et al. (2009)
	Fruit development		Galla et al. (2009)

8 Where to look for further information

Further researches and updates about plant flowering and reproduction are found in scientific journals such as *Plant Reproduction* (https://link.springer.com/journal/497), *Plant Physiology* (http://www.plantphysiol.org/), *The Plant Cell* (http://www.plantcell.org/), *The Plant Journal* (https://onlinelibrary.wiley.com/journal/1365313x), *Journal of Experimental Botany* (https://academic.oup.com/jxb), *New Phytologist* (https://nph.onlinelibrary.wiley.com/journal/14698137), *Plant & Cell Physiology* (https://academic.oup.com/pcp), *Planta* (https://link.springer.com/journal/425), *Frontiers in Plant Science* (https://www.frontiersin.org/journals/plant-science#), *BMC Plant Biology* (https://bmcplantbiol.biomedcentral.com/), *Plant Science* (https://www.journals.elsevier.com/plant-science), *Plant, Cell & Environment* (https://onlinelibrary.wiley.com/journal/13653040), *Plant Cell Reports* (https://link.springer.com/journal/299), *Plant Molecular Biology* (https://link.springer.com/journal/11103), *Horticulture Research* (https://www.nature.com/hortres/), *Journal of the American Society for Horticultural Science* (http://journal.ashspublications.org/), *The Horticulture Journal* (http://hortj.org/), *The Journal of Horticultural Science and Biotechnology* (https://www.tandfonline.com/toc/thsb20/current), *Scientia Horticulturae* (https://www.journals.elsevier.com/scientia-horticulturae), *Functional Plant Biology* (http://www.publish.csiro.au/fp), *Plant Biotechnology* (http://www.jspcmb.jp/english/sub04.html), *Tree Physiology* (https://academic.oup.com/treephys) and so on. Some journals contain special issues, for example *Focus Issue on Flowering and Reproduction* of *Plant Physiology*: 173 (1) (http://www.plantphysiol.org/content/173/1) and *The Flowering Newsletter* in *Journal of Experimental Botany* (https://academic.oup.com/jxb/pages/flowering_newsletter), which focus on plant reproductive development.

9 References

Abdul-Rahman, A., Goh, H. H., Loke, K. K., et al. (2017). RNA-seq analysis of mangosteen (*Garcinia mangostana* L.) fruit ripening. *Genomics Data* 12: 159–60. doi:10.1016/j.gdata.2017.05.013.

Ainalidou, A., Tanou, G., Belghazi, M., et al. (2016). Integrated analysis of metabolites and proteins reveal aspects of the tissue-specific function of synthetic cytokinin in kiwifruit development and ripening. *Journal of Proteomics* 143: 318–33. doi:10.1016/j.jprot.2016.02.013.

Akagi, T., Henry, I. M., Tao, R., et al. (2014). A Y-chromosome-encoded small RNA acts as a sex determinant in persimmons. *Science* 346: 646–50.

Akagi, T., Henry, I. M., Ohtani, H., et al. (2018). A Y-encoded suppressor of feminization arose via lineage-specific duplication of a cytokinin response regulator in kiwifruit. *Plant Cell* 30: 780–95.

Alagna, F., D'Agostino, N., Torchia, L., et al. (2009). Comparative 454 pyrosequencing of transcripts from two olive genotypes during fruit development. *BMC Genomics* 10: 399. doi:10.1186/1471-2164-10-399.

Alba, R., Payton, P., Fei, Z. J., et al. (2005). Transcriptome and selected metabolite analyses reveal multiple points of ethylene control during tomato fruit development. *Plant Cell* 17: 2954–65. doi:10.1105/tpc.105.036053.

Aleza, P., Juarez, J., Cuenca, J., et al. (2010). Recovery of citrus triploid hybrids by embryo rescue and flow cytometry from 2x × 2x sexual hybridisation and its application to extensive breeding programs. *Plant Cell Reports* 29: 1023–34.

Aleza, P., Cuenca, J., Juarez, J., et al. (2016). Inheritance in doubled-diploid clementine and comparative study with SDR unreduced gametes of diploid clementine. *Plant Cell Reports* 35: 1573–86.

Almeida, A. M., Urra, C., Moraga, C., et al. (2016). Proteomic analysis of a segregant population reveals candidate proteins linked to mealiness in peach. *Journal of Proteomics* 131: 71–81. doi:10.1016/j.jprot.2015.10.011.

Ampomah-Dwamena, C., Morris, B. A., Sutherland, P., et al. (2002). Down-regulation of *TM29*, a tomato *SEPALLATA* homolog, causes parthenocarpic fruit development and floral reversion. *Plant Physiology* 130: 605–17.

Asif, M. H., Lakhwani, D., Pathak, S., et al. (2014). Transcriptome analysis of ripe and unripe fruit tissue of banana identifies major metabolic networks involved in fruit ripening process. *BMC Plant Biology* 14: 316. doi:10.1186/s12870-014-0316-1.

Azuma, A. (2018). Genetic and environmental impacts on the biosynthesis of anthocyanins in grapes. *Horticulture Journal* 87: 1–17.

Bai, S. L., Sun, Y. W., Qian, M. J., et al. (2017). Transcriptome analysis of bagging-treated red Chinese sand pear peels reveals light-responsive pathway functions in anthocyanin accumulation. *Scientific Reports* 7: 63. doi:10.1038/s41598-017-00069-z.

Baldassarre, V., Cabassi, G., Spadafora, N. D., et al. (2015). Wounding tomato fruit elicits ripening-stage specific changes in gene expression and production of volatile compounds. *Journal of Experimental Botany* 66: 1511–26. doi:10.1093/jxb/eru516.

Balic, I., Vizoso, P., Nilo-Poyanco, R., et al. (2018). Transcriptome analysis during ripening of table grape berry cv. Thompson Seedless. *PLoS One* 13: e0190087. doi:10.1371/journal.pone.0190087.

Barsan, C., Sanchez-Bel, P., Rombaldi, C., et al. (2010). Characteristics of the tomato chromoplast revealed by proteomic analysis. *Journal of Experimental Botany* 61: 2413–31. doi:10.1093/jxb/erq070.

Barsan, C., Zouine, M., Maza, E., et al. (2012). Proteomic analysis of chloroplast-to-chromoplast transition in tomato reveals metabolic shifts coupled with disrupted thylakoid biogenesis machinery and elevated energy-production components. *Plant Physiology* 160: 708–25. doi:10.1104/pp.112.203679.

Bemer, M., Karlova, R., Ballester, A. R., et al. (2012). The tomato FRUITFULL homologs TDR4/FUL1 and MBP7/FUL2 regulate ethylene-independent aspects of fruit ripening. *Plant Cell* 24: 4437–51. doi:10.1105/tpc.112.103283.

Benlloch, R., Berbel, A., Ali, L., et al. (2015). Genetic control of inflorescence architecture in legumes. *Frontiers in Plant Science* 6: 543.

Bermejo, A., Granero, B., Mesejo, C., et al. (2018). Auxin and gibberellin interact in citrus fruit set. *Journal of Plant Growth Regulation* 37: 491–501.

Bianco, L., Lopez, L., Scalone, A. G., et al. (2009). Strawberry proteome characterization and its regulation during fruit ripening and in different genotypes. *Journal of Proteomics* 72: 586–607. doi:10.1016/j.jprot.2008.11.019.

Bombarely, A., Merchante, C., Csukasi, F., et al. (2010). Generation and analysis of ESTs from strawberry (*Fragaria xananassa*) fruits and evaluation of their utility in genetic and molecular studies. *BMC Genomics* 11: 503. doi:10.1186/1471-2164-11-503.

Boualem, A., Fergany, M., Fernandez, R., et al. (2008). A conserved mutation in an ethylene biosynthesis enzyme leads to andromonoecy in melons. *Science* 321: 836–8.

Boualem, A., Troadec, C., Camps, C., et al. (2015). A cucurbit androecy gene reveals how unisexual flowers develop and dioecy emerges. *Science* 350: 688–91.

Boualem, A., Lemhemdi, A., Sari, M. A., et al. (2016). The andromonoecious sex determination gene predates the separation of *Cucumis* and *Citrullus* genera. *PLoS One* 11: e0155444.

Bradley, D., Carpenter, R., Copsey, L., et al. (1996). Control of inflorescence architecture in *Antirrhinum. Nature* 379: 791–7.

Bradley, D., Ratcliffe, O., Vincent, C., et al. (1997). Inflorescence commitment and architecture in *Arabidopsis. Science* 275: 80–3.

Brownfield, L. and Kohler, C. (2011). Unreduced gamete formation in plants: Mechanisms and prospects. *Journal of Experimental Botany* 62: 1659–68.

Bustamante, C. A., Budde, C. O., Borsani, J., et al. (2012). Heat treatment of peach fruit: Modifications in the extracellular compartment and identification of novel extracellular proteins. *Plant Physiology and Biochemistry* 60: 35–45. doi:10.1016/j.plaphy.2012.07.021.

Calonje, M., Cubas, P., Martinez-Zapater, J. M., et al. (2004). Floral meristem identity genes are expressed during tendril development in grapevine. *Plant Physiology* 135: 1491–501.

Carbonell-Bejerano, P., Maria, E. S., Torres-Perez, R., et al. (2013). Thermotolerance responses in ripening berries of *Vitis vinifera* L. cv Muscat Hamburg. *Plant and Cell Physiology* 54: 1200–16. doi:10.1093/pcp/pct071.

Carmona, M. J., Cubas, P. and Martinez-Zapater, J. M. (2002). *VFL*, the grapevine *FLORICAULA/LEAFY* ortholog, is expressed in meristematic regions independently of their fate. *Plant Physiology* 130: 68–77.

Carmona, M. J., Calonje, M. and Martinez-Zapater, J. M. (2007). The *FT/TFL1* gene family in grapevine. *Plant Molecular Biology* 63: 637–50.

Castede, S., Campoy, J. A., Le Dantec, L., et al. (2015). Mapping of candidate genes involved in bud dormancy and flowering time in sweet cherry (*Prunus avium*). *PLoS One* 10: e0143250.

Chen, M., Liu, X., Huan, L., et al. (2017a). Genome-wide analysis of Dof family genes and their expression during bud dormancy in peach (*Prunus persica*). *Scientia Horticulturae* 214: 18–26.

Chen, W. X., Xiong, Y. L., Xu, L. Q., et al. (2017b). An integrated analysis based on transcriptome and proteome reveals deastringency-related genes in CPCNA persimmon. *Scientific Reports* 7: 44671. doi:10.1038/srep44671.

Cherian, S., Figueroa, C. R. and Nair H. (2014). 'Movers and shakers' in the regulation of fruit ripening: A cross-dissection of climacteric versus non-climacteric fruit. *Journal of Experimental Botany* 65: 4705–22.

Cockerham, L. E. and Galletta, G. J. (1976). Survey of pollen characteristics in certain *Vaccinium* species. *Journal of the American Society for Horticultural Science* 101: 671–6.

Cooke, J. E. K., Eriksson, M. E. and Junttila, O. (2012). The dynamic nature of bud dormancy in trees: Environmental control and molecular mechanisms. *Plant Cell and Environment* 35: 1707–28.

Costa, F., Alba, R., Schouten, H., et al. (2010). Use of homologous and heterologous gene expression profiling tools to characterize transcription dynamics during apple fruit maturation and ripening. *BMC Plant Biology* 10: 229. doi:10.1186/1471-2229-10-229.

Cramer, G. R., Ghan, R., Schlauch, K. A., et al. (2014). Transcriptomic analysis of the late stages of grapevine (*Vitis vinifera* cv. Cabernet Sauvignon) berry ripening reveals significant induction of ethylene signaling and flavor pathways in the skin. *BMC Plant Biology* 14: 370. doi:10.1186/s12870-014-0370-8.

D'Esposito, D., Ferriello, F., Dal Molin, A., et al. (2017). Unraveling the complexity of transcriptomic, metabolomic and quality environmental response of tomato fruit. *BMC Plant Biology* 17: 66. doi:10.1186/s12870-017-1008-4.

da Costa, J. H. P., Rodriguez, G. R., Picardi, L. A., et al. (2018). Genome-wide expression analysis at three fruit ripening stages for tomato genotypes differing in fruit shelf life. *Scientia Horticulturae* 229: 125–31. doi:10.1016/j.scienta.2017.11.004.

Dautt-Castro, M., Ochoa-Leyva, A., Contreras-Vergara, C. A., et al. (2015). Mango (*Mangifera indica* L.) cv. Kent fruit mesocarp *de novo* transcriptome assembly identifies gene families important for ripening. *Frontiers in Plant Science* 6: 62. doi:10.3389/fpls.2015.00062.

Deluc, L. G., Grimplet, J., Wheatley, M. D., et al. (2007). Transcriptomic and metabolite analyses of Cabernet Sauvignon grape berry development. *BMC Genomics* 8: 429. doi:10.1186/1471-2164-8-429.

Dermen, H. (1937). Detection of polyploidy by pollen-grain size: Investigations with peaches and apricots. *Proceedings of the American Society for Horticultural Science* 35: 96–103.

Deshpande, A. B., Anamika, K., Jha, V., et al. (2017). Transcriptional transitions in Alphonso mango (*Mangifera indica* L.) during fruit development and ripening explain its distinct aroma and shelf life characteristics. *Scientific Reports* 7: 8711 doi:10.1038/s41598-017-08499-5.

Deytieux, C., Geny, L., Lapaillerie, D., et al. (2007). Proteome analysis of grape skins during ripening. *Journal of Experimental Botany* 58: 1851–62. doi:10.1093/jxb/erm049.

Dheilly, E., Le Gall, S., Guillou, M. C., et al. (2016). Cell wall dynamics during apple development and storage involves hemicellulose modifications and related expressed genes. *BMC Plant Biology* 16: 201. doi:10.1186/s12870-016-0887-0.

Diaz-Riquelme, J., Lijavetzky, D., Martinez-Zapater, J. M., et al. (2009). Genome-wide analysis of MIKCC-type MADS box genes in grapevine. *Plant Physiology* 149: 354–69.

Ding, J. G., Chen, B. W., Xia, X. J., et al. (2013). Cytokinin-induced parthenocarpic fruit development in tomato is partly dependent on enhanced gibberellin and auxin biosynthesis. *PLoS One* 8: e70080.

Du, L. N., Song, J., Forney, C., et al. (2016). Proteome changes in banana fruit peel tissue in response to ethylene and high-temperature treatments. *Horticulture Research* 3: 16012. doi:10.1038/hortres.2016.12.

Einset, J. (1952). Spontaneous polyploidy in cultivated apples. *Proceedings of the American Society for Horticultural Science* 59: 291–302.

El-Sharkawy, I., Liang, D. and Xu, K. (2015). Transcriptome analysis of an apple (*Malus* x *domestica*) yellow fruit somatic mutation identifies a gene network module highly associated with anthocyanin and epigenetic regulation. *Journal of Experimental Botany* 66: 7359–76. doi:10.1093/jxb/erv433.

Endo, T., Shimada, T., Fujii, H., et al. (2005). Ectopic expression of an *FT* homolog from *Citrus* confers an early flowering phenotype on trifoliate orange (*Poncirus trifoliata* L. Raf.). *Transgenic Research* 14: 703–12.

Esen, A. and Soost, R. K. (1971). Unexpected triploids in *Citrus* – their origin, identification, and possible use. *Journal of Heredity* 62: 329–33.

Espley, R. V., Hellens, R. P., Putterill, J., et al. (2007). Red colouration in apple fruit is due to the activity of the MYB transcription factor, MdMYB10. *Plant Journal* 49: 414–27.

Espley, R. V., Brendolise, C., Chagne, D., et al. (2009). Multiple repeats of a promoter segment causes transcription factor autoregulation in red apples. *Plant Cell* 21: 168–83.

Esteve, C., D'Amato, A., Marina, M. L., et al. (2013). In-depth proteomic analysis of banana (*Musa* spp.) fruit with combinatorial peptide ligand libraries. *Electrophoresis* 34: 207–14. doi:10.1002/elps.201200389.

Estrada-Johnson, E., Csukasi, F., Pizarro, C. M., et al. (2017). Transcriptomic analysis in strawberry fruits reveals active auxin biosynthesis and signaling in the ripe receptacle. *Frontiers in Plant Science* 8: 889. doi:10.3389/fpls.2017.00889.

Esumi, T., Tao, R. and Yonemori, K. (2007). Relationship between floral development and transcription levels of *LEAFY* and *TERMINAL FLOWER 1* homologs in Japanese pear (*Pyrus pyrifolia* Nakai) and quince (*Cydonia oblonga* Mill.). *Journal of the Japanese Society for Horticultural Science* 76: 294–304.

Fabi, J. P., Seymour, G. B., Graham, N. S., et al. (2012). Analysis of ripening-related gene expression in papaya using an *Arabidopsis*-based microarray. *BMC Plant Biology* 12: 242. doi:10.1186/1471-2229-12-242.

Fan, S., Bielenberg, D. G., Zhebentyayeva, T. N., et al. (2010). Mapping quantitative trait loci associated with chilling requirement, heat requirement and bloom date in peach (*Prunus persica*). *New Phytologist* 185: 917–30.

Fang, Z. Z., Zhou, D. R., Ye, X. F., et al. (2016). Identification of candidate anthocyanin-related genes by transcriptomic analysis of 'Furongli' plum (*Prunus salicina* Lindl.) during fruit ripening using RNA-seq. *Frontiers in Plant Science* 7: 1338. doi:10.3389/fpls.2016.01338.

Farcuh, M., Li, B., Rivero, R. M., et al. (2017). Sugar metabolism reprogramming in a non-climacteric bud mutant of a climacteric plum fruit during development on the tree. *Journal of Experimental Botany* 68: 5813–28. doi:10.1093/jxb/erx391.

Farinati, S., Rasori, A., Varotto, S., et al. (2017). Rosaceae fruit development, ripening and post-harvest: An epigenetic perspective. *Frontiers in Plant Science* 8: 1247.

Faurobert, M., Mihr, C., Bertin, N., et al. (2007). Major proteome variations associated with cherry tomato pericarp development and ripening. *Plant Physiology* 143: 1327–46. doi:10.1104/pp.106.092817.

Faust, M. (1989). *Physiology of Temperate Zone Fruit Trees*. New York: John Wiley & Sons.

Fernandez, L., Torregrosa, L., Terrier, N., et al. (2007). Identification of genes associated with flesh morphogenesis during grapevine fruit development. *Plant Molecular Biology* 63: 307–23. doi:10.1007/s11103-006-9090-2.

Fernandez, L., Torregrosa, L., Segura, V., et al. (2010). Transposon-induced gene activation as a mechanism generating cluster shape somatic variation in grapevine. *Plant Journal* 61: 545–57.

Fernandez, L., Chaib, J., Martinez-Zapater, J. M., et al. (2013). Mis-expression of a *PISTILLATA*-like MADS box gene prevents fruit development in grapevine. *Plant Journal* 73: 918–28.

Fernandez, L., Le Cunff, L., Tello, J., et al. (2014). Haplotype diversity of *VvTFL1A* gene and association with cluster traits in grapevine (*V. vinifera*). *BMC Plant Biology* 14: 209.

Ficcadenti, N., Sestili, S., Pandolfini, T., et al. (1999). Genetic engineering of parthenocarpic fruit development in tomato. *Molecular Breeding* 5: 463–70.

Foucher, F., Morin, J., Courtiade, J., et al. (2003). *DETERMINATE* and *LATE FLOWERING* are two *TERMINAL FLOWER1/CENTRORADIALIS* homologs that control two distinct phases of flowering initiation and development in pea. *Plant Cell* 15: 2742–54.

Fraige, K., Gonzalez-Fernandez, R., Carrilho, E., et al. (2015). Metabolite and proteome changes during the ripening of Syrah and Cabernet Sauvignon grape varieties cultured in a nontraditional wine region in Brazil. *Journal of Proteomics* 113: 206–25. doi:10.1016/j.jprot.2014.09.021.

Freiman, A., Shlizerman, L., Golobovitch, S., et al. (2012). Development of a transgenic early flowering pear (*Pyrus communis* L.) genotype by RNAi silencing of *PcTFL1-1* and *PcTFL1-2*. *Planta* 235: 1239–51.

Freiman, Z. E., Doron-Faigenboim, A., Dasmohapatra, R., et al. (2014). High-throughput sequencing analysis of common fig (*Ficus carica* L.) transcriptome during fruit ripening. *Tree Genetics & Genomes* 10: 923–35. doi:10.1007/s11295-014-0732-2.

Fujii, H., Shimada, T., Sugiyama, A., et al. (2007). Profiling ethylene-responsive genes in mature mandarin fruit using a citrus 22K oligoarray. *Plant Science* 173: 340–8. doi:10.1016/j.plantsei.2007.06.006.

Fujisawa, M., Nakano, T., Shima, Y., et al. (2013). A large-scale identification of direct targets of the tomato MADS box transcription factor RIPENING INHIBITOR reveals the regulation of fruit ripening. *Plant Cell* 25: 371–86. doi:10.1105/tpc.112.108118.

Fujisawa, M., Shima, Y., Nakagawa, H., et al. (2014). Transcriptional regulation of fruit ripening by tomato FRUITFULL homologs and associated MADS box proteins. *Plant Cell* 26: 89–101. doi:10.1105/tpc.113.119453.

Galla, G., Barcaccia, G., Ramina, A., et al. (2009). Computational annotation of genes differentially expressed along olive fruit development. *BMC Plant Biology* 9: 128. doi:10.1186/1471-2229-9-128.

Gao, Z., Zhang, C. J., Luo, M., et al. (2016). Proteomic analysis of pear (*Pyrus pyrifolia*) ripening process provides new evidence for the sugar/acid metabolism difference between core and mesocarp. *Proteomics* 16: 3025–41. doi:10.1002/pmic.201600108.

Gao, L., Zhao, S., Lu, X., et al. (2018). Comparative transcriptome analysis reveals key genes potentially related to soluble sugar and organic acid accumulation in watermelon. *PLoS One* 13: e0190096. doi:10.1371/journal.pone.0190096.

Giovannoni, J., Nguyen, C., Ampofo, B., et al. (2017). The epigenome and transcriptional dynamics of fruit ripening. *Annual Review of Plant Biology* 68: 61–84.

Giribaldi, M., Geny, L., Delrot, S., et al. (2010). Proteomic analysis of the effects of ABA treatments on ripening *Vitis vinifera* berries. *Journal of Experimental Botany* 61: 2447–58. doi:10.1093/jxb/erq079.

Giribaldi, M., Purrotti, M., Pacifico, D., et al. (2011). A multidisciplinary study on the effects of phloem-limited viruses on the agronomical performance and berry quality of *Vitis vinifera* cv. Nebbiolo. *Journal of Proteomics* 75: 306–15. doi:10.1016/j.jprot.2011.08.006.

Grassi, S., Piro, G., Lee, J. M., et al. (2013). Comparative genomics reveals candidate carotenoid pathway regulators of ripening watermelon fruit. *BMC Genomics* 14: 781. doi:10.1186/1471-2164-14-781.

Grimplet, J., Romieu, C., Audergon, J. M., et al. (2005). Transcriptomic study of apricot fruit (*Prunus armeniaca*) ripening among 13 006 expressed sequence tags. *Physiologia Plantarum* 125: 281–92. doi:10.1111/j.1399-3054.2005.00563.x.

Grimplet, J., Martinez-Zapater, J. M. and Carmona, M. J. (2016). Structural and functional annotation of the MADS-box transcription factor family in grapevine. *BMC Genomics* 17: 80.

Guarino, C., Arena, S., De Simone, L., et al. (2007). Proteomic analysis of the major soluble components in Annurca apple flesh. *Molecular Nutrition & Food Research* 51: 255–62. doi:10.1002/mnfr.200600133.

Guillaumie, S., Fouquet, R., Kappel, C., et al. (2011). Transcriptional analysis of late ripening stages of grapevine berry. *BMC Plant Biology* 11: 165. doi:10.1186/1471-2229-11-165.

Guo, S. G., Liu, J. A., Zheng, Y., et al. (2011). Characterization of transcriptome dynamics during watermelon fruit development: Sequencing, assembly, annotation and gene expression profiles. *BMC Genomics* 12: 454. doi:10.1186/1471-2164-12-454.

Guo, S. G., Sun, H. H., Zhang, H. Y., et al. (2015). Comparative transcriptome analysis of cultivated and wild watermelon during fruit development. *PLoS One* 10: e0130267. doi:10.1371/journal.pone.0130267.

Gupta, V., Estrada, A. D., Blakley, I., et al. (2015). RNA-Seq analysis and annotation of a draft blueberry genome assembly identifies candidate genes involved in fruit ripening, biosynthesis of bioactive compounds, and stage-specific alternative splicing. *Gigascience* 4: 5. doi:10.1186/s13742-015-0046-9.

Habu, T., Yamane, H., Naito, I., et al. (2016). Differences in physiological characteristics and gene expression levels in fruits between Japanese persimmon (*Diospyros kaki* Thunb.) 'Hiratanenashi' and its small fruit mutant 'Totsutanenashi'. *Horticulture Journal* 85: 306–14.

Hanano, S. and Goto, K. (2011). *Arabidopsis* TERMINAL FLOWER1 is involved in the regulation of flowering time and inflorescence development through transcriptional repression. *Plant Cell* 23: 3172–84.

Hartl, K., Denton, A., Franz-Oberdorf, K., et al. (2017). Early metabolic and transcriptional variations in fruit of natural white-fruited *Fragaria vesca* genotypes. *Scientific Reports* 7: 45113. doi:10.1038/srep45113.

Hattasch, C., Flachowsky, H., Kapturska, D., et al. (2008). Isolation of flowering genes and seasonal changes in their transcript levels related to flower induction and initiation in apple (*Malus domestica*). *Tree Physiology* 28: 1459–66.

Honsho, C., Yamamura, E., Tsuruta, K., et al. (2012). Unreduced 2n pollen production in 'Nishiuchi Konatsu' hyuganatsu as inferred by pollen characteristics and progeny ploidy level. *Journal of the Japanese Society for Horticultural Science* 81: 19–26.

Honsho, C., Sakata, A., Tanaka, H., et al. (2016). Single-pollen genotyping to estimate mode of unreduced pollen formation in *Citrus tamurana* cv. Nishiuchi Konatsu. *Plant Reproduction* 29: 189–97.

Hu, W., Wang, L. Z., Tie, W. W., et al. (2016). Genome-wide analyses of the bZIP family reveal their involvement in the development, ripening and abiotic stress response in banana. *Scientific Reports* 6: 30203. doi:10.1038/srep30203.

Huang, G., Li, T., Li, X., et al. (2014). Comparative transcriptome analysis of climacteric fruit of Chinese pear (*Pyrus ussuriensis*) reveals new insights into fruit ripening. *PLoS One* 9: e107562. doi:10.1371/journal.pone.0107562.

Iezzoni, A. F. and Hancock, A. M. (1984). A comparison of pollen size in sweet and sour cherry. *Hortscience* 19: 560–2.

Ireland, H. S., Yao, J. L., Tomes, S., et al. (2013). Apple *SEPALLATA1/2*-like genes control fruit flesh development and ripening. *Plant Journal* 73: 1044–56.

Isuzugawa, K., Murayama, H. and Nishio, T. (2014). Characterization of a giant-fruit mutant exhibiting fruit-limited polyploidization in pear (*Pyrus communis* L.). *Scientia Horticulturae* 170: 196–202.

Jiang, J. M., Lin, Y. X., Chen, Y. Y., et al. (2015). Proteomics approach reveals mechanism underlying susceptibility of loquat fruit to sunburn during color changing period. *Food Chemistry* 176: 388–95. doi:10.1016/j.foodchem.2014.12.076.

Joldersma, D. and Liu, Z. (2018). The making of virgin fruit: The molecular and genetic basis of parthenocarpy. *Journal of Experimental Botany* 69: 955–62.

Jung, J., Choi, S. C., Jung, S., et al. (2017). A transcriptome approach toward understanding fruit softening in persimmon. *Frontiers in Plant Science* 8: 1556. doi:10.3389/fpls.2017.01556.

Kambiranda, D., Katam, R., Basha, S. M., et al. (2014). iTRAQ-based quantitative proteomics of developing and ripening muscadine grape berry. *Journal of Proteome Research* 13: 555–69. doi:10.1021/pr400731p.

Kambiranda, D., Basha, S. M., Singh, R. K., et al. (2016). In depth proteome analysis of ripening muscadine grape berry cv. carlos reveals proteins associated with flavor and aroma compounds. *Journal of Proteome Research* 15: 2910–23. doi:10.1021/acs.jproteome.5b01064.

Kang, R. Y., Zhang, L., Jiang, L., et al. (2016). Effect of postharvest nitric oxide treatment on the proteome of peach fruit during ripening. *Postharvest Biology and Technology* 112: 277–89. doi:10.1016/j.postharvbio.2015.08.017.

Karagiannis, E., Tanou, G., Samiotaki, M., et al. (2016). Comparative physiological and proteomic analysis reveal distinct regulation of peach skin quality traits by altitude. *Frontiers in Plant Science* 7: 1689. doi:10.3389/fpls.2016.01689.

Karlova, R., Rosin, F. M., Busscher-Lange, J., et al. (2011). Transcriptome and metabolite profiling show that APETALA2a is a major regulator of tomato fruit ripening. *Plant Cell* 23: 923–41. doi:10.1105/tpc.110.081273.

Karlova, R., Chapman, N., David, K., et al. (2014). Transcriptional control of fleshy fruit development and ripening. *Journal of Experimental Botany* 65: 4527–41.

Katz, E., Fon, M., Lee, Y. J., et al. (2007). The citrus fruit proteome: Insights into citrus fruit metabolism. *Planta* 226: 989–1005. doi:10.1007/s00425-007-0545-8.

Katz, E., Fon, M., Eigenheer, R. A., et al. (2010). A label-free differential quantitative mass spectrometry method for the characterization and identification of protein changes during citrus fruit development. *Proteome Science* 8: 68. doi:10.1186/1477-5956-8-68.

Kilambi, H. V., Kumar, R., Sharma, R., et al. (2013). Chromoplast-specific carotenoid-associated protein appears to be important for enhanced accumulation of carotenoids in *hp1* tomato fruits. *Plant Physiology* 161: 2085–101. doi:10.1104/pp.112.212191.

Klocko, A. L., Borejsza-Wysocka, E., Brunner, A. M., et al. (2016). Transgenic suppression of *AGAMOUS* genes in apple reduces fertility and increases floral attractiveness. *PLoS One* 11: e0159421.

Kobayashi, Y., Kaya, H., Goto, K., et al. (1999). A pair of related genes with antagonistic roles in mediating flowering signals. *Science* 286: 1960–2.

Kobayashi, S., Goto-Yamamoto, N. and Hirochika, H. (2004). Retrotransposon-induced mutations in grape skin color. *Science* 304: 982.

Kok, E. J., Franssen-van Hal, N. L. W., Winnubst, L. N. W., et al. (2007). Assessment of representational difference analysis (RDA) to construct informative cDNA microarrays for gene expression analysis of species with limited transcriptome information, using red and green tomatoes as a model. *Journal of Plant Physiology* 164: 337–49. doi:10.1016/j.jplph.2006.02.013.

Kok, E. J., Lehesranta, S. J., van Dijk, J. P., et al. (2008). Changes in gene and protein expression during tomato ripening – Consequences for the safety assessment of new crop plant varieties. *Food Science and Technology International* 14: 503–18. doi:10.1177/1082013208100771.

Konozy, E. H. E., Rogniaux, H., Causse, M., et al. (2013). Proteomic analysis of tomato (*Solanum lycopersicum*) secretome. *Journal of Plant Research* 126: 251–66. doi:10.1007/s10265-012-0516-4.

Kotoda, N., Wada, M., Komori, S., et al. (2000). Expression pattern of homologues of floral meristem identity genes *LFY* and *AP1* during flower development in apple. *Journal of the American Society for Horticultural Science* 125: 398–403.

Kotoda, N., Wada, M., Kusaba, S., et al. (2002). Overexpression of *MdMADS5*, an *APETALA1*-like gene of apple, causes early flowering in transgenic *Arabidopsis*. *Plant Science* 162: 679–87.

Kotoda, N., Iwanami, H., Takahashi, S., et al. (2006). Antisense expression of *MdTFL1*, a *TFL1*-like gene, reduces the juvenile phase in apple. *Journal of the American Society for Horticultural Science* 131: 74–81.

Kotoda, N., Hayashi, H., Suzuki, M., et al. (2010). Molecular characterization of *FLOWERING LOCUS T*-like genes of apple (*Malus × domestica* Borkh.). *Plant Cell Physiology* 51: 561–75.

Koyama, K., Sadamatsu, K. and Goto-Yamamoto, N. (2010). Abscisic acid stimulated ripening and gene expression in berry skins of the Cabernet Sauvignon grape. *Functional & Integrative Genomics* 10: 367–81. doi:10.1007/s10142-009-0145-8.

Kuang, J. F., Chen, J. Y., Liu, X. C., et al. (2017). The transcriptional regulatory network mediated by banana (*Musa acuminata*) dehydration-responsive element binding (MaDREB) transcription factors in fruit ripening. *New Phytologist* 214: 762–81. doi:10.1111/nph.14389.

Kumar, R., Sharma, M. K., Kapoor, S., et al. (2012). Transcriptome analysis of *rin* mutant fruit and in silico analysis of promoters of differentially regulated genes provides insight into LeMADS-RIN-regulated ethylene-dependent as well as ethylene-independent aspects of ripening in tomato. *Molecular Genetics and Genomics* 287: 189–203. doi:10.1007/s00438-011-0671-7.

Kumar, G., Arya, P., Gupta, K., et al. (2016). Comparative phylogenetic analysis and transcriptional profiling of MADS-box gene family identified *DAM* and *FLC*-like genes in apple (*Malus × domestica*). *Scientific Reports* 6: 20695.

Lai, B., Hu, B., Qin, Y. H., et al. (2015). Transcriptomic analysis of *Litchi chinensis* pericarp during maturation with a focus on chlorophyll degradation and flavonoid biosynthesis. *BMC Genomics* 16: 225. doi:10.1186/s12864-015-1433-4.

Lara, M. V., Borsani, J., Budde, C. O., et al. (2009). Biochemical and proteomic analysis of 'Dixiland' peach fruit (*Prunus persica*) upon heat treatment. *Journal of Experimental Botany* 60: 4315–33. doi:10.1093/jxb/erp267.

Lecourieux, F., Kappel, C., Pieri, P., et al. (2017). Dissecting the biochemical and transcriptomic effects of a locally applied heat treatment on developing cabernet sauvignon grape berries. *Frontiers in Plant Science* 8: 53. doi:10.3389/fpls.2017.00053.

Lee, S., Chung, E. J., Joung, Y. H., et al. (2010). Non-climacteric fruit ripening in pepper: Increased transcription of *EIL*-like genes normally regulated by ethylene. *Functional & Integrative Genomics* 10: 135–46. doi:10.1007/s10142-009-0136-9.

Lee, J. M., Joung, J. G., McQuinn, R., et al. (2012). Combined transcriptome, genetic diversity and metabolite profiling in tomato fruit reveals that the ethylene response factor *SlERF6* plays an important role in ripening and carotenoid accumulation. *Plant Journal* 70: 191–204. doi:10.1111/j.1365-313X.2011.04863.x.

Leng, F., Lin, Q., Wu, D., et al. (2016). Comparative transcriptomic analysis of grape berry in response to root restriction during developmental stages. *Molecules* 21: 1431. doi:10.3390/molecules21111431.

Li, C., Xie, H., Zhang, L., et al. (2012a). Molecular characterization of the *PpMADS1* gene from peach. *Tree Genetics & Genomes* 8: 831–40.

Li, X. Y., Sun, H. Y., Pei, J. B., et al. (2012b). *De novo* sequencing and comparative analysis of the blueberry transcriptome to discover putative genes related to antioxidants. *Gene* 511: 54–61. doi:10.1016/j.gene.2012.09.021.

Li, L., Song, J., Kalt, W., et al. (2013). Quantitative proteomic investigation employing stable isotope labeling by peptide dimethylation on proteins of strawberry fruit at different ripening stages. *Journal of Proteomics* 94: 219–39. doi:10.1016/j.jprot.2013.09.004.

Li, D. D., Li, L., Luo, Z. S., et al. (2015a). Comparative transcriptome analysis reveals the influence of abscisic acid on the metabolism of pigments, ascorbic acid and folic acid during strawberry fruit ripening. *PLoS One* 10: e0130037. doi:10.1371/journal.pone.0130037.

Li, J. M., Huang, X. S., Li, L. T., et al. (2015b). Proteome analysis of pear reveals key genes associated with fruit development and quality. *Planta* 241: 1363–79. doi:10.1007/s00425-015-2263-y.

Li, W. B., Liu, Y. F., Zeng, S. H., et al. (2015c). Gene expression profiling of development and anthocyanin accumulation in kiwifruit (*Actinidia chinensis*) based on transcriptome sequencing. *PLoS One* 10: e0136439. doi:10.1371/journal.pone.0136439.

Li, J. Y., Tao, X. Y., Li, L., et al. (2016a). Comprehensive RNA-seq analysis on the regulation of tomato ripening by exogenous auxin. *PLoS One* 11: e0156453. doi:10.1371/journal.pone.0156453.

Li, L. L., Zhang, H. H., Liu, Z. S., et al. (2016b). Comparative transcriptome sequencing and *de novo* analysis of *Vaccinium corymbosum* during fruit and color development. *BMC Plant Biology* 16: 223. doi:10.1186/s12870-016-0866-5.

Li, Y., Li, H. Y., Wang, F. D., et al. (2016c). Comparative transcriptome analysis reveals effects of exogenous hematin on anthocyanin biosynthesis during strawberry fruit ripening. *International Journal of Genomics* 2016: 6762731. doi:10.1155/2016/6762731.

Li, J. Y., Tao, X. Y., Bu, J. W., et al. (2017a). Global transcriptome profiling analysis of ethylene-auxin interaction during tomato fruit ripening. *Postharvest Biology and Technology* 130: 28–38. doi:10.1016/j.postharvbio.2017.03.021.

Li, L., Wang, X., Zhang, X., et al. (2017b). Unraveling the target genes of RIN transcription factor during tomato fruit ripening and softening. *Journal of the Science of Food and Agriculture* 97: 991–1000. doi:10.1002/jsfa.7825.

Lijavetzky, D., Carbonell-Bejerano, P., Grimplet, J., et al. (2012). Berry flesh and skin ripening features in *Vitis vinifera* as assessed by transcriptional profiling. *PLoS One* 7: e39547. doi:10.1371/journal.pone.0039547.

Lin-Wang, K., Bolitho, K., Grafton, K., et al. (2010). An R2R3 MYB transcription factor associated with regulation of the anthocyanin biosynthetic pathway in Rosaceae. *BMC Plant Biology* 10: 50.

Liu, X. M., Anderson, J. M. and Pijut, P. M. (2010). Cloning and characterization of *Prunus serotina AGAMOUS*, a putative flower homeotic gene. *Plant Molecular Biology Reporter* 28: 193–203.

Liu, Z. X., Zhang, D. D., Liu, D., et al. (2013). Exon skipping of *AGAMOUS* homolog *PrseAG* in developing double flowers of *Prunus lannesiana* (Rosaceae). *Plant Cell Reports* 32: 227–37.

Liu, X. T., Zhai, R., Feng, W. T., et al. (2014). Proteomic analysis of 'Zaosu' pear (*Pyrus bretschneideri* Rehd.) and its early-maturing bud sport. *Plant Science* 224: 120–35. doi:10.1016/j.plantsci.2014.04.012.

Liu, L. H., Shao, Z. Y., Zhang, M., et al. (2015a). Regulation of carotenoid metabolism in tomato. *Molecular Plant* 8: 28–39.

Liu, M. C., Pirrello, J., Chervin, C., et al. (2015b). Ethylene control of fruit ripening: Revisiting the complex network of transcriptional regulation. *Plant Physiology* 169: 2380–90.

Liu, R. L., Wang, Y. Y., Qin, G. Z., et al. (2016). iTRAQ-based quantitative proteomic analysis reveals the role of the tonoplast in fruit senescence. *Journal of Proteomics* 146: 80–9. doi:10.1016/j.jprot.2016.06.031.

Liu, K., Feng, S., Jiang, Y., et al. (2017). Identification and expression analysis of seven MADS-box genes from *Annona squamosa*. *Biologia Plantarum* 61: 24–34.

Liu, L., Wang, Z., Liu, J., et al. (2018). Histological, hormonal and transcriptomic reveal the changes upon gibberellin-induced parthenocarpy in pear fruit. *Horticulture Research* 5: 1.

Llorente, B., Martinez-Garcia, J. F., Stange, C., et al. (2017). Illuminating colors: Regulation of carotenoid biosynthesis and accumulation by light. *Current Opinion in Plant Biology* 37: 49-55.

Lu, X. P., Cao, X. J., Li, F. F., et al. (2016). Comparative transcriptome analysis reveals a global insight into molecular processes regulating citrate accumulation in sweet orange (*Citrus sinensis*). *Physiologia Plantarum* 158: 463–82. doi:10.1111/ppl.12484.

Lu, X. P., Li, F. F., Xiong, J., et al. (2017). Transcriptome and metabolome analyses provide insights into the occurrence of peel roughing disorder on satsuma mandarin (*Citrus unshiu* Marc.) fruit. *Frontiers in Plant Science* 8: 1907. doi:10.3389/fpls.2017.01907.

Lucker, J., Laszczak, M., Smith, D., et al. (2009). Generation of a predicted protein database from EST data and application to iTRAQ analyses in grape (*Vitis vinifera* cv. Cabernet Sauvignon) berries at ripening initiation. *BMC Genomics* 10: 50. doi:10.1186/1471-2164-10-50.

Malladi, A. and Hirst, P. M. (2010). Increase in fruit size of a spontaneous mutant of 'Gala' apple (*Malus × domestica* Borkh.) is facilitated by altered cell production and enhanced cell size. *Journal of Experimental Botany* 61: 3003–13.

Manganaris, G. A., Rasori, A., Bassi, D., et al. (2011). Comparative transcript profiling of apricot (*Prunus armeniaca* L.) fruit development and on-tree ripening. *Tree Genetics & Genomes* 7: 609–16. doi:10.1007/s11295-010-0360-4.

Marondedze, C. and Thomas, L. A. (2012). Apple hypanthium firmness: New insights from comparative proteomics. *Applied Biochemistry and Biotechnology* 168: 306–26. doi:10.1007/s12010-012-9774-9.

Martens, M. H. R., Reisch, B. I. and Mauro, M. C. (1989). Pollen size variability within genotypes of *Vitis*. *Hortscience* 24: 659–62.

Martin, A., Troadec, C., Boualem, A., et al. (2009). A transposon-induced epigenetic change leads to sex determination in melon. *Nature* 461: 1135–8.

Martinez-Esteso, M. J., Vilella-Anton, M. T., Pedreno, M. A., et al. (2013). iTRAQ-based protein profiling provides insights into the central metabolism changes driving grape berry development and ripening. *BMC Plant Biology* 13: 167. doi:10.1186/1471-2229-13-167.

Martinez-Lopez, L. A., Ochoa-Alejo, N. and Martinez, O. (2014). Dynamics of the chili pepper transcriptome during fruit development. *BMC Genomics* 15: 143. doi:10.1186/1471-2164-15-143.

Massonnet, M., Fasoli, M., Tornielli, G. B., et al. (2017). Ripening transcriptomic program in red and white grapevine varieties correlates with berry skin anthocyanin accumulation. *Plant Physiology* 174: 2376–96. doi:10.1104/pp.17.00311.

Matas, A. J., Agusti, J., Tadeo, F. R., et al. (2010). Tissue-specific transcriptome profiling of the citrus fruit epidermis and subepidermis using laser capture microdissection. *Journal of Experimental Botany* 61: 3321–30. doi:10.1093/jxb/erq153.

Matsuda, N., Ikeda, K., Kurosaka, M., et al. (2009). Early flowering phenotype in transgenic pears (*Pyrus communis* L.) expressing the *CiFT* gene. *Journal of the Japanese Society for Horticultural Science* 78: 410–16.

Matsumoto, D. and Tao R. (2016). Distinct self-recognition in the *Prunus* S-RNase-based gametophytic self-incompatibility system. *Horticulture Journal* 85: 289–305.

Matsuo, S., Kikuchi, K., Fukuda, M., et al. (2012). Roles and regulation of cytokinins in tomato fruit development. *Journal of Experimental Botany* 63: 5569–79.

Medina-Puche, L., Blanco-Portales, R., Molina-Hidalgo, F. J., et al. (2016). Extensive transcriptomic studies on the roles played by abscisic acid and auxins in the development and ripening of strawberry fruits. *Functional & Integrative Genomics* 16: 671–92. doi:10.1007/s10142-016-0510-3.

Megalos, B. S. and Ballington, J. R. (1988). Unreduced pollen frequencies versus hybrid production in diploid-tetraploid *Vaccinium* crosses. *Euphytica* 39: 271–8.

Mesejo, C., Reig, C., Martinez-Fuentes, A., et al. (2010). Parthenocarpic fruit production in loquat (*Eriobotrya japonica* Lindl.) by using gibberellic acid. *Scientia Horticulturae* 126: 37–41.

Mesejo, C., Yuste, R., Reig, C., et al. (2016). Gibberellin reactivates and maintains ovary-wall cell division causing fruit set in parthenocarpic *Citrus* species. *Plant Science* 247: 13–24.

Mimida, N., Kidou, S., Iwanami, H., et al. (2011a). Apple FLOWERING LOCUS T proteins interact with transcription factors implicated in cell growth and organ development. *Tree Physiology* 31: 555–66.

Mimida, N., Ureshino, A., Tanaka, N., et al. (2011b). Expression patterns of several floral genes during flower initiation in the apical buds of apple (*Malus* × *domestica* Borkh.) revealed by in situ hybridization. *Plant Cell Reports* 30: 1485–92.

Mohamed, R., Wang, C. T., Ma, C., et al. (2010). Populus *CEN/TFL1* regulates first onset of flowering, axillary meristem identity and dormancy release in *Populus*. *Plant Journal* 62: 674–88.

Mori, K., Shirasawa, K., Nogata, H., et al. (2017). Identification of *RAN1* orthologue associated with sex determination through whole genome sequencing analysis in fig (*Ficus carica* L.). *Scientific Reports* 7: 46784.

Moussa, H. R. and Salem, A. A. K. (2010). Parthenocarpy of watermelon cultivars induced by gamma-irradiation. *Russian Journal of Plant Physiology* 57: 574–81.

Moyle, R., Fairbairn, D. J., Ripi, J., et al. (2005). Developing pineapple fruit has a small transcriptome dominated by metallothionein. *Journal of Experimental Botany* 56: 101–12. doi:10.1093/jxb/eri015.

Muccilli, V., Licciardello, C., Fontanini, D., et al. (2009). Proteome analysis of *Citrus sinensis* L. (Osbeck) flesh at ripening time. *Journal of Proteomics* 73: 134–52. doi:10.1016/j.jprot.2009.09.005.

Nakagawa, T., Nakatsuka, A., Yano, K., et al. (2008). Expressed sequence tags from persimmon at different developmental stages. *Plant Cell Reports* 27: 931–8. doi:10.1007/s00299-008-0518-9.

Nashima, K., Takahashi, H., Nakazono, M., et al. (2013). Transcriptome analysis of giant pear fruit with fruit-specific DNA reduplication on a mutant branch. *Journal of the Japanese Society for Horticultural Science* 82: 301–11. doi:10.2503/jjshs1.82.301.

Negri, A. S., Prinsi, B., Scienza, A., et al. (2008b). Analysis of grape berry cell wall proteome: A comparative evaluation of extraction methods. *Journal of Plant Physiology* 165: 1379–89. doi:10.1016/j.jplph.2007.10.011.

Negri, A. S, Robotti, E., Prinsi, B., et al. (2011). Proteins involved in biotic and abiotic stress responses as the most significant biomarkers in the ripening of Pinot Noir skins. *Functional & Integrative Genomics* 11: 341–55. doi:10.1007/s10142-010-0205-0.

Nguyen, C. V., Vrebalov, J. T., Gapper, N. E., et al. (2014). Tomato *GOLDEN2-LIKE* transcription factors reveal molecular gradients that function during fruit development and ripening. *Plant Cell* 26: 585–601. doi:10.1105/tpc.113.118794.

Nham, N. T., de Freitas, S. T., Macnish, A. J., et al. (2015). A transcriptome approach towards understanding the development of ripening capacity in 'Bartlett' pears (*Pyrus communis* L.). *BMC Genomics* 16: 762. doi:10.1186/s12864-015-1939-9.

Nilo, R., Campos-Vargas, R. and Orellana, A. (2012). Assessment of *Prunus persica* fruit softening using a proteomics approach. *Journal of Proteomics* 75: 1618–38. doi:10.1016/j.jprot.2011.11.037.

Nishikawa, F., Endo, T., Shimada, T., et al. (2007). Increased *CiFT* abundance in the stem correlates with floral induction by low temperature in Satsuma mandarin (*Citrus unshiu* Marc.). *Journal of Experimental Botany* 58: 3915–27.

Nishikawa, F., Endo, T., Shimada, T., et al. (2010). Transcriptional changes in *CiFT*-introduced transgenic trifoliate orange (*Poncirus trifoliata* L. Raf.). *Tree Physiology* 30: 431–9.

Nishikawa, F., Iwasaki, M., Fukamachi, H., et al. (2012). Fruit bearing suppresses citrus *FLOWERING LOCUS T* expression in vegetative shoots of satsuma mandarin (*Citrus unshiu* Marc.). *Journal of the Japanese Society for Horticultural Science* 81: 48–53.

Nishikawa, F., Iwasaki, M., Fukamachi, H., et al. (2017). Predicting the number of flowers in satsuma mandarin (*Citrus unshiu* Marc.) trees based on citrus *FLOWERING LOCUS T* mRNA levels. *Horticulture Journal* 86: 305–10.

Negri, A. S., Prinsi, B., Rossoni, M., et al. (2008a). Proteome changes in the skin of the grape cultivar Barbera among different stages of ripening. *BMC Genomics* 9: 378. doi:10.1186/1471-2164-9-378.

Nishitani, C., Shimizu, T., Fujii, H., et al. (2010). Oligoarray analysis of gene expression in ripening Japanese pear fruit. *Scientia Horticulturae* 124: 195–203. doi:10.1016/j.scienta.2009.12.032.

Nishiyama, S., Onoue, N., Kono, A., et al. (2018). Characterization of a gene regulatory network underlying astringency loss in persimmon fruit. *Planta* 247: 733–43. doi:10.1007/s00425-017-2819-0.

Nogueira, S. B., Labate, C. A., Gozzo, F. C., et al. (2012). Proteomic analysis of papaya fruit ripening using 2DE-DIGE. *Journal of Proteomics* 75: 1428–39. doi:10.1016/j.jprot.2011.11.015.

Okada, K. (2015). DNA markers and the molecular mechanism of self-(in)compatibility in Japanese pear (*Pyrus pyrifolia* Nakai). *Horticulture Journal* 84: 183–94.

Ortiz, R. (1997). Occurrence and inheritance of 2*n* pollen in *Musa*. *Annals of Botany* 79: 449–53.

Ortiz, R., Vorsa, N., Bruederle, L. P., et al. (1992). Occurrence of unreduced pollen in diploid blueberry species, *Vaccinium* sect. *Cyanococcus*. *Theoretical and Applied Genetics* 85: 55–60.

Page, D., Gouble, B., Valot, B., et al. (2010). Protective proteins are differentially expressed in tomato genotypes differing for their tolerance to low-temperature storage. *Planta* 232: 483–500. doi:10.1007/s00425-010-1184-z.

Pan, Z. Y., Liu, Q., Yun, Z., et al. (2009). Comparative proteomics of a lycopene-accumulating mutant reveals the important role of oxidative stress on carotenogenesis in sweet orange (*Citrus sinensis* [L.] osbeck). *Proteomics* 9: 5455–70. doi:10.1002/pmic.200900092.

Pan, Z. Y., Zeng, Y. L., An, J. Y., et al. (2012). An integrative analysis of transcriptome and proteome provides new insights into carotenoid biosynthesis and regulation in sweet orange fruits. *Journal of Proteomics* 75: 2670–84. doi:10.1016/j.jprot.2012.03.016.

Pan, X. Q., Zhu, B. Z., Zhu, H. L., et al. (2014). iTRAQ protein profile analysis of tomato green-ripe mutant reveals new aspects critical for fruit ripening. *Journal of Proteome Research* 13: 1979–93. doi:10.1021/pr401091n.

Pandolfini, T. (2009). Seedless fruit production by hormonal regulation of fruit set. *Nutrients* 1: 168–77.

Pastore, C., Zenoni, S., Tornielli, G. B., et al. (2011). Increasing the source/sink ratio in *Vitis vinifera* (cv Sangiovese) induces extensive transcriptome reprogramming and modifies berry ripening. *BMC Genomics* 12: 631. doi:10.1186/1471-2164-12-631.

Pastore, C., Zenoni, S., Fasoli, M., et al. (2013). Selective defoliation affects plant growth, fruit transcriptional ripening program and flavonoid metabolism in grapevine. *BMC Plant Biology* 13: 30. doi:10.1186/1471-2229-13-30.

Pastore, C., Dal Santo, S., Zenoni, S., et al. (2017). Whole plant temperature manipulation affects flavonoid metabolism and the transcriptome of grapevine berries. *Frontiers in Plant Science* 8: 929. doi:10.3389/fpls.2017.00929.

Patel, M., Manvar, T., Apurwa, S., et al. (2014). Comparative *de novo* transcriptome analysis and metabolic pathway studies of *Citrus paradisi* flavedo from naive stage to ripened stage. *Molecular Biology Reports* 41: 3071–80. doi:10.1007/s11033-014-3166-x.

Pena, L., Martin-Trillo, M., Juarez, J., et al. (2001). Constitutive expression of *Arabidopsis LEAFY* or *APETALA1* genes in citrus reduces their generation time. *Nature Biotechnology* 19: 263–7.

Peng, T. and Moriguchi, T. (2013). The molecular network regulating the coloration in apple. *Scientia Horticulturae* 163: 1–9.

Pignataro, V., Canton, C., Spadafora, A., et al. (2010). Proteome from lemon fruit flavedo reveals that this tissue produces high amounts of the Cit s1 germin-like isoforms. *Journal of Agricultural and Food Chemistry* 58: 7239–44. doi:10.1021/jf1006825.

Pillet, J., Yu, H. W., Chambers, A. H., et al. (2015). Identification of candidate flavonoid pathway genes using transcriptome correlation network analysis in ripe strawberry (*Fragaria x ananassa*) fruits. *Journal of Experimental Botany* 66: 4455–67. doi:10.1093/jxb/erv205.

Pillitteri, L. J., Lovatt, C. J. and Walling, L. L. (2004). Isolation and characterization of *LEAFY* and *APETALA1* homologues from *Citrus sinensis* L. Osbeck 'Washington'. *Journal of the American Society for Horticultural Science* 129: 846–56.

Pirona, R., Vecchietti, A., Lazzari, B., et al. (2013). Expression profiling of genes involved in the formation of aroma in two peach genotypes. *Plant Biology* 15: 443–51. doi:10.1111/j.1438-8677.2012.00666.x.

Pnueli, L., Carmel-Goren, L., Hareven, D., et al. (1998). The *SELF-PRUNING* gene of tomato regulates vegetative to reproductive switching of sympodial meristems and is the ortholog of *CEN* and *TFL1*. *Development* 125: 1979–89.

Pnueli, L., Gutfinger, T., Hareven, D., et al. (2001). Tomato SP-interacting proteins define a conserved signaling system that regulates shoot architecture and flowering. *Plant Cell* 13: 2687–702.

Prinsi, B., Negri, A. S., Espen, L., et al. (2016). Proteomic comparison of fruit ripening between 'Hedelfinger' sweet cherry (*Prunus avium* L.) and its somaclonal variant 'HS'. *Journal of Agricultural and Food Chemistry* 64: 4171–81. doi:10.1021/acs.jafc.6b01039.

Qi, Y. W., Lei, Q., Zhang, Y. J., et al. (2017). Comparative transcripome data for commercial maturity and physiological maturity of 'Royal Gala' apple fruit under room temperature storage condition. *Scientia Horticulturae* 225: 386–93. doi:10.1016/j.scienta.2017.07.024.

Qin, G. Z., Meng, X. H., Wang, Q., et al. (2009a). Oxidative damage of mitochondrial proteins contributes to fruit senescence: A redox proteomics analysis. *Journal of Proteome Research* 8: 2449–62. doi:10.1021/pr801046m.

Qin, G. Z., Wang, Q., Liu, J., et al. (2009b). Proteomic analysis of changes in mitochondrial protein expression during fruit senescence. *Proteomics* 9: 4241–53. doi:10.1002/pmic.200900133.

Quecini, V., Torres, G. A. M., de Rosa, V. E., et al. (2007). *In silico* analysis of phytohormone metabolism and communication pathways in citrus transcriptome. *Genetics and Molecular Biology* 30: 713–33. doi:10.1590/s1415-47572007000500002.

Riebel, M., Fronk, P., Distler, U., et al. (2017). Proteomic profiling of German Dornfelder grape berries using data-independent acquisition. *Plant Physiology and Biochemistry* 118: 64–70. doi:10.1016/j.plaphy.2017.06.003.

Rienth, M., Torregrosa, L., Luchaire, N., et al. (2014). Day and night heat stress trigger different transcriptomic responses in green and ripening grapevine (*vitis vinifera*) fruit. *BMC Plant Biology* 14: 108. doi:10.1186/1471-2229-14-108.

Rienth, M., Torregrosa, L., Sarah, G., et al. (2016). Temperature desynchronizes sugar and organic acid metabolism in ripening grapevine fruits and remodels their transcriptome. *BMC Plant Biology* 16: 164. doi:10.1186/s12870-016-0850-0.

Ring, L., Yeh, S. Y., Huecherig, S., et al. (2013). Metabolic interaction between anthocyanin and lignin biosynthesis is associated with peroxidase FaPRX27 in strawberry fruit. *Plant Physiology* 163: 43–60. doi:10.1104/pp.113.222778.

Rinne, P. L., Welling, A., Vahala, J., et al. (2011). Chilling of dormant buds hyperinduces *FLOWERING LOCUS T* and recruits GA-inducible 1,3-β-glucanases to reopen signal conduits and release dormancy in *Populus*. *Plant Cell* 23: 130–46.

Rios, G., Leida, C., Conejero, A., et al. (2014). Epigenetic regulation of bud dormancy events in perennial plants. *Frontiers in Plant Science* 5: 247.

Rocco, M., D'Ambrosio, C., Arena, S., et al. (2006). Proteomic analysis of tomato fruits from two ecotypes during ripening. *Proteomics* 6: 3781-91. doi:10.1002/pmic.200600128.

Rogiers, S. Y., Coetzee, Z. A., Walker, R. R., et al. (2017). Potassium in the grape (*Vitis vinifera* L.) berry: Transport and function. *Frontiers in Plant Science* 8: 1629.

Rosianski, Y., Doron-Faigenboim, A., Freiman, Z. E., et al. (2016). Tissue-specific transcriptome and hormonal regulation of pollinated and parthenocarpic fig (*Ficus carica* L.) fruit suggest that fruit ripening is coordinated by the reproductive part of the syconium. *Frontiers in Plant Science* 7: 1696. doi:10.3389/fpls.2016.01696.

Rouiss, H., Cuenca, J., Navarro, L., et al. (2017a). Tetraploid citrus progenies arising from FDR and SDR unreduced pollen in 4x X 2x hybridizations. *Tree Genetics & Genomes* 13: 10.

Rouiss, H., Cuenca, J., Navarro, L., et al. (2017b). Unreduced megagametophyte production in lemon occurs via three meiotic mechanisms, predominantly second-division restitution. *Frontiers in Plant Science* 8: 1211.

Ru, L., Osorio, S., Wang, L., et al. (2017). Transcriptomic and metabolomics responses to elevated cell wall invertase activity during tomato fruit set. *Journal of Experimental Botany* 68: 4263–79. doi:10.1093/jxb/erx219.

Ruiu, F., Picarella, M. E., Imanishi, S., et al. (2015). A transcriptomic approach to identify regulatory genes involved in fruit set of wild-type and parthenocarpic tomato genotypes. *Plant Molecular Biology* 89: 263–78.

Ruiz-May, E. and Rose, J. K. C. (2013). Progress toward the tomato fruit cell wall proteome. *Frontiers in Plant Science* 4: 159. doi:10.3389/fpls.2013.00159.

Sanchez-Sevilla, J. F., Vallarino, J. G., Osorio, S., et al. (2017). Gene expression atlas of fruit ripening and transcriptome assembly from RNA-seq data in octoploid strawberry (*Fragaria x ananassa*). *Scientific Reports* 7: 13737. doi:10.1038/s41598-017-14239-6.

Sarry, J. E., Sommerer, N., Sauvage, F. X., et al. (2004). Grape berry biochemistry revisited upon proteomic analysis of the mesocarp. *Proteomics* 4: 201–15. doi:10.1002/pmic.200300499.

Sassa, H. (2016). Molecular mechanism of the S-RNase-based gametophytic self-incompatibility in fruit trees of Rosaceae. *Breeding Science* 66: 116–21.

Savoi, S., Wong, D. C. J., Arapitsas, P., et al. (2016). Transcriptome and metabolite profiling reveals that prolonged drought modulates the phenylpropanoid and terpenoid pathway in white grapes (*Vitis vinifera* L.). *BMC Plant Biology* 16: 67. doi:10.1186/s12870-016-0760-1.

Savoi, S., Wong, D. C. J., Degu, A., et al. (2017). Multi-omics and integrated network analyses reveal new insights into the systems relationships between metabolites, structural genes, and transcriptional regulators in developing grape berries (*Vitis vinifera* L.) exposed to water deficit. *Frontiers in Plant Science* 8: 1124. doi:10.3389/fpls.2017.01124.

Schaeffer, S., Harper, A., Raja, R., et al. (2014). Comparative analysis of predicted plastid-targeted proteomes of sequenced higher plant genomes. *PLoS One* 9: e112870. doi:10.1371/journal.pone.0112870.

Segonne, S. M., Bruneau, M., Celton, J. M., et al. (2014). Multiscale investigation of mealiness in apple: An atypical role for a pectin methylesterase during fruit maturation. *BMC Plant Biology* 14: 375. doi:10.1186/s12870-014-0375-3.

Serrano, A., Espinoza, C., Armijo, G., et al. (2017). Omics approaches for understanding grapevine berry development: Regulatory networks associated with endogenous processes and environmental responses. *Frontiers in Plant Science* 8: 1486. doi:10.3389/fpls.2017.01486.

Serrano-Mislata, A., Fernandez-Nohales, P., Domenech, M. J., et al. (2016). Separate elements of the *TERMINAL FLOWER 1 cis*-regulatory region integrate pathways to control flowering time and shoot meristem identity. *Development* 143: 3315–27.

Shangguan, L., Mu, Q., Fang, X., et al. (2017). RNA-sequencing reveals biological networks during table grapevine ('Fujiminori') fruit development. *PLoS One* 12: e0170571. doi:10.1371/journal.pone.0170571.

Shen, Y. H., Lu, B. G., Feng, L., et al. (2017). Isolation of ripening-related genes from ethylene/1-MCP treated papaya through RNA-seq. *BMC Genomics* 18: 671. doi:10.1186/s12864-017-4072-0.

Shin, A. Y., Kim, Y. M., Koo, N., et al. (2017). Transcriptome analysis of the oriental melon (*Cucumis melo* L. var. *makuwa*) during fruit development. *Peer Journal* 5: e2834. doi:10.7717/peerj.2834.

Shinozaki, Y., Nicolas, P., Fernandez-Pozo, N., et al. (2018). High-resolution spatiotemporal transcriptome mapping of tomato fruit development and ripening. *Nature Communications* 9: 364.

Smita, S., Rajwanshi, R., Lenka, S. K., et al. (2013). Comparative analysis of fruit transcriptome in tomato (*Solanum lycopersicum*) genotypes with contrasting lycopene contents. *Plant Molecular Biology Reporter* 31: 1384–96. doi:10.1007/s11105-013-0613-0.

Song, G. Q., Walworth, A., Zhao, D. Y., et al. (2013). The *Vaccinium corymbosum FLOWERING LOCUS T*-like gene (*VcFT*): A flowering activator reverses photoperiodic and chilling requirements in blueberry. *Plant Cell Reports* 32: 1759–69.

Song, H. W., Zhao, X. X., Hu, W. C., et al. (2016). Comparative transcriptional analysis of loquat fruit identifies major signal networks involved in fruit development and ripening process. *International Journal of Molecular Sciences* 17: 1837. doi:10.3390/ijms17111837.

Sotelo-Silveira, M., Marsch-Martinez, N. and de Folter, S. (2014). Unraveling the signal scenario of fruit set. *Planta* 239: 1147–58.

Sreekantan, L. and Thomas, M. R. (2006). *VvFT* and *VvMADS8*, the grapevine homologues of the floral integrators *FT* and *SOC1*, have unique expression patterns in grapevine and hasten flowering in *Arabidopsis*. *Functional Plant Biology* 33: 1129–39.

Srinivasan, C., Dardick, C., Callahan, A., et al. (2012). Plum (*Prunus domestica*) trees transformed with poplar *FT1* result in altered architecture, dormancy requirement, and continuous flowering. *PLoS One* 7: e40715.

Srivastava, A., Chung, S. H., Fatima, T., et al. (2007). Polyamines as anabolic growth regulators revealed by transcriptome analysis and metabolite profiles of tomato fruits engineered to accumulate spermidine and spermine. *Plant Biotechnology* 24: 57–70. doi:10.5511/plantbiotechnology.24.57.

Srivastava, S., Singh, R. K., Pathak, G., et al. (2016). Comparative transcriptome analysis of unripe and mid-ripe fruit of *Mangifera indica* (var. 'Dashehari') unravels ripening associated genes. *Scientific Reports* 6: 32557. doi:10.1038/srep32557.

Sugiura, A., Ohkuma, T., Choi, Y. A., et al. (2000). Production of nonaploid (2n=9x) Japanese persimmons (*Diospyros kaki*) by pollination with unreduced (2n=6x) pollen and embryo rescue culture. *Journal of the American Society for Horticultural Science* 125: 609–14.

Sugiyama, K. and Morishita, M. (2000). Production of seedless watermelon using soft-X-irradiated pollen. *Scientia Horticulturae* 84: 255–64.

Sugiyama, K., Morishita, M. and Nishino, E. (2002). Seedless watermelons produced via soft-X-irradiated pollen. *Hortscience* 37: 292–5.

Suh, S. J., Lee, S. H., Lee, D. H., et al. (2013). Transcriptome analysis of a spontaneous reddish mutant in Miyagawa Wase satsuma mandarin. *Journal of the Korean Society for Applied Biological Chemistry* 56: 391–9. doi:10.1007/s13765-013-3085-x.

Sun, L. M., Zhang, J. Z., Mei, L., et al. (2014). Molecular cloning, promoter analysis and functional characterization of *APETALA 1*-like gene from precocious trifoliate orange (*Poncirus trifoliata* L. Raf.). *Scientia Horticulturae* 178: 95–105.

Sun, T. H., Yuan, H., Cao, H. B., et al. (2018). Carotenoid metabolism in plants: The role of plastids. *Molecular Plant* 11: 58–74.

Sung, S. K., Yu, G. H. and An, G. H. (1999). Characterization of *MdMADS2*, a member of the *SQUAMOSA* subfamily of genes, in apple. *Plant Physiology* 120: 969–78.

Sung, S. K., Yu, G. H., Nam, J., et al. (2000). Developmentally regulated expression of two MADS-box genes, *MdMADS3* and *MdMADS4*, in the morphogenesis of flower buds and fruits in apple. *Planta* 210: 519–28.

Sweetman, C., Wong, D. C. J., Ford, C.M., et al. (2012). Transcriptome analysis at four developmental stages of grape berry (*Vitis vinifera* cv. Shiraz) provides insights into regulated and coordinated gene expression. *BMC Genomics* 13: 691. doi:10.1186/1471-2164-13-691.

Szymanski, J., Levin, Y., Savidor, A., et al. (2017). Label-free deep shotgun proteomics reveals protein dynamics during tomato fruit tissues development. *Plant Journal* 90: 396–417. doi:10.1111/tpj.13490.

Tafolla-Arellano, J. C., Zheng, Y., Sun, H. H., et al. (2017). Transcriptome analysis of mango (*Mangifera indica* L.) fruit epidermal peel to identify putative cuticle-associated genes. *Scientific Reports* 7: 46163. doi:10.1038/srep46163.

Takos, A. M., Jaffe, F. W., Jacob, S. R., et al. (2006). Light-induced expression of a *MYB* gene regulates anthocyanin biosynthesis in red apples. *Plant Physiology* 142: 1216–32.

Tanaka, N., Ureshino, A., Shigeta, N., et al. (2014). Overexpression of *Arabidopsis FT* gene in apple leads to perpetual flowering. *Plant Biotechnology* 31: 11–20.

Tanaka, N., Tanaka-Moriya, Y., Mimida, N., et al. (2016). The analysis of transgenic apples with down-regulated expression of *MdPISTILLATA*. *Plant Biotechnology* 33: 395–401.

Tang, X. F., Tang, Z. Z., Huang, S. X., et al. (2013). Whole transcriptome sequencing reveals genes involved in plastid/chloroplast division and development are regulated by the *HP1/DDB1* at an early stage of tomato fruit development. *Planta* 238: 923–36. doi:10.1007/s00425-013-1942-9.

Tang, N., Deng, W., Hu, G., et al. (2015). Transcriptome profiling reveals the regulatory mechanism underlying pollination dependent and parthenocarpic fruit set mainly mediated by auxin and gibberellin. *PLoS One* 10: e0125355.

Tang, W., Zheng, Y., Dong, J., et al. (2016). Comprehensive transcriptome profiling reveals long noncoding RNA expression and alternative splicing regulation during fruit development and ripening in kiwifruit (*Actinidia chinensis*). *Frontiers in Plant Science* 7: 335. doi:10.3389/fpls.2016.00335.

Terrier, N., Glissant, D., Grimplet, J., et al. (2005). Isogene specific oligo arrays reveal multifaceted changes in gene expression during grape berry (*Vitis vinifera* L.) development. *Planta* 222: 832–47. doi:10.1007/s00425-005-0017-y.

Theissen, G., Melzer, R. and Rumpler, F. (2016). MADS-domain transcription factors and the floral quartet model of flower development: Linking plant development and evolution. *Development* 143: 3259–71.

Thouet, J., Quinet, M., Ormenese, S., et al. (2008). Revisiting the involvement of *SELF-PRUNING* in the sympodial growth of tomato. *Plant Physiology* 148: 61–4.

Tohge, T., de Souza, L. P. and Fernie, A. R. (2017). Current understanding of the pathways of flavonoid biosynthesis in model and crop plants. *Journal of Experimental Botany* 68: 4013–28.

Toledo, T. T., Nogueira, S. B., Cordenunsi, B. R., et al. (2012). Proteomic analysis of banana fruit reveals proteins that are differentially accumulated during ripening. *Postharvest Biology and Technology* 70: 51–8. doi:10.1016/j.postharvbio.2012.04.005.

Trainotti, L., Bonghi, C., Ziliotto, F., et al. (2006). The use of microarray μPEACH1.0 to investigate transcriptome changes during transition from pre-climacteric to climacteric phase in peach fruit. *Plant Science* 170: 606–13. doi:10.1016/j.plantsci.2005.10.015.

Trankner, C., Lehmann, S., Hoenicka, H., et al. (2010). Over-expression of an *FT*-homologous gene of apple induces early flowering in annual and perennial plants. *Planta* 232: 1309–24.

Ueta, R., Abe, C., Watanabe, T., et al. (2017). Rapid breeding of parthenocarpic tomato plants using CRISPR/Cas9. *Scientific Reports* 7: 507.

Vecchietti, A., Lazzari, B., Ortugno, C., et al. (2009). Comparative analysis of expressed sequence tags from tissues in ripening stages of peach (*Prunus persica* L. Batsch). *Tree Genetics & Genomes* 5: 377–91. doi:10.1007/s11295-008-0193-6.

Vega, A., Gutierrez, R. A., Pena-Neira, A., et al. (2011). Compatible GLRaV-3 viral infections affect berry ripening decreasing sugar accumulation and anthocyanin biosynthesis in *Vitis vinifera*. *Plant Molecular Biology* 77: 261–74. doi:10.1007/s11103-011-9807-8.

Velazquez, K., Aguero, J., Vives, M. C., et al. (2016). Precocious flowering of juvenile citrus induced by a viral vector based on *Citrus leaf blotch virus*: A new tool for genetics and breeding. *Plant Biotechnology Journal* 14: 1976–85.

Wang, N. and Kao, T. H. (2012). Self-incompatibility in *Petunia*: A self/nonself-recognition mechanism employing S-locus F-box proteins and S-RNase to prevent inbreeding. *Wiley Interdisciplinary Reviews-Developmental Biology* 1: 267–75.

Wang, W., Bianchi, L., Scali, M., et al. (2009). Proteomic analysis of β-1,3-glucanase in grape berry tissues. *Acta Physiologiae Plantarum* 31: 597–604. doi:10.1007/s11738-008-0269-9.

Wang, X. C., Kayesh, E., Han, J., et al. (2014). Microarray analysis of differentially expressed genes engaged in fruit development between table and wine grape. *Molecular Biology Reports* 41: 4397–412. doi:10.1007/s11033-014-3311-6.

Wang, L., Yin, X., Cheng, C., et al. (2015). Evolutionary and expression analysis of a MADS-box gene superfamily involved in ovule development of seeded and seedless grapevines. *Molecular Genetics and Genomics* 290: 825–46.

Wang, R. H., Yuan, X. Y., Meng, L. H., et al. (2016). Transcriptome analysis provides a preliminary regulation route of the ethylene signal transduction component, *SlEIN2*, during tomato ripening. *PLoS One* 11: e0168287. doi:10.1371/journal.pone.0168287.

Wang, J. H., Liu, J. J., Chen, K. L., et al. (2017a). Comparative transcriptome and proteome profiling of two *Citrus sinensis* cultivars during fruit development and ripening. *BMC Genomics* 18: 984. doi:10.1186/s12864-017-4366-2.

Wang, S. M., Li, W. J., Liu, Y. X., et al. (2017b). Comparative transcriptome analysis of shortened fruit mutant in woodland strawberry (*Fragaria vesca*) using RNA-Seq. *Journal of Integrative Agriculture* 16: 828–44. doi:10.1016/s2095-3119(16)61448-x.

Wang, W., Khalil-Ur-Rehman, M., Feng, J., et al. (2017c). RNA-seq based transcriptomic analysis of CPPU treated grape berries and emission of volatile compounds. *Journal of Plant Physiology* 218: 155–66. doi:10.1016/j.jplph.2017.08.004.

Wang, X. C., Zhao, M. Z., Wu, W. M., et al. (2017d). Comparative transcriptome analysis of berry-sizing effects of gibberellin (GA_3) on seedless *Vitis vinifera* L. *Genes & Genomics* 39: 493–507. doi:10.1007/s13258-016-0500-9.

Wang, Z., Du, H., Zhai, R., et al. (2017e). Transcriptome analysis reveals candidate genes related to color fading of 'Red Bartlett' (*Pyrus communis* L.). *Frontiers in Plant Science* 8: 455. doi:10.3389/fpls.2017.00455.

Wang, Z. R., Cui, Y. Y., Vainstein, A., et al. (2017f). Regulation of fig (*Ficus carica* L.) fruit color: Metabolomic and transcriptomic analyses of the flavonoid biosynthetic pathway. *Frontiers in Plant Science* 8: 1990. doi:10.3389/fpls.2017.01990.

Waters, D. L. E., Holton, T. A., Ablett, E. M., et al. (2006). The ripening wine grape berry skin transcriptome. *Plant Science* 171: 132–8. doi:10.1016/j.plantsci.2006.03.002.

Wei, H., Chen, X., Zong, X., et al. (2015). Comparative transcriptome analysis of genes involved in anthocyanin biosynthesis in the red and yellow fruits of sweet cherry (*Prunus avium* L.). *PLoS One* 10: e0121164. doi:10.1371/journal.pone.0121164.

Weigel, D., Alvarez, J., Smyth, D. R., et al. (1992). *LEAFY* controls floral meristem identity in Arabidopsis. *Cell* 69: 843–59.

Wells, C. E., Vendramin, E., Tarodo, S. J., et al. (2015). A genome-wide analysis of MADS-box genes in peach [*Prunus persica* (L.) Batsch]. *BMC Plant Biology* 15: 41.

Wen, L. H., Zhong, W. J., Huo, X. M., et al. (2016). Expression analysis of ABA- and GA-related genes during four stages of bud dormancy in Japanese apricot (*Prunus mume* Sieb. et Zucc). *Journal of Horticultural Science & Biotechnology* 91: 362–9.

Wisniewski, M., Norelli, J. and Artlip, T. (2015). Overexpression of a peach CBF gene in apple: A model for understanding the integration of growth, dormancy, and cold hardiness in woody plants. *Frontiers in Plant Science* 6: 85.

Wong, D. C. J., Gutierrez, R. L., Dimopoulos, N., et al. (2016). Combined physiological, transcriptome, and cis-regulatory element analyses indicate that key aspects of ripening, metabolism, and transcriptional program in grapes (*Vitis vinifera* L.) are differentially modulated accordingly to fruit size. *BMC Genomics* 17: 416. doi:10.1186/s12864-016-2660-z.

Wu, B. H., Cao, Y. G., Guan, L., et al. (2014a). Genome-wide transcriptional profiles of the berry skin of two red grape cultivars (*Vitis vinifera*) in which anthocyanin synthesis is sunlight-dependent or -independent. *PLoS One* 9: e105959. doi:10.1371/journal.pone.0105959.

Wu, H. X., Jia, H. M., Ma, X. W., et al. (2014b). Transcriptome and proteomic analysis of mango (*Mangifera indica* Linn) fruits. *Journal of Proteomics* 105: 19–30. doi:10.1016/j.jprot.2014.03.030.

Wu, J. X., Xu, Z. L., Zhang, Y. J., et al. (2014c). An integrative analysis of the transcriptome and proteome of the pulp of a spontaneous late-ripening sweet orange mutant and its wild type improves our understanding of fruit ripening in citrus. *Journal of Experimental Botany* 65: 1651–71. doi:10.1093/jxb/eru044.

Wu, Y. Y., Fu, T., Wang, Z. H, et al. (2015). Differential gene expression analysis of early-ripening mutants of grape (*Vitis vinifera* L.). *Scientia Horticulturae* 194: 7–17. doi:10.1016/j.scienta.2015.07.022.

Wu, X. Q., Jiang, L., Yu, M. L., et al. (2016). Proteomic analysis of changes in mitochondrial protein expression during peach fruit ripening and senescence. *Journal of Proteomics* 147: 197–211. doi:10.1016/j.jprot.2016.06.005.

Xia, Y., Kuan, C., Chiu, C. H., et al. (2016). Global transcriptomic analysis of targeted silencing of two paralogous ACC oxidase genes in banana. *International Journal of Molecular Sciences* 17: 1632. doi:10.3390/ijms17101632.

Xie, M., Huang, Y., Zhang, Y. P., et al. (2013). Transcriptome profiling of fruit development and maturation in Chinese white pear (*Pyrus bretschneideri* Rehd). *BMC Genomics* 14: 823. doi:10.1186/1471-2164-14-823.

Xue, Y. B., Carpenter, R., Dickinson, H. G., et al. (1996). Origin of allelic diversity in antirrhinum S locus RNases. *Plant Cell* 8: 805–14.

Yakushiji, H., Kobayashi, S., Goto-Yamamoto, N., et al. (2006). A skin color mutation of grapevine, from black-skinned Pinot Noir to white-skinned Pinot Blanc, is caused by deletion of the functional *VvmybA1* allele. *Bioscience, Biotechnology, and Biochemistry* 70: 1506–8.

Yamada, A. and Tao, R. (2007). Controlled pollination with sorted reduced and unreduced pollen grains reveals unreduced embryo sac formation in *Diospyros kaki* Thunb. 'Fujiwaragosho'. *Journal of the Japanese Society for Horticultural Science* 76: 133–8.

Yamada, A., Tao, R. and Sugiura, A. (2005). Influence of low temperature before flowering on the occurrence of unreduced pollen in Japanese persimmon (*Diospyros kaki* thunb.). *Hortscience* 40: 24–8.

Yamagishi, N., Sasaki, S., Yamagata, K., et al. (2011). Promotion of flowering and reduction of a generation time in apple seedlings by ectopical expression of the *Arabidopsis thaliana FT* gene using the *Apple latent spherical virus* vector. *Plant Molecular Biology* 75: 193–204.

Yamane, H. (2014). Regulation of bud dormancy and bud break in Japanese apricot (*Prunus mume* Siebold & Zucc.) and peach [*Prunus persica* (L.) Batsch]: A summary of recent studies. *Journal of the Japanese Society for Horticultural Science* 83: 187–202.

Yamane, H., Ichiki, M., Tao, R., et al. (2008). Growth characteristics of a small-fruit dwarf mutant arising from bud sport mutation in Japanese persimmon (*Diospyros kaki* thunb.). *Hortscience* 43: 1726–30.

Yan, G. J., Ferguson, A. R., McNeilage, M. A., et al. (1997). Numerically unreduced (2*n*) gametes and sexual polyploidization in *Actinidia*. *Euphytica* 96: 267–72.

Yano, R., Nonaka, S. and Ezura, H. (2018). Melonet-DB, a grand RNA-seq gene expression atlas in melon (*Cucumis melo* L.). *Plant and Cell Physiology* 59: e4. doi:10.1093/pcp/pcx193.

Yao, J. L., Dong, Y. H., Kvarnheden, A., et al. (1999). Seven MADS-box genes in apple are expressed in different parts of the fruit. *Journal of the American Society for Horticultural Science* 124: 8–13.

Yao, J. L., Dong, Y. H. and Morris, B. A. M. (2001). Parthenocarpic apple fruit production conferred by transposon insertion mutations in a MADS-box transcription factor. *Proceedings of the National Academy of Sciences of the United States of America* 98: 1306–11.

Ye, J, Hu, T. X., Yang, C. M., et al. (2015). Transcriptome profiling of tomato fruit development reveals transcription factors associated with ascorbic acid, carotenoid and flavonoid biosynthesis. *PLoS One* 10: e0130885. doi:10.1371/journal.pone.0130885.

Yeats, T. H., Howe, K. J., Matas, A. J., et al. (2010). Mining the surface proteome of tomato (*Solanum lycopersicum*) fruit for proteins associated with cuticle biogenesis. *Journal of Experimental Botany* 61: 3759–71. doi:10.1093/jxb/erq194.

Yin, X. R., Xie, X. L., Xia, X. J., et al. (2016). Involvement of an ethylene response factor in chlorophyll degradation during citrus fruit degreening. *Plant Journal* 86: 403–12.

Yordanov, Y. S., Ma, C., Strauss, S. H., et al. (2014). EARLY BUD-BREAK 1 (*EBB1*) is a regulator of release from seasonal dormancy in poplar trees. *Proceedings of the National Academy of Sciences of the United States of America* 111: 10001–6.

Yu, K. Q., Xu, Q., Da, X. L., et al. (2012). Transcriptome changes during fruit development and ripening of sweet orange (*Citrus sinensis*). *BMC Genomics* 13: 10. doi:10.1186/1471-2164-13-10.

Yuan, H., Zhang, J. X., Nageswaran, D., et al. (2015). Carotenoid metabolism and regulation in horticultural crops. *Horticulture Research* 2: 15036.

Zeng, Y. L., Pan, Z. Y., Ding, Y. D., et al. (2011). A proteomic analysis of the chromoplasts isolated from sweet orange fruits *Citrus sinensis* (L.) Osbeck. *Journal of Experimental Botany* 62: 5297–309. doi:10.1093/jxb/err140.

Zeng, J. W., Gao, C. Y., Deng, G. M., et al. (2012). Transcriptome analysis of fruit development of a citrus late-ripening mutant by microarray. *Scientia Horticulturae* 134: 32–9. doi:10.1016/j.scienta.2011.10.018.

Zeng, Y. L., Pan, Z. Y., Wang, L., et al. (2014). Phosphoproteomic analysis of chromoplasts from sweet orange during fruit ripening. *Physiologia Plantarum* 150: 252–70. doi:10.1111/ppl.12080.

Zhang, X. Z., Liu, G. J. and Zhang, D. M. (1998). Occurrence and cytogenetic development of unreduced pollen in *Vitis*. *Vitis* 37: 63–5.

Zhang, J. W., Ma, H. Q., Feng, J. D., et al. (2008). Grape berry plasma membrane proteome analysis and its differential expression during ripening. *Journal of Experimental Botany* 59: 2979–990. doi:10.1093/jxb/ern156.

Zhang, Y. J., Wang, X. J., Wu, J. X., et al. (2014). Comparative transcriptome analyses between a spontaneous late-ripening sweet orange mutant and its wild type suggest the functions of ABA, sucrose and JA during citrus fruit ripening. *PLoS One* 9: e116056. doi:10.1371/journal.pone.0116056.

Zhang, H., Wang, H. S., Yi, H. P., et al. (2016a). Transcriptome profiling of *Cucumis melo* fruit development and ripening. *Horticulture Research* 3: 16014. doi:10.1038/hortres.2016.14.

Zhang, M. Y., Xue, C., Xu, L. L., et al. (2016b). Distinct transcriptome profiles reveal gene expression patterns during fruit development and maturation in five main cultivated species of pear (*Pyrus* L.). *Scientific Reports* 6: 28130. doi:10.1038/srep28130.

Zheng, Q. F., Song, J., Campbell-Palmer, L., et al. (2013). A proteomic investigation of apple fruit during ripening and in response to ethylene treatment. *Journal of Proteomics* 93: 276–94. doi:10.1016/j.jprot.2013.02.006.

Zheng, C., Halaly, T., Acheampong, A. K., et al. (2015). Abscisic acid (ABA) regulates grape bud dormancy, and dormancy release stimuli may act through modification of ABA metabolism. *Journal of Experimental Botany* 66: 1527–42.

Zhou, Y. Z., Xu, Z. D., Yong, X., et al. (2017). SEP-class genes in *Prunus mume* and their likely role in floral organ development. *BMC Plant Biology* 17: 10.

Zhu, X. Y., Chen, J. Y., Qiu, K., et al. (2017a). Phytohormone and light regulation of chlorophyll degradation. *Frontiers in Plant Science* 8: 1911.

Zhu, Q. L., Gao, P., Liu, S., et al. (2017b). Comparative transcriptome analysis of two contrasting watermelon genotypes during fruit development and ripening. *BMC Genomics* 18: 3. doi:10.1186/s12864-016-3442-3.

Ziliotto, F., Begheldo, M., Rasori, A., et al. (2008). Transcriptome profiling of ripening nectarine (*Prunus persica* L. Batsch). fruit treated with 1-MCP. *Journal of Experimental Botany* 59: 2781–91. doi:10.1093/jxb/ern136.

Chapter 5

Advances in fruit genetics

Ksenija Gasic and Christopher Saski, Clemson University, USA

1 Introduction

Environmental challenges, changes in production systems and human preferences are driving the need for delivery of new cultivars for economical and sustainable production (Byrne, 2005; Bliss, 2006). The latest innovations in science and technology are integrated in plant breeding to provide sustained solutions to production challenges and market demands. Climate-change resilience is becoming one of the most topical and concerning issues among plant scientists (Kole et al., 2015). Elastic climatic changes and environmental instabilities are imparting global ecosystem disruptions, and the sustainability and oftentimes feasibility of fruit production is being challenged by increased abiotic and biotic stresses. Therefore, understanding the genetics of the key traits of importance for fruit tree production and sustainability is the main generator of advances in breeding and development of new cultivars that can address changing climate and needs of the fruit industry (Badenes et al., 2016).

Characterization and utilization of germplasm collections, with an emphasis on preservation of genetic diversity found in nature, is in demand by fruit breeders who are required to address climate challenges by developing cultivars that are productive, nutritious, tolerant to biotic and abiotic stresses, and efficient in use of water and fertilizer (Byrne et al., 2018). Emerging abiotic and biotic threats to fruit production sustainability often can be addressed only

http://dx.doi.org/10.19103/AS.2018.0040.05

by introgression of alleles from wild relatives (Li and Liu, 2016; Migicovsky and Myles, 2017), which is time consuming and typically difficult to achieve.

Availability of high-quality genome sequences of many fruit crops (Chagné, 2015; Sharma et al., 2016) provides an opportunity through a new lens to investigate the effects and occurrence of whole-genome duplication and the evolutionary significance of ploidy (van der Peer et al., 2017). Many fruit crops have undergone various scales of genome duplications and subsequent fractionation leaving unanswered questions about the copy number and functional significance of genes involved in the traits of interest. Understanding which genes undergo duplication and are preserved following duplication may provide insights into their importance and biological role (Li et al., 2016). The complexity of fruit genomes is often perceived as an obstacle in the discovery and implementation of new molecular tools and technologies.

Transcriptome sequencing is now a common approach to measure and compare gene expression profiles at a given time under specific conditions (e.g. developmental or treatment time course), and has been an avenue for researchers to understand the functionality behind the existing or newly developed genomic resources and the genetic mechanisms underlying trait genetics (Ong et al., 2012; Socquet-Juglard et al., 2013b; Wang et al., 2013; Li et al., 2015; González et al., 2016; Tang et al., 2016; Jazi et al., 2017; Kose et al., 2017; Guo et al., 2018). Functional and comparative genomics studies leverage conservation of genomic collinearity among species and genera, and often are utilized in the development of new cultivars with desirable traits via marker saturation and candidate gene searches (Arús et al., 2005; Jung et al., 2006, 2009; Tang et al., 2008). With the demonstrated power of comparative genomics, we expect to see a proportional increase in comparative studies as the resources on both genomic and transcriptome sequences increase.

Routine DNA-informed breeding has become a reality via translation of new discoveries into practical breeding-friendly tools and knowledge (Peace, 2017). Challenges in understanding genetics of fruit quality and disease resistance, and application of genomics tools available for cultivar development, will be addressed in more detail. Most of the examples will cover Rosaceae species presented in this book; however, notable discoveries in other fruit species will be mentioned wherever possible.

2 Genomic mapping

Some of the major obstacles preventing more progressive advances in genomic mapping in fruit species include perennial nature, long juvenile period and large genome size. In the last decade, development of molecular resources has become less time consuming, labour intensive and costly, thus allowing for a steep increase in accumulation of genetic resources in fruit crops (Jung and Main, 2014; Jung et al., 2018) (www.rosaceae.org; www.citrusgenomedb.org; www.vaccinium.org). Genotyping by sequencing (GBS) through massively parallel next-generation sequencing (NGS) technologies (Elshire et al., 2011), initially applied to field crops (Barabaschi et al., 2016), have enabled much faster and more affordable discoveries in fruit genetics (Ward et al., 2013; Bielenberg et al., 2015; Gurcan et al., 2016; Salazar et al., 2017; Goonetilleke et al., 2018).

The last few years have been really productive in the release of cornerstone genomic sequences of fruit crops. Since the initial summary of the available genome sequences for fruit crops (Chagné, 2015; Sharma et al., 2016), several new fruit genomes have been released or reported as being at the final stage of assembly (Table 1). The availability of high-quality genome sequences of many fruit crops provided the needed resources for discovery of markers necessary for linkage and QTL mapping (Klagges et al., 2013; Martínez-García et al., 2013a,b; Chen et al., 2015; Scaglione et al., 2015; Schlautman et al., 2017), development of high-throughput genotyping platforms (Chagné et al., 2012b; Peace et al., 2012; Verde et al., 2012; Bianco et al., 2014; Bassil et al., 2015) and practical application via DNA-informed breeding (Cuenca et al., 2016; Meneses et al., 2016; Wang et al., 2016; Peace, 2017).

Rapidly expanding genome sequencing also facilitated the candidate gene discovery approach and comparative studies within and between genera that have translational potential to benefit fruit crops without extensive genetic resources (Dirlewanger et al., 2004; Bernet et al., 2010; Illa et al., 2011; Koepke et al., 2013; Cao et al., 2014; Goldy et al., 2017; Schlautman et al., 2018).

Large single nucleotide polymorphism (SNP) arrays, such as Illumina (Peace et al., 2012; Verde et al., 2012; Bianco et al., 2014) and Affymetrix Axiom® SNP arrays (Bassil et al., 2015; Bianco et al., 2016), are becoming the standard genotyping tool in the development of high-density linkage maps. Those high-density linkage maps, usually based on the bi-parental populations, could then be used to develop consensus linkage maps providing higher marker density and precision in downstream applications (Di Pierro et al., 2016; Khan et al., 2012; da Silva Linge et al., 2018). Consensus linkage maps may also serve as a reference map for the species to enable estimation of genetic position of unmapped markers. This is especially useful in newly emerging mapping approaches where only information about marker genetic position is needed without the actual linkage mapping effort.

Pedigree-based (Bink et al., 2012, 2014; Allard et al., 2016; Fresnedo-Ramirez et al., 2016b; Mangandi et al., 2017) and/or genome-wide (Korte and Farlow, 2013; Navarro et al., 2017) mapping strategies have recently surfaced as complementary techniques in QTL and association mapping that mitigate constraints of bi-parental linkage and QTL mapping where recombination

Table 1 Whole-genome sequences of fruit crops

Common name	Genus	Species	Cultivar/accession	Ploidy	Genome size/ Assembled (Mb)	References
Almond	*Prunus*	*dulcis*	Nonpareil	2n = 16		Fresnedo (*pers. comm.*)
Almond	*Prunus*	*dulcis*	Texas	2n = 16		Alioto et al. (2016); www.rosaceae.org
Apple	*Malus*	× *domestica*	Golden Delicious_DH	2n = 34	750/604	Velasco et al. (2010)
Banana	*Musa*	*acuminata*	Pahang_DH	2n = 22	523/472	D'Hont et al. (2012)
Banana	*Musa*	*balbisiana*	Pisang Klutuk Wulung (PKW)			Davey et al. (2013)
Black raspberry	*Rubus*	*occidentalis*	ORUS 4115-3 (PI 672644)	2n = 14	283/243	VanBuren et al. (2016)
Blueberry	*Vaccinium*	*corymbosum*	W8520	2n = 24	500/-	Gupta et al. (2015)
Chinese plum	*Prunus*	*mume*	BJFU1210120008	2n = 16	280/237	Zhang et al. (2012)
Citrus	*Citrus*	× *clementina*	Clemenules	2n = 18	367/301	
Cocoa	*Theobroma*	*cacao*	B97-61/B2	2n = 20	430/327	Argout et al. 2011
Cocoa	*Theobroma*	*cacao*	Matina 1-6		445/346	
Coffee	*Coffea*	*canephora*	DH200-94	2n = 22	710/568	Denoeud et al. (2014)
Cranberry	*Vaccinium*	*macrocarpon*	Ben Lear clone CNJ99-125-1	2n = 24	470/420	Polashock et al. (2014)
Date palm	*Phoenix*	*dactylifera*	Khalas	2n = 36	660/558	Al-Dous et al. (2011); Al-Mssallem et al. (2013)
Grape	*Vitis*	*vinifera*	Pinot noir (PN40024)	2n = 38	475/481	Jaillon et al. (2007); Velasco et al. (2007)
Hazelnut	*Corylus*	*avellana*	Jefferson	2n = 22	379/345	Rowley et al. (2012)

Jujube	*Ziziphus*	*jujuba*	Dongzao	2n = 24	444/437	Liu et al. (2014)
Kiwi	*Actinidia*	*chinensis*	Hongyang	2n = 58	758/616	Huang et al. (2013)
Macadamia	*Macadamia*	*integrifolia*	Mauka	2n = 28	652/518	Nock et al. (2014)
Mango	*Magnifera*	*indica*	Amrapali	4n = 40	439/432	Singh et al. (2016)
Mulberry	*Morus*	*notabilis*	NA	2n = 14	357/330	He et al. (2013)
Papaya	*Carica*	*papaya*	SunUp	2n = 18	372/370	Ming et al. (2008)
Peach	*Prunus*	*persica*	Lovell_DH	2n = 16	265/223	Verde et al. (2013, 2017)
European pear	*Pyrus*	*communis*	Bartlett	2n = 34	600/577	Chagné et al. (2014)
Pear	*Pyrus*	× *bretschneiderei*	Dangshansuli	2n = 34	527/512	Wu et al. (2012)
Persian walnut	*Juglans*	*regia*	Chandler	2n = 32	606/667	Martínez-Garcia et al. (2016)
Pineapple	*Ananas*	*comosus*	F153	2n = 50	526/375	Zhang et al. (2014)
Strawberry	*Fragaria*	× *ananassa*	Reikou	2n = 8× = 56	720/689	Hirakawa et al. (2013)
Strawberry	*Fragaria*	*vesca*	Hawaii-4 (PI551572)	2n = 14	240/209.8	Shulaev et al. (2010)
Sweet cherry	*Prunus*	*avium*	Satonishiki	2n = 16	353/272.4	Shirasawa et al. (2017)
Sweet cherry	*Prunus*	*avium*	Regina	2n = 16	353/277	Dirlewanger et al. (2018)
Sweet orange	*Citrus*	*sinensis*	Valencia	2n = 18	367/320	Xu et al. (2013)

events are provided by only two parental genotypes. Combination of mapping approaches provides necessary depth and precision to understand the genetics behind the traits of interest in any material available, segregating population, pedigreed germplasm or unrelated germplasm (Picanol et al., 2013).

Genome-wide association study (GWAS) has been successfully applied in animal breeding and is gaining momentum in plants, especially for highly quantitative traits (that are under the influence of many genes with small effects) where other mapping techniques fail to provide results (Brachi et al., 2011; Khan and Korban, 2012). Decreases in genotyping prices, and the availability of many different genotyping platforms/techniques and service providers, have eliminated bottlenecks in obtaining cost-effective genotyping information for large germplasm collections. Perennial fruit crops often have decades of phenotypic data that can now be used in GWAS to reconnect particular traits with their underlying genetics across the genome. This approach has been recently applied to fruit crops in understanding fruit quality traits in citrus (*Citrus* sp.) (Minamikawa et al., 2017; Imai et al., 2018), peach (*Prunus persica*) (Cao et al., 2016) and apple (*Malus* × *domestica* Borkh.) (Kumar et al., 2013, 2015), as well as flowering and ripening periods in apple (Urrestarazu et al., 2017) and diversity and genetic structure in peach (Micheletti et al., 2015). As association mapping in fruit crops becomes more popular, the need to understand the limitations of the different models and minimum threshold necessary for number of markers and individuals for analyses is increasing (Khan and Korban, 2012).

Genomic research offers unprecedented discovery potential driven by the massive influx of molecular information of ultra-high resolution. With accessibility and affordability of latest NGS technologies, SNPs are becoming the genotyping system of choice in genetic studies. However, being typically biallelic, individually they provide little resolution for understanding SNP-trait relationships. Therefore, haplotypes constructed based on linkage disequilibrium (LD) (Qian et al., 2017) or pedigree (Cai et al., 2017) can complement both quantitative trait analysis and genomic selection. Within the RosBREED project (www.rosbreed.org), efforts to provide haploblock maps for apple, cherry and peach are under way (Vanderzande et al., 2019). Preliminary results suggest only 104 haploblocks in 8 chromosomes of the peach genome (Fig. 1). The low number of haploblocks observed in the peach genome is in agreement with the recognized narrow genetic base of North American peach germplasm (Scorza and Sherman, 1996). When overlaid with the detected QTLs, haploblock maps could provide the overview of hot spots (regions where multiple QTLs are mapped) and bottlenecks (regions where low or no recombination exist). This information then can be used to attempt manipulation via breeding, for example introgression, and or biotechnology, for example genetic engineering or genome editing.

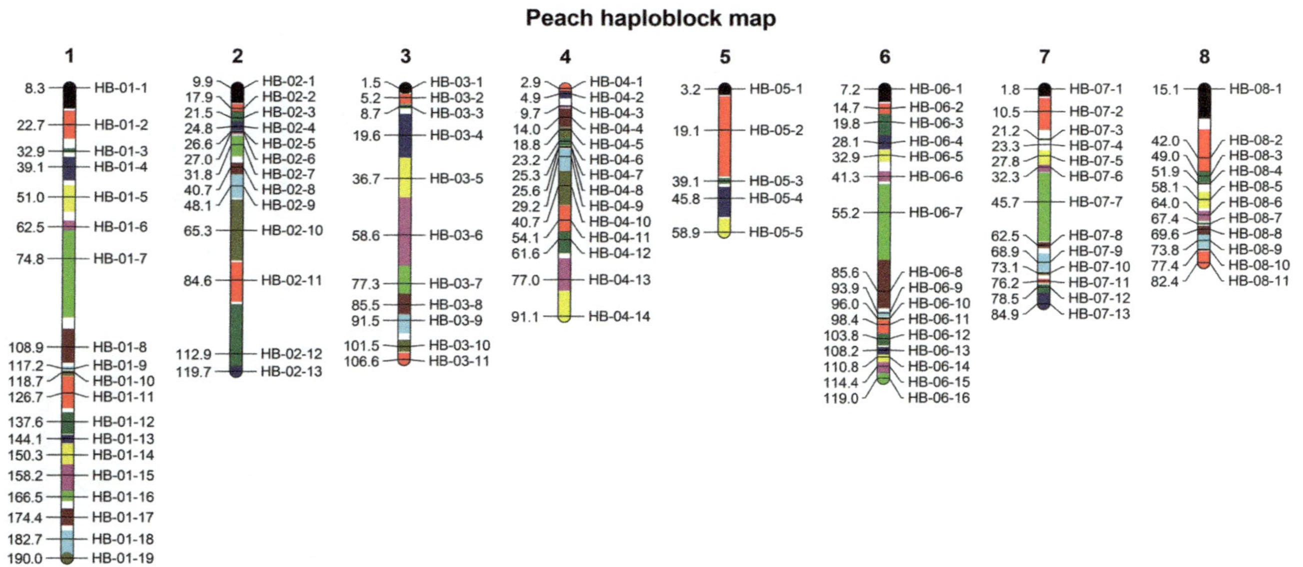

Figure 1 A haploblock map of the peach genome based on the pedigreed material genotyped with a 9K peach SNP array (Verde et al., 2012) in the RosBREED project (www.rosbreed.org). Haploblocks are colour coded; white area designates regions of recombination; HB – haploblock, followed by the chromosome number (e.g. 01) and block number (e.g. 1), for example HB-01-1. Physical position of haploblock mid-point in the peach genome in Mbp is shown on the left.

3 Marker-assisted breeding

Marker-assisted breeding (MAB) has always been the Holy Grail of plant breeders. The ability to assess the phenotype before it is expressed and in the early stages of the plant's life cycle is of ultimate importance since perennial fruit crops can take five or more years before the phenotype can be assessed in the field. Many QTLs for fruit quality and disease resistance are reported in fruit crops (Hancock, 2008; Bus et al., 2009; Verma et al., 2017) and are housed in genomic and genetic databases (Jung and Main, 2014; Jung et al., 2018) (www.rosaceae.org; www.citrusgenomedb.org/). Several collaborative attempts to translate that vast information into user-friendly informative DNA tests have been reported or are ongoing (www.rosbreed.org; www.fruitbreedomics.com). The promises, progress and prospects of using DNA-based information in breeding of Rosaceae and other horticultural crops have been recently summarized (Edge-Garza et al., 2015; Ru et al., 2015; Peace, 2017) and steps to translate promising QTL information to a trait-predictive DNA test have been proposed (Vanderzande et al., 2018).

Most of the reported DNA tests that are applied routinely in breeding programs are either associated with disease resistance (Jänsch et al., 2015; Baumgartner et al., 2016; Noh et al., 2018) and/or fruit quality (Longhi et al., 2013; Mitani et al., 2014; Stegmeir et al., 2015; Sandefur et al., 2016, 2017; Wang et al., 2016; Cruz-Rus et al., 2017; Noh et al., 2017). The development and validation of DNA tests is a slow process and very often results are not published, but rather are used in-house. There are various reasons for slow progress in the translation of marker-trait knowledge into application in breeding, some being the cost of application, and/or breeders' lack of trust and willingness to cull seedlings based on the DNA test (Edge-Garza et al., 2015; Peace, 2017).

Genome-scale predictions and their application in genomic selection are being explored in fruit breeding for their potential to overcome the limitations of marker-assisted selection based on major effect genes (Iwata et al., 2013; Viana et al., 2016; Biscarini et al., 2017; Gezan et al., 2017; Minamikawa et al., 2017; O'Connor et al., 2018). Genomic selection in fruit tree crops promises to enhance breeding efficiency via increased prediction accuracy and selection intensity, and decrease in generational interval (Muranty et al., 2015). This exciting possibility is being explored in many fruit crops where genomic assisted breeding can shorten the breeding cycle by aiding in faster selection of superior parental genotypes and maximizing genetic gain (Iwata et al., 2013; Minamikawa et al., 2017; Navarro et al., 2017).

Application of genomic prediction in plant breeding is still a fairly new approach and requires further investigation in the most appropriate model(s) to apply, and the most optimal number of markers and plants to use (Daetwyler et al., 2012; Desta and Ortiz, 2014; Stich and van Inghelandt, 2018). Preliminary reports on the factors affecting accuracy of genomic prediction in fruit crops suggested importance of relatedness between the reference and validation individuals, with prediction accuracy increasing in highly related material (Muranty et al., 2015; Gezan et al., 2017). In addition, studies in strawberry (*Fragaria* × *ananassa*) revealed potential of the replicated advanced selection trials to serve as reference/training populations for predicting parental performance in the next breeding cycle (Gezan et al., 2017; Osorio et al., 2018).

Genetic architecture of the trait of interest influences the size of the reference population needed to accurately estimate the SNP effects. The smaller the largest SNP effects are, the larger the reference population is needed (Muranty et al., 2015). Modelling for optimal number of markers in apricot (*Prunus armeniaca*) (Nisibi et al., 2018) and strawberry (Osorio et al., 2018) suggested that prediction accuracy of 88%–97% is achieved with ~500–1000 informative, non-redundant, randomly distributed markers. In addition, multicycle analysis in strawberry showed a steady increase of prediction accuracy in all traits as data were aggregated across cycles in the reference population (Gezan et al., 2017; Osorio et al., 2018).

4 Biotechnology

Conventional approaches to fruit tree breeding and genetic enhancement are time consuming and offer challenges that can quickly become compounded with long breeding cycles as a primary variable. Because genetic diversity is the key source for trait improvement in plants and time is not on our side, biotechnology and genome engineering approaches have an undeniable role as tools for the addition of novel desired traits to genetic resources (Badenes et al., 2016; Igarashi et al., 2016; Limera et al., 2017). These approaches hold great potential, and promise large strides in progress because of (1) the relatively short time necessary to observe an event and (2) the minimal impact on other beneficial traits.

As of this writing, the CRISPR/CAS9 technique is regarded as the most advanced tool for precision genome editing, and is considered an indispensable tool in biological research (Ma et al., 2016; Liu et al., 2017; Fernandez i Marti and Dodd, 2018). Compared to classical genetic engineering approaches, which require the use of selective agents (typically antibiotic or herbicide selection), the CRISPR/CAS9 system can modify a genome in a precise and predictable manner using targeted nucleases which do not leave plasmid sequences and selectable markers in the target genome. Precision genome editing is now emerging as a new plant breeding technology (NPBT) and has attracted the attention of politicians, stakeholders and citizens due to the revolutionary impact on agriculture. In fact, this approach holds a tremendous potential for fruit trees, such as peach, plum (*Prunus domestica* L.), cherry (*Prunus avium*), almond (*Prunus dulcis*) and others that utilize a rootstock system (Malyska et al., 2016; Dalla Costa et al., 2017). Because of the dual genetic system, rootstock and scion, the fruit on the scion of a rootstock endowed with genetic modification for enhancement regarding soil abiotic and biotic pressures would not be considered a genetically modified fruit, hence circumventing regulatory measures.

As noted above, breeding fruit trees with a combination of desirable traits, such as disease resistance and high-quality fruit, can take most of a career (e.g. between 15 and 20 years). Using gene knock-in genetic engineering, researchers were able to insert a gene native to the California Poplar (*Populus trichocarpa*), PtFt1, into plum and the birch $MADS_4$ transcription factor (TF), into apple to induce early and continued flowering (Srinivasan et al., 2012; Scorza et al., 2013; Schlathölter et al., 2018). Similarly RNAi silencing of apple *MdTFL1* in pear (*Pyrus communis* L.) reduced expression of both *PcTFL1-1* and *PcTFL1-2* in one transgenic pear line, resulting in a shorter juvenile period (Freiman et al., 2012). This transgenic system, called FasTrak or rapid cycle breeding, offers breeders the opportunity to make crosses all year long, and coupled with molecular marker technology, this system can reduce cultivar development time to less than half that of conventional breeding (Srinivasan et al., 2012).

Advances in improvement of fruit crops through genetic engineering for traits such as abiotic stress tolerance, disease resistance, reduction of the generation time, improving rooting ability, fruit quality and post-harvest shelf life have been summarized recently (Mir and Patel, 2018; Upadhyay, 2018). For example, abiotic pressures, such as drought, heat and salinity, are regulated by multiple signalling pathways that control common genes and biochemical pathways. TF genes are commonly used to endow abiotic resilience. In apple, improved drought tolerance was achieved by overexpressing the MdATG18a TF (Sun et al., 2018). Also in apple, expression of the MdMyb10 TF improved disease resistance by increasing production of flavonoids (Rihani et al., 2017). Genome editing has been used in woodland strawberry (*Fragaria vesca*) to determine gene function of auxin biosynthesis and sensing genes, in addition to understanding flower development (Martín-Pizarro and Posé, 2018; Zhou et al., 2018).

5 Genetics of fruit quality and disease resistance

The major aim in many genetic studies in fruit crops is to understand the genetic and physiological control of major fruit quality and disease resistance traits with a common goal to translate new discoveries into tools for breeding new varieties (www.fruitbreedomics.com; www.rosbreed.org). The recent rapid increase in availability of genomic and transcriptomic data for fruit crops has facilitated a better understanding of the genetics of agronomically important traits. As a result, the importance and role of TFs in the genetics of many fruit quality and disease resistance traits has been revealed (Degenhardt et al., 2005; Chagné et al., 2012a; Bianchi et al., 2015; Zhou et al., 2015; Kan et al., 2017; Allan and Espley, 2018; Zhang et al., 2018; Zuriaga et al., 2018).

Breeding perspectives on utility of *Prunus* TFs have recently been summarized by Bianchi et al. (2015). The authors provided an inventory of peach TFs at the whole genome level and discussed implications of this information on marker-assisted selection strategies for traits of agronomical interest in *Prunus*. Synteny and collinearity between genomes and transcriptomes of fruit crops, such as those in the *Rosaceae* family, aid in discovery and understanding of underlying genetics in fruit quality and disease resistance (Dirlewanger et al., 2004; Jung et al., 2006, 2009; Tang et al., 2008; Schlautman et al., 2018), as detailed in the case study below.

5.1 Fruit quality

Historically, fruit quality has been the most important trait for consumers and industry, therefore it has the highest priority in breeding programs (Crisosto et al., 2006; Gallardo et al., 2018; Gilbert et al., 2014; Olmstead et al., 2015). Fruit quality traits (e.g. appearance, taste, size) for many fruit crops are similar and have enabled collaborative efforts in understanding their genetics (www.rosbreed.org; www.fruitbreedomics.com). Genetics of fruit quality, in particular sweetness, acidity and phenolic compound content, in *Rosaceae* species has recently been investigated via mapping of candidate genes in peach (Le Dantec et al., 2010). Eight candidate genes were mapped in peach and 14 in strawberry, and co-localization between some of the candidate genes and QTLs for fruit quality has been identified.

Colouration of skin and/or flesh has been a topic of many genetics studies, especially since various market analyses suggested that consumers are drawn to higher and brighter coloured fruits. Skin and/or flesh colouration in the *Rosaceae* is governed by MYB TFs, as reported in apple (Chagné et al., 2007, 2012a,b; Telias et al., 2011; Zhang et al., 2014; Moriya et al., 2017), cherry (Sooriyapathirana et al., 2010; Stegmeir et al., 2015; Sandefur et al., 2016) and peach (Frett et al., 2014; Zhou et al., 2015; Sandefur et al., 2017). MYB TFs control not only pigmentation but are also involved in flavour and texture, and often influence health benefits of the crop. Recently, Allan and Espley (2018) summarized the potential of MYBs to influence the development of novel consumer traits in fruits and vegetables and discussed how breeding techniques may manipulate or create new alleles of key MYB TFs.

Fruit size is another very important trait that determines fruit market value, the bigger the size, the higher the market value. Despite its importance, the genetics of fruit size for many fruit crops is still not fully understood. The earliest insights into fruit size genetics suggested that cell number and size are the most important factors that determine the potential for achieving large size in optimal growing conditions (Scorza et al., 1991; Zhang et al., 2010; Johnson et al., 2011). The copy number regulator (CNR) gene family has been indicated in the genetics of plant and organ size, with the *FW2.2* gene controlling a major QTL for tomato (*Solanum lycopersicum*) fruit size (Guo and Simmons, 2011). Investigation of this gene in peach led to the discovery of 23 orthologous genes, two of which have been located within fruit size QTLs mapped in sweet cherry (De Franceschi et al., 2013; Cai et al., 2017; Qi et al., 2017).

In addition to appearance and size, taste and flavour have been among the most sought-after traits in fruit crops. Taste is often associated with level of sugar(s) or soluble solids concentration (SSC) and ratio between sweetness and acidity (Crisosto et al., 2006). Flavour, on the other hand, involves the combination of sugars, acids and a set of about 20 volatile chemicals (Klee, 2010). The complexity of these traits, difficulty in obtaining relevant phenotypic data and their mutual dependency require the coordinated approach of many disciplines (Sanchez et al., 2013).

Volatile profiles have been reported for apple (Rowan et al., 2009; Costa et al., 2013; Kumar et al., 2015), peach (Wang et al., 2009; Sanchez et al., 2012, 2013; Eduardo et al., 2013) and strawberry (Schwieterman et al., 2014; Cruz-Rus et al., 2017; Zhang et al., 2018). Strawberry fruit aroma has been researched extensively, with detection of some key flavour compounds that give sweet cherry and fresh peach-like notes to strawberry fruits (Zorrilla-Fontanesi et al., 2012; Sánchez-Sevilla et al., 2014; Ulrich et al., 2018; Yan et al., 2018).

Texture is also a very important and heavily researched trait in apple and peach (King et al., 2001; Haji et al., 2005; Longhi et al., 2013; Serra et al., 2017). Fruit texture, firmness and crispness are associated with the ethylene production pathway and polygalactorunase (Costa et al., 2010; Gu et al., 2016; Qian et al., 2016; Morgutti et al., 2017; Fernandez i Marti et al., 2018; Zhang et al., 2018). Understanding the role of ethylene in ripening and softening of climacteric fruits (Payasi and Sanwal, 2010) has benefited from the discovery of bud sport mutants in Japanese plum (*Prunus salicina*) (Minas et al., 2015; Fernandez i Marti et al., 2018). This mutant series is providing unique resources for understanding ripening and fruit softening in tree fruit species and could serve as a model for other *Prunus* species (Haji et al., 2005; Hayama et al., 2006; Begheldo et al., 2008; Pan et al., 2015).

5.2 Disease resistance

Genetic solutions for many diseases in fruit crops are becoming the only sustainable long-term option. Sources of resistance (R) genes quite often reside in uncultivated germplasm, extending the effort and time needed for introgression of R-genes into the acceptable background. This becomes an even bigger problem in perennial fruit species for which the generation cycle is very long. Traditional breeding has been slowly incorporating major R alleles into newly released cultivars. Genetic studies of disease responses are expanding in parallel with the improvement in understanding the plant disease interaction and our ability to phenotype plant response.

Nucleotide binding site leucine-rich repeat (NBS-LRR) disease resistance proteins have been the most extensively researched R-genes in plants (Arya et al., 2014; Jia et al., 2015; Zhong et al., 2015, 2018; Pandolfi et al., 2017). Evaluation of the NBS-LRRs in the five Rosaceae species (apple, pear, peach, Japanese apricot [*Prunus mume*] and strawberry) indicated species-specific duplication causing expansion of the NBS-LRR genes (Zhong et al., 2015). Genome-wide identification and expression analyses of NBS-LRR genes in apple identified a family of 1015 genes and indicated both tandem and segmental duplication events responsible for the expansion of the apple NBS-LRR gene family (Arya et al., 2014). Study of evolutionary processes of NBS-LRR genes in six *Fragaria* genomes revealed a total of 1134 NBS-LRRs with 866 multi-genes and 184 gene families, and implied that the NBS-LRR genes duplicated prior to divergence of the six species (Zhong et al., 2018).

The focus of many disease resistance genetic studies in temperate tree fruits and berries has been directed towards finding markers associated with disease response and providing breeding programmes with efficient tools for early selection. In total, 425 QTLs associated with biotic stress have been reported in *Rosaceae* and housed in GDR (www.rosaceae.org); apricot (52), apple (167), peach (156), pear (12), plum (3), sweet cherry (7), black (1) and red raspberries (*Rubus occidentalis* and *Rubus idaeus* subsp. *idaeus*, respectively) (1), and strawberry (10). Similar to the fruit quality genetic advances, the majority of new discoveries in disease resistance genetics in *Rosaceae* have not yet been incorporated in the database.

Great progress on understanding the genetics of disease resistance in stone fruits has been achieved, with 237 QTLs reported in apricot, peach, plum, sweet cherry and *Prunus* sp. The most researched are resistance to plum pox virus (Vilanova et al., 2003; Decroocq et al., 2005; Lalli et al., 2008; Rubio et al., 2008; Cirilli et al., 2016; Passaro et al., 2017; Decroocq et al., 2018), brown rot (*Monilinia* sp.) (Martínez-García et al., 2013a; Pacheco et al., 2014; Rungjindamai et al., 2014; Fresnedo-Ramirez et al., 2016a; Fu et al., 2018; Quilot-Turion et al., 2018), bacterial spot (*Xanthomonas arboricola* pv. *pruni*) (Socquet-Juglard et al., 2013a,b; Yang et al., 2013; Lamichhane, 2014; Gervasi et al., 2018) and powdery mildew (*Sphaerotheca pannosa* var. *persicae*) (Dirlewanger et al., 1996; Foulongne et al., 2002, 2003; Quilot et al., 2004; Salazar et al., 2016).

In addition to scion diseases, rootstock resistance to nematodes or peach tree short life (PTSL) syndrome has also been important (Lu et al., 1998; Yamamoto and Hayashi, 2002; Arús et al., 2004; Claverie et al., 2004; Lecouls et al., 2004; Gillen and Bliss, 2005; Blenda et al., 2007; Gainza et al., 2015). PTSL is a complex disease syndrome caused by different biotic, climatic and edaphic factors. The PTSL disease aetiology is associated with the presence of high population densities of ring nematode (*Criconemoides xenoplax* Raski syn. *Mesocriconema xenoplax* (Raski) Loof and de Grisse). The resulting injury and physiological response to nematode feeding increases susceptibility of peach trees to bacterial canker (*Pseudomonas syringae* pv. *syringae* van Hall) or cold injury, or an interaction of both. Recent findings suggest regions in the peach genome associated with bacterial spot resistance on chromosomes 1, 4 and 6 (Yang et al., 2013; Gasic et al., 2015) being also involved in PTSL. These regions have not been identified before in association with nematode response (Gasic, *unpublished data*). We believe that our recent findings indicate regions in the peach genome associated with a general response to bacterial infection, and this will be explored further in the case study below.

The genetics of disease resistance has been well-researched in pome fruit as well, with major efforts in understanding apple resistance to powdery mildew (*Podosphaera leucotricha*) (Stankiewicz-Kosyl et al., 2005; Jensen et al., 2014; Jänsch et al., 2015; Baumgartner et al., 2016), fire blight (*Erwinia amylovora*) (Jänsch et al., 2015; Kellerhals et al., 2017; van de Weg et al., 2017; Emeriewen et al., 2018; Schlathölter et al., 2018; Schropfer et al., 2018) and scab (*Venturia inaequalis*) (Terakami et al., 2006; Padmarasu et al., 2014; Won et al., 2014; Perchepied et al., 2015; Baumgartner et al., 2016; Khajuria et al., 2018). Apple scab resistance, derived from the major gene Rvi6 from *Malus floribunda*, has been the basis of resistance in many breeding programmes. Development of *Venturia* strains capable of overcoming Rvi6 resistance exposed weaknesses of a single gene, vertical resistance, and thus breeding focus has shifted towards pyramiding of multiple resistance genes to ensure durable resistance (Khajuria et al., 2018).

The sequencing of several *Fragaria* genomes (Zhong et al., 2018), as well as the development of genotyping platforms (Bassil et al., 2015), has enabled comparison studies within *Fragaria* and between *Rosaceae* facilitating rapid advances in understanding disease genetics in strawberry (Jia et al., 2015; Mangandi et al., 2017; Noh et al., 2018). Strawberry is one of the first fruit crops for which genomic selection has been explored (Gezan et al., 2017; Osorio et al., 2018; Verma et al., 2018), paving the path for other *Rosaceae* species to follow. Research on other temperate zone berries, such as red raspberry, black raspberry, cranberry (*Vaccinium macrocarpon* Ait.), blackberry (*Rubus fruticosus*) and blueberry (*Vaccinium angustifolium*), are slowly catching up (Woodhead et al., 2010; Castro et al., 2013; Longhi et al., 2014; Bushakra et al., 2015; Gupta et al., 2015; Stephens et al., 2016). With their genomes sequenced

(Polashock et al., 2014; Gupta et al., 2015; VanBuren et al., 2016; Schlautman et al., 2018), we anticipate rapid expansion of new discoveries in the near future.

6 Case study: hot spot on chromosome 4 in *Prunus*

The advancement of genomic technologies coupled with the drastic reduction in sequencing costs have yielded unprecedented opportunities to rapidly understand the linkage between genetics, environment and the presentation of traits in rosaceous crops. The development of the peach reference genome assembly has provided the necessary foundation for quantum leaps in progress in understanding and mapping trait genetics in peach, plum, cherry and other closely related rosaceous species. Many quantitative trait loci have been mapped and converted into simple DNA tests that are prognostic for horticultural and disease traits (Cai et al., 2017). Further work in peach and cherry, surveying related pedigrees and other exotic non-related germplasm using dense SNP-based genotyping, has facilitated the ability to phase the genome into discrete haplotype blocks, where meiotic recombination events are seemingly rare. Defining discrete haplotype blocks as hereditary units increases power in approaches to transgressively stack favourable alleles and predictive breeding. Furthermore, the high degree of relatedness and conservation of macro/micro synteny among rosaceous species affords opportunities to translate horticultural and disease trait information across species.

In peach, chromosome 4 harbours many fruit quality and disease resistance QTLs that cluster along the chromosome as 'hot spots' (Mézard, 2006). In sweet cherry, the same region is important for fruit traits. The recent release of the sweet cherry reference genome assembly (Shirasawa et al., 2017) provides an opportunity to study these important traits as chromosome 4 'hot spots' at the genomic level in detail between peach and cherry. In peach, there are four main hot spots on chromosome four that span the genomic interval between 7 and 20 Mbp (Fig. 2).

In peach, this clustered trait hot spot includes fruit quality traits such as stone adhesion to the flesh (freestone or clingstone) (Peace et al., 2006; Gu et al., 2016), flesh texture (melting or non-melting) (Peace et al., 2005; Martinez-Garcia et al., 2013b), fruit weight/size (da Silva Linge et al., 2015; Hernández Mora et al., 2017), fruit development period (Hernández Mora et al., 2017), bloom date (days to bloom) (Bielenberg et al., 2015; Hernández Mora et al., 2017), ripening date (Serra et al., 2017), maturity (Pirona et al., 2013), and individual sugars and soluble solids (Zeballos et al., 2016; Hernández Mora et al., 2017).

This region also includes QTLs for resistance to powdery mildew (Foulongne et al., 2002), brown rot (Martínez-García et al., 2013a; Pacheco et al., 2014; Fu et al., 2018), bacterial spot (Yang et al., 2013) and peach tree short life (Gasic, unpublished) (Fig. 2). Sizes of these regions in peach range from 52 kilobases to several megabases in genomic length, with corresponding regions of microsynteny in sweet cherry of similar length (Table 2). The number of genes within these windows in cherry spanned as few as 13 and as many as 557 (Table 2).

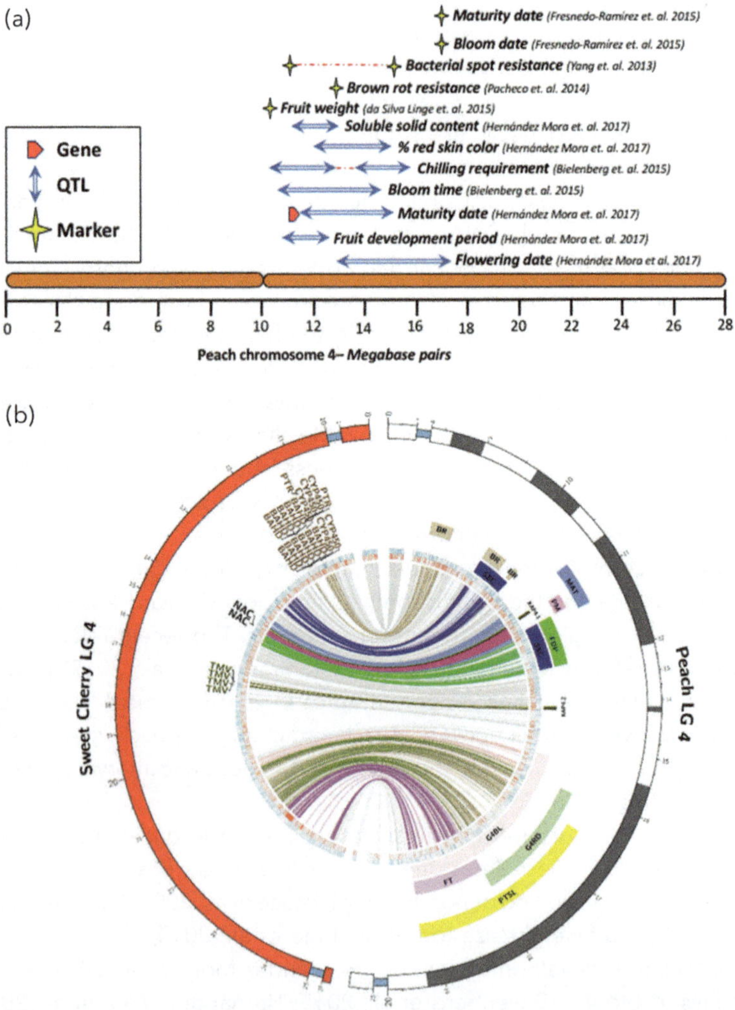

Figure 2 Chromosome 4 trait hot spots. (a) Linear representation of the peach chromosome 4 with known locations of markers, QTLs and genes. (b) A circular whole chromosome alignment of chromosome 4 in peach with pseudomolecule 4 in sweet cherry. Chromosomal hot spot regions are marked as shaded grey regions on the peach ideogram. Horticultural and disease traits are highlighted by different colours as mapped according to the peach reference genome. Coloured links from peach to cherry indicate direct syntenic relationships of the QTL regions. Grey links indicate stretches of general collinearity outside of the encoded hot spots.

Table 2 Number of genes detected in chromosome 4 7-20 Mbp syntenic region between peach and sweet cherry

Peach chromosome	Peach start	Peach stop	Length in peach (bp)	Trait	Cherry chromosome	No. genes in cherry	Cherry start	Cherry stop	Length in cherry
Pp04	8 581 271	8 977 975	396 704	BRT	Pa04	119	11 036 495	11 758 129	721 634
Pp04	9 647 311	10 015 773	368 462	BRT	Pa04	25	12 534 379	12 759 528	225 149
Pp04	9 647 311	10 280 095	632 784	SSC	Pa04	80	12 534 379	13 144 230	609 851
Pp04	10 194 038	10 280 095	86 057	BRT	Pa04	12	13 056 611	13 144 230	87 619
Pp04	10 696 489	11 298 736	602 247	MAT	Pa04	103	13 585 417	14 286 450	701 033
Pp04	10 923 464	10 976 364	52 900	XAP4_1	Pa04	18	13 865 184	13 948 849	83 665
Pp04	10 981 971	11 208 347	226 376	PM	Pa04	27	13 956 995	14 183 757	226 762
Pp04	11 208 347	12 107 191	898 844	FDP	Pa04	103	14 188 495	14 986 433	797 938
Pp04	11 208 347	12 107 191	898 844	SSC	Pa04	103	14 188 495	14 986 433	797 938
Pp04	14 074 784	14 137 816	63 032	XAP4_2	Pa04	13	17 498 020	17 592 902	94 882
Pp04	15 646 308	19 206 580	3 560 272	G4BL	Pa04	557	20 543 936	24 261 542	3 717 606
Pp04	16 084 694	17 984 794	1 900 100	G4RD	Pa04	249	20 779 040	22 341 967	1 562 927
Pp04	16 511 312	19 206 580	2 695 268	PTSL	Pa04	557	20 543 936	24 261 542	3 717 606
Pp04	17 984 794	19 206 580	1 221 786	AD	Pa04	268	22 370 707	24 261 542	1 890 835
Pp04	17 984 794	19 206 580	1 221 786	FT	Pa04	268	22 370 707	24 261 542	1 890 835

Abbreviations: Pp - *Prunus persica*; Pa - *Prunus avium*; BRT - Brown Rot; SSC - Soluble solids; MAT - maturity; XAP - bacterial spot; PM - powdery mildew; FDP - fruit development period; G4BL - bloom time; G4RD - ripening date; PTSL - peach tree short life; AD - adhesion; FT - flesh texture.

Of the traits described in Fig. 2a, only one trait, maturity time, has a published gene that was discovered in peach (Pirona et al., 2013). In peach, this gene falls almost in the centre of the QTL window and contains a No Apical Meristem (NAM/NAC) domain which is functionally involved in developmental processes such as shoot apical meristem formation, floral organs and lateral shoots, as well as plant hormonal control and defence. Using the synteny map described in Fig. 2b and reciprocal BLAST-based orthology analysis between peach and cherry, we determined the cherry ortholog of ppa008301m (peach) to be Pav_sc0000029.1_g070, which was annotated as a NAC protein. In fact, just 7.1 kilobases downstream, another NAC domain-encoding protein is annotated as Pav_sc0000029.1_g090 with the same reciprocal match to peach, suggesting a possible recent duplication event of this gene in cherry.

Brown rot is a critical fungal disease in stone fruit caused by *Monilinia* spp. (Oliveira Lino et al., 2016). Resistance to brown rot was encoded among three QTLs in peach (Martínez-García et al., 2013a; Pacheco et al., 2014). There are no mapped QTLs in cherry for brown rot and progress in peach has been recent, therefore comparison between peach and sweet cherry genomic regions might shed light on the genetics of brown rot resistance in *Prunus*. Alignment of the regions mapped in peach with the sweet cherry genome revealed similar-sized regions (~1 Mbp total), with a total of 156 predicted coding genes (Fig. 2b and Table 2).

Further investigation of the function of the predicted genes in cherry revealed clusters of cytochrome P450 genes, BAHD acyltransferase genes and pentatricopeptide genes. Cytochrome p450 genes are one of the largest super-family of genes in plants that produce enzymes to metabolize potentially toxic compounds (Bak et al., 2011). In plant defence responses, these genes are involved in the biosynthetic pathway of major phytoalexins, such as antibacterial and antifungal compounds (Durst and Benveniste, 1993). The BAHD acyltransferase family of genes regulate salicylic acid (SA), which is an important regulator of plant resistance to biotrophic and hemi-biotrophic pathogens (Zheng et al., 2009). Quilot-Turion et al. (2018) recently reported genetic variability in peach skin phenolics and triterpenoids as a natural defence mechanism in peach against brown rot. Further investigation of these genes in sweet cherry will be required for validation of their role in brown rot resistance.

Bacterial spot caused by *Xanthomons arboricola* pv. *pruni* is another devastating disease that affects stone fruits (Palacio-Bielsa et al., 2010; Tjou-Tam-Sin et al., 2012; Shen et al., 2013; Lamichhane, 2014). Two major QTLs associated with leaf resistance were mapped on chromosome 4 in peach (Yang et al., 2013) (Fig. 2b), in a region that contains only 31 predicted genes in cherry (Table 3). These regions in cherry are encoded for pentatricopeptide repeat-containing proteins, cytochrome p450, leucine-rich repeat (LRR) proteins and a tandem array of tobacco mosaic virus (TMV) resistance-like genes (Table 3, Fig. 2b). The annotation of these genes implicates possible resistance or defence

Table 3 Candidate genes in sweet cherry residing in the syntenic region of bacterial spot peach QTLs

Chromosome	Cherry start	Cherry stop	Gene	Annotation	QTL
Pa04	13920777	13921489	Pav_sc0002828.1_g220.1.mk	Pentatricopeptide repeat-containing protein At5g66520-like	4.1
Pa04	13935032	13935244	Pav_sc0002828.1_g190.1.br	Anoctamin-like protein At1g73020 isoform X3	
Pa04	13921625	13922107	Pav_sc0002828.1_g210.1.mk	Cytochrome P450 704C1-like	
Pa04	13916200	13916955	Pav_sc0002828.1_g230.1.br	Hypothetical protein PRUPE_ppa000565mg	
Pa04	13868641	13869483	Pav_sc0002828.1_g310.1.mk	Pav_sc0002828.1_g310.1.mk	
Pa04	13865215	13867293	Pav_sc0002828.1_g320.1.mk	HrpN-interacting protein from *Malus*	
Pa04	13892641	13893365	Pav_sc0002828.1_g280.1.mk	MADS-box transcription factor 50-like	
Pa04	13885956	13892166	Pav_sc0002828.1_g290.1.mk	Nucleolar GTP-binding protein 2	
Pa04	13893387	13896080	Pav_sc0002828.1_g270.1.mk	Nucleolar GTP-binding protein 2	
Pa04	13925874	13934453	Pav_sc0002828.1_g200.1.mk	Plant intracellular Ras-group-related LRR protein 6	
Pa04	13937888	13939888	Pav_sc0002828.1_g170.1.br	Putative ribonuclease H protein At1g65750	
Pa04	13939917	13941300	Pav_sc0002828.1_g160.1.br	Putative ribonuclease H protein At1g65750	
Pa04	13906779	13908579	Pav_sc0002828.1_g240.1.br	Uncharacterized protein DDB_G0271670-like	
Pa04	13944829	13945560	Pav_sc0002828.1_g150.1.mk	Uncharacterized protein LOC103326227	
Pa04	13936661	13937380	Pav_sc0002828.1_g180.1.mk	Uncharacterized protein LOC103326337	
Pa04	13899856	13901772	Pav_sc0002828.1_g260.1.br	Uncharacterized protein LOC103327533	
Pa04	13869556	13883222	Pav_sc0002828.1_g300.1.mk	Uncharacterized protein LOC103338760	
Pa04	13906228	13906761	Pav_sc0002828.1_g250.1.br	Wall-associated receptor kinase-like 4 isoform X2	

(Continued)

Table 3 (*Continued*)

Chromosome	Cherry start	Cherry stop	Gene	Annotation	QTL
Pa04	17526852	17527283	Pav_sc0001293.1_g150.1.br	TMV resistance protein N-like	4.2
Pa04	17529288	17529617	Pav_sc0001293.1_g160.1.br	TMV resistance protein N-like	
Pa04	17530377	17530850	Pav_sc0001293.1_g170.1.br	TMV resistance protein N-like	
Pa04	17581007	17583956	Pav_sc0001293.1_g220.1.br	TMV resistance protein N-like	
Pa04	17524050	17525300	Pav_sc0001293.1_g140.1.br	ABC transporter B family member 11-like	
Pa04	17502924	17503586	Pav_sc0001293.1_g100.1.br	Pav_sc0001293.1_g100.1.br	
Pa04	17505648	17506133	Pav_sc0001293.1_g110.1.br	Pav_sc0001293.1_g110.1.br	
Pa04	17507195	17512816	Pav_sc0001293.1_g120.1.mk	Hypothetical protein PRUPE_ppa001040mg	
Pa04	17532198	17532699	Pav_sc0001293.1_g180.1.br	Pav_sc0001293.1_g180.1.br	
Pa04	17543619	17543903	Pav_sc0001293.1_g200.1.br	Pav_sc0001293.1_g200.1.br	
Pa04	17534610	17535699	Pav_sc0001293.1_g190.1.mk	Probable calcium-binding protein CML29	
Pa04	17514893	17520466	Pav_sc0001293.1_g130.1.mk	Putative BPI/LBP family protein At1g04970	
Pa04	17549996	17570508	Pav_sc0001293.1_g210.1.mk	U3 small nucleolar RNA-associated protein 6 homolog	

Abbreviation: Pa - *Prunus avium*.

mechanisms, and further studies are necessary to validate a functional role in bacterial spot resistance. In addition, with the anticipated release of almond and plum genomic sequences, further comparison of this syntenic region among *Prunus* will provide insight into the genetics of bacterial spot in stone fruit.

In our recent effort to understand the genetics of PTSL in peach rootstock and facilitate development of predictive DNA test(s), we used genotyping by sequencing (Elshire et al., 2011) to refine QTLs associated with PTSL previously mapped in peach (Blenda et al., 2007). A meta-QTL approach revealed PTSL-associated QTLs on chromosomes 1, 4 and 6 (Gasic and Saski, unpublished). Peach rootstock Guardian®, a source of resistance to PTSL (Beckman et al., 1997; Parker et al., 1997) in mapping studies, was subjected to *de novo* genome sequencing and assembly using long-read single-molecule sequencing. Comparative alignments were performed with the susceptible Lovell haploid reference assembly (Verde et al., 2017) and PTSL QTL boundaries for two QTLs mapped on chromosome 4, *PTSL.meta_4.1* and *PTSL.meta_4.2* (Fig. 3).

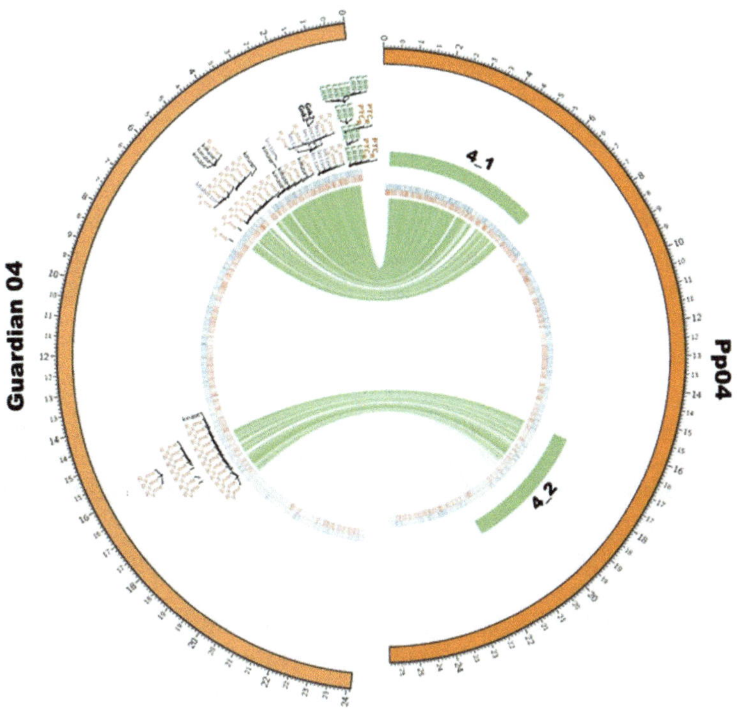

Figure 3 Alignment of the peach reference chromosome 4 (Lovell) to the *de novo* Guardian® pseudo-chromosome 4. Green histograms labelled 4_1 and 4_2 are two newly mapped PTSL QTLs, and the green links indicate syntenic alignments. Genes listed as text on the Guardian chromosome 4 are unique resistance genes found in Guardian®. The Guardian® ideogram is offset to illustrate significant R-gene expansion.

The alignment of the Guardian® chromosome 4 to the reference peach chromosome 4 did not reveal any large-scale structural variants. The most interesting discovery is localized expansions of disease resistance genes (R-genes) within the QTL intervals. We identified a total of 66 and 22 resistance-related genes in *PTSL.meta_4.1* and *PTSL.meta_4.2*, respectively, which appear to have distinct functional clusters (Fig. 3 and Table 4). For example, PTSL. meta_4.1 in Guardian® encodes tandem arrays of LRR genes, intermixed with proteins that encode Gnk2 domains and Bulb-type lectin domains. Gnk2 proteins have been shown to be involved in response to environmental stress, such as oxidative, salt and drought stress, as well as responses to both fungal and bacterial pathogen infection (Czernic et al., 1999; Ohtake et al., 2000; Acharya et al., 2007). Plant lectins are critical components of the plant's innate immune system (Lannoo and van Damme, 2014).

The PTSL.meta_ 4.2 region in Guardian® contains a general expansion of R-genes in comparison to the peach reference chromosome 4. Taken together, these gene expansion events and localized functions extend our current understanding of genetic resistance to the PTSL disease complex, while advances in technology are driving our ability to collect and analyse massive amounts of data at the single genome level.

7 Future trends and conclusion

The future of fruit production, fruit quality, plant resilience and ultimately sustainability heavily depends on continued advances in trait genetics, our understanding of environmental dependencies and interactions, availability and continued maintenance of diverse germplasm, and progress in tissue culture and *in vitro* systems alongside genetic engineering. Discovery genomics is being fuelled by cheap comprehensive genome sequencing at the population level, which endows scientists with unprecedented opportunities to precisely map quantitative traits, and conduct in-depth experiments to mechanistically disentangle complex traits into the underlying biochemical pathways, genes, epistatic interactions, regulatory elements and so on, to recapitulate the system driving the trait on a molecular level.

The philosophies and approaches to classical plant breeding are still required for the sustainable future of agriculture; however, a shift in selection approaches to molecular genotyping and DNA-informed breeding efforts have proven useful and are now economically feasible on much larger scales. Furthermore, the promises of genome editing have been realized in a number of fruit tree species and continue to hold the potential for quantum gains in progress through designer trait genetics and refined breeding approaches. The future is bright for durable genetic solutions to disease and abiotic stress resilience as well as sustainability of fruit tree crops.

Table 4 Resistance (R) genes unique to Guardian® peach rootstock detected in PTSL.meta_4.1 and PTSL.meta_4.2 QTL regions

Chromosome	Gene	Gene start	Gene stop	Transcription	QTL	Description
Pp_G_04	Pp_G_0055	233535	234130	-	4.1	Pentatricopeptide repeat
Pp_G_04	Pp_G_0067	268005	268331	-	4.1	Leucine-rich repeat-containing N-terminal, plant type
Pp_G_04	Pp_G_0068	270991	272494	+	4.1	Leucine-rich repeat
Pp_G_04	Pp_G_0069	272820	273395	-	4.1	Leucine-rich repeat
Pp_G_04	Pp_G_0070	275307	275760	+	4.1	Leucine-rich repeat-containing N-terminal, plant type
Pp_G_04	Pp_G_0073	289634	291837	+	4.1	Leucine-rich repeat
Pp_G_04	Pp_G_0074	292454	292945	-	4.1	Leucine-rich repeat
Pp_G_04	Pp_G_0075	293547	294101	+	4.1	Leucine-rich repeat-containing N-terminal, plant type
Pp_G_04	Pp_G_0081	324804	326426	+	4.1	Leucine-rich repeat
Pp_G_04	Pp_G_0082	327570	328622	+	4.1	Leucine-rich repeat 4
Pp_G_04	Pp_G_0083	329980	331999	+	4.1	Leucine-rich repeat 4
Pp_G_04	Pp_G_0087	345596	346814	-	4.1	Leucine-rich repeat
Pp_G_04	Pp_G_0092	371880	374507	-	4.1	Leucine-rich repeat
Pp_G_04	Pp_G_0108	422455	425185	-	4.1	Pentatricopeptide repeat
Pp_G_04	Pp_G_0110	436386	438098	+	4.1	Pentatricopeptide repeat
Pp_G_04	Pp_G_0112	443474	446965	+	4.1	Pentatricopeptide repeat
Pp_G_04	Pp_G_0120	461967	462275	+	4.1	Leucine-rich repeat
Pp_G_04	Pp_G_0121	462654	463295	-	4.1	Leucine-rich repeat
Pp_G_04	Pp_G_0123	473853	478526	+	4.1	Leucine-rich repeat

(Continued)

Table 4 (*Continued*)

Chromosome	Gene	Gene start	Gene stop	Transcription	QTL	Description
Pp_G_04	Pp_G_0134	512394	514723	-	4.1	Leucine-rich repeat
Pp_G_04	Pp_G_0139	531050	533523	+	4.1	Leucine-rich repeat-containing N-terminal, plant type
Pp_G_04	Pp_G_0399	1660901	1663453	+	4.1	R-Gene
Pp_G_04	Pp_G_0400	1664303	1668731	+	4.1	Gnk2-homologous domain
Pp_G_04	Pp_G_0406	1692840	1694191	+	4.1	Gnk2-homologous domain
Pp_G_04	Pp_G_0407	1696421	1700728	+	4.1	R-Gene
Pp_G_04	Pp_G_0413	1712244	1712462	+	4.1	R-Gene
Pp_G_04	Pp_G_0444	1847775	1855621	+	4.1	Domain of unknown function DUF4371
Pp_G_04	Pp_G_0445	1856533	1857087	-	4.1	Bulb-type lectin domain
Pp_G_04	Pp_G_0451	1883445	1887233	+	4.1	R-Gene
Pp_G_04	Pp_G_0457	1908635	1910419	-	4.1	Bulb-type lectin domain
Pp_G_04	Pp_G_0458	1914486	1916002	+	4.1	Bulb-type lectin domain
Pp_G_04	Pp_G_0460	1919780	1921119	-	4.1	Bulb-type lectin domain
Pp_G_04	Pp_G_0461	1924581	1925279	-	4.1	R-Gene
Pp_G_04	Pp_G_0462	1925951	1926634	-	4.1	Bulb-type lectin domain
Pp_G_04	Pp_G_0463	1930321	1932261	+	4.1	Bulb-type lectin domain
Pp_G_04	Pp_G_0465	1936410	1936547	-	4.1	Bulb-type lectin domain
Pp_G_04	Pp_G_0638	2664290	2665150	-	4.1	R-Gene
Pp_G_04	Pp_G_0738	3058463	3060427	+	4.1	Serine-threonine/tyrosine-protein kinase, catalytic domain
Pp_G_04	Pp_G_0781	3249952	3256406	+	4.1	Bulb-type lectin domain

Pp_G_04	Pp_G_0806	3396424	3398893	-	4.1	Protein kinase domain
Pp_G_04	Pp_G_0812	3428861	3429067	+	4.1	Ribonuclease T2-like
Pp_G_04	Pp_G_0927	3944412	3944759	-	4.1	Chlorophyll A-B binding protein
Pp_G_04	Pp_G_0930	3953146	3958645	+	4.1	Protein kinase domain
Pp_G_04	Pp_G_0931	3964593	3965798	+	4.1	Serine-threonine/tyrosine-protein kinase, catalytic domain
Pp_G_04	Pp_G_1010	4336296	4341571	+	4.1	Serine-threonine/tyrosine-protein kinase, catalytic domain
Pp_G_04	Pp_G_1086	4778316	4779042	+	4.1	R-Gene
Pp_G_04	Pp_G_1092	5071027	5071530	-	4.1	R-Gene
Pp_G_04	Pp_G_1093	5072719	5074127	-	4.1	Protein kinase domain
Pp_G_04	Pp_G_1095	5081305	5084530	+	4.1	R-Gene
Pp_G_04	Pp_G_1097	5093524	5095200	+	4.1	Protein kinase domain
Pp_G_04	Pp_G_1099	5098569	5101543	-	4.1	Wall-associated receptor kinase, galacturonan-binding domain
Pp_G_04	Pp_G_1102	5115303	5120145	+	4.1	Wall-associated receptor kinase, galacturonan-binding domain
Pp_G_04	Pp_G_1105	5130310	5131888	+	4.1	R-Gene
Pp_G_04	Pp_G_1106	5132191	5139006	-	4.1	Protein kinase domain
Pp_G_04	Pp_G_1107	5139765	5142653	+	4.1	EGF-like calcium-binding domain
Pp_G_04	Pp_G_1161	5510037	5513785	-	4.1	R-Gene
Pp_G_04	Pp_G_1162	5514991	5515962	-	4.1	Malectin
Pp_G_04	Pp_G_1164	5517392	5518124	-	4.1	R-Gene
Pp_G_04	Pp_G_1165	5521803	5528258	+	4.1	R-Gene

(*Continued*)

Table 4 (*Continued*)

Chromosome	Gene	Gene start	Gene stop	Transcription	QTL	Description
Pp_G_04	Pp_G_1195	5669923	5670347	+	4.1	Argonaute, linker 1 domain
Pp_G_04	Pp_G_1205	5699517	5700569	-	4.1	R-Gene
Pp_G_04	Pp_G_1206	5703769	5704068	+	4.1	R-Gene
Pp_G_04	Pp_G_1214	5734976	5744082	+	4.1	R-Gene
Pp_G_04	Pp_G_1220	5761372	5761734	-	4.1	R-Gene
Pp_G_04	Pp_G_1290	6154729	6161259	+	4.1	R-Gene
Pp_G_04	Pp_G_1400	6746038	6746382	+	4.1	R-Gene
Pp_G_04	Pp_G_2781	15155373	15160841	-	4.2	Protein kinase domain
Pp_G_04	Pp_G_2850	15576450	15576645	-	4.2	R-Gene
Pp_G_04	Pp_G_2906	15957158	15957706	-	4.2	R-Gene
Pp_G_04	Pp_G_2909	15973112	15973431	+	4.2	R-Gene
Pp_G_04	Pp_G_2913	15982920	15983348	-	4.2	R-Gene
Pp_G_04	Pp_G_2914	15987935	15996860	+	4.2	R-Gene
Pp_G_04	Pp_G_2920	16009333	16009386	-	4.2	R-Gene
Pp_G_04	Pp_G_2921	16012653	16013208	+	4.2	R-Gene
Pp_G_04	Pp_G_2934	16099777	16101036	+	4.2	R-Gene
Pp_G_04	Pp_G_2937	16179710	16180158	-	4.2	R-Gene
Pp_G_04	Pp_G_2957	16286228	16287036	+	4.2	R-Gene
Pp_G_04	Pp_G_2961	16293750	16294892	+	4.2	R-Gene
Pp_G_04	Pp_G_2971	16461596	16464226	+	4.2	Domain of unknown function DUF4216
Pp_G_04	Pp_G_3000	16721115	16722323	+	4.2	R-Gene

Pp_G_04	Pp_G_3014	16890761	16891510	-	4.2	R-Gene
Pp_G_04	Pp_G_3015	16891703	16892008	-	4.2	R-Gene
Pp_G_04	Pp_G_3029	17153718	17154992	+	4.2	R-Gene
Pp_G_04	Pp_G_3035	17165553	17166099	-	4.2	R-Gene
Pp_G_04	Pp_G_3070	17482889	17483053	-	4.2	R-Gene
Pp_G_04	Pp_G_3073	17493660	17494675	-	4.2	R-Gene
Pp_G_04	Pp_G_3080	17560845	17561680	-	4.2	R-Gene
Pp_G_04	Pp_G_3090	17601862	17602769	+	4.2	R-Gene

+/− indicates orientation of transcription.

8 Where to look for further information

Additional information about advances in fruit genetics can be found at websites of projects such as:

- FruitBreedomics (http://fruitbreedomics.com/);
- RosBREED (www.rosbreed.org).

There are a number of databases that house up-to-date information on newly detected marker-trait associations and DNA tests, and use various tools to explore the vast amount of information available for fruit crops, including:

- Genome Database for *Rosaceae* (www.rosaceae.org);
- Genome Database for *Vaccinium* (www.vaccinium.org/); and
- Genome Database for *Citrus* (www.citrusgenomedb.org).

Newer approaches to mapping on the whole-genome level, such as GWAS, can be confusing and hard to understand. For novice readers, a review of the advantages and limitations of trait analysis with GWAS provides a good start (Korte and Farlow, 2013). Genomic prediction is becoming the next hot topic in fruit breeding, and many reports of testing this approach in plant breeding programmes are available (Daetwyler et al., 2012; Gezan et al., 2017; Stich and Van Inghelandt, 2018).

Additional excellent sources of the newest advances in fruit genetics, some of which have been mentioned in this chapter, include the following:

- Plant & Animal Genome Conference (www.intlpag.org);
- International Horticultural Congress (www.ihc2018.org); and
- International Rosaceae Genomic Conference (www.rgc9.org).

9 References

Acharya, B. R., Raina, S., Maqbool, S. B., Jagadeeswaran, G., Mosher, S. L., Appel, H. M., et al. (2007). Overexpression of CRK13, an Arabidopsis cysteine-rich receptor-like kinase, results in enhanced resistance to *Pseudomonas syringae*. *The Plant Journal: For Cell and Molecular Biology* 50(3): 488–99.

Al-Dous, E. K., George, B., Al-Mahmoud, M. E., Al-Jaber, M. Y., Wang, H., Salameh, Y. M., et al. (2011). *De novo* genome sequencing and comparative genomics of date palm (*Phoenix dactylifera*). *Nature Biotechnology* 29(6): 521–84.

Al-Mssallem, I. S., Hu, S., Zhang, X., Lin, Q., Liu, W., Tan, J., et al. (2013). Genome sequence of the date palm *Phoenix dactylifera* L. *Nature Communications* 4.

Alioto, T. S., Cruz, F., Gomez-Garrido, J., Frias, L., Ribeca, P., Gut, M., et al. (2016). A draft sequence of the almond genome. *Plant and Animal Genome conference XXIV*.

Allan, A. C. and Espley, R. V. (2018). MYBs drive novel consumer traits in fruits and vegetables. *Trends in Plant Science* 23(8): 693–705.

Allard, A., Bink, M. C. A. M., Martinez, S., Kelner, J., Legave, J., di Guardo, M., et al. (2016). Detecting QTLs and putative candidate genes involved in budbreak and flowering time in an apple multiparental population. *Journal of Experimental Botany* 67(9): 2875–88.

Argout, X., Salse, J., Aury, J.-M., Guiltinan, M. J., Droc, G., Gouzy, J., et al. (2011). The genome of *Theobroma cacao*. *Nature Genetics* 43(2): 101–8. http://dx.doi.org/10.1038/ng.736.

Arús, P., Mnejja, M., Esmenjaud, D., Bosselut, N. and Dirlewanger, E. (2004). High marker density around the peach nematode resistance genes. *Acta Horticulturae* 658: 567–71.

Arús, P., Yamamoto, T., Dirlewanger, E. and Abbott, A. G. (2005). Synteny in the Rosaceae. *Plant Breeding Reviews* 27: 175–211.

Arya, P., Kumar, G., Acharya, V. and Singh, A. K. (2014). Genome-wide identification and expression analysis of NBS-encoding genes in *Malus × domestica* and expansion of NBS genes family in Rosaceae. *PLoS ONE* 9(9): e107987.

Badenes, M. L., Fernandez i Marti, A., Rios, G. and Rubio-Cabetas, M. J. (2016). Application of genomic technologies to the breeding of trees. *Frontiers in Genetics* 7: 198.

Bak, S., Beisson, F., Bishop, G., Hamberger, B., Höfer, R., Paquette, S., et al. (2011). Cytochromes p450. *The Arabidopsis Book* 9: e0144.

Barabaschi, D., Tondelli, A., Desiderio, F., Volante, A., Vaccino, P., Vale, G., et al. (2016). Next generation breeding. *Plant Science: From Genomics to Breeding* 242: 3–13.

Bassil, N. V., Davis, T. M., Zhang, H., Ficklin, S., Mittmann, M., Webster, T., et al. (2015). Development and preliminary evaluation of a 90K Axiom® SNP array for the allo-octoploid cultivated strawberry *Fragaria × ananassa*. *BMC Genomics* 16(1): 155.

Baumgartner, I. O., Kellerhals, M., Costa, F., Dondini, L., Pagliarani, G., Gregori, R., et al. (2016). Development of SNP-based assays for disease resistance and fruit quality traits in apple (*Malus × domestica* Borkh.) and validation in breeding pilot studies. *Tree Genetics & Genomes* 12(3): 35.

Beckman, T. G., Okie, W. R., Nyczepir, A. P., Reighard, G. L., Zehr, E. I. and Newall, W. C. (1997). History current status, and future potential of Guardian (BY520-9) peach rootstock. *Acta Horticulturae* 451(451): 251–8.

Begheldo, M., Manganaris, G. A., Bonghi, C. and Tonutti, P. (2008). Different postharvest conditions modulate ripening and ethylene biosynthetic and signal transduction pathways in Stony Hard peaches. *Postharvest Biology and Technology* 48(1): 84–91.

Bernet, G. P., Fernandez-Ribacoba, J., Carbonell, E. A. and Asins, M. J. (2010). Comparative genome-wide segregation analysis and map construction using a reciprocal cross design to facilitate citrus germplasm utilization. *Molecular Breeding* 25(4): 659–73.

Bianchi, V. J., Rubio, M., Trainotti, L., Verde, I., Bonghi, C. and Martínez-Gómez, P. (2015). *Prunus* transcription factors: Breeding perspectives. *Frontiers in Plant Science* 6: 443.

Bianco, L., Cestaro, A., Sargent, D. J., Banchi, E., Derdak, S., Di Guardo, M., et al. (2014). Development and validation of a 20K single nucleotide polymorphism (SNP) whole genome genotyping array for apple (*Malus × domestica* Borkh). *PLoS ONE* 9(10): e110377.

Bianco, L., Cestaro, A., Linsmith, G., Muranty, H., Denancé, C., Théron, A., et al. (2016). Development and validation of the Axiom®Apple480K SNP genotyping array. *The Plant Journal: For Cell and Molecular Biology* 86(1): 62–74.

Bielenberg, D. G., Rauh, B., Fan, S., Gasic, K., Abbott, A. G., Reighard, G. L., et al. (2015). Genotyping by sequencing for SNP-based linkage map construction and QTL analysis of chilling requirement and bloom date in peach [*Prunus persica* (L.) Batsch]. *PLoS ONE* 10(10): e0139406.

Bink, M. C., Totir, R. L., ter Braak, C. J. F., Winkler, C. R., Boer, M. P. and Smith, O. S. (2012). QTL linkage analysis of connected populations using ancestral marker and pedigree information. *Theoretical and Applied Genetics* 24(6): 1097–113.

Bink, M. C., Jansen, J., Madduri, M., Voorrips, R. E., Durel, C. E., Kouassi, A. B., et al. (2014). Bayesian QTL analyses using pedigreed families of an outcrossing species, with application to fruit firmness in apple. *Theoretical and Applied Genetics* 127(5): 1073–90.

Biscarini, F., Nazzicari, N., Bink, M., Arús, P., Aranzana, M. J., Verde, I., et al. (2017). Genome-enabled predictions for fruit weight and quality from repeated records in European peach progenies. *BMC Genomics* 18: 432. doi:10.1186/s12864-017-3781-8.

Blenda, A. V., Verde, I., Georgi, L. L., Reighard, G. L., Forrest, S. D., Munoz-Torres, M., et al. (2007). Construction of a genetic linkage map and identification of molecular markers in peach rootstocks for response to peach tree short life syndrome. *Tree Genetics & Genomes* 3: 341–50.

Bliss, F. (2006). Plant breeding in the U.S. private sector. *HortScience* 41(1): 45–7.

Brachi, B., Morris, G. P. and Borevitz, J. O. (2011). Genome-wide association studies in plants: The missing heritability is in the field. *Genome Biology* 12(10): 232.

Bus, V. G. M., Esmenjaud, D., Buck, E. and Laurens, F. (2009). Application of genetic markers in rosaceous crops. In: Folta, K. M. and Gardiner, S. E. (Ed.), *Genetics and Genomics of Rosaceae*. Springer-Verlag, New York, NY, pp. 563–99.

Bushakra, J. M., Bryant, D. W., Dossett, M., Vining, K. J., VanBuren, R., Gilmore, B. S., et al. (2015). A genetic linkage map of black raspberry (*Rubus occidentalis*) and the mapping of Ag(4) conferring resistance to the aphid *Amphorophora agathonica*. *Theoretical and Applied Genetics* 128(8): 1631–46.

Byrne, D. H. (2005). Trends in stone fruit cultivar development. *HortTechnology* 15: 494–500.

Byrne, P. F., Volk, G. M., Gardner, C., Gore, M. A., Simon, P. W. and Smith, S. (2018). Sustaining the future of plant breeding: The critical role of the USDA-ARS national plant germplasm system. *Crop Science* 58(2): 451–68.

Cai, L., Voorrips, R. E., van de Weg, E., Peace, C. and Iezzoni, A. (2017). Genetic structure of a QTL hotspot on chromosome 2 in sweet cherry indicates positive selection for favorable haplotypes. *Molecular Breeding* 37(7): 85.

Cao, K., Zheng, Z., Wang, L., Liu, X., Zhu, G., Fang, W., et al. (2014). Comparative population genomics reveals the domestication history of the peach, *Prunus persica*, and human influences on perennial fruit crops. *Genome Biology* 15(7): 1.

Cao, K., Zhou, Z., Wang, Q., Guo, J., Zhao, P., Zhu, G., et al. (2016). Genome-wide association study of 12 agronomic traits in peach. *Nature* 7: 13246.

Castro, P., Stafne, E. T., Clark, J. R. and Lewers, K. S. (2013). Genetic map of the primocane-fruiting and thornless traits of tetraploid blackberry. *Theoretical and Applied Genetics* 126(10): 2521–32.

Chagné, D. (2015). Whole genome sequencing of fruit tree species. In: Plomion, C. and Adam-Blondon, A.-F. (Ed.), *Advances in Botanical Research*. Elsevier Ltd., Vol. 74, pp. 1–37. doi:10.1016/bs.abr.2015.04.004.

Chagné, D., Carlisle, C. M., Blond, C., Volz, R. K., Oraguzie, N. C., Crowhurst, R. N., et al. (2007). Mapping a candidate gene (*MdMYB10*) for red flesh and foliage colour in apple. *BMC Genomics* 8: 212.

Chagné, D., Lin-Wang, K., Espley, R. V., Volz, R. K., How, N. M., Rouse, S., et al. (2012a). An ancient duplication of apple MYB transcription factors is responsible for novel red fruit-flesh phenotypes. *Plant Physiology* 161: 225–39.

Chagné, D., Crowhurst, R. N., Troggio, M., Davey, M. W., Gilmore, B., Lawley, C., et al. (2012b). Genome-wide SNP detection, validation, and development of an 8K SNP array for apple. *PLoS ONE* 7(2): e31745.

Chagné, D., Crowhurst, R. N., Pindo, M., Thrimawithana, A., Deng, C., Ireland, H., et al. (2014). The draft genome sequence of European pear (*Pyrus communis* L. 'Bartlett'). *PLoS ONE* 9(4): 12.

Chen, H., Song, Y., Li, L., Khan, M. A., Li, X., Korban, S. S., et al. (2015). Construction of a high-density simple sequence repeat consensus genetic map for pear (*Pyrus* spp.). *Plant Molecular Biology Reporter* 33(2): 316–25.

Cirilli, M., Geuna, F., Babini, A. R., Bozhkova, V., Catalano, L., Cavagna, B., et al. (2016). Fighting sharka in peach: Current limitations and future perspectives. *Frontiers in Plant Science* 7: 1290.

Claverie, M., Dirlewanger, E., Cosson, P., Bosselut, N., Lecouls, A. C., Voisin, R., et al. (2004). High-resolution mapping and chromosome landing at the root-knot nematode resistance locus *Ma* from Myrobalan plum using a large-insert BAC DNA library. *Theoretical and Applied Genetics* 109(6): 1318–27.

Costa, F., Peace, C. P., Stella, S., Serra, S., Musacchi, S., Bazzani, M., et al. (2010). QTL dynamics for fruit firmness and softening around an ethylene-dependent polygalacturonase gene in apple (*Malus × domestica* Borkh.). *Journal of Experimental Botany* 61(11): 3029–39.

Costa, F., Cappellin, L., Zini, E., Patocchi, A., Kellerhals, M., Komjanc, M., et al. (2013). QTL validation and stability for volatile organic compounds (VOCs) in apple. *Plant Science* 211: 1–7.

Crisosto, C. H., Crisosto, G. and Neri, F. (2006). Understanding tree fruit quality based on consumer acceptance. *Acta Horticulturae (ISHS)* 712: 183–90.

Cruz-Rus, E., Sesmero, R., Angel-Perez, J. A., Sanchez-Sevilla, J. F., Ulrich, D. and Amaya, I. (2017). Validation of a PCR test to predict the presence of flavor volatiles mesifurane and gamma-decalactone in fruits of cultivated strawberry (*Fragaria × ananassa*). *Molecular Breeding* 37(10): 131.

Cuenca, J., Aleza, P., Garcia-Lor, A., Ollitrault, P. and Navarro, L. (2016). Fine mapping for identification of citrus alternaria brown spot candidate resistance genes and development of new SNP markers for marker-assisted selection. *Frontiers in Plant Science* 7: 1948.

Czernic, P., Visser, B., Sun, W., Savouré, A., Deslandes, L., Marco, Y., et al. (1999). Characterization of an *Arabidopsis thaliana* receptor-like protein kinase gene activated by oxidative stress and pathogen attack. *The Plant Journal: For Cell and Molecular Biology* 18(3): 321–7.

D'Hont, A., Denoeud, F., Aury, J.-M., Baurens, F.-C., Carreel, F., Garsmeur, O., et al. (2012). The banana (*Musa acuminata*) genome and the evolution of monocotyledonous plants. *Nature* 488(7410): 213–17. http://dx.doi.org/10.1038/nature11241.

Da Silva Linge, C., Bassi, D., Bianco, L., Pacheco, I., Pirona, R. and Rossini, L. (2015). Genetic dissection of fruit weight and size in an F2 peach (*Prunus persica* (L.) Batsch) progeny. *Molecular Breeding* 35(2): 71.

Da Silva Linge, C., Antanaviciute, L., Abdelghafar, A., Arus, P., Bassi, D., Rossini, L., et al. (2019). High-density multi-population consensus genetic linkage map for peach. *PLoS ONE* 13(11): e0207724. doi:10.1371/journal.pone.0207724.

Daetwyler, H. D., Calus, M. P. L., Pong-Wong, R., de Los Campos, G. and Hickey, J. M. (2012). Genomic prediction in animals and plants: Simulation of data, validation, reporting, and benchmarking. *Genetics* 193(2): 347–65.

Dalla Costa, L., Malnoy, M. and Gribaudo, I. (2017). Breeding next generation tree fruits: Technical and legal challenges. *Horticulture Research* 4: 17067.

Davey, M., Gudimella, R., Harikrishna, J. A., Sin, L. W., Khalid, N. and Keulemans, J. (2013). A draft *Musa balbisiana* genome sequence for molecular genetics in polyploid, inter- and intra-specific *Musa* hybrids. *BMC Genomics* 14(1): 683.

De Franceschi, P., Stegmeir, T., Cabrera, A., van der Knaap, E., Rosyara, U. R., Sebolt, A. M., et al. (2013). Cell number regulator genes in *Prunus* provide candidate genes for the control of fruit size in sweet and sour cherry. *Molecular Breeding* 32(2): 311–26.

Decroocq, V., Foulongne, M., Lambert, P., Gall, O., Mantin, C., Pascal, T., et al. (2005). Analogues of virus resistance genes map to QTLs for resistance to sharka disease in *Prunus davidiana*. *Molecular Genetics and Genomics* 272(6): 680–9.

Decroocq, V., Decroocq, S., Tricon, D., Flutre, T., Pascal, T., Tuero, C., et al. (2018). Resistance to sharka in stone fruit trees: Genetic and genomic technologies for new breeding strategies. *Proceeding of 9th International Rosaceae Genomic Conference*, Nanjing, China, p. 39.

Degenhardt, J., Al-Masri, A. N., Kurkcuoglu, S., Szankowski, I. and Gau, A. E. (2005). Characterization by suppression subtractive hybridization of transcripts that are differentially expressed in leaves of apple scab-resistant and susceptible cultivars of *Malus domestica*. *Molecular Genetics and Genomics* 273(4): 326–35.

Denoeud, F., Carretero-Paulet, L., Dereeper, A., Droc, G., Guyot, R., Pietrella, M., et al. (2014). The coffee genome provides insight into the convergent evolution of caffeine biosynthesis. *Science* 345(6201): 1181–4.

Desta, Z. A. and Ortiz, R. (2014). Genomic selection: Genome-wide prediction in plant improvement. *Trends in Plant Science* 19(9): 592–601.

Di Pierro, E. A., Gianfranceschi, L., Di Guardo, M., Koehorst-van Putten, H. J. J., Kruisselbrink, J. W., Longhi, S., et al. (2016). A high-density, multi-parental SNP genetic map on apple validates a new mapping approach for outcrossing species. *Horticulture Research* 3: 16057.

Dirlewanger, E., Pascal, T., Zuger, C. and Kervella, J. (1996). Analysis of molecular markers associated with powdery mildew resistance genes in peach [*Prunus persica* (L) Batsch] × *Prunus davidiana* hybrids. *Theoretical and Applied Genetics* 93(5–6): 909–19.

Dirlewanger, E., Fouche, M., Le Dantec, L., Zaracho Echague, N., Flutre, T., Barreneche, T. et al. (2018). Sweet cherry: New genomic tools for the creation of varieties adapted to future conditions. *Proceedings of 9th Rosaeae Genomic Conference*, 26–30 June, Nanjing, China, p. 31.

Dirlewanger, E., Graziano, E., Joobeur, T., Garriga-Caldere, F., Cosson, P., Howad, W., et al. (2004). Comparative mapping and marker-assisted selection in Rosaceae fruit crops. *Proceedings of the National Academy of Sciences* 101(26): 9891–6.

Durst, F. and Benveniste, I. (1993). Cytochrome P450 in plants. In: Schenkman, J. B. and Greim, H. (Eds), *Cytochrome P450*. Springer Berlin Heidelberg, Berlin, Heidelberg, pp. 293–310.

Edge-Garza, D. A., Luby, J. J. and Peace, C. (2015). Decision support for cost-efficient and logistically feasible marker-assisted seedling selection in fruit breeding. *Molecular Breeding* 35(12): 223.

Eduardo, I., Chietera, G., Pirona, R., Pacheco, I., Troggio, M., Banchi, E., et al. (2013). Genetic dissection of aroma volatile compounds from the essential oil of peach fruit: QTL analysis and identification of candidate genes using dense SNP maps. *Tree Genetics & Genomes* 9(1): 189–204.

Elshire, R. J., Glaubitz, J. C., Sun, Q., Poland, J. A., Kawamoto, K., Buckler, E. S., et al. (2011). A robust, simple genotyping-by-sequencing (GBS) approach for high diversity species. *PLoS ONE* 6(5): e19379.

Emeriewen, O. F., Richter, K., Piazza, S., Micheletti, D., Broggini, G. A. L., Berner, T., et al. (2018). Towards map-based cloning of FB_Mfu10: Identification of a receptor-like kinase candidate gene underlying the *Malus fusca* fire blight resistance locus on linkage group 10. *Molecular Breeding* 38(8): 106.

Fernandez i Marti, A. and Dodd, R. S. (2018). Using CRISPR as a gene editing tool for validating adaptive gene function in tree landscape genomics. *Frontiers in Ecology and Evolution* 6.

Fernandez i Marti, A., Saski, C. A., Manganaris, G. A., Gasic, K. and Crisosto, C. H. (2018). Genomic sequencing of Japanese plum (*Prunus salicina* Lindl.) mutants provides a new model for Rosaceae fruit ripening studies. *Frontiers in Plant Science* 9: 21.

Foulongne, M., Pascal, T., Pfeiffer, F. and Kervella, J. (2002). Introgression of a polygenic resistance to powdery mildew from a wild species *Prunus davidiana* into peach [*Prunus persica* (L.) Batsch], a case study of marker assisted selection in fruit tree. *Acta Horticulturae* 592: 259–65.

Foulongne, M., Pascal, T., Pfeiffer, F. and Kervella, J. (2003). QTLs for powdery mildew resistance in peach × *Prunus davidiana* crosses: Consistency across generations and environments. *Molecular Breeding* 12(1): 33–50.

Freiman, A., Shlizerman, L., Golobovitch, S., Yablovitz, Z., Korchinsky, R., Cohen, Y., et al. (2012). Development of a transgenic early flowering pear (*Pyrus communis* L.) genotype by RNAi silencing of PcTFL1-1 and PcTFL1-2. *Planta* 235(6): 1239–51.

Fresnedo-Ramirez, J., Famula, T. R. and Gradziel, T. M. (2016a). Application of a Bayesian ordinal animal model for the estimation of breeding values for the resistance to *Monilinia fruticola* (G.Winter) Honey in progenies of peach [*Prunus persica* (L.) Batsch]. *Breeding Science* 67(2): 110–22.

Fresnedo-Ramirez, J., Frett, T. J., Sandefur, P. J., Salgado-Rojas, A., Clark, J. R., Gasic, K., et al. (2016b). QTL mapping and breeding value estimation through pedigree-based analysis of fruit size and weight in four diverse peach breeding programs. *Tree Genetics & Genomes* 12(2): 25.

Frett, T. J., Reighard, G. L., Okie, W. R. and Gasic, K. (2014). Mapping quantitative trait loci associated with blush in peach [*Prunus persica* (L.) Batsch]. *Tree Genetics & Genomes* 10(2): 367–81.

Fu, W., Da Silva Linge, C., Burrell, R. I., Schnabel, G. and Gasic, K. (2018). A genome wide association study for brown rot tolerance in peach. *Proceedings of 9th International Genomic Conference*, Nanjing, China, p. 104.

Gainza, F., Opazo, I., Guajardo, V., Meza, P., Ortiz, M., Pinochet, J., et al. (2015). Rootstock breeding in *Prunus* species: Ongoing efforts and new challenges. *Chilean Journal of Agricultural Research* 75: 6–16.

Gallardo, R. K., Zhang, Q., Dossett, M., Polashock, J. J., Rodriguez-Saona, C., Vorsa, N., et al. (2018). Breeding trait priorities of the blueberry industry in the United States and Canada. *HortScience* 53(7): 1021–8.

Gasic, K., Reighard, G., Okie, W., Clark, J., Gradziel, T., Byrne, D., et al. (2015). Bacterial spot resistance in peach: Functional allele distribution in breeding germplasm. *Acta Horticulturae* 1084: 69–74.

Gervasi, F., Ferrante, P., Dettori, M. T., Scortichini, M. and Verde, I. (2018). Transcriptome reprogramming of resistant and susceptible peach genotypes during *Xanthomonas arboricola* pv. *pruni* early leaf infection. *PLoS ONE* 13(4): e0196590.

Gezan, S. A., Osorio, L. F., Verma, S. and Whitaker, V. M. (2017). An experimental validation of genomic selection in octoploid strawberry. *Horticulture Research* 4: 16070.

Gilbert, J. L., Olmstead, J. W., Colquhoun, T. A., Levin, L. A., Clark, D. G. and Moskowitz, H. R. (2014). Consumer-assisted selection of blueberry fruit quality traits. *HortScience* 49(7): 864–73.

Gillen, A. M. and Bliss, F. A. (2005). Identification and mapping of markers linked to the *Mi* gene for root-knot nematode resistance in peach. *Journal of the American Society for Horticultural Science* 130(1): 24–33.

Goldy, C., Svetaz, L. A., Bustamante, C. A., Allegrini, M., Valentini, G. H., Drincovich, M. F., et al. (2017). Comparative proteomic and metabolomic studies between *Prunus persica* genotypes resistant and susceptible to *Taphrina deformans* suggest a molecular basis of resistance. *Plant Physiology and Biochemistry* 118: 245–55.

González, M., Maldonado, J., Salazar, E., Silva, H. and Carrasco, B. (2016). *De novo* transcriptome assembly of 'Angeleno' and 'Lamoon' Japanese plum cultivars (*Prunus salicina*). *Genomics Data* 9: 35–6.

Goonetilleke, S. N., March, T. J., Wirthensohn, M. G., Arús, P., Walker, A. R. and Mather, D. E. (2018). Genotyping by sequencing in almond: SNP discovery, linkage mapping, and marker design. *G3: Genes|Genomes|Genetics* 8(1): 161.

Gu, C., Wang, L., Wang, W., Zhou, H., Ma, B., Zheng, H., et al. (2016). Copy number variation of a gene cluster encoding endopolygalacturonase mediates flesh texture and stone adhesion in peach. *Journal of Experimental Botany* 67(6): 1993–2005.

Guo, M. and Simmons, C. R. (2011). Cell number counts – the *fw2.2* and *CNR* genes and implications for controlling plant fruit and organ size. *Plant Science: An International Journal of Experimental Plant Biology* 181(1): 1–7.

Guo, J., Cao, K., Li, Y., Yao, J., Deng, C., Wang, Q., et al. (2018). Comparative transcriptome and microscopy analyses provide insights into flat shape formation in peach (*Prunus persica*). *Frontiers in Plant Science* 8: 2215.

Gupta, V., Estrada, A. D., Blakley, I., Reid, R., Patel, K., Meyer, M. D., et al. (2015). RNA-Seq analysis and annotation of a draft blueberry genome assembly identifies candidate genes involved in fruit ripening, biosynthesis of bioactive compounds, and stage-specific alternative splicing. *GigaScience* 4(1): 1–22.

Gurcan, K., Teber, S., Ercisli, S. and Yilmaz, K. U. (2016). Genotyping by sequencing (GBS) in apricots and genetic diversity assessment with GBS-derived single-nucleotide polymorphisms (SNPs). *Biochemical Genetics* 54(6): 854–85.

Haji, T., Yaegaki, H. and Yamaguchi, M. (2005). Inheritance and expression of fruit texture melting, non-melting and stony hard in peach. *Scientia Horticulturae* 105(2): 241–8.

Hancock, J. F. (2008). *Temperate Fruit Crop Breeding*. Springer, Dordrecht.
Hayama, H., Tatsuki, M., Ito, A. and Kashimura, Y. (2006). Ethylene and fruit softening in the stony hard mutation in peach. *Postharvest Biology and Technology* 41(1): 16–21.
He, N., Zhang, C., Qi, X., Zhao, S., Tao, Y., Yang, G., et al. (2013). Draft genome sequence of the mulberry tree *Morus notabilis*. *Nature Communications* 4.
Hernández Mora, J. R., Micheletti, D., Bink, M., Weg Vd, Eric, Cantín, C., Nazzicari, N., et al. (2017). Integrated QTL detection for key breeding traits in multiple peach progenies. *BMC Genomics* 18(1): 404–15.
Hirakawa, H., Shirasawa, K., Kosugi, S., Tashiro, K., Nakayama, S., Yamada, M., et al. (2013). Dissection of the octoploid strawberry genome by deep sequencing of the genomes of *Fragaria* species. *DNA Research: An International Journal for Rapid Publication of Reports on Genes and Genomes* 21(2): 169–81.
Huang, S. X., Ding, J., Deng, D. J., Tang, W., Sun, H. H., Liu, D. Y., et al. (2013). Draft genome of the kiwifruit *Actinidia chinensis*. *Nature Communications* 4: 9.
Igarashi, M., Hatsuyama, Y., Harada, T. and Fukasawa-Akada, T. (2016). Biotechnology and apple breeding in Japan. *Breeding Science* 66(1): 18–33.
Illa, E., Sargent, D., Girona, E., Bushakra, J., Cestaro, A., Crowhurst, R., et al. (2011). Comparative analysis of rosaceous genomes and the reconstruction of a putative ancestral genome for the family. *BMC Evolutionary Biology* 11(1): 9.
Imai, A., Nonaka, K., Kuniga, T., Yoshioka, T. and Hayashi, T. (2018). Genome-wide association mapping of fruit-quality traits using genotyping-by-sequencing approach in citrus landraces, modern cultivars, and breeding lines in Japan. *Tree Genetics & Genomes* 14(2): 24.
Iwata, H., Hayashi, T., Terakami, S., Takada, N., Sawamura, Y. and Yamamoto, T. (2013). Potential assessment of genome-wide association study and genomic selection in Japanese pear *Pyrus pyrifolia*. *Breeding Science* 63(1): 125–40.
Jaillon, O., Aury, J.-M., Noel, B., Policriti, A., Clepet, C., Casagrande, A., et al. (2007). The grapevine genome sequence suggests ancestral hexaploidization in major angiosperm phyla. *Nature* 449(7161), 463–5.
Jänsch, M., Broggini, G. A. L., Weger, J., Bus, V. G. M., Gardiner, S. E., Bassett, H., et al. (2015). Identification of SNPs linked to eight apple disease resistance loci. *Molecular Breeding* 35(1): 45.
Jazi, M. M., Seyedi, S. M., Ebrahimie, E., Ebrahimi, M., De Moro, G. and Botanga, C. (2017). A genome-wide transcriptome map of pistachio (*Pistacia vera* L.) provides novel insights into salinity-related genes and marker discovery. *BMC Genomics* 18: 627.
Jensen, P. J., Fazio, G., Altman, N., Praul, C. and McNellis, T. W. (2014). Mapping in an apple (*Malus × domestica*) F1 segregating population based on physical clustering of differentially expressed genes. *BMC Genomics* 15(1): 261.
Jia, Y., Yuan, Y., Zhang, Y., Yang, S. and Zhang, X. (2015). Extreme expansion of NBS-encoding genes in Rosaceae. *BMC Genetics* 16(1): 48.
Johnson, L. K., Malladi, A. and NeSmith, D. S. (2011). Differences in cell number facilitate fruit size variation in Rabbiteye blueberry genotypes. *Journal of the American Society for Horticultural Science* 136(1): 10–15.
Jung, S. and Main, D. (2014). Genomics and bioinformatics resources for translational science in Rosaceae. *Plant Biotechnology Reports* 8(2): 49–64.
Jung, S., Main, D., Staton, M., Cho, I., Zhebentyayeva, T., Arus, P., et al. (2006). Synteny conservation between the *Prunus* genome and both the present and ancestral *Arabidopsis* genomes. *BMC Genomics* 7(1): 81.

Jung, S., Jiwan, D., Cho, I., Lee, T., Abbott, A., Sosinski, B., et al. (2009). Synteny of *Prunus* and other model plant species. *BMC Genomics* 10: 76.

Jung, S., Lee, T., Cheng, C.-H., Buble, K., Zheng, P., Yu, J., et al. (2018). 15 years of GDR: New data and functionality in the genome database for Rosaceae. *Nucleic Acid Research*. doi:10.1093/nar/gky1000.

Kan, J., Liu, T., Ma, N., Li, H., Li, X., Wang, J., et al. (2017). Transcriptome analysis of Callery pear (*Pyrus calleryana*) reveals a comprehensive signalling network in response to *Alternaria alternata*. *PLoS ONE* 12(9): e0184988.

Kellerhals, M., Schutz, S. and Patocchi, A. (2017). Breeding for host resistance to fire blight. *Journal of Plant Pathology* 99: 37–43.

Khajuria, Y. P., Kaul, S., Wani, A. A. and Dhar, M. K. (2018). Genetics of resistance in apple against *Venturia inaequalis* (Wint.) Cke. *Tree Genetics & Genomes* 14(2): 16.

Khan, M. A. and Korban, S. S. (2012). Association mapping in forest trees and fruit crops. *Journal of Experimental Botany* 63(11): 4045–60.

Khan, M. A., Han, Y., Zhao, Y. F., Troggio, M. and Korban, S. S. (2012). A multi-population consensus genetic map reveals inconsistent marker order among maps likely attributed to structural variations in the apple genome. *PLoS ONE* 7(11): e47864.

King, G. J., Lynn, J. R., Dover, C. J., Evans, K. M. and Seymour, G. B. (2001). Resolution of quantitative trait loci for mechanical measures accounting for genetic variation in fruit texture of apple (*Malus pumila* Mill.). *Theoretical and Applied Genetics* 102(8): 1227–35.

Klagges, C., Antonio Campoy, J., Quero-Garcia, J., Guzman, A., Mansur, L., Gratacos, E., et al. (2013). Construction and comparative analyses of highly dense linkage maps of two sweet cherry intra-specific progenies of commercial cultivars. *PLoS ONE* 8(1): e54743.

Klee, H. J. (2010). Improving the flavor of fresh fruits: Genomics, biochemistry, and biotechnology. *New Phytologist* 187(1): 44–56.

Koepke, T., Schaeffer, S., Harper, A., Dicenta, F., Edwards, M., Henry, R. J., et al. (2013). Comparative genomics analysis in *Prunoideae* to identify biologically relevant polymorphisms. *Plant Biotechnology Journal* 11(7): 883–93.

Kole, C., Muthamilarasan, M., Henry, R., Edwards, D., Sharma, R., Abberton, M., et al. (2015). Application of genomics-assisted breeding for generation of climate resilient crops: Progress and prospects. *Frontiers in Plant Science* 6: 563.

Korte, A. and Farlow, A. (2013). The advantages and limitations of trait analysis with GWAS: A review. *Plant Methods* 9(1): 29.

Kose, M. A., Cetinsag, N. and Gurcan, K. (2017). *De novo* transcriptome assembly and SSR marker development in apricot (*Prunus armeniaca*). *Turkish Journal of Agriculture and Forestry* 41(4): 305–15.

Kumar, S., Garrick, D., Bink, M., Whitworth, C., Chagné, D. and Volz, R. (2013). Novel genomic approaches unravel genetic architecture of complex traits in apple. *BMC Genomics* 14(1): 393.

Kumar, S., Rowan, D., Hunt, M., Chagné, D., Whitworth, C. and Souleyre, E. (2015). Genome-wide scans reveal genetic architecture of apple flavour volatiles. *Molecular Breeding* 35(5): 118.

Lalli, D., Abbott, A., Zhebentyayeva, T., Badenes, M., Damsteegt, V., Polák, J., et al. (2008). A genetic linkage map for an apricot (*Prunus armeniaca* L.) BC1 population mapping plum pox virus resistance. *Tree Genetics & Genomes* 4(3): 481–93.

Lamichhane, J. (2014). *Xanthomonas arboricola* diseases of stone fruit, almond and walnut trees: Progress toward understanding and management. *Plant Disease* 98(12): 1600–10.

Lannoo, N. and van Damme, Els J. M. (2014). Lectin domains at the frontiers of plant defense. *Frontiers in Plant Science* 5: 397.

Le Dantec, L., Cardinet, G., Bonet, J., Fouché, M., Boudehri, K., Monfort, A., et al. (2010). Development and mapping of peach candidate genes involved in fruit quality and their transferability and potential use in other *Rosaceae* species. *Tree Genetics & Genomes* 6(6): 995–1012.

Lecouls, A.-C., Bergougnoux, V., Rubio-Cabetas, M. J., Bosselut, N., Voisin, R., Poessel, J. L., et al. (2004). Marker-assisted selection for the wide-spectrum resistance to root-knot nematodes conferred by the Ma gene from myrobalan plum (*Prunus cerasifera*) in interspecific *Prunus* material. *Molecular Breeding* 13(2): 113–24.

Li, D. and Liu, Y. (2016). Gene introgression from wild relatives. In: Testolin, R., Huang, H. and Ferguson, A. R. (Eds), *The Kiwifruit Genome*. Springer International Publishing, Cham, pp. 237–48.

Li, J. F., Gao, Z., Lou, Y. S., Luo, M., Song, S. R., Xu, W. P., et al. (2015). *De novo* assembly, functional annotation, and marker development of Asian pear (*Pyrus pyrifolia*) fruit transcriptome through massively parallel sequencing. *Genetics and Molecular Research* 14(4): 18344–55.

Li, Z., Defoort, J., Tasdighian, S., Maere, S., van de Peer, Y. and De Smet, R. (2016). Gene duplicability of core genes is highly consistent across all angiosperms. *The Plant Cell*. doi:10.1105/tpc.15.00877

Limera, C., Sabbadini, S., Sweet, J. B. and Mezzetti, B. (2017). New biotechnological tools for the genetic improvement of major woody fruit species. *Frontiers in Plant Science* 8: 1418.

Liu, M.-J., Zhao, J., Cai, Q.-L., Liu, G.-C., Wang, J.-R., Zhao, Z.-H., et al. (2014). The complex jujube genome provides insights into fruit tree biology. *Nature Communications* 5.

Liu, X., Wu, S., Xu, J., Sui, C. and Wei, J. (2017). Application of CRISPR/Cas9 in plant biology. *Acta Pharmaceutica Sinica B* 7(3): 292–302.

Longhi, S., Cappellin, L., Guerra, W. and Costa, F. (2013). Validation of a functional molecular marker suitable for marker-assisted breeding for fruit texture in apple (*Malus × domestica* Borkh.). *Molecular Breeding* 32(4): 841–52.

Longhi, S., Giongo, L., Buti, M., Surbanovski, N., Viola, R., Velasco, R., et al. (2014). Molecular genetics and genomics of the Rosoideae: State of the art and future perspectives. *Horticulture Research* 1: 1. doi:10.1038/hortres.2014.1

Lu, Z. X., Sosinski, B., Reighard, G. L., Baird, V. W. and Abbott, A. G. (1998). Construction of a genetic linkage map and identification of AFLP markers for resistance to root-knot nematodes in peach rootstocks. *Genome* 41(2): 199–207.

Ma, X., Zhu, Q., Chen, Y. and Liu, Y. (2016). CRISPR/Cas9 platforms for genome editing in plants: Developments and applications. *Molecular Plant* 9(7): 961–74.

Malyska, A., Bolla, R. and Twardowski, T. (2016). The role of public opinion in shaping trajectories of agricultural biotechnology. *Trends Biotechnology* 34(7): 530–4.

Mangandi, J., Verma, S., Osorio, L., Peres, N. A., van de Weg, E. and Whitaker, V. M. (2017). Pedigree-based analysis in a multiparental population of octoploid strawberry reveals QTL alleles conferring resistance to *Phytophthora cactorum*. *G3: Genes|Genomes|Genetics* 7(6): 1707–19.

Martín-Pizarro, C. and Posé, D. (2018). Genome editing as a tool for fruit ripening manipulation. *Frontiers in Plant Science* 9: 1415. doi:10.3389/fpls.2018.01415.

Martínez-García, P. J., Parfitt, D. E., Bostock, R. E., Fresnedo-Ramírez, J., Vazquez-Lobo, A., Ogundiwin, E. A., et al. (2013a). Application of genomic and quantitative genetic tools to identify candidate resistance genes for brown rot resistance in peach. *PLoS ONE* 8(11): e78634.

Martínez-García, P., Parfitt, D., Ogundiwin, E., Fass, J., Chan, H., Ahmad, R., et al. (2013b). High density SNP mapping and QTL analysis for fruit quality characteristics in peach (*Prunus persica* L.). *Tree Genetics & Genomes* 9(1): 19–36.

Martínez-García, P. J., Crepeau, M. W., Puiu, D., Gonzalez-Ibeas, D., Whalen, J., Stevens, K. A., et al. (2016). The walnut (*Juglans regia*) genome sequence reveals diversity in genes coding for the biosynthesis of non-structural polyphenols. *The Plant Journal: For Cell and Molecular Biology* 87(5): 507–32.

Meneses, C., Ulloa-Zepeda, L., Cifuentes-Esquivel, A., Infante, R., Cantin, C. M., Batlle, I., et al. (2016). A codominant diagnostic marker for the slow ripening trait in peach. *Molecular Breeding* 36(6): 77.

Mézard, C. (2006). Meiotic recombination hotspots in plants. *Biochemical Society Transactions* 34(Pt 4): 531–4.

Micheletti, D., Dettori, M. T., Micali, S., Aramini, V., Pacheco, I., Linge, C. D. S., et al. (2015). Whole-genome analysis of diversity and SNP-major gene association in peach germplasm. *PLoS ONE* 10(9): e0136803.

Migicovsky, Z. and Myles, S. (2017). Exploiting wild relatives for genomics-assisted breeding of perennial crops. *Frontiers in Plant Science* 8: 460.

Minamikawa, M. F., Nonaka, K., Kaminuma, E., Kajiya-Kanegae, H., Onogi, A., Goto, S., et al. (2017). Genome-wide association study and genomic prediction in citrus: Potential of genomics-assisted breeding for fruit quality traits. *Scientific Reports* 7: 4721.

Minas, I. S., Font i Forcada, C., Dangl, G. S., Gradziel, T. M., Dandekar, A. M. and Crisosto, C. H. (2015). Discovery of non-climacteric and suppressed climacteric bud sport mutations originating from a climacteric Japanese plum cultivar (*Prunus salicina* Lindl.). *Frontiers in Plant Science* 6: 316.

Ming, R., Hou, S., Feng, Y., Yu, Q., Dionne-Laporte, A., Saw, J. H., et al. (2008). The draft genome of the transgenic tropical fruit tree papaya (*Carica papaya* Linnaeus). *Nature* 452(7190): 991–6. http://dx.doi.org/10.1038/nature06856.

Mir, H. and Patel, V. B. (2018). Genetic engineering of temperate fruit crops. In: Rout, G. R. and Peter, K. V. (Eds), *Genetic Engineering of Horticultural Crops*. Academic Press, pp. 89–119. doi:10.1016/B978-0-12-810439-2.00005-2.

Mitani, N., Kono, A., Yamada, M., Sato, A., Kobayashi, S., Ban, Y., et al. (2014). Application of marker-assisted selection in persimmon breeding of PCNA offspring using SCAR markers among the population from the cross between Non-PCNA 'Taigetsu' and PCNA 'Kanshu'. *HortScience* 49(9): 1132–5.

Morgutti, S., Negrini, N., Ghiani, A., Baldin, F., Bassi, D. and Cocucci, M. (2017). *Endopolygalacturonase* gene polymorphisms: Asset of the locus in different peach accessions. *American Journal of Plant Sciences* 8(4): 941–57.

Moriya, S., Kunihisa, M., Okada, K., Shimizu, T., Honda, C., Yamamoto, T., et al. (2017). Allelic composition of *MdMYB1* drives red skin color intensity in apple (*Malus* × *domestica* Borkh.) and its application to breeding. *Euphytica* 213(4): 78.

Muranty, H., Troggio, M., Sadok, I. B., Rifai, M. A., Auwerkerken, A., Banchi, E., et al. (2015). Accuracy and responses of genomic selection on key traits in apple breeding. *Horticulture Research* 2: 15060.

Navarro, J. A. R., Phillips-Mora, W., Arciniegas-Leal, A., Mata-Quiros, A., Haiminen, N., Mustiga, G., et al. (2017). Application of genome wide association and genomic prediction for improvement of cacao productivity and resistance to black and frosty pod diseases. *Frontiers in Plant Science* 8: 1905.

Nock, C. J., Elphinstone, M. S., Ablett, G., Kawamata, A., Hancock, W., Hardner, C. M., et al. (2014). Whole genome shotgun sequences for microsatellite discovery and application in cultivated and wild Macadamia (*Proteaceae*). *Applications in Plant Sciences* 2(4).

Noh, Y., Lee, S., Whitaker, V. M., Cearley, K. R. and Cha, J. (2017). A high-throughput marker-assisted selection system combining rapid DNA extraction high-resolution melting and simple sequence repeat analysis: Strawberry as a model for fruit crops. *Journal of Berry Research* 7(1): 23–31.

Noh, Y., Oh, Y., Mangandi, J., Verma, S., Zurn, J. D., Lu, Y., et al. (2018). High-throughput marker assays for *FaRPc2*-mediated resistance to Phytophthora crown rot in octoploid strawberry. *Molecular Breeding* 38(8): 104.

Nsibi, M., Confolent, C., Gouble, B., Bureau, S., Blanc, A., Roch, G., Lambert, P., et al. (2018). Genomic selection – which prospects in *Prunus armeniaca*? preliminary results issues from fruit quality traits and phenology. *Proceedings of 9th International Genomic Conference*, Nanjing, China, p. 61.

O'Connor, K., Hayes, B. and Topp, B. (2018). Prospects for increasing yield in macadamia using component traits and genomics. *Tree Genetics & Genomes* 14(1): 7.

Ohtake, Y., Takahashi, T. and Komeda, Y. (2000). Salicylic acid induces the expression of a number of receptor-like kinase genes in Arabidopsis thaliana. *Plant & Cell Physiology* 41(9): 1038–44.

Oliveira Lino, L., Pacheco, I., Mercier, V., Faoro, F., Bassi, D., Bornard, I., et al. (2016). Brown rot strikes *Prunus* fruit: An ancient fight almost always lost. *Journal of Agricultural and Food Chemistry* 64(20): 4029–47.

Olmstead, M. A., Gilbert, J. L., Colquhoun, T. A., Clark, D. G., Kluson, R. and Moskowitz, H. R. (2015). In pursuit of the perfect peach: Consumer-assisted selection of peach fruit traits. *HortScience* 50(8): 1202–12.

Ong, W. D., Voo, L. C. and Kumar, V. S. (2012). *De Novo* assembly, characterization and functional annotation of pineapple fruit transcriptome through massively parallel sequencing. *PLoS ONE* 7(10): e46937.

Osorio, L. F., Gezan, S., Verma, S. and Whitaker, V. M. (2018). Genomic prediction in strawberry breeding at the University of Florida. *Proceedings of 9th International Rosaceae Genomic Conference*, Nanjing, China, p. 75.

Pacheco, I., Bassi, D., Eduardo, I., Ciacciulli, A., Pirona, R., Rossini, L., et al. (2014). QTL mapping for brown rot (*Monilinia fructigena*) resistance in an intraspecific peach (*Prunus persica* L. Batsch) F1 progeny. *Tree Genetics & Genomes* 10(5): 1223–42.

Padmarasu, S., Sargent, D. J., Jaensch, M., Kellerhals, M., Tartarini, S., Velasco, R., et al. (2014). Fine-mapping of the apple scab resistance locus *Rvi12* (*Vb*) derived from 'Hansen's baccata #2'. *Molecular Breeding* 34(4): 2119–29.

Palacio-Bielsa, A., Roselló, M., Cambra, M. A. and López, M. M. (2010). First report on almond in Europe of bacterial spot disease of stone fruits caused by *Xanthomonas arboricola* pv. *pruni*. *Plant Disease* 94(6): 786.

Pan, L., Zeng, W., Niu, L., Lu, Z., Wang, X., Liu, H., et al. (2015). *PpYUC11*, a strong candidate gene for the stony hard phenotype in peach (*Prunus persica* L. Batsch), participates in IAA biosynthesis during fruit ripening. *Journal of Experimental Botany* 66(22): 7031–44.

Pandolfi, V., Neto, J. R. C. F., da Silva, M. D., Amorim, L. L. B., Wanderley-Nogueira, A. C., de Oliveira Silva, R. L., et al. (2017). Resistance (R) genes: Applications and prospects for plant biotechnology and breeding. *Current Protein & Peptide Science* 18(4): 323–34.

Parker, M. L., Ritchie, D. and Nyczepir, A. (1997). Peach rootstock performance of BY-520-9 and Lovell in a peach tree short life replant site. *HortScience* 32(3): 497.

Passaro, M., Geuna, F., Bassi, D. and Cirilli, M. (2017). Development of a high-resolution melting approach for reliable and cost-effective genotyping of *PPVres locus* in apricot (*P. armeniaca*). *Molecular Breeding* 37(6): 74.

Payasi, A. and Sanwal, G. (2010). Ripening of climacteric fruits and their control. *Journal of Food Biochemistry* 34(4): 679–710.

Peace, C. P. (2017). DNA-informed breeding of rosaceous crops: Promises, progress and prospects. *Horticulture Research* 4: 17006.

Peace, C. P., Crisosto, C. H. and Gradziel, T. M. (2005). Endopolygalacturonase: A candidate gene for freestone and melting flesh in peach. *Molecular Breeding* 16(1): 21–31.

Peace, C. P., Crisosto, C. H., Garner, D. T., Dandekar, A. M., Gradziel, T. M. and Bliss, F. A. (2006). Genetic control of internal breakdown in peach. *Acta Horticulturae* 713: 489–96.

Peace, C., Bassil, N., Main, D., Ficklin, S., Rosyara, U. R., Stegmeir, T., et al. (2012). Development and evaluation of a genome-wide 6K SNP array for diploid sweet cherry and tetraploid sour cherry. *PLoS ONE* 7(12): e48305.

Perchepied, L., Leforestier, D., Ravon, E., Guérif, P., Denancé, C., Tellier, M., et al. (2015). Genetic mapping and pyramiding of two new pear scab resistance QTLs. *Molecular Breeding* 35(10): 197.

Picanol, R., Eduardo, I., Aranzana, M. J., Howad, W., Batlle, I., Iglesias, I., et al. (2013). Combining linkage and association mapping to search for markers linked to the flat fruit character in peach. *Euphytica* 190(2): 279–88.

Pirona, R., Eduardo, I., Pacheco, I., Da Silva Linge, C., Miculan, M., Verde, I., et al. (2013). Fine mapping and identification of a candidate gene for a major locus controlling maturity date in peach. *BMC Plant Biology* 13(1): 166.

Polashock, J., Zelzion, E., Fajardo, D., Zalapa, J., Georgi, L., Bhattacharya, D., et al. (2014). The American cranberry: First insights into the whole genome of a species adapted to bog habitat. *BMC Plant Biology* 14(1): 165.

Qi, X., Liu, C., Song, L., Li, Y. and Li, M. (2017). PaCYP78A9, a Cytochrome P450, regulates fruit size in sweet cherry (*Prunus avium* L.). *Frontiers in Plant Science* 8: 2076.

Qian, M., Zhang, Y., Yan, X., Han, M., Li, J., Li, F., et al. (2016). Identification and expression analysis of Polygalacturonase family members during peach fruit softening. *International Journal of Molecular Sciences* 17(11): 1933–50.

Qian, L., Hickey, L. T., Stahl, A., Werner, C. R., Hayes, B., Snowdon, R. J., et al. (2017). Exploring and harnessing haplotype diversity to improve yield stability in crops. *Frontiers in Plant Science* 8: 1534.

Quilot, B., Wu, B. H., Kervella, J., Génard, M., Foulongne, M. and Moreau, K. (2004). QTL analysis of quality traits in an advanced backcross between *Prunus persica* cultivars and the wild relative species *P. davidiana*. *Theoretical and Applied Genetics* 109(4): 884–97.

Quilot-Turion, B., Scariotto, S., Corre, M., Oliveira Lino Leandro, Signoret, V., Dini Vinoly, M., et al. (2018). Exploring the genetic variability of peach skin phenolics and triterpenoids as natural defenses against brown rot. *Proceedings of 9th International Rosaceae Genomic Conference*, Nanjing, China, p. 38.

Rihani, K. A. L., Jacobsen, H., Hofmann, T., Schwab, W. and Hassan, F. (2017). Metabolic engineering of apple by overexpression of the *MdMyb10* gene. *Journal of Genetic Engineering and Biotechnology* 15(1): 263–73.

Rowan, D. D., Hunt, M. B., Dimouro, A., Alspach, P. A., Weskett, R., Volz, R. K., et al. (2009). Profiling fruit volatiles in the progeny of a 'Royal Gala' × 'Granny Smith' Apple (Malus × domestica) cross. *Journal of Agricultural and Food Chemistry* 57(17): 7953–61.

Rowley, E. R., Fox, S. E., Bryant, D. W., Sullivan, C. M., Priest, H. D., Givan, S. A., et al. (2012). Assembly and characterization of the European hazelnut 'Jefferson' transcriptome. *Crop Science* 52: 2679–86. doi:10.2135/cropsci2012.02.0065.

Ru, S., Main, D., Evans, K. and Peace, C. (2015). Current applications, challenges, and perspectives of marker-assisted seedling selection in Rosaceae tree fruit breeding. *Tree Genetics & Genomes* 11(1): 8.

Rubio, M., Ruiz, D., Egea, J., Martínez-Gómez, P. and Dicenta, F. (2008). Evaluation of apricot resistance to *Plum pox virus* (Sharka) in controlled greenhouse and natural field conditions. *Scientia Horticulturae* 116(2): 176–9.

Rungjindamai, N., Jeffries, P. and Xu, X. (2014). Epidemiology and management of brown rot on stone fruit caused by *Monilinia laxa*. *European Journal of Plant Pathology* 140(1): 3.

Salazar, J. A., Batnini, M., Trifi-Farah, N., Ruiz, D., Martínez-Gómez, P. and Rubio, M. (2016). Quantitative trait loci (QTLs) identification and the transmission of resistance to powdery mildew in apricot. *Euphytica* 211(2): 245–54.

Salazar, J. A., Pacheco, I., Shinya, P., Zapata, P., Silva, C., Aradhya, M., et al. (2017). Genotyping by sequencing for SNP-based linkage analysis and identification of QTLs linked to fruit quality traits in Japanese plum (*Prunus salicina* Lindl.). *Frontiers in Plant Science* 8: 476.

Sanchez, G., Besada, C., Badenes, M. L., Monforte, A. J. and Granell, A. (2012). A non-targeted approach unravels the volatile network in peach fruit. *PLoS ONE* 7(6): e38992.

Sanchez, G., Venegas-Caleron, M., Salas, J., Monforte, A., Badenes, M. and Granell, A. (2013). An integrative 'omics' approach identifies new candidate genes to impact aroma volatiles in peach fruit. *BMC Genomics* 14(1): 343.

Sánchez-Sevilla, J. F., Cruz-Rus, E., Valpuesta, V., Botella, M. A. and Amaya, I. (2014). Deciphering gamma-decalactone biosynthesis in strawberry fruit using a combination of genetic mapping, RNA-Seq and eQTL analyses. *BMC Genomics* 15(1): 218.

Sandefur, P., Oraguzie, N. and Peace, C. (2016). A DNA test for routine prediction in breeding of sweet cherry fruit color, Pav-R_f-SSR. *Molecular Breeding* 36(3): 33.

Sandefur, P., Frett, T., Clark, J., Gasic, K. and Peace, C. (2017). A DNA test for routine prediction in breeding of peach blush, Ppe-R_f-SSR. *Molecular Breeding* 37(1): 1–15.

Scaglione, D., Fornasiero, A., Pinto, C., Cattonaro, F., Spadotto, A., Infante, R., et al. (2015). A RAD-based linkage map of kiwifruit (*Actinidia chinensis* Pl.) as a tool to improve the genome assembly and to scan the genomic region of the gender determinant for the marker-assisted breeding. *Tree Genetics & Genomes* 11(6): 115.

Schlathölter, I., Jänsch, M., Flachowsky, H., Broggini, G. A. L., Hanke, M. and Patocchi, A. (2018). Generation of advanced fire blight-resistant apple (*Malus* × *domestica*) selections of the fifth generation within 7 years of applying the early flowering approach. *Planta* 247(6): 1475–88.

Schlautman, B., Covarrubias-Pazaran, G., Diaz-Garcia, L., Iorizzo, M., Polashock, J., Grygleski, E., et al. (2017). Construction of a high-density American cranberry (*Vaccinium macrocarpon* Ait.) composite map using genotyping-by-sequencing for multi-pedigree linkage mapping. *G3: Genes|Genomes|Genetics* 7(4): 1177–89.

Schlautman, B., Diaz-Garcia, L., Covarrubias-Pazaran, G., Schlautman, N., Vorsa, N., Polashock, J., et al. (2018). Comparative genetic mapping reveals synteny and collinearity between the American cranberry and diploid blueberry genomes. *Molecular Breeding* 38(1): 9.

Schropfer, S., Bottcher, C., Wohner, T., Richter, K., Norelli, J., Rikkerink, E. H. A., et al. (2018). A single effector protein, AvrRpt2EA, from *Erwinia amylovora* can cause fire blight disease symptoms and induces a salicylic acid-dependent defense response. *Molecular Plant-Microbe Interactions* 31: 1179–91.

Schwieterman, M. L., Colquhoun, T. A., Jaworski, E. A., Bartoshuk, L. M., Gilbert, J. L., Tieman, D. M., et al. (2014). Strawberry flavor: Diverse chemical compositions, a seasonal influence, and effects on sensory perception. *PLoS ONE* 9(2): e88446.

Scorza, R. and Sherman, W. B. (1996). Peaches. In: Janick, J. and Moore, J. N. (Eds), *Fruit Breeding*. Wiley, New York, NY, pp. 325–440.

Scorza, R., May, L. G., Purnell, B. and Upchurch, B. (1991). Difference in number and area of mesocarp cells between small- and large-fruited peach cultivars. *J. Am. Soc. Hort. Sci.* 5(116): 861–4.

Scorza, R., Callahan, A., Dardick, C., Ravelonandro, M., Polak, J., Malinowski, T., et al. (2013). Genetic engineering of *Plum pox* virus resistance: 'HoneySweet' plum–from concept to product. *Plant Cell, Tissue and Organ Culture (PCTOC)* 115(1): 1–12.

Serra, O., Gina Bordonaba, J., Eduardo, I., Bonany, J., Echeverria, G., Larrigaudiere, C., et al. (2017). Genetic analysis of the slow-melting flesh character in peach. *Tree Genetics & Genomes* 13: 77.

Sharma, N., Singh, S. K., Lal, S. and Singh, N. K. (2016). Genome sequence information in fruit crops: Current status. *Transcriptomics* 4: 130. doi:10.4172/2329-8936.1000130.

Shen, Y. M., Huang, T. C., Chao, C. H. and Liu, H. L. (2013). First report of bacterial spot caused by *Xanthomonas arboricola* pv. *pruni* on Japanese plum in Taiwan. *Plant Disease* 97(6): 835.

Shirasawa, K., Isuzugawa, K., Ikenaga, M., Saito, Y., Yamamoto, T., Hirakawa, H., et al. (2017). The genome sequence of sweet cherry (*Prunus avium*) for use in genomics-assisted breeding. *DNA Research: An International Journal for Rapid Publication of Reports on Genes and Genomes* 24(5): 499–508.

Shulaev, V., Sargent, D. J., Crowhurst, R. N., Mockler, T. C., Folkerts, O., Delcher, A. L., et al. (2010). The genome of woodland strawberry (*Fragaria vesca*). *Nature Genetics* 43(2): 109–18.

Singh, N. K., Mahato, A. K., Jayaswal, P. K., Singh, A., Singh, S., Singh, N., et al. (2016). Origin, diversity and genome sequence of mango (*Magnifera indica* L.). *Indian Journal of History of Science* 51(2.2): 355–68.

Socquet-Juglard, D., Duffy, B., Pothier, J. F., Christen, D., Gessler, C. and Patocchi, A. (2013a). Identification of a major QTL for *Xanthomonas arboricola* pv. *pruni* resistance in apricot. *Tree Genetics & Genomes* 9: 409–21.

Socquet-Juglard, D., Kamber, T., Pothier, J. F., Christen, D., Gessler, C., Duffy, B., et al. (2013b). Comparative RNA-Seq analysis of early-infected peach leaves by the invasive phytopathogen *Xanthomonas arboricola* pv. *pruni*. *PLoS ONE* 8(1): e54196.

Sooriyapathirana, S., Khan, A., Sebolt, A., Wang, D., Bushakra, J., Lin-Wang, K., et al. (2010). QTL analysis and candidate gene mapping for skin and flesh color in sweet cherry fruit (*Prunus avium* L.). *Tree Genetics & Genomes*: 1–12.

Srinivasan, C., Dardick, C., Callahan, A. and Scorza, R. (2012). Plum (*Prunus domestica*) trees transformed with poplar FT1 result in altered architecture, dormancy requirement, and continuous flowering. *PLoS ONE* 7(7): e40715.

Stankiewicz-Kosyl, M., Pitera, E. and Gawronski, S. W. (2005). Mapping QTL involved in powdery mildew resistance of the apple clone U 211. *Plant Breeding* 124(1): 63–6.

Stegmeir, T., Cai, L., Basundari, F. R. A., Sebolt, A. M. and Iezzoni, A. F. (2015). A DNA test for fruit flesh color in tetraploid sour cherry (*Prunus cerasus* L.). *Molecular Breeding* 35(7): 149.

Stephens, M. J., Buck, E. J. and Tahir, J. (2016). Mapping a potential resistance gene for Raspberry bushy dwarf virus in red raspberry. *XI International Rubus and Ribes Symposium* 1133: 121–7.

Stich, B. and Van Inghelandt, D. (2018). Prospects and potential uses of genomic prediction of key performance traits in tetraploid potato. *Frontiers in Plant Science* 9: 159.

Sun, X., Wang, P., Jia, X., Huo, L., Che, R. and Ma, F. (2018). Improvement of drought tolerance by overexpressing *MdATG18a* is mediated by modified antioxidant system and activated autophagy in transgenic apple. *Plant Biotechnology Journal* 16(2): 545–57.

Tang, H., Bowers, J. E., Wang, X., Ming, R., Alam, M. and Paterson, A. H. (2008). Synteny and collinearity in plant genomes. *Science* 320(5875): 486–8.

Tang, W., Zheng, Y., Dong, J., Yu, J., Yue, J., Liu, F., et al. (2016). Comprehensive transcriptome profiling reveals long noncoding RNA expression and alternative splicing regulation during fruit development and ripening in kiwifruit (*Actinidia chinensis*). *Frontiers in Plant Science* 7: 335.

Telias, A., Lin-Wang, K., Stevenson, D. E., Cooney, J. M., Hellens, R. P., Allan, A. C., et al. (2011). Apple skin patterning is associated with differential expression of *MYB10*. *BMC Plant Biology* 11: 93.

Terakami, S., Shoda, M., Adachi, Y., Gonai, T., Kasumi, M., Sawamura, Y., et al. (2006). Genetic mapping of the pear scab resistance gene *Vnk* of Japanese pear cultivar Kinchaku. *Theoretical and Applied Genetics* 113(4): 743–52.

Tjou-Tam-Sin, N. N. A., van de Bilt, J. L. J., Bergsma-Vlami, M., Koenraadt, H., Naktuinbouw, J. W., van Doorn, J., et al. (2012). First report of *Xanthomonas arboricola* pv. *pruni* in Ornamental *Prunus laurocerasus* in the Netherlands. *Plant Disease* 96(5): 759.

Ulrich, D., Kecke, S. and Olbricht, K. (2018). What do we know about the chemistry of strawberry aroma? *Journal of Agricultural and Food Chemistry* 66(13): 3291–301.

Upadhyay, A. (2018). Transgenic research in fruit crops. In: Rout, G. R. and Peter, K. V. (Eds), *Genetic Engineering of Horticultural Crops*. Academic Press, pp. 63–87. doi:10.1016/B978-0-12-810439-2.00004-0.

Urrestarazu, J., Muranty, H., Denance, C., Leforestier, D., Ravon, E., Guyader, A., et al. (2017). Genome-wide association mapping of flowering and ripening periods in apple. *Frontiers in Plant Science* 8: 1923.

van de Weg, E., Di Guardo, M., Jänsch, M., Socquet-Juglard, D., Costa, F., Baumgartner, I., et al. (2017). Epistatic fire blight resistance QTL alleles in the apple cultivar

'Enterprise' and selection X-6398 discovered and characterized through pedigree-informed analysis. *Molecular Breeding* 38(1): 5.

Van der Peer, Y., Mizrachi, E. and Marchal, K. (2017). The evolutionary significance of polyploidy. *Nature Reviews Genetics* 18: 411.

VanBuren, R., Bryant, D., Bushakra, J. M., Vining, K. J., Edger, P. P., Rowley, E. R., et al. (2016). The genome of black raspberry (*Rubus occidentalis*). *The Plant Journal: For Cell and Molecular Biology* 87(6): 535–47.

Vanderzande, S., Piaskowski, J. L., Luo, F., Edge Garza, D. A., Klipfel, J., Schaller, A., et al. (2018). Crossing the finish line: How to develop diagnostic DNA tests as breeding tools after QTL discovery. *Journal of Horticulture* 5(1): 1–6.

Vanderzande, S., Cai, L., Howard, N. P., Da Silva Linge, C., Bink, Marko C. A. M., Gasic, K., et al. (2019). *A Data Curation Workflow Providing Access to High-Quality, Genome-Wide SNP Genotypic Data for Pedigreed Germplasm of Diploid Outbreeding Species as Demonstrated in Apple, Peach, and Sweet Cherry (PLoS ONE submitted).*

Velasco, R., Zharkikh, A., Troggio, M., Cartwright, D. A., Cestaro, A., Pruss, D., et al. (2007). A high quality draft consensus sequence of the genome of a heterozygous grapevine variety. *PLoS ONE* 2(12): e1326.

Velasco, R., Zharkikh, A., Affourtit, J., Dhingra, A., Cestaro, A., Kalyanaraman, A., et al. (2010). The genome of the domesticated apple (*Malus* × *domestica* Borkh.). *Nature Genetics* 42(10): 833–9.

Verde, I., Bassil, N., Scalabrin, S., Gilmore, B., Lawley, C. T., Gasic, K., et al. (2012). Development and evaluation of a 9K SNP array for peach by internationally coordinated SNP detection and validation in breeding germplasm. *PLoS ONE* 7(4): e35668.

Verde, I., Abbott, A. G., Scalabrin, S., Jung, S., Shu, S., Marroni, F., et al. (2013). The high-quality draft genome of peach (*Prunus persica*) identifies unique patterns of genetic diversity, domestication and genome evolution. *Nature Genetics* 45(5): 487–94.

Verde, I., Jenkins, J., Dondini, L., Micali, S., Pagliarani, G., Vendramin, E., et al. (2017). The Peach v2.0 release: High-resolution linkage mapping and deep resequencing improve chromosome-scale assembly and contiguity. *BMC Genomics* 18(1): 225.

Verma, S., Zurn, J. D., Salinas, N., Mathey, M. M., Denoyes, B., Hancock, J. F., et al. (2017). Clarifying sub-genomic positions of QTLs for flowering habit and fruit quality in U.S. strawberry (*Fragaria* × *ananassa*) breeding populations using pedigree-based QTL analysis. *Horticulture Research* 4: 17062.

Verma, S., Osorio, L. F., Lee, S., Bassil, N. V. and Whitaker, V. M. (2018). Genome assisted breeding in the octoploid strawberry. In: Hytonen, T., Graham, J. and Harrison, R. (Eds), *The Genomes of Rosaceous Berries and their Wild Relatives*. Springer, New York, pp. 161–84.

Viana, A. P., Vilela de Resende, M. D., Riaz, S. and Walker, M. A. (2016). Genome selection in fruit breeding: Application to table grapes. *Scientia Agricola* 73(2): 142–9.

Vilanova, S., Romero, C., Abbott, A. G., Llacer, G. and Badenes, M. L. (2003). An apricot (*Prunus armeniaca* L.) F2 progeny linkage map based on SSR and AFLP markers, mapping plum pox virus resistance and self-incompatibility traits. *Theoretical Applied Genetics* 107: 239–47.

Wang, Y., Yang, C., Li, S., Yang, L., Wang, Y., Zhao, J., et al. (2009). Volatile characteristics of 50 peaches and nectarines evaluated by HP-SPME with GC-MS. *Food Chemistry* 116(1): 356–64.

Wang, L., Zhao, S., Gu, C., Zhou, Y., Zhou, H., Ma, J., et al. (2013). Deep RNA-Seq uncovers the peach transcriptome landscape. *Plant Molecular Biology* 83(4–5): 365–77.

Wang, Q., Wang, L., Zhu, G., Cao, K., Fang, W., Chen, C., et al. (2016). DNA marker-assisted evaluation of fruit acidity in diverse peach (*Prunus persica*) germplasm. *Euphytica* 210(3): 413–26.

Ward, J., Bhangoo, J., Fernandez-Fernandez, F., Moore, P., Swanson, J. D., Viola, R., et al. (2013). Saturated linkage map construction in *Rubus idaeus* using genotyping by sequencing and genome-independent imputation. *BMC Genomics* 14(1): 2.

Won, K., Bastiaanse, H., Kim, Y. K., Song, J. H., Kang, S. S., Lee, H. C., et al. (2014). Genetic mapping of polygenic scab (*Venturia pirina*) resistance in an interspecific pear family. *Molecular Breeding* 34(4): 2179–89.

Woodhead, M., Weir, A., Smith, K., McCallum, S., MacKenzie, K. and Graham, J. (2010). Functional markers for red raspberry. *Journal of the American Society for Horticultural Science* 135(5): 418–27.

Wu, J., Wang, Z., Shi, Z., Zhang, S., Ming, R., Zhu, S., et al. (2012). The genome of the pear (*Pyrus bretschneideri* Rehd.). *Genome Research* 23(2): 396–408.

Xu, Q., Chen, L.-L., Ruan, X., Chen, D., Zhu, A., Chen, C., et al. (2013). The draft genome of sweet orange (*Citrus sinensis*). *Nature Genetics*, 45(1), 59–66. http://dx.doi.org/10.1038/ng.2472.

Yamamoto, T. and Hayashi, T. (2002). New root-knot nematode resistance genes and their STS markers in peach. *Scientia Horticulturae*, 96(1–4): 81–90.

Yan, J., Ban, Z., Lu, H., Li, D., Poverenov, E., Luo, Z., et al. (2018). The aroma volatile repertoire in strawberry fruit: A review. *Journal of the Science of Food and Agriculture* 98(12): 4395–402.

Yang, N., Reighard, G., Ritchie, D., Okie, W. and Gasic, K. (2013). Mapping quantitative trait loci associated with resistance to bacterial spot *(Xanthomonas arboricola* pv. *pruni)* in peach. *Tree Genetics & Genomes* 9(2): 573–86.

Zeballos, J. L., Abidi, W., Gimenez, R., Monforte, A. J., Angeles Moreno, M. and Gogorcena, Y. (2016). Mapping QTLs associated with fruit quality traits in peach [*Prunus persica* (L.) Batsch] using SNP maps. *Tree Genetics & Genomes* 12(3): 37.

Zhang, J., Liu, J. and Ming, R. (2014). Genomic analyses of the CAM plant pineapple. *Journal of Experimental Botany*, 65(13), 3395–404.

Zhang, G., Sebolt, A., Sooriyapathirana, S., Wang, D., Bink, M., Olmstead, J., et al. (2010). Fruit size QTL analysis of an F_1 population derived from a cross between a domesticated sweet cherry cultivar and a wild forest sweet cherry. *Tree Genetics & Genomes* 6(1): 25–36.

Zhang, Q., Chen, W., Sun, L., Zhao, F., Huang, B., Yang, W., et al. (2012). The genome of *Prunus mume*. *Nature* Communications, 3, 1318.

Zhang, X. J., Wang, L. X., Chen, X. X., Liu, Y. L., Meng, R., Wang, Y. J., et al. (2014). A and *MdMYB1* allele-specific markers controlling apple (*Malus × domestica* Borkh.) skin color and suitability for marker-assisted selection. *Genetics and Molecular Research* 13(4): 9103–14.

Zhang, Y., Yin, X., Xiao, Y., Zhang, Z., Li, S., Liu, X., et al. (2018). An ethylene response factor-MYB transcription complex regulates furaneol biosynthesis by activating *Quinone Oxidoreductase* expression in strawberry. *Plant Physiology* 178: 189–201.

Zheng, Z., Qualley, A., Fan, B., Dudareva, N. and Chen, Z. (2009). An important role of a BAHD acyl transferase-like protein in plant innate immunity. *The Plant Journal: For Cell and Molecular Biology* 57(6): 1040–53.

Zhong, Y., Yin, H., Sargent, D. J., Malnoy, M. and Cheng, Z. M. (2015). Species-specific duplications driving the recent expansion of NBS-LRR genes in five *Rosaceae* species. *BMC Genomics* 16: 77.

Zhong, Y., Zhang, X. and Cheng, Z. (2018). Lineage-specific duplications of NBS-LRR genes occurring before the divergence of six *Fragaria* species. *BMC Genomics* 19(1): 128.

Zhou, H., Lin-Wang, K., Wang, H., Gu, C., Dare, A. P., Espley, R. V., et al. (2015). Molecular genetics of blood-fleshed peach reveals activation of anthocyanin biosynthesis by NAC transcription factors. *The Plant Journal: For Cell and Molecular Biology* 82(1): 105–21.

Zhou, J., Wang, G. and Liu, Z. (2018). Efficient genome editing of wild strawberry genes, vector development and validation. *Plant Biotechnology Journal*. https://doi.org/10.1111/pbi.12922.

Zorrilla-Fontanesi, Y., Rambla, J., Cabeza, A., Medina, J. J., Sánchez-Sevilla, J. F., Valpuesta, V., et al. (2012). Genetic analysis of strawberry fruit aroma and identification of O-*methyltransferase FaOMT* as the locus controlling natural variation in mesifurane content. *Plant Physiology* 159(2): 851–70.

Zuriaga, E., Romero, C., Miguel Blanca, J. and Luisa Badenes, M. (2018). Resistance to *Plum Pox Virus* (PPV) in apricot (*Prunus armeniaca* L.) is associated with down-regulation of two *MATHd* genes. *BMC Plant Biology* 18: 25.

Part 2

Optimizing sustainable cultivation and quality

Chapter 6

Optimizing production of quality nursery plants for fruit tree cultivation

Stefano Musacchi, Washington State University and Tree Fruit Research and Extension Center, USA; and Davide Neri, Università Politecnica delle Marche, Italy

1 Introduction

Grafting of high-quality fruiting scion genotypes onto rootstocks is a necessary practice in fruit-tree production to assure high yield efficiency and fruit quality. While the use of scion-rootstock combinations has made it possible to solve many orchard problems, it also has created other problems on the biological-functionality of the tree. Several critical areas of ongoing investigation regarding rootstock-scion interactions exist: structural and molecular analyses of tissues grown together at the graft union; new vascular connections; xylem and phloem flow; xylem transport of nutrients and amino acids; canopy gas exchange; water consumption and use efficiency; and canopy-to-root ratio (Andrews and Serrano-Marquez, 1993).

Historically, rootstocks were simply a seedling of the same species as the scion variety. Modern pome and stone fruit orchards are designed to increase plant density, tree efficiency, light interception and early cropping with less use of labour per yield unit (Neri et al., 2010). For these reasons, rootstock production by vegetative propagation ensures uniform trees with similar cropping characteristics (Musacchi, 2008a; Musacchi et al., 2015; Musacchi and Green, 2017). Clonal rootstocks for pome and stone crops can provide a broad range of tree size control (Howard, 1987; Musacchi, 2008a). Further, researchers

http://dx.doi.org/10.19103/AS.2018.0040.06

are evaluating the propagation of self-rooted scion varieties (Ossani, 1991; De Paoli, 1996; Neri et al., 1989) as has been done in olives (*Olea europaea*) (Cioccolanti et al., 2012), kiwi (*Actinidia deliciosa*), nuts (Cozzolino and Neri, 2017), and pear (*Pyrus communis*) cultivars William and Abbe' Fetel (Colombo, 2003). This is convenient when self-rooting is easy, and no specific advantages (e.g. yield efficiency, pest resistance and fruit quality improvement) are induced by available rootstocks. Also, in this case, young potted trees (less than 1-year-old) can be prepared to meet the demands of different planting systems. While many fruit industries have shifted towards alternative rootstock genetic sources, seedling rootstocks are still widely used for peach and pear orchards.

High-quality nursery trees are desired by fruit growers to obtain early returns on orchard investments. For this reason, in recent years the nursery industry has developed different types of trees congruent with the desired orchard training system design (Musacchi, 2007a, 2008b; Neri et al., 2015). Nearly all trees planted in modern fruit orchards are a combination of two genotypes: rootstock and scion. Rootstocks are produced utilizing different techniques such as stooling, layering, cuttings, micropropagation ('in vitro') and seedlings. The grafting material (i.e. bud or bud-stick) of the scion is produced in special orchards called mother plant blocks. The most innovative nurseries utilize virus-free sources for both rootstocks and scion. In many countries, such as Italy, Holland and France, a national or private certification program has been created to provide material clean of viruses and infectious phytoplasma (Catalano and Martinelli, 2005).

There are several types of nursery tree that can be produced. In fact, nurseries can offer traditional chip-budded trees in a two-year cycle, two-year 'knip' plants obtained from early summer ('June') budding (1-year-old scion), bench-budded plants in a 1-year cycle, all of which are generally suitable for the spindle training system. The innovative 'bi-axis' system is planted with nursery-created, split-leader trees created with the 'Bibaum®' technique of double budding or bench grafting (Lugli and Musacchi, 1993; Musacchi, 2008b).

The common goal for these nursery tree types is to create lateral branches ('feathers') that speed up the early formation of fruiting branches with wide crotch angles (Neri, 2004; Neri et al., 2001). Depending on the cultivar, this makes it possible to rapidly bring an orchard into production (Sansavini and Zocca, 1965; Mitov and Zocca, 1966). Only in cases where the tree produced by the nursery is completely devoid of feathers (i.e. a 'whip') does the orchardist need to head the stem to 50–60 cm to produce lateral shoots and initiate a new leader. Some contemporary orchard training systems are better suited to whip nursery trees than feathered trees, such as open vase, quad-V or dual-V training of peaches, or Kym Green Bush (KGB), Upright Fruiting Offshoots (UFO) or Steep Leader training of sweet cherries.

For some high-density (HD) systems trained to be exploited with continuous mechanical plantation and harvest (e.g. olive, almond), the young tree is developed with several lateral shoots that are able to create a continuous fruiting wall in only 2 years. In this case, the nursery prepares a young feathered scion which must grow very rapidly when it is planted in the field. Pruning must adapt to the HD architectural characteristics of the trees (Lodolini et al., 2017). The very narrow canopy fruiting wall architecture of UFO-trained sweet cherries, as well as apricots, plums, apples or pears which bear exclusively on spurs or very short shoots, is best established with single or bi-axe whip (unfeathered) nursery trees (Long et al., 2015; G. Lang, pers. comm.).

Organic farming is spreading worldwide and organic fruit production is becoming significant in certain areas: in the EU, Italy, Spain, Germany and Poland are rapidly growing, while in the United States, California and Washington are leading in organic vegetable and fruit production (Kirby and Granatstein, 2017). World organic tree fruits are now produced on more than 701 600 ha, including 250 000 ha of temperate fruit crops (FAO, 2016). In Europe, to obtain organic certification, it is necessary to plant organically certified trees. In Italy, the data bank about the availability of organic-certified trees is managed by CREA (National Research Centre for Agricultural Research and Economic Studies, http://scs.entecra.it/biologico-indice/biologic.html). In September 2018, there were more than 135 organically certified genotypes available from 12 accredited nurseries, comprised of 17 fruit species including apple (*Malus × domestica* Borkh.), pear (*Pyrus communis* L.), peach (*Prunus persica* L.) and blueberry (*Vaccinium corymbosum* L.). Organic-certified trees are limited in comparison to the extremely high number of new varieties released every year by breeding programs. In fact, the laws of different countries (like Italy and the United States) authorize the organic sector to use trees that are not certified organic if they are not available. In this case, it is possible to buy conventional trees from a nursery as an exception to the rule. After three years of organic transition (which correspond to the growing stage phase of the orchard tree), it is possible to became fully organic-certified. Another challenge for some organic nurseries is to meet health requirements for organically certified fruit tree production.

2 Rootstock production techniques

2.1 Seed propagation

Propagation by seed is of great importance for genetic improvement. This technique is utilized mainly for peach, cherry and pear rootstocks (Fig. 1a and b). In the case of peach, the seed-containing pits are collected from fields planted to mother seedlings that are isolated from other peaches to exclude

Figure 1 (a) Germination of a peach seed. (b) Germination of a pear seed.

cross-pollination. To obtain seedlings, the pits are stratified at the end of September and start to germinate the following March and April. Immediately after germination, the seedlings are transplanted in the nursery. At the end of the summer, the rootstocks are ready to be grafted or budded. If the grafted/budded rootstocks are dug out in the winter and planted directly in the orchard, they are called 'sleeping eye' or 'sleeping bud' trees. These trees have very low costs. Seedlings can remain in the nursery for another year to allow the scion bud to develop into a leader and form a feathered nursery tree.

2.2 Vegetative propagation

Vegetative propagation requires the induction of adventitious roots on a shoot of the rootstock, which may be accomplished several ways: stooling, layering, cutting and micropropagation *in vitro*. The term 'adventitious' indicates that such organs arise in tissue unrelated to the normal hypocotyl and radicle of meristems of the seedling plant (Howard, 1987). Vegetative propagation systems fall into two basic classes of either division (in which a potential complete plant develops while attached to the stock plant) or cuttings (in which a potential complete plant develops after detachment from the stock plant).

2.2.1 Stooling

Also known as mound or stool layering, this division technique is primarily used to propagate apple rootstocks like the Malling and Malling-Merton series (e.g. 'M9') (Howard, 1987). Geneva series apple rootstocks also may be propagated by stooling (Fazio et al., 2005). In this technique, a 'stool' is established by planting a rootstock for one season to develop a strong root system before cutting back aerial shoots to encourage new shoot development from the stump (Fig. 2) (Howard, 1987). Virus-free starting material is recommended to guarantee that the rootstocks produced will be free of contamination. Planting time usually is fall if rooted plants are used, while micropropagated plants can be transplanted in spring.

The planting distance to establish each stool is a function of the level of mechanization used in the nursery. Typical distance between the stool rows varies from 130 cm to 160 cm, while plant distance within the row may range from 25 cm to 30 cm, depending on the vigour of the stump at maturity. If the operations of planting and establishment in the first year are excluded, it is possible to mechanize nearly the entire cycle of stool-propagated rootstocks.

The development of vigorous shoots (new rootstocks) in the field can be promoted by covering the base of the shoots with soil to increase etiolation of the tissue, which together with higher humidity promotes adventitious root formation (Baldini, 1986; Hartmann and Kester, 2002). Choosing a soil with a light structure is strongly recommended (Fig. 3a). In fact, heavy soils (with stagnant water) can make the winter operation of cutting the rootstocks from the stump and harvesting them quite difficult. Replanting must be avoided due to possible phytosanitary issues. In sandy soils, light organic substrates (such as rice husks, sawdust, or wood shavings) may be added at the base of the stump and during the subsequent shoot's growth to maintain a good level of humidity and promote rooting (Fig. 3b). Soil may be added periodically following shoot elongation and to maintain etiolated shoot bases (Figs. 3c and d). Before harvesting the rootstocks, the soil close to the root system needs

Figure 2 Shoots growing from the stool bed in spring.

Figure 3 (a) Stool bed during winter. (b) Details of adventitious root formed in a stool bed. (c) Stool bed completely developed in summer. (d) Stool bed in autumn. Notice the amount of soil at the base of the shoots.

to be removed using a jet of air that leaves the shoots and roots completely exposed and ready to be cut with a rotating saw (Fig. 4a) or new equipment based on a cutting blade shaped like a U (i.e. Oliver model SP96, Fig. 4b) (Ade and Musacchi, 2003).

Figure 4 (a) Harvest of the stool bed with a rotative saw copy. (b) Harvest of the stool bed with a new equipment.

In the second year after planting, yields typically increase to 30% of the expected mature total and by the fourth or fifth year, the stool will be in full production. In order for the stump to produce a large number of shoots, high levels of soil fertility and careful weed control must be maintained. If these conditions are met, it is possible to keep production at full potential for 15–20 years, though the period in which material can be certified virus-free may be limited to 15 years, depending on the country (Lugli and Musacchi, 1993). After harvest of the rootstocks (late winter), they can be graded in classes based on the stem size (Lugli and Musacchi, 1993).

2.2.2 Layering

The layering technique, much like stooling, has been applied primarily to the propagation of rootstocks of pome fruit species, especially apple and quince (*Cydonia oblonga*) (Lugli and Musacchi, 1993). The two methods are slightly different, but they work according to the same principle of inducing root formation on the shoots while they are still connected to the mother plant. To establish a layering, the rootstocks are initially planted at an inclined angle of 30° relative to the horizon and bent down the following spring to build a permanent continuous horizontal cordon that will produce shoots every year. The cordon is covered with 3–4 cm of soil to allow the axillary buds of the layered shoots to push through. The base of the shoot is etiolated following the same method described for stooling. The distance of planting of the layering is also a function of the degree of mechanization, and of the vigour of the rootstock. The distance between the rows varies from 130 cm to 160 cm, with 35 cm to 40 cm between the plants (Lugli and Musacchi, 1993).

Layering presents some disadvantages during harvest because the permanent cordon can be damaged by the rotative saw, and over time the gaps in the cordon can decrease the production of rootstocks. A possible solution to minimize this problem is to leave some shoots attached to the cordon and bend them down to cover the gaps. This additional labour is very expensive and is the major limitation of this propagation technique.

2.2.3 Cuttings

For this technique, a piece of shoot (1-year-old wood) or a branch 15–30 cm long (Fig. 5) is separated from the mother plant and placed in a suitable environment to root. Propagation by cuttings exploits the ability of certain woody plants to develop adventitious roots from already differentiated tissues. Usually the pieces to root are planted in a moist substrate, with the portion of the cutting covered by the substrate forming roots while the aerial portion generates the crown. Depending on the starting material, there are several

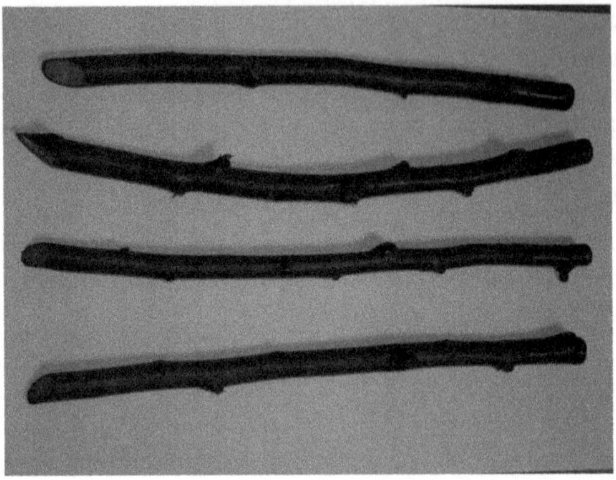

Figure 5 Hardwood cutting of Mirabolan 29.

types of cuttings, the most common being: hardwood cuttings, semi-woody cuttings, herbaceous cuttings and radical cuttings. Not all of these methods are important for rootstock production, so only the hardwood cutting technique will be described in detail.

2.2.3.1 Hardwood cuttings

Hardwood cuttings developed for easy-to-root genotypes, especially those capable of developing pre-formed root initials on shoots during the growing season. This technique is a viable alternative to stooling and layering, as it is less expensive (Lugli and Musacchi, 1993). This technique has been used for apple rootstocks such as M26, MM106, MM111 and M27, cherry rootstocks such as Colt, some plum (*Prunus salicina, Prunus domestica*) rootstocks (Howard, 1987; Cherubini and Preka, 2001) and for certain kiwi varieties.

The material used needs to be collected from a virus-free mother plant block during dormancy or better just before bud swell (Cherubini and Preka, 2001), and can be performed directly in the field or in a rooting area of a nursery. The length of the cutting varies depending on the growth rate of the genotype, typically ranging from 15 cm to 30 cm. Root formation occurs from or near the cambium and wounding the tissue can improve the results. In fact, slicing the shoot base through its diameter for at least 2 cm can enhance root formation. Cuttings require auxin treatment with synthetic plant growth regulators (PGRs) such as indolebutyric acid (IBA) and naphthaleneacetic acid (NAA) to induce a fibrous root system (Howard et al., 1983; Pastyrik and Ivanicka, 1977; Cappellini and Nicotra, 1973). IBA can be dissolved in an organic solvent such as ethanol or acetone for application to the base of cuttings. It can also be applied as the potassium salt form dissolved in water or as a powder formulation using talc as a carrier for coating the base of the cutting (Howard, 1987). Liquid treatments of 50% water and organic solvent are preferred because they increase PGR penetration into the tissue. Treatment must be done within 24 h after harvesting the shoots, usually by dipping 1 cm of the base in the solution for 5 s. The concentration of IBA for plum and apple is 2500 ppm, while for other species such as cherry and quince, this can be reduced to 1000 ppm (Howard et al., 1983).

The timing for propagation by cuttings is determined by the capacity of the species to form roots. In some species, a higher percentage of cuttings produce roots when taken in February before bud swell. However, peach rootstocks formed more roots when cuttings were collected in October (Cherubini and Preka, 2001). Species that develop pre-formed 'root initials' while growth can be collected, stored and planted directly in the field, such as with the cherry rootstock 'Colt' (*Prunus avium* × *P. pseudocerasus*) that can be planted from leaf fall through bud burst. Other species, such as plum and quince, perform better when planted in autumn directly in the field. Apple rootstocks require a combined treatment of IBA and base heat in the rooting medium. In this case, the best time to take cuttings is late winter through early spring. Cherubini and Preka (2001) compared several rootstocks from different species (peach, plum, cherry and quince), finding a broad range of responses related to the genotypes utilized. The peach rootstock 'GF677' showed 79.3% of cuttings with root formation compared to 'Barrier' with only 32.3%. Among the plum genotypes tested, Marianna 2624 was highest with 79.3%, while only 47% of Myrocal cuttings formed roots.

2.2.3.2 Leafy summer cuttings

Semi-woody and herbaceous (also called 'softwood') cuttings utilize young shoots with leaves. This technique requires specialized areas of the nursery with mist irrigation to reduce water lost by transpiration until functional roots form. For this reason, softwood cuttings are not widely adopted by tree fruit nurseries (Howard, 1987), but it is commonly used for olive and sometimes for the GiSeLa® series of cherry rootstocks (*P. cerasus* × *P. canescens*).

2.2.4 Micropropagation in vitro

Micropropagation *in vitro* allows production from small cuttings multiplied under aseptic and controlled environmental conditions on an artificial medium (Fig. 6). Following the discovery of plant hormones, their influence on plant development and refinement of growing media formulations in the second half of the last century (Murashige and Skoog, 1962), *in vitro* propagation techniques

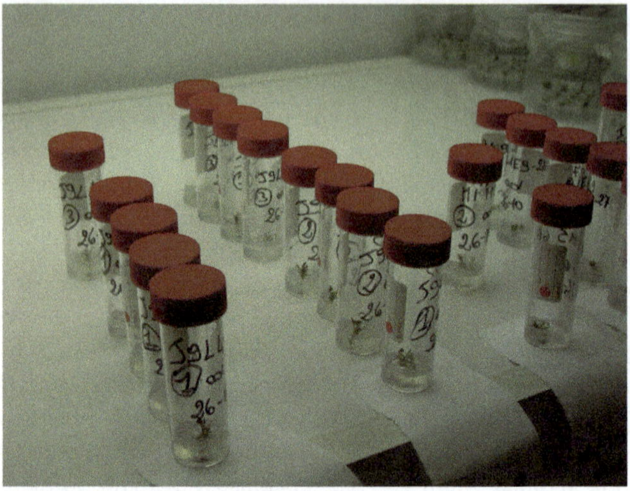

Figure 6 Explant of shoots to start micropropagation.

have improved rapidly (Bonga and Von Aderkas, 1992). Micropropagation utilizes the ability of axillary buds in the shoot-tip meristem to produce small lateral shoots. These shoots can be separated repeatedly (subcultured), then cultured at intervals of a few weeks to increase the potential shoot multiplication rate without seasonal limitations (Bonga and Von Aderkas, 1992; D'Oria, 2012).

Micropropagation began in the 1960s, and during the 1980s, a large number of species (Table 1), both herbaceous and woody, began to be propagated commercially (Zimmerman, 1992). This initial expansion has been followed by a widespread reduction in commercial activities. In France, micropropagated plant production has declined from 40 million at the end of the 1980s to only 8–10 million in 2008 (Grapin, 2009; D'Oria, 2012). This trend is mainly due to difficulty in automation and an increase in production costs (Rugini et al., 2009). The production steps, advantages and drawbacks for micropropagation are discussed below. Medium composition is not detailed due to variability in the needs of individual species, which require a dedicated chapter to cover the subject properly.

2.2.4.1 Phases of micropropagation

The micropropagation process is divided in four phases: the first three are performed *in vitro* (Phase I, II, III), while the last (Phase IV) is carried out under controlled environmental conditions (Razdan, 2003; D'Oria, 2012).

Table 1 The main tree fruit and small fruit rootstocks and cultivars produced in Italy and the United States by micropropagation

Species		Production
Apple	Rootstocks:	Geneva® 'G.11' (*Malus* hybrid)
		Geneva® 'G.202' (*Malus* hybrid)
		Geneva® 'G.210' (*Malus* hybrid)
		Geneva® 'G.214' (*Malus* hybrid)
	(USA)	Geneva® 'G.222' (*Malus* hybrid)
		Geneva® 'G.41' (*Malus* hybrid)
		Geneva® 'G.890' (*Malus* hybrid)
		Geneva® 'G.935' (*Malus* hybrid)
		Geneva® 'G-969' (*Malus* hybrid) Budagovski® 'B.10' (*Malus* hybrid)
Apricot	Rootstocks:	Myrobalan 29C (*P. cerasifera*)
		MRS 2/5 (Selection of *P. cerasifera x P. spinosa*)
		Torine® Avifel (*P. domestica*)
Kiwifruit	Rootstocks:	D1-Vitroplant e Bruno
	Cultivar:	*Actinidia deliciosa*: ♀ Hayward, Hayward Clone 8 (Testolin selection), Hayward K (Vitroplant's selection), Green Light, Tuscia, Jenny, Summer 3373
		♂ Tomuri, M3M, NPP1, MAU3, P1 Vitroplant
		Actinidia chinensis: ♀ Diploids (Zespri Gold - Hort 16A); Tetraploids: Jintao e Sorci
		♂ Belen
		Actinidia arguta: ♀ ornamentals: Issai, Green, Jumbo verde
		For production: Sel. R8P1 (Green Bunch Vitroplant), Sel. R8P23 (Green Delice Vitroplant), Sel. R10P20 (Reti Delice Vitroplant)
		♂ Cornell, Sel. R9P18 e R9P16 (Vitroplant)
Cherry	Rootstocks:	Adara (*P. cerasifera*)
		CAB 11E (*P. cerasus*)
		CAB 6P (*P. cerasus*)
		Gisela® 3 (*P. cerasus x. P. canescens*)
		Gisela® 5 (*P. cerasus x. P. canescens*)
		Gisela® 6 (*P. cerasus x. P. canescens*)
		MaxMa Delbard® 14 Brokforest (*P. mahaleb x P. avium*)
		MaxMa Delbard® 60 Broksec (*P. mahaleb x P. avium*)
		Maxma Delbard® 97 Brokgrove (*P. mahaleb x P. avium*)
		Piku 1 (*P. avium x [P. canescens x P. tomentosa]*)
		Santa Lucia 64 (*P. mahaleb*)
		PHLC (*P. avium x P. cerasus*)

(Continued)

Table 1 (*Continued*)

Species		Production
Cherry/plum	Rootstocks:	Colt cherry (*P. avium x P. pseudocerasus*)
		Krymsk® 1 peach/apricot/plum (*P. tomentosa x P. cerasifera*)
		F 12/1 Mazzard cherry (*P. avium*)
		Damil® Gm 61/1 cherry (*P. dawyckensis x [P. canescens x P. dielsiana]*)
		Krymsk® 5 cherry (*P. fruticosa x P. serrulata var. lannesiana*)
	(USA)	Krymsk® 6 cherry (*P. cerasus x [P. cerasus x P. maackii]*)
		Krymsk® 7 cherry (*P. lannesiana*)
		Maxma® 14 cherry (*P. mahaleb x P. avium*)
		Weiroot™ 158 cherry (*P. cerasus*)
		Tabel™ Edabriz cherry (*P. cerasus*)
		Marielle® 'Mar 4001' plum/almond (*P. cerasifera x P. munsoniana*)
Fig	Cultivar:	Brogiotto, Fico di Troia, Dottato, Fiorone, Melanzana, Signora, Turca, Verdone
		Brown Turkey, Celeste
Jujube	Cultivar:	Li, Lang
Pomegranate	Cultivar:	Wonderful Pomegranate
Raspberry	Cultivar:	Alpengold, Autumn Bliss, Bristol, Golden Bliss, Heritage, Rossana, Tulameen, Willamette
	(USA)	Kokanee Raspberry, Cascade Bounty Raspberry, Cascade Dawn Raspberry, Cascade Delight Raspberry, Cascade Harvest Raspberry, Caroline, Meeker, Munger, Nova Red, Nantahala, Amity, Bristol, Jewel, Tulameen, Heritage, Autumn Bliss, Canby, Crimson Giant, Fall Gold, Honey Queen, Vintage, Crimson Night Raspberry, Double Gold, Lewis
Blueberry	Cultivar:	Berkeley, Blueray, Blue Crop, Bluetta, Brigitta, Coville, Duke, Early Blue, Elliot, Late Blue, Legaci, O'Neal, Toro, Spartan, Patriot, Pilgrim
	(USA)	Berkeley, Biloxi, Bluechip, Bluecrop, Blue Gold, Bluehaven, Bluejay, Blueray, Brigitta, Duke, Earliblue, Elliott, Hardyblue, Jersey, Jubilee, Legacy, Misty, O'Neal, Patriot, Reka, Rubel, Sharpblue, Southland, Spartan, Summit, Sunshine Blue, Toro, Bluetta, Brightwell, Powderblue, Chandler, Native Blue, Pink Lemonade, Sweet Heart
Blackberry	Cultivar:	Chester, Hull Thornless, Lochness, Thorn Free
	(USA)	Prime-Ark® Freedom, Prime-Ark® Traveler, Columbia Giant Blackberry, Columbia Sunrise Blackberry, Prime-Ark® 45, Apache, Arapaho, Boysenberry, Chester, Hull Thornless, Kotata, Lochness, Loganberry, Marionberry, Natchez, Navaho, Obsidian, Ouachita, Sylvan, Tayberry, Black Satin 3037-2, Thornless Boysen 2202-20, Triple Crown, Waldo, Olallie Trailing, Tupi, Black Diamond, Newberry (Orus 1324-1), Wild Treasure, Osage, Onyx (Orus 1523-4), Columbia Star

Species		Production
Hazelnut	Cultivar:	Tonda Romana, Purpurea, Webb
	(USA)	McDonald, Wepster, Lewis, Sacajawea, Yamhill, Epsilon (Pollinizer), Zeta (Pollinizer), Jefferson (703.007), Theta (1001.008) Pollinizer, York Hazelnut (878.048), Dorris Hazelnut (876.041), Felix (941.016) Pollinizer
Walnut	Rootstocks:	Vlach (*Juglans* hybrid)
		VX211 (*Juglans* hybrid)
	(USA)	RX1 (*Juglans* hybrid)
Olive tree	Cultivar:	Arbequina, Galega, Koroneiki, Carbuncion, Correggiolo, Nostrana di Brisighella, Novellare del Belice, Kalamata
Pistachio	Rootstocks:	UCB-1 (*Pistacia vera*)
Peach	Rootstocks:	GF677 (*P. persica x P. davidiana*)
		Cadaman (*P. persica x P. davidiana*)
		Isthara ([*P. cerasifera x P. salicina*] x [*P. cerasifera x P. persica*])
		Garnem (*P. amygdalus x P. persica*)
		Adesoto (Puebla de Soto) (*P. insititia*)
		Barrier 1® (*P. persica x P. davidiana*)
		Penta® (*P. domestica*)
		S. Giuliano GF 655/2 (selection of *P. insititia*)
		Tetra® (*P. domestica*)
Peach/ Almond	Rootstocks:	BH® 106 Cv. Lillian (*Prunus*)
		BH® 5 (*P. dulcis x P. persica*)
		Cadaman® peach/almond (*P. persica x P. davidiana*)
		Hansen 536 peach/almond (([*P. davidiana x P. persica*] x *P. persica 'Okinawa'*) x *P. amygdalus*)
		Krymsk® 86 peach/plum (*P. persica x P. cerasifera*)
	(USA)	Marianna 26-24 plum/almond (*P. cerasifera x P. munsoniana*)
		Marianna M-40 plum/almond (*P. cerasifera x P. munsoniana*)
		Paramount® 'GF677' peach/almond (*P. amygdalus x P. persica*)
		Rootpac®-40 ([*P. dulcis x P. persica*] x [*P. dulcis x P. persica*])
		Rootpac®-70 ([*P. persica x P. davidiana*] x [*P. dulcis x P. persica*])
		Rootpac®-90 ([*P. persica x P. davidiana*] x [*P. dulcis x P. persica*])
		Tempropac peach/almond (*P. persica x P. dulcis*)
Plum/Almond	Rootstocks:	Rootpac®-R (PAC 941) (*P. cerasifera x P. dulcis*)
Plum	Rootstocks:	Marianna 2624 (*P. cerasifera x P. munsoniana*)
		Marianna GF 8/1 (*P. cerasifera x P. munsoniana*)
		Wavit® (*P. domestica* Wangenheim)
		S. Giuliano GF 655/2 (selection of *P. insititia*)

(*Continued*)

Table 1 (*Continued*)

Species		Production
Pear	Rootstocks:	Farold® 40 Daygon and Farold® 69 Daymir (Old Home x Farmingdale)
		Fox 9 and Fox 11 (Volpina)
		Pyrodwarf® (Old Home x Buona Luisa d'Avranches)
		OHxF 87 (*P. communis* 'Old Home' x 'Farmingdale')
		OHxF 97 (*P. communis* 'Old Home' x 'Farmingdale')
	(USA)	OHxF 513 (*P. communis* 'Old Home' x 'Farmingdale')
		Pyro™ 2-33 (*P. communis* 'Old Home' x 'Bonne Louise')
		OHxF 69 (*P. communis*)
	Cultivar:	Abbé Fétel, Bartlett (William), Conference, Bosc (Kaiser)
Currant	Cultivar:	Titania, Junifer, Rivada, Blanka
Gooseberry	Cultivar:	Hinnomaki, Invicta, Rokula, White Smith
Grape	Rootstocks:	1103 P (*V. berlandieri x V. rupestris*)
		140 R (*V. berlandieri x V. rupestris*)
Serviceberry	Cultivar:	Rainbow Pillar® (*Amelanchier canadensis* 'Glenn Form')
		Prince William Shadblow (*Amelanchier canadensis* 'Prince William')
	(USA)	Princess Diana(*Amelanchier x Grandiflora* 'Princess Diana')
		Robin Hill (*Amelanchier x Grandiflora* 'Robin Hill')
		Autumn Brilliance® (*Amelanchier x Grandiflora*)

Source: Dept. Tree Fruit, University of Bologna (Italy); Vivai Battistini, Vitroplant, Microplant, Apice Pinte (nurseries in Italy) (Modified from Lambardi & Ozudogru, 2010)
North American Plants LLC

2.2.4.1.1 Phase I – selection of explants and establishment in aseptic culture

As a critical starting point in the micropropagation process, selection of the optimum explant is of utmost importance. Explant age, size, position, sampling season, phenological and sanitary state must be taken into account. Due to higher reactivity, younger tissues or organs may give better results during explant establishment compared to older ones (Smith, 2000). After the explant tissue has been collected, a disinfection process is required, since pathogens like moulds, yeasts, fungi and bacteria may be present either on the explant's surface or in its vascular system (D'Oria, 2012). Several products are available to sterilize the tissue surface, such as alcohol (e.g. ethyl or methyl alcohol, usually at 80%), bleach, (e.g. calcium or sodium hypochlorite at 5.25–6% active ingredient) and others such as silver nitrate, hydrogen peroxide and benzalkonium chloride (D'Oria, 2012). Disinfection protocols are time and concentration dependent.

Internal contaminations are difficult to detect, and pathogen identification may be difficult when infected explants are symptomless (Hartmann and Kester, 2002).

After completing disinfection, the explants are stabilized on the substrate (Fig. 6). This requires several (3–4) months through different subculture steps (Smith, 2000). To enhance axillary shoot formation, a 0.5–1.0 mg/L concentration of cytokinin (e.g. kinetin, benzyladenine [BA], 6-benzylaminopurine [BAP] or 2-isopentenyladenine [2iP]) is required, while auxins (e.g. IBA, NAA or 2,4-dichlorophenoxyacetic acid [2,4-D]) need to be used at low concentrations (e.g. 0.01 mg/L) or may even be avoided entirely for some genotypes (Hartmann and Kester, 2002).

2.2.4.1.2 Phase II - multiplication

During this phase, explant elongation and axillary shoot formation is emphasized with a goal of shoot proliferation (Razdan, 2003). Shoot formation is promoted by cytokinin concentration but can be affected indirectly by auxins (D'Oria, 2012). Further, the effect of most plant growth regulators on *in vitro* development is highly genotype dependent (Dobránszki and da Silva, 2010).

Depending on species and the cultural conditions, one explant may produce up to 25 microshoots (Fig. 7). Once newly-formed microshoots reach the optimum size (i.e. capable of being productive in the next subculture), they can be separated and placed on new media. Explants can be subcultured every four to eight weeks. Depending on the length of the shoots, different numbers

Figure 7 Shoot of pear in multiplication stage.

of explants may be obtained. Such subcultures can be repeated multiple times, until the desired number of microshoots has been achieved to be moved to the rooting stage (D'Oria, 2012). The number of subcultures will depend on genotype ability to avoid epigenetic alteration and deterioration during the culturing process (Hartmann and Kester, 2002; Razdan, 2003; D'Oria, 2012).

2.2.4.1.3 Phase III – root induction and formation

Microshoots must be transferred to a root induction substrate for rooting and preparation for eventual transplanting (Razdan, 2003; D'Oria, 2012). Some species are capable of rooting during the multiplication step, but recalcitrant genotypes require changes in media composition and environmental conditions. The rooting process will be enhanced by adding auxins to the substrate and reducing or eliminating cytokinins (Razdan, 2003; D'Oria, 2012). Due to differences in plant rooting ability (noted above), the rooting induction process may be carried out either *in vitro or ex vitro*.

In vitro rooting is carried out in a sterile environment (Fig. 8). Microshoots are transferred to a new medium characterized by a high auxin concentration and a reduced or absent cytokinin level (D'Oria, 2012). *Ex vitro* rooting occurs outside culture vessels or test tubes. Microshoots require two different auxin pre-treatments before going into the transplant area. They can be soaked in an auxin solution (quick-dip) or transferred for 5–10 days on a substrate with a high auxin concentration (D'Oria, 2012). Afterwards, they can be transferred to a suitable greenhouse rooting medium. Rooting explants *ex vitro*, besides

Figure 8 Root formation in vitro.

reducing production costs, leads to practical advantages as well. *Ex vitro* rooting seems to confer a better structured vascular system that leads to more successful adaptation compared to *in vitro* roots (Hartmann and Kester, 2002).

2.2.4.1.4 Phase VI - acclimatization

Rooted or still un-rooted plantlets from the previous steps are transferred to, and acclimatized in, a greenhouse (Fig. 9a and b) or growth chambers (D'Oria, 2012). Plantlets can be moved either in small pots containing a suitable growing substrate or even placed in a soilless system.

Plants grown *in vitro* often show unusual morphological characters. The high level of humidity can induce less epicuticular wax production and malfunctioning stomata that fail to close once transferred. For this reason, plantlets readily become water stressed and can fail to acclimate (Razdan, 2003; D'Oria, 2012). Further plant physiological changes are required since the plantlets must shift from a heterotrophic to an autotrophic condition, increasing plantlet susceptibility to death during transplant (Hartmann and Kester, 2002). To minimize this problem, plantlets must be cultivated in high humidity conditions. Air temperature and light intensity need to be controlled by computerized systems and specialized software. During this process, anti-transpirants or wax-stimulating products may be applied (Razdan, 2003). The interaction between the root system and the substrate is also important to induce a better root formation and water uptake (Neri et al., 2017).

2.2.4.2 Advantages and drawbacks of micropropagation

Micropropagation may be a good alternative for certain species or genotypes that cannot be propagated using traditional methods such as stooling and

Figure 9 (a) Acclimatization stage. Plantlets are very delicate and susceptible to stress. (b) Plant completely acclimatized in the green house.

cuttings (Pasqual et al., 2002; Razdan, 2003; D'Oria, 2012). Additional potential advantages of micropropagation include:

- *Rapid clonal mass propagation* – Micropropagation is faster than traditional propagation methods (Dolcet-Sanjuan et al., 2008; Mencuccini, 2009). Each subculture takes almost only one month, allowing laboratories to, in theory, produce more than 1 million plantlets per year (D'Oria, 2012). Such a high production level requires the ability to sell the material produced continuously, or the possibility to store the unsold stocks (Razdan, 2003; D'Oria, 2012).
- *Year-round nursery production* – Micropropagation is performed in a controlled environment, so production can be programmed and scheduled throughout the year in accordance with market requirements (Razdan, 2003).
- *Pathogen-free plant production* – Micropropagated explants can be grown and kept under aseptic conditions (Smith, 2000; Zilka et al., 2002). Once the starting material is tested and certified to be free of critical pathogens, all subsequent explants are assured of also being certifiable. Thermotherapy and meristem-tip culture are typical micropropagation techniques for isolating pathogen-free explants from starting materials.
- *In vitro grafting (micro-grafting)* – This technique facilitates the grafting of rootstocks and scions before they exit sterile conditions (Musacchi, 1996; Espen et al., 2002) (Fig. 10).
- *Endangered plant conservation* – Micropropagation can be used to maintain repositories of live genotypes that are not cultivated through slow growth storage at a temperature range of 2–5°C, or through cryopreservation in liquid nitrogen (De Carlo et al., 2009).
- *Germplasm preservation* – In this case, plant tissues can be cryopreserved - conserved until needed, then thawed and micropropagated (Razdan, 2003). Since its development in the 1960s, plant cryopreservation is considered an extraordinary method of safe, long-term conservation of biological material, as it does not induce genetic alterations, and it preserves the regeneration potential of the stored material. It is based on storage of explants at cryogenic temperatures, such as that of liquid nitrogen (−196°C), in which metabolism within the cells is suspended; thus, metabolism for these cells is theoretically 'stopped'. Cryopreservation is particularly important for embryogenic cultures, as they require periodic subculturing for their maintenance, and this, in turn, increases the risk of losing the material, as well as its embryogenic potential. Periodic re-initiation of embryogenic cultures is possible; however, it is labour-intensive, expensive and particularly difficult when working with species for which embryogenic explants are available only during a limited period

Figure 10 *In vitro* grafting to study graft-incompatibility in sterile conditions.

of the year. Among various methods of cryopreservation available for embryogenic cultures, slow cooling is still the most common approach, especially in callus cultures from softwood species (Ozudogru and Lambardi, 2016).

- On the other hand, some limitations of micropropagation include:
- *Variability* - plantlets can develop somaclonal variation, which can be genetic, phenotypic or epigenetic. This can be induced by various factors, such as the high multiplication rates and repeated subcultures, growth regulator imbalance and culture aging (Razdan, 2003; D'Oria, 2012).
- *Hyperhydricity (vitrification)* - this physiological disorder is induced by morphological and metabolic changes (Fig. 11). Plantlets subjected to this disorder grow less and have a lower survival rate. Leaves appear translucent and glassy, and stems can be deformed. This disorder is mainly due to a high multiplication rate and an excess uptake of water. Schloupt et al. (1994) showed that vitrification appears mainly when using substrate with a low agar concentration or in liquid substrates. Other factors inducing vitrification are high relative humidity, low nitrate-ammonium ratio, and high cytokinin concentrations (D'Oria, 2012).
- *Species difficult to root (recalcitrant)* - some species can be difficult to root even with micropropagation (e.g. many gymnosperms) (Razdan, 2003).
- *Substrate browning* - this problem is frequent in species characterized by a high level of phenolic compounds. When these compounds are released in the media, they oxidize to brown and have a subsequent toxic effect that leads to a decrease in explant growth. The oxidation process

Figure 11 Vitrification of shoots.

may be reduced by using antioxidants like ascorbic acid, citric acid and cysteine or by keeping the culture in the dark (D'Oria, 2012). Charcoal or polyvinylpyrrolidone can be used to absorb phenolic compounds and prevent their accumulation (D'Oria, 2012).

- *Contamination* - many pathogens can contaminate the explant starting materials and be very difficult to eradicate, or can infect the plantlets during subculturing, especially fungi, bacteria (Fig. 12), viruses or virus-like organisms. Some bacterial pathogens (e.g. *Bacillus* spp., *Pseudomonas* spp.) and viral or virus-like contaminants are very resistant to sterilization procedures and are able to persist latently through many subcultures (Hartmann and Kester, 2002).
- *Shoot-tip necrosis* - there are two main causes of this physiological disorder: the first may be due to a lack of calcium during shoot growth, requiring a change in media nutrient composition to correct (Hartmann and Kester, 2002) and the second may be related to a deficiency of cytokinin in the substrate, therefore requiring a change in media hormone composition.
- *Economical sustainability* - challenging economic factors include currently low automation of this technique, as well as high costs for laboratory equipment and skilled work staff (Razdan, 2003).

2.2.4.3 Micropropagation technology innovations

The phases of micropropagation are well established, and innovation has only been incremental. The substrates and PGRs utilized have changed little over the

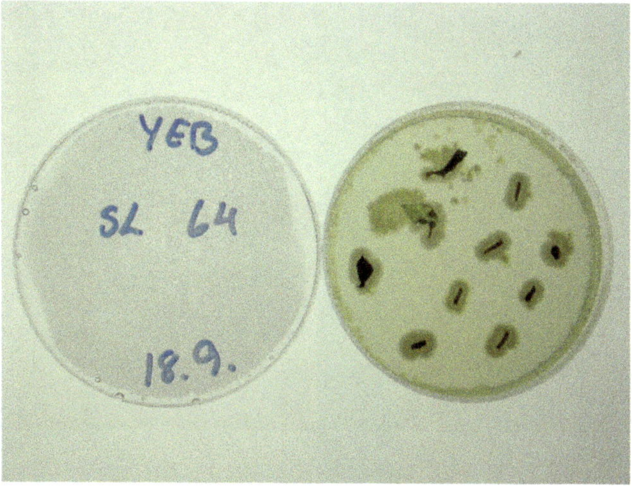

Figure 12 Explant of Santa Lucia 64 cultivated on YEB substrate to enhance endogenous bacteria growth.

years, with minor variations in composition for each phase specific to different species and cultivars. A new cytokinin-related PGR, thidiazuron (a phenylurea), has shown positive results for shoot proliferation (Arndt et al., 1976). Another important innovation is the use of lightweight disposable (or recyclable) plastic containers as opposed to glass jars (Fig. 13). However, the high cost of conversion to new systems has limited adoption of this technology. A different container advancement that facilitates better gas exchange (i.e. VitroVent®, Microbox®, Eco2®, StarPac®) has not been widely adopted for a similar reason. A temporary immersion (RITA®) system (Fig. 14a and b) based on alternating cycles of shoot immersion with dryer periods has proven to be successful (Lambardi and Ozudogru, 2010). With the rapid innovations in LED lighting technologies, there may be future advances in micropropagation light source and quality on the horizon, including for the acclimation phase (Lambardi and Ozudogru, 2010).

2.2.4.4 Effects of micropropagation on rejuvenation

Trees that develop from seeds typically exhibit a juvenile phase. In apple, this phase can last 4–8 years, and in olive up to 20 years, characterized by total sterility that is manifested as the inability to flower. Other morphological peculiarities include branches that may have thorns and a generally more open orientation than that of adult forms. Juvenile branches may have a distinct phyllotaxis, vigorous buds, small leaves and/or dimorphism, and less developed cortical

Figure 13 Plastic jar.

Figure 14 (a) RITA®, temporary immersion system. (b) Shoots cultivated in the RITA®, temporary immersion system.

tissues, with wood that is characterized by a greater quantity of fibres, a smaller quantity of parenchymatous cells and a low degree of porosity (Baldini, 1986). From a biochemical point of view, juvenile branches are poorer in reserves and richer in lignin and hemicellulose. Importantly for propagation, young tissues have a higher rhizogenic capacity and capacity to produce buds (Monteuuis, 2004).

As this juvenile phase ends, the tree goes through a transition phase in which the juvenile characteristics disappear and the ability to flower and fruit increases, progressing at different rates according to species. The meristems and plants reach a mature stage that is stable, which maintain growth and reproductive characteristics for the rest of the life of the tree, but there are some events, interventions, and propagation techniques that can promote a temporary return to the juvenile characteristics along with the absence of flowering (Hackett, 1985). In several species, heavy and repeated pruning, grafts on juvenile rootstocks, hormonal treatments (in particular, gibberellins) and high temperatures accompanied or not by low luminosity induce (in part or in total) the reversion of the adult phase to the juvenile one. This is not necessarily rejuvenation, in fact; some of these methods do not lead to a real change of phase, but to a simple reversion of the physiological age and reinvigoration. The reinvigoration, however, may be considered as the first step towards rejuvenation (Neri, 1990a).

The propagation technique with the highest risk of resulting in a rejuvenated nursery tree is micropropagation. In several species, this leads to a delay in production and greater vigour than plants propagated by traditional methods. Hackett (1985) has reported that this is greatly accentuated, and the symptoms can become those of a true rejuvenation, when the following techniques are used: creation of a very large number of transplants and subcultures; maintenance of the *in vitro* material for a long time; application of high concentrations of cytokinin; use of low concentrations of sugars; and starting with vigorous suckers or watersprouts.

Franclet (1979) stated that it is impossible to distinguish a rejuvenated micropropagated plant from a plant obtained by seed. The only difference found was that when plants produced *in vitro* were transplanted in the field, the juvenile features persisted for less time than did the same plant developed by seed (Neri, 1996; Brand and Lineberger, 1992). Moreover, the rejuvenated material also showed morphological changes. For example, the leaves may show lobes, indentations and laminae different from normal (mature) leaves (Sriskandarajah et al., 1982) and there may be a reappearance of torn formations (Sansavini et al., 2004). Another noteworthy feature is the rhizogenic capacity of the shoots is increased (Takeno et al., 1982/83), which can be linked to a reduction in the production of endogenous gibberellins and cytokinins. This feature could be favourable for rootstock propagation, but micropropagated 'M9' apple rootstocks also exhibited more suckers and burrknots as well as more roots as a consequence of increasing the number of subcultures carried out *in vitro* (Webster and Jones, 1992). This is undesirable because it requires additional labour for cleaning procedures and creates wounds that expose the plant to pathogens and insects.

2.2.4.5 Epigenetic effects

The return to a juvenile phase is a temporary event that, when it ends, leaves the plant with its appearance and its original characters, so changes are not real mutations. Mutations can affect the entire genome, parts of chromosomes or even simply single nucleotide sequences, and except for some rare cases (e.g. when they are due to histones), are stable over time. Additionally, the effect of the environment can be excluded because the juvenile characteristics occur in any environment in which the plant is grown. The reasons for the modification are therefore due to other causes, so-called epigenetic factors.

In this way, the modification of gene expression (sometimes heritable) is defined as not due to mutations in the nucleotide sequence, but due to other factors, such as a change or strengthening of the links of proteins associated with the DNA sequence, DNA methylation, histone modifications and/or chromatin remodelling (Liu and Wendel, 2003). These factors can be caused by agronomic interventions (Hackett, 1985), by micropropagation techniques or through the normal morpho-physiological development of plants as well.

The most studied of these effects is undoubtedly the DNA methylation that occurs on cytosine (Martienssen and Colot, 2001). In this case, a methyl group is bound by the action of the enzyme DNA methyltransferase, at a carbon at position five of the nitrogenous base. The sequence that is methylated is 5′-CG-3′ containing the cytosine also in the opposite strand. The DNA can be defined as hemi-methylated when only one of the two cytosines is methylated, or completely methylated when both cytosines are methylated. Methylation in plants is generally high in heterochromatin areas and is also associated with gene silencing, while genes that are actively transcribed have hypomethylation (Martienssen and Colot, 2001).

Poething (1990) suggested that the reacquisition of juvenile morphology and physiology in plants could depend on the change in the rate of DNA methylation, and in fact Finnegan (2001) and Wolffe and Matzke (1999) demonstrated that a planned control of gene expression is essential to ensure proper growth and adequate development.

3 Scion grafting and budding techniques

The execution of the graft, with the consequent union of two or more parts of tissues to form a single individual, is an agamic propagation technique known since the time of Theophrastus (371–287 BC) and Virgil (70–19 BC) (Andrews and Serrano Marquez, 1993). The advantages deriving from graft unions are well known: possible control of tree vegetative growth and advancement of initial production; ability to adapt to different types of terrain (e.g. low porosity, calcareous soils); possible improvement of growth on replant sites (e.g. peach on GF677 rootstock); influence on fruit quality; reduction of the juvenile period of seedlings; study of viral diseases; and replacement of plant structural components (Musacchi, 1996; Hartmann and Kester, 2002).

3.1 Histogenesis of the graft

According to the scheme proposed by McCully (1983) for various herbaceous species such as *Nicotiana* and *Coleus*, the first phase of the histological repair process in grafting (even in the presence of a perfect affinity) leads to the formation of necrotic areas. These are produced by the walls and by the cells damaged by the graft cut, elongation of intact cells that start a division towards free space between the two bionts, and the formation of callus that joins the surfaces in contact. In this way, a precarious merging of the tissue (cohesion) is achieved between the rootstock and the scion cultivar. Subsequently, differentiation from the cells of the parenchyma and the callus, of tracheal elements from the wound of cribrose vessels, and of a meristematic tissue (cambiform), which connects the primary changes of the two bionts, is observed. After this phase, the cambiform resume their activity with the production of new xylem and phloem and normally lead to reconnection of the graft (McCully, 1983; Baldini, 1986; Stoddard and McCully, 1980).

3.2 Grafting and budding techniques

There are several possible fruit tree grafting types, which can be classified according to the type of grafting applied and the degree of lignification of the bionts to be used, that is herbaceous grafts (if the two genotypes to be joined are not yet lignified), semi-woody (when the rootstock is lignified and the cultivar is not) and woody grafts (if both rootstock and cultivar are lignified). According to the type of graft performed, there are three basic groups: approximation grafts, bud grafts and bud-stick grafts. With approximation grafts, the separation of the new composite tree from the mother (scion source) plant takes place after grafting. These can be performed even just by tying together two branches, after removing the cortex (exposing the cambium) in the contact area. Bud-shaped grafts are comprised of a scion bud and a portion of smaller or larger rootstock wood that can have lignified bark. The most common bud graft techniques are T-budding and chip budding. Bud-stick grafts are performed with a scion stick having one or more buds. The main grafting techniques are cleft graft, bark graft, side-veneer graft, whip and tongue graft, bridge graft and inarch graft. In modern nursery practice, three main techniques are used: chip budding (Fig. 15a–e), side graft (Fig. 16a and b) and whip and tongue (Fig. 17a–c).

3.3 New grafting technique: micro- and mini-grafting and herbaceous grafting of lignified rootstocks

There are different micro-grafting techniques, determined by the size of the rootstock and cultivar to be grafted. The two main types of micro-grafts are *in*

Figure 15 (a) Preparation of the rootstock for the chip budding. (b) Chip budding. (c) Protection of the chip budding with plastic tape. (d) Perfect merging of the cultivar with the rootstock. (e) Graft union obtained with the bud chip budding technique (1-year-old tree).

vitro and *in vivo*. The herbaceous grafting of woody plants refers to the grafting of micropropagated material in the herbaceous state after acclimatization (*ex vitro*) (Neri, 1990b). *In vitro* micro-grafting occurs in the laboratory under aseptic conditions. The technique is also known as shoot-tip grafting and consists of excising a very small bud and grafting it with a portion of a meristematic apex of equally small size (0.1–1 mm). This graft is often performed under the microscope. The steps involved in grafting are as follows: (1) preparation of the rootstock and of the *in vitro* graft, (2) the *in vitro* grafting procedure, (3) the *in vitro* activation of growth and (4) the transfer of the grafted plant to *in vivo* conditions. This type of grafting is used mainly for the remediation of viruses of infected plants because the meristems used for this type of graft are usually free from viruses, as the meristem grows more rapidly than the virus replicates to infect the meristematic cells. Other applications of this type of graft are to study graft incompatibility (histology and physiology) (Navarro, 1988).

Figure 16 (a) Side graft executed with a double bud stick. (b) Graft union obtained with the side graft (1-year-old tree).

Figure 17 (a) Whip and tongue graft in apple. (b) Preparation of the whip and tongue graft. (c) Details of the graft union obtained whip and tongue technique (1-year-old tree) copy.

The procedure for micro-grafts with *Prunus* have been described by Gebhardt and Goldbach (1988) and Amiri (2006), among others. However, this technique makes it possible to carry out only a few grafts per worker and day. Thus, it is used primarily for research purposes and for cleaning from viruses and phytoplasmas, but it is not used for the commercial production of grafted seedlings (Neri, 1990b).

In the case of *vivo* micro-grafting, an *in vitro* shoot of the scion cultivar is taken when it reaches dimensions suitable for grafting. The rootstock is not in aseptic conditions and is grafted with the *in vitro*-derived scion cultivar (split grafting). This technique is no longer suitable for virus recovery but can be used for other purposes such as the study of graft incompatibility (Musacchi et al., 2004). Preka and Cherubini (2001) developed a method to graft potted rootstocks with herbaceous material *ex vitro*, using the split grafting technique performed in a controlled environment.

4 Biochemical and physiological aspects of graft incompatibility

Grafting is an important technique for the propagation of fruit trees, providing the potential to adapt a specific fruiting cultivar to multiple cultivation areas and planting densities by using different rootstocks. The two grafting partners may belong to the same species, but it is also possible to have phylogenetically more distant combinations. In some cases, the graft may not be successful, and the two genotypes show symptoms of incompatibility (Fig. 18a and b) (Pina and Errea, 2005). While grafts of different cultivars of the same species to a certain rootstock may be successful, often grafts of different species within the same genus to a certain rootstock fail. Even less probable is the success of graft combinations belonging to different genera of the same family, and the success of graft combinations between different families are very unlikely (Hartmann and Kester, 2002). According to Hartmann and Kester (2002), affinity occurs if two plants can yield a permanent union after grafting and developing as a single plant. Feucht (1988) defined the term incompatibility as a phenomenon of premature senescence caused by physiological and biochemical processes.

Causes of graft incompatibility, however, are not completely understood although different hypotheses about the presence or absence of cell-cell recognition have been postulated. Yeoman and Brown (1976) posited, with Solanaceae, a cell-cell recognition mechanism based on the formation of a catalyst complex (protein-based) in compatible unions, but they did not identify these proteins. Incompatible unions fail to synthesize this complex, and a cell-cell recognition mechanism based on lectins causes a rejection of the two members (Heslop-Harrison, 1978). On the contrary, Moore (1984a,b) questioned this mechanism and proved that pear callus cells grafted on quince

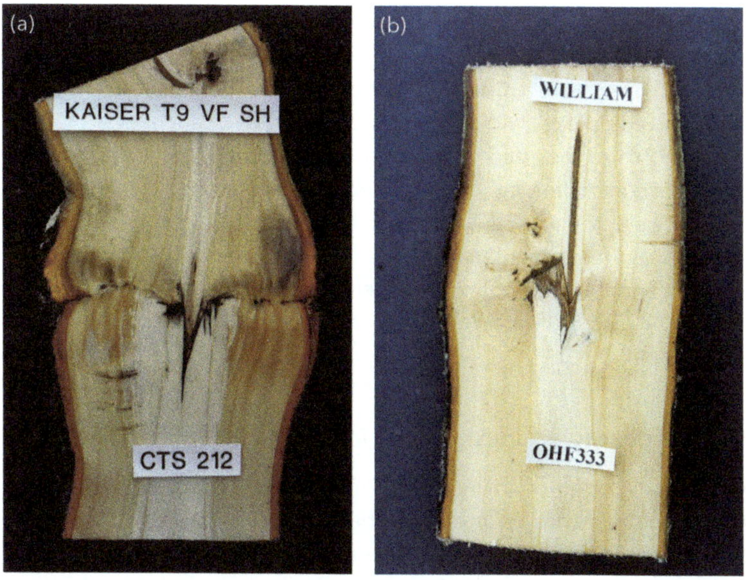

Figure 18 (a) Graft incompatible union between Beurre' Bosc and Quince CTS212. (b) Compatible union of William grafted on OHF 333.

callus cells underwent cellular senescence independently of direct contact. Moore and Walker (1981a,b) demonstrated that cohesion between scion and rootstock is due to the wound response and the subsequent polymerization of cell wall material. Musacchi (1996) simulated *in vitro* a pear/quince graft union by placing two portions of callus side-by-side in the same Petri dishes and detected changes in the protein profile due to cell-cell recognition. These results have been confirmed by Nocito et al. (2010), who also identified a high level of oxidation in the incompatible unions. An early study on pear/quince incompatibility by Gur et al. (1968) reported the role of prunasin (cyanogenic glycoside) in this phenomenon. This compound present in quince rootstock was translocated and decomposed in a narrow zone above the graft union of the pear scion with formation of hydrocyanic acid. The anatomical malformations in the tissues adjacent to the interface were caused by cambium inactivation during the growing season due to prunasin hydrolysis compounds. This could modify the metabolism and create a metabolic block in the graft interface where the cells were merged (Feucht and Treutter, 1991). Treutter and Feucht (1988) reported prunin accumulation in incompatible combinations of *Prunus avium* on *Prunus cerasus* above the graft union due to histo-anatomical disorders (Errea, 1998).

Phenolic compounds are widely present in higher plants where they are produced mainly by the shikimic acid pathway (Harborne, 1980; Feucht and

Treutter, 1989). Moreover, polyphenols are involved in the chemical defence strategy of plants, especially for plant-pathogen interactions (Harbonne, 1980; Haslam, 1989). They have a repellent taste to potential predators and are able to form insoluble complexes with proteins (Haslam, 1989). Feucht et al. (1985) and Edwards (1992) also reported polyphenol accumulation as a wound response. Polyphenol accumulation at the graft point of incompatible combinations with quince was mainly due to (-)-epicatechin. This flavan-3-ol increases in concentration at the graft union compared with 2 cm above, according to graft combination, and the incompatibility levels for quince and certain cultivars are 'Beurré Hardy' (+56%), 'William' (+172%) and 'Beurré Bosc' (+353%) (Musacchi et al., 2000).

4.1 General characteristics of graft incompatibility

The symptoms of incompatibility differ from species to species, and in general, internal symptoms usually precede the manifestation of external ones. Furthermore, normally the phloem appears to be more damaged than the xylem. The symptomatology of graft incompatibility includes several events that also involve the cortex (Breen, 1974). The following are the main external symptoms identified by various authors in graft incompatibility: formation of anomalous xylem (Copes, 1980a); lack of parenchyma in phloem and cell necrosis (Copes, 1980b); and increased peroxidase activity in both bionts and deposition of lignin and polyphenols in the grafting area, which prevents the formation of vascular (and relative connections) translocation (Deloire and Hebant, 1983).

When these vascular conductivity problems occur, high quantities of starch, free sugars and sugar alcohols accumulate in the scion cultivar. High quantities of inorganic elements such as N, P, K, Ca and Mg were found in the rootstock (Breen, 1975; Salesses and Al Käi, 1985). Breen and Muraoka (1975) found that modified nutrient levels were associated with carbohydrate level changes and proposed that this anomalous distribution of starch was related to vascular discontinuity, refuting the hypothesis that it is the cause of incompatibility (Rogers and Beakbane, 1957; Garner, 1979). The external symptoms of incompatibility observed were: greater development of one of the two bionts (Garner, 1979); tissue hypertrophy or hyperplasia above and below the graft union (these symptoms are not necessarily attributable to incompatibility); easy breakage of the shaft at the point of engagement (Garner, 1979); delay in bud break (Nelson, 1968); abnormal morphology and premature leaf senescence (Copes, 1980a); reduction of growth associated with the death of the shoots and the plant (Hartmann and Kester, 2002); loss of leaf turgidity and reduced transpiration rate during the day (Schmid et al., 1988); and loss of leaf colour due to vascular discontinuity in the graft area (Breen, 1974).

The presence of hyperplasia in the graft area is considered a symptom of incompatibility but is not always a reliable indicator (Bhattacharya and Dutta, 1952; Lapins, 1959; Garner, 1979; Westwood, 1988). The tissue enlargement can occur both above the joint area (e.g. 'Colt' and 'GiSeLa®6' cherry rootstocks) and below (e.g. 'M9' apple rootstock), and often is accompanied by the presence of vascular discontinuity. Hyperplasia also occurs in relation to genetic differences in the vigour of the two bionts. Rupturing of the graft area is considered as the clearest symptom of incompatibility (Garner, 1979; Eleftheriou, 1985). This implies the total absence of (or the presence of few) vascular connections between cultivar and rootstock able to confer any significant mechanical resistance (Andrews and Serrano Marquez, 1993).

4.2 Classification of graft incompatibility

Herrero (1951) and Mosse (1962) proposed two classifications of graft incompatibility based on how symptoms develop: translocated and localized. The former, caused by toxic compounds transported from one graft partner to the other, is associated with leaf ageing, alterations at the root and degeneration of the sieve tubes and companion cells. *Localized* graft incompatibility occurs at the level of graft interface with the appearance of histo-anatomical disorders caused by cell-cell recognition processes at the contact point between the two graft components. A third classification, delayed incompatibility, has also been described.

4.2.1 Translocated incompatibility

Translocated incompatibility occurs when symptoms appear away from the graft union due to transport of some factors of a biochemical nature, such as toxins, from one component of the graft to another. The interposition of a related intermediate does not overcome the problem of incompatibility. Distinctive factors of translocated incompatibility are phloem degeneration (the main symptom); normal vascular continuity in the area of union; accumulation of starch in the cultivar; different behaviour of reciprocal grafts; and blockage at the area of the union. Early symptoms are evident during the growing season, but it is rare to observe the death of the plant (Andrews and Serrano Marques, 1993).

The most important factor for identifying this incompatibility type is the translocation of a compound responsible for incompatibility from one genotype to another, which the use of an interstem cannot overcome (Andrews and Serrano Marquez, 1993). Examples of this type of incompatibility include different cultivars of almond (*Prunus dulcis*) (Kester et al., 1965; Heuser, 1987) and peach grafted on plum 'Marianna 2624' (*Prunus cerasifera* × *Prunus munsoniana*). In peach and almond incompatible combinations, trees showed degeneration of the phloem and the failure of the interstock even if the latter is compatible with both. The incompatibility of the almond is due to the production of a factor that is translocated through the phloem of the grafting area to the rootstock, thus producing a toxic reaction of the latter. This can also occur when one biont of the graft is sensitive to a virus that has infected the other biont, as with hypersensitivity of some cherry rootstocks to prune dwarf virus or prunus necrotic ringspot virus in infected cherry scions (Lang et al., 1997; Lang and Howell, 2001). This virus-induced translocated incompatibility is sometimes mistaken for localized incompatibility between genotypes, but it cannot be overcome with an interstem since eventually the viral genetic material will infect the interstem as well. This underscores the importance of using virus-free rootstocks and budwood for grafting.

4.2.2 Localized incompatibility

Localized incompatibility occurs at the graft interface and requires contact between the tissues of the two bionts (Mosse, 1962). In this case, the use of a compatible interstem genotype can overcome the problem. The main characteristics of this type of incompatibility include necrosis of the initial cells at the graft union interface and discontinuity of the vascular tissues with rupture of the grafting zone (Fig. 18a and b); gradual depletion of the root system's nutritional level; and an appearance of external symptoms proportional to the degree of union discontinuity. The slow appearance of external symptoms is due to the reduced translocation of the metabolites through the union that presents anatomical connection problems (Andrews and Serrano Marquez, 1993). An example of localized incompatibility is 'William' pear grafted on quince. The use of a pear-like cultivar (e.g. 'Old Home') as an interstem overcomes this incompatibility (Mosse, 1958).

4.2.3 Delayed incompatibility

Delayed incompatibility describes combinations that seem to grow well for a few years and then quickly break at the graft union or begin to deteriorate (Argles, 1937; Mosse, 1960; Randhawa and Bajwa, 1958). This term has been criticized because many occurrences of delayed incompatibility have more recently been identified with diseases attributable to phytoplasmas, as in the case of pear decline (Hibino et al., 1971) or to viruses, as discussed above. Additional examples of virus-induced delayed incompatibility include apple graft union necrosis caused by tomato ring spot virus (Cummins and Gonsalves, 1982) and tristeza virus of citrus (Andrews and Serrano Marquez, 1993).

4.3 Cell-cell recognition

The ability to respond to environmental stimuli and the control of entry and exit of molecules through the plasma membrane are fundamental characteristics for any type of cell (Lewin, 1994). Cell recognition mechanisms include all those phenomena in which cells select or specifically reject other types of cells that are either similar to them or complementary (Curtis, 1978). Yeoman et al. (1978) and Jeffree and Yeoman (1983) describe in the graft a recognition mechanism between the opposing cells of the two bionts that dissolve, with formation of holes in the cell wall that allow the contact of the plasmalemma of the cells of the two bionts and formation of plasmodesmata. The basis of this system is the liberation of recognized protein molecules by the plasmalemma, depending on whether the bionts are compatible or incompatible. When the union is good, the two sets of molecules stimulate transcription of nuclear mRNA, which does not occur when the graft combinations are incompatible.

Lectins are compounds that can be involved in specific phenomena of cell recognition and can interact with specific complementary carbohydrates (Andrews and Serrano Marquez, 1993). Feucht and Schmid (1988) considered lectins to not be connected with graft incompatibility between *Prunus avium/ Prunus cerasus*, since various graft combinations showed a remarkable variability in life span rather than exhibiting a similar mortality rate for all combinations. The timing of the appearance of symptoms did not give any indication of the metabolism of the lectins, and hence their role remains unclear (Andrews and Serrano Marquez, 1993).

4.4 Biochemical compounds involved in graft incompatibility

4.4.1 Cyanogenic glycosides (Gur hypothesis)

Moore (1984a,b) postulated that graft incompatibility occurs when compounds (auxins) that promote the success of the union are overcome by toxins that promote incompatibility. The toxins would prevent callus growth and inhibit vascular re-differentiation, thus preventing proper support for the scion. Researchers have previously investigated toxins produced by physiological processes of grafting and many consider these metabolites to be the cause of incompatibility. The movement of substances that can negatively affect the graft or that may alter the tissues has long been a controversial topic of the relationship between scion and rootstock.

The substances most often cited belong to the group of cyanogenic glycosides, which are present in many species (about 1000) and are derived from different amino acids (Jones, 1961). The mechanism of action of these compounds was described for the graft combinations of pear on quince and peach on almond, respectively (Gur et al., 1968; Gur and Blum, 1973). In the first case, an explanation of incompatibility was proposed: a cyanogenic glucoside (amygdalin) which is produced by quince, is translocated in pear and comes into contact with enzymes (β-glucosidase and mandelonitrile-lyase) that catabolize it by releasing two molecules of glucose, benzoic aldehyde and hydrogen cyanide. The latter is toxic to the cells, causing death and disorders in the graft union, with the consequent appearance of incompatibility symptoms (Gur et al., 1968). Hydrogen cyanide stops translocation and destroys phloem and xylem; this causes a reduction in the carbohydrates available for quince and thus increases the destruction of the phloem tissue. So, the different level of compatibility of pear cultivars utilized as interstock with quince, for example, 'Beurre' Hardy' and 'Doyenne' du Comice', can be explained by a different activity of enzymes like β-glucosidase and mandelonitrile-lyase capable of catabolizing amygdalin. A different explanation of this phenomenon can be related to the existence of a plant defence mechanism that was demonstrated in herbaceous species where the amino acid L-cysteine can incorporate the cyanidric group and form L-asparagine (Goodwin and Mercer, 1983).

Gur and Blum (1973) also studied incompatibility between some peach cultivars and almond, identifying cyanogenic glycosides as responsible for the incompatibility and finding these compounds in both bionts. However, the incompatible peach cultivars had a higher content of these compounds compared to the compatible cultivars relative to the same rootstock (Moing et al., 1987; Moing and Salesses, 1988).

4.4.2 Polyphenols and their role in the lignification of the graft union

The description of the models of morpho-functional graft events highlights the role of lignification in graft success. According to Buchloh (1962), the processes involved in the lignification of the cell wall are the main processes responsible for a solid union, and the compounds that inhibit the formation of lignin and median lamella in the cells of the contact zone can be jointly responsible for the appearance of symptoms of incompatibility. Lignin is a high molecular weight compound that is deposited on cell walls where its hydroxyl groups can join, covalently or with bonds, to hydrogen, to carbohydrates present, and above all, to cellulose (Buchloh, 1960; Guerrjn, 1970; Gross, 1979; Alpi et al., 1992). Various chemical molecules can influence the synthesis of lignin, among these are (a) the phenols, flavanones (prunin), coumarins, flavanols (especially catechins) and (b) the peroxidases. Some of these substances seem to have a positive role in the union, while others are more related to incompatibility.

Among the secondary products present in higher plants, phenols are involved in most of the metabolic processes: 60% of plant biomass production is associated with phenolic compounds (Errea, 1991). There are phenol monomers (cinnamic acid) and polymers (tannins and lignin) (Alpi et al., 1992). The main route of synthesis of these compounds is the 'shikimate pathway' (Stafford, 1988; Harborne, 1980) whose most important product is lignin (Rhodes, 1985; Gross, 1979). The phenols also regulate auxin activity. Auxins are important in the early stages of the union by acting on vascular re-differentiation. The phenolic compounds become part of the regulation of auxin catabolism, acting on IAA-oxidase, an enzyme that degrades auxins and whose activity is regulated by phenols. In particular, monophenols increase IAA_oxidase activity while the diphenols, such as chlorogenic acid, decrease it and then act as protectors (Nitsch, 1962; Stenlid, 1976; Kefeli and Kutacek, 1977; Pöessel, 1983).

In sweet cherry callus cultures, high concentrations of prunin were found to limit cell proliferation and differentiation (Feucht, 1988; Feucht and Schmid, 1988). Other experiments performed by Tretter et al. (1985), on trees formed by combinations of sweet and sour cherry, showed accumulation of prunin in the phloem above the graft union before the onset of external symptoms of incompatibility. In a subsequent work, the same authors proposed, for the same combination, to use prunin as a marker of incompatibility for early selection of the compatible genotypes (Treutter and Feucht, 1988). Prunin, in fact, is a growth inhibitor that promotes IAA degradation by creating an imbalance in growth regulators and also inhibits oxidative phosphorylation (Treutter and Feucht, 1988).

4.4.2.1 Coumarin

Yu and Carlson (1975) studied sweet cherry grafted onto *Prunus mahaleb*, finding a noticeable difference in the content of phenolic compounds. More precisely, sweet cherry contains high concentrations of *para*-coumaric acid and caffeic acid, which are absent in *P. mahaleb*. The coumarin content, on the other hand, is greater in the bark and roots of *P. mahaleb*. As already mentioned, these phenolic compounds regulate the synthesis of IAA-oxidase. More precisely, coumaric acid and hydroxybenzoic acid (monophenol) inhibit plant development because they activate auxin oxidation, while caffeic acid inhibits auxin oxidation and therefore promotes plant growth. The translocation capacity of these phenolic compounds was tested in various species such as in *Vicia faba* and in *Prunus* species (Favre-Bonvin et al., 1968); one can therefore hypothesize that the translocation of *para*-coumaric acid from sweet cherry (scion) to *P. mahaleb* (rootstock) favours decarboxylation of the auxins, reducing the vigour of the rootstock and favouring the onset of incompatibility. It is known that coumarins exert an inhibitory effect on the germination and growth of roots and shoots. The cultivars of sweet cherry grafted on *P. mahaleb* that are characterized by a high level of coumarins could lead to a slowing of growth (similar to what happens in seeds, sprouts and roots) and the appearance of incompatibility symptoms.

4.4.2.2 Flavanols (catechin)

Flavanols seem to have ambivalent roles in graft incompatibility. First, they are natural protectors of auxins (Feucht and Nachit, 1977), which regulate the synthesis of the lignin responsible for a successful union. In fact, in *in vitro* growth tests starting from sweet cherry bud segments, it was seen that the application of exogenous catechins increases the production of callus. There are not many studies on the role of catechins, but it was shown that these compounds increase cell division in *Prunus*; overall, therefore, catechins seem to play an active role in the incompatibility of grafting. However, it was proved that flavanols also can undergo enzymatic oxidation, with the consequent formation of quinones which, in contrast to catechins, can exert an inhibitory action on lignin synthesis, with negative effects on graft union stability (Freudenberg and Weinges, 1962; Pöessel et al., 1980; Feucht and Treutter, 1989).

4.4.2.3 Role of peroxidase

Peroxidases are enzymes that use hydrogen peroxide (H_2O_2) to oxidize a large number of hydrogen donors, including phenolic compounds. They are involved in various physiological processes and also play a role in lignin synthesis and graft incompatibility (Andrews and Serrano Marquez, 1993). Plant peroxidases catalyse the last enzymatic step in the biosynthesis of lignin, with the conversion of cinnamyl alcohols into their free-radical forms using hydrogen peroxide (Higuchi, 1985). Peroxidases supply hydrogen peroxide necessary for the oxidation of cinnamic acid and convert ferulic acid into diferulic which acts as a bridge to hemicellulose, joining the cinnamic acid to the proteins and carbohydrates of the wall which enhance consolidation (Bruce and West, 1989). Harkin and Obst (1973) demonstrated the presence of peroxidase in lignified cells; perhaps peroxidase is the only enzyme that polymerizes alcohols to lignin.

The role of these peroxidases in incompatibility is not yet clear. One view is that increased peroxidase activity is positive because it is accompanied by an accumulation of lignin, which is generally considered a positive factor. Peroxidases were studied in apricot (*Prunus armeniaca*) where it was found that compatible cultivars have a greater affinity for the rootstock of those that are incompatible. The lack of compatibility could be due to a lack of lignification at the graft union (Quessada and Macheix, 1984). A negative effect of peroxidase activity was observed in incompatible sweet cherry/sour cherry combinations in which the increase of the peroxidase activity coincided with tissue browning (Schmid and Feucht, 1984; Schmid and Feucht, 1985; Schmid and Freitag, 1988).

5 Tree production in the nursery

Fruit tree nursery production cycles vary according to species, but in general are rather short duration (from one to three years) and are characterized by high labour and management costs. The cycles usually begin with the preparation of the rootstock, followed by grafting or budding and bud development (Fig. 19). The timing of the graft operation determines the length of the cycle. The production cycle usually ends with a tree ready to be planted (Musacchi, 2007b). The current trend is to shorten the production cycles (sometimes also creating new types of products) and to move part of the production cycle from the open field to a container. This helps to reduce production costs and better respond to market needs (faster introduction of new cultivars) (Sansavini et al., 2006).

The traditional production cycle for apple nursery trees is a two-year cycle: the first year involves spring planting of the rootstock (obtained through one of the previously described techniques) and chip budding – a graft during the summer. In the following spring, the bud grows and at the end of the second growing season, the tree is ready for digging (Lugli and Musacchi, 1993).

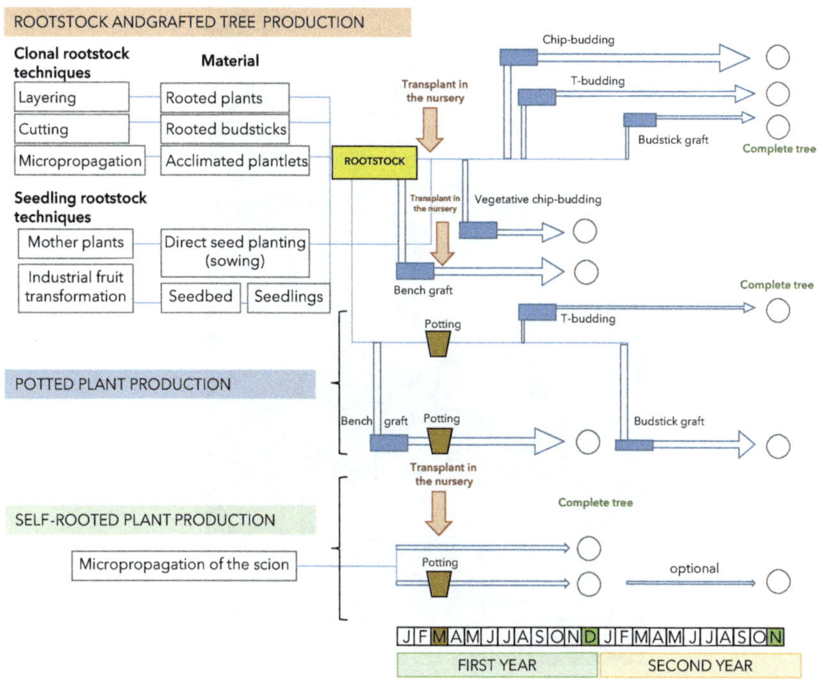

Figure 19 Plant production cycles in nursery.

An equally valid cycle for the production of a 'knip' tree in a 2-year cycle: the first year involves spring planting of the rootstock which has been grafted during the winter. During the first growing season, a tree with a short leader is obtained. This is cut during the winter, which then regrows vigorously during the second growing season and is conducive to forming numerous lateral shoots ('feathers'), resulting in a well-feathered 'knip' tree (Fig. 20) at the end of the second year (Catalano and Martinelli, 2005). A less common technique is to market the plant after the first year, because the tree produced in this case is usually of rather poor quality.

5.1 Two-year nursery cycle

The 2-year nursery tree production cycle begins in late February to March with the planting of rootstocks. The first year is dedicated to growing the root system and a rootstock shank to the prime size for budding. The second year focuses on growth of the scion to form the structural foundation for the future orchard canopy (Lugli and Musacchi, 1993).

Figure 20 Details of the knip tree. Notice that different age of the bottom part and the upper part of the tree.

Vegetative buds of the desired scion cultivar must be collected from the mother field of mature trees that (in most cases) are certified to be virus-free. The shoots from which they are taken are defoliated by hand or machine. In the case of T-budding, a short section of the petiole must be saved to make handling the buds during budding operations easier (Hartmann and Kester, 2002). It is better to use buds located in the middle part of the shoot because basal buds do not swell as well or may be floral (as in the case of cherry), and apical buds may still be immature.

Budding occurs from late July to mid-September, depending on the cambial activity of the rootstocks and the weather conditions. The cambium must be active enough to generate a good callus, which will heal the wound and rebuild tissues between the rootstock and the scion. Budding on the rootstock is done 10–15 cm above the soil level. Budding at 5 cm or lower may increase the possibility of scion rooting after the tree is planted. Placement of the bud onto the rootstock is performed by T-budding (a traditional technique) or with the more innovative practice of chip budding. Chip budding, developed in the 1970s, increases the time frame when budding is possible, allows for better contact between the two bionts, promotes a faster merging and more solid union, creates better tree uniformity, leads to a greater number of feathers because of accelerated growth and has a union that is less sensitive to cold and fungal spores (Howard et al., 1974; Ligonniere, 1981). Chip budding also expands the time that trees can be propagated because 'slipping' of the bark is not necessary. The wound is healed after 10–20 days of budding, and the bud remains dormant until the following spring. The rootstock's canopy above the budding is mechanically removed during winter. The stem just above the bud is cut at an oblique angle to minimize the effect of wood necrosis in the graft union to make the union stronger (Lugli and Musacchi, 1993).

In the spring of the second year, the shoot from the grafted bud usually grows very fast, achieving a length of 50–60 cm after one month. During this vegetative stage, PGR branching agents (cytokinins and gibberellins) can be applied to induce feather formation, depending on where the first scaffold of branches is desired. Treatments to promote feathers can be repeated 4–6 times as the stem grows, resulting in a pre-formed tree with 5-6 good feathers and many short shoots in the upper zone of the tree. All the shoots at heights of 50 cm or less from the soil surface must be removed when they are still herbaceous because they will not produce branches. In the fall of the second year, trees are ready to be dug and stored in a cold room until the orchard is ready to be planted.

5.2 Two-year nursery cycle with side graft

This type of tree production differs from that described above because a side graft is used to place the scion cultivar on the rootstock in the second year, starting in February or later depending on the climatic zone. This method is useful to recover rootstocks for which budding was unsuccessful the previous year. Side grafting utilizes a bud-stick with one or two buds. One end of the bud-stick is shaped like a triangle and it interlocks with the rootstock cut to receive it. The wood necessary for this kind of graft must be collected during winter and cold-stored to maintain dormancy (Lugli and Musacchi, 1993).

5.3 Bench grafts

Bench grafting is an important technique to produce a tree in 1 year. It is performed during the dormant season and scion wood is grafted on a bare root rootstock that may or may not be dormant. Scion and rootstock must be of equal diameter, and middle and upper diameter classes (9–11 mm and over 11 mm) are preferred. A deep V shape is machine cut on each of the bionts (Fig. 21a) and the two pieces are grafted together (Fig. 21b). This graft also can be performed by hand using the whip and tongue technique (Lugli and Musacchi, 1993). After grafting, the wound is covered with paper tape and dipped in melted wax to cover all the wounded surfaces (Fig. 22). The newly-grafted trees must be stored at 15–20°C for 10–15 days to allow the wounds to heal (Fig. 23). Afterwards, the grafted trees should be stored at 4–6°C and transferred to the orchard when the risk of a hard freeze is low.

5.4 'Knip' trees (apple, pear and cherry)

The 'knip' tree (Fig. 24) is the best choice for a traditional spindle training system, since it allows production in the same year of planting if all the management

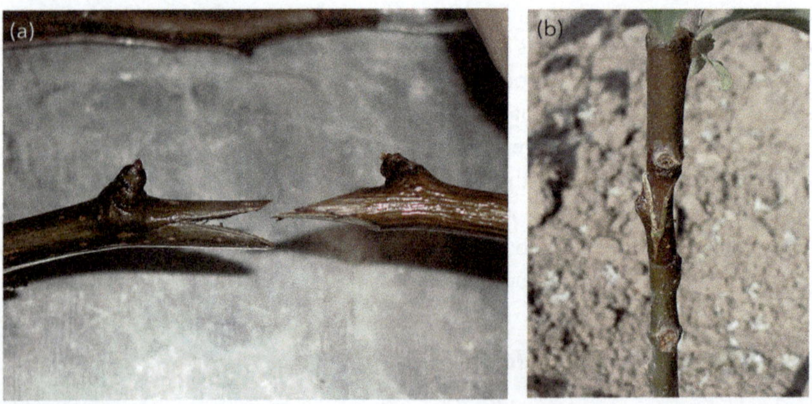

Figure 21 (a) Graft done with machine V shape. (b) Perfect merging of the two bionts of the V graft.

Figure 22 Sealing of the wound with paraffin.

techniques are done correctly. By the second year in the orchard, it is possible to obtain a significant crop of 10–15 t/ha. 'Knip' tree has been developed for the first time in the Netherlands where unsold trees were planted and headed back in the nursery and allowed to grow a new leader (Ravaglia and Corelli Grappadelli, 1990). 'Knip' is a Dutch word that means 'cut'. Originally it was a 3-year cycle of production which explains why the basal zone of the tree is older than the upper zone. The Italian nurseries developed a 2-year cycle to produce 'knip' trees; where bench grafted 6-month-old trees are headed back to 50–70 cm depending where the first scaffold of branches is desired. A single leader shoot is selected and treated with PGRs to induce feather formation. These trees usually develop feathers with a wider crotch angle and early bearing.

Figure 23 Bench graft ready to be planted in the field.

Figure 24 Knip trees.

5.5 Bi-axis trees (apple, pear, cherry and peach)

Bi-axis nursery trees are created with the 'Bibaum®' technique: two buds of the same scion cultivar are grafted opposite each other at the same level on the rootstock to develop two leaders (axes) during the growing season in the nursery (Fig. 25). Another method to achieve a bi-axis nursery tree is to bench graft a bar root rootstock with two buds. In this case, the insertion point of the two axes will be at different heights and may make the resulting tree unbalanced, especially if management in the nursery is not performed correctly, resulting in the axes having different diameters (Musacchi, 2008b).

5.6 Trees with an interstem

This cycle is necessary for those species, such as pear grafted on quince, characterized by localized graft incompatibility. To overcome this problem, it is necessary to utilize an interstem - usually a genotype compatible with both rootstock and scion. Traditionally, it would take three years to produce a complete interstem tree. In the first year, the rootstock is planted in the nursery

Figure 25 Bi-axis pear tree in nursery.

and summer-budded with the interstem. In the second year, the scion cultivar is budded onto the interstem during summer and will develop in the following (third) year. To shorten the cycle to two years, concomitant with planting and growing the rootstock during the first year, several buds of the scion cultivar can be budded on the interstem mother tree in the nursery, producing a sort of double bud-stick that can be grafted on the established rootstock at the beginning of the second year. A further possibility is to bench graft the scion cultivar and the interstem to create a double bud-stick and perform a side graft on the rootstock after one year in the nursery.

5.7 Mini-chip budding

The mini-chip budding technique facilitates the grafting of 'small plants' in pots. Buds are recovered from young weak shoots at the beginning of autumn or at the end of winter, while rootstocks are propagated *in vitro* and acclimatized. The budded rootstocks, after a period of healing and acclimatization in the greenhouse, can be transplanted in nursery for the final stage of growth or can be sold as pot plants after budding for direct transplant in the field (Fig. 26a

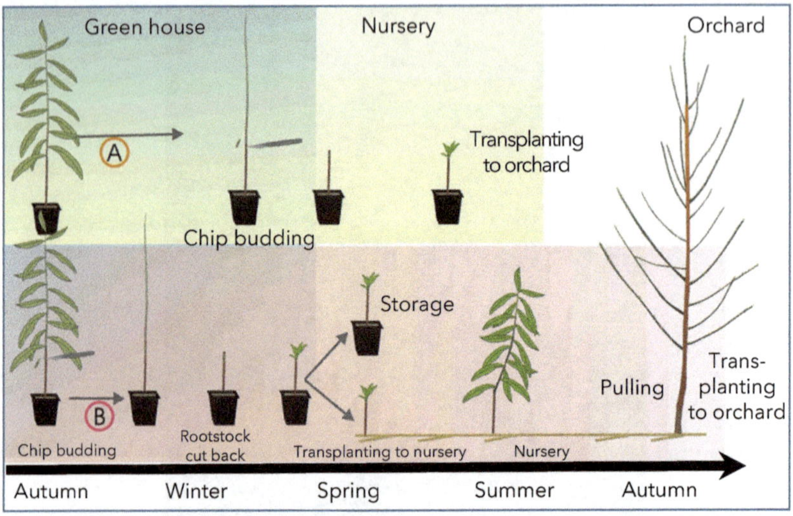

Figure 26 Short growth cycle of peach tree in 1 year: (A) chip budding done in January-February followed by a rapid sale in March-April immediately after the bud swelling; (B) chip budding done in September-October and tree are sold in February-March of the following year (Neri et al., 2008).

and b). The cycle is reduced to 6 months if the grafting is in early autumn and the sale in March or even 3 months if the grafting is at the end of winter and the sale in May. The mini-chip budding creates plants with a young and explorative root system that are easy to establish in the field (Fig. 27).

Figure 27 Left: Mini-chip budding in the fall; center: young growing shoot in late winter or early spring; right: end of the first year in the field.

This technique is increasing in the Mediterranean basin and California to produce trees (especially stone species) in very short production cycles, sometimes only a few months (Massetani et al., 2008a,b). The grafting technique differs from that of traditional chip budding only in the size of the plant material used. In fact, both the rootstocks and the buds are much smaller than the traditional technique. It has numerous advantages over dormant bud grafting. The healing of the wound and callus formation is faster. A disadvantage is that the bandage of the budding must be particularly accurate (better results are obtained with polyethylene tapes). The constant introduction of new stone fruit varieties required by the market makes it necessary to reduce the length of the production cycles in the nursery, and this also facilitates a more organized supply chain. The changeover from a biennial to an annual cycle or even to potted tree production in repeated cycles during the year can also be important for reducing costs.

5.8 Physiological aspects of tree growth in the nursery

There is evidence that the emergence of sylleptic shoots is highly probable when a threshold growth rate is exceeded. This means that in some conditions, apical dominance is not able to inhibit the growth of lateral meristems, which thus originate sylleptic (neo-formed) shoots rather than buds (Neri and Massetani, 2011) (Fig. 28). Generally, lateral vegetative buds

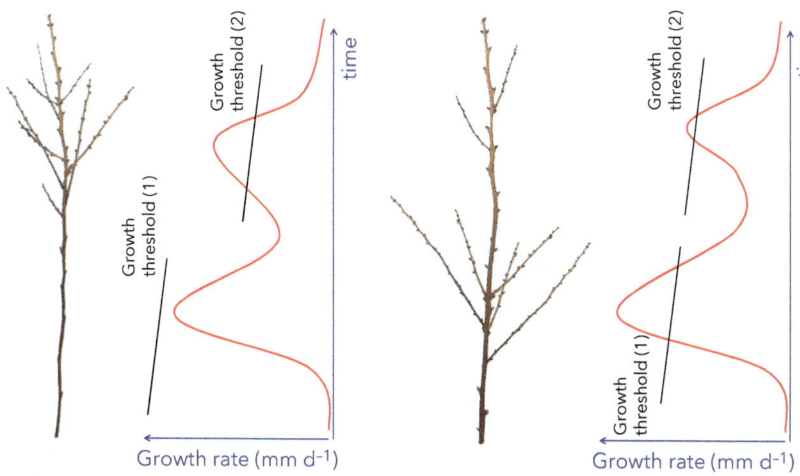

Figure 28 Shoot growth rate and sylleptic shoot formation in apricot. This model assumes that the growth rate threshold for inducing sylleptic growth decreases during the season. Left: the threshold was overcome only at the end of the season. Right: the critical growth rate was overcome twice in the season.

(once formed) become paradormant rapidly and will only grow if apical dominance is removed, as by heading. As the season progresses, these transition to endodormancy (Fig. 29), which means they will only grow in the following spring after a specific amount of chilling, emerging as proleptic shoots. However, exceeding the threshold growth rate releases some of the meristems from paradormancy, resulting in sylleptic branching. It is also possible that, depending on growing conditions, sylleptic shoot formation can take place more than once as the leader grows. Furthermore, the greater the distance between the apical meristem and the sylleptic branches, the greater the epinastic effect on wide crotch angles (Figs. 30 and 31). Thus, buds that are formed on sylleptic shoots at different times compared to proleptic shoots may have a different fruiting performance and time of flowering (generally later). This is frequently reported by growers (Neri and Massetani, 2011).

In high-density plantings, early limb (branch) formation is critical for filling the orchard's canopy fruiting area and obtaining a viable crop in year three or four following planting (Sansavini et al., 2008). It is well known that precocity is most easily achieved when trees possess a high number of horizontal laterals and develop well in the first years after planting. The growth of axillary meristems into sylleptic (lateral) shoots is 'controlled' by the apical meristem of growing shoots (Tromp, 1996) (Fig. 30). However, pinching cherry shoots to release the inhibition of axillary meristem can be inconsistent (Neri et al., 2004). Also, scion quality is negatively affected by pinching because of bending in the resulting axis. On the other hand, PGRs seem to be even less reliable for promotion of branching in sweet cherry

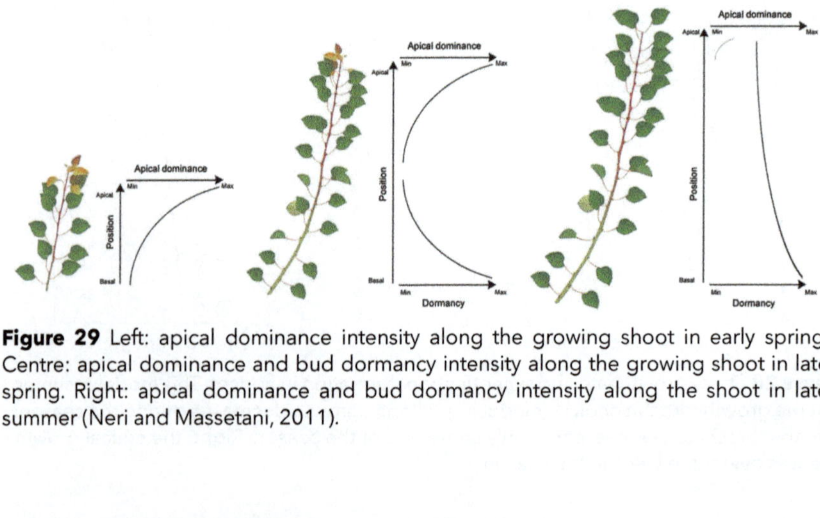

Figure 29 Left: apical dominance intensity along the growing shoot in early spring. Centre: apical dominance and bud dormancy intensity along the growing shoot in late spring. Right: apical dominance and bud dormancy intensity along the shoot in late summer (Neri and Massetani, 2011).

Figure 30 Apical dominance (left) and epinasty (right). Apical dominance 'ApD' is the control exerted along the stem by apical meristem on the axillary meristems which cannot elongate while shoot apex is growing. Epinasty: diffused control by upper parts (it determines the sylleptic shoot angle).

nursery trees. In any case, the absence of a well-shaped central leader, and the development of only a low number of lateral shoots, which may be too upright and vigorous, reduces the suitability of such trees for high-density planting.

In the apple nursery, stimulating fast growth has been used to obtain a desirable number of lateral shoots with wide crotch angles (Tromp, 1996; Neri et al., 2001). Moreover, removal, or spraying, of young leaves with benzyladenine (BA) formulations at the tip of the scion leader increased the tendency to form sylleptic laterals. Sweet cherry sylleptic shoot formation was obtained in the nursery by deblading apical leaflets, application of cytokinins, and avoiding pinching of the scion terminal shoot (Figs. 32–34). 'Kordia', 'Celeste', 'Summit' and 'Sunburst' were treated during the first growing season after budding onto Gisela 5, Colt or Mazzard rootstocks. Results indicate that laterals could be initiated successfully by this technique. Three to five lateral shoots, with low vigour and wide crotch angles, were consistently formed, producing quality 1-year-old well-feathered plants suitable for spindle training in high-density cherry orchards (Neri et al., 2004).

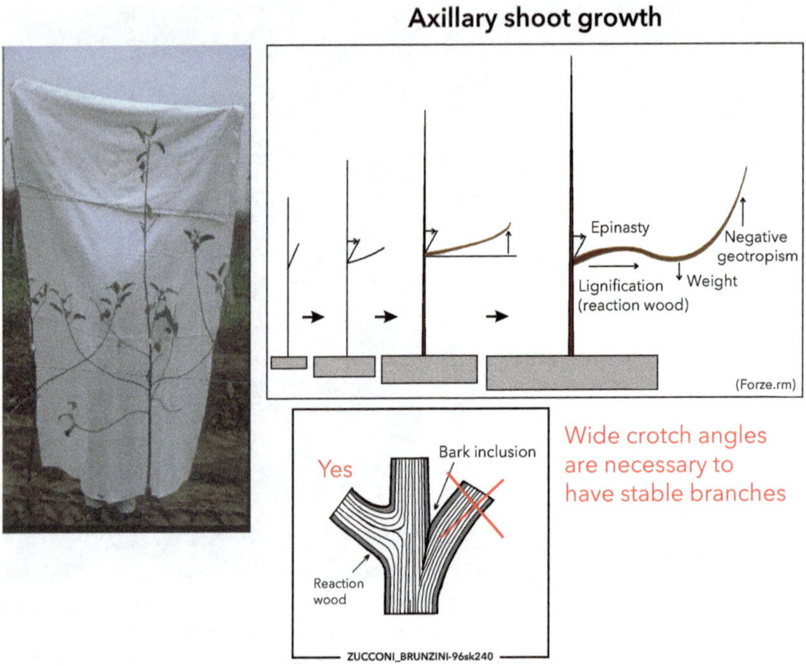

Figure 31 Sylleptic shoot growth. Wide crotch angle formation in the nursery of Florina apple scion grafted on M9.

5.9 Storage of the scions

Trees of pome and stone species commonly planted in the orchard may be kept safely in cold stores at +2°C, with a high level of humidity (almost 100%). The storage period can span from December to June, depending the growing conditions. Trees must be fully dormant to store them safely (Lugli and Musacchi, 1993).

6 Future trends

Orchards that are fully mechanized from planting to harvest is a logical next fruit production goal. To achieve this, simplified training systems must be developed, starting with economic, efficient and easy to manage nursery tree scions. Well-feathered single- or bi-axis scions, bare root planted and trained as a continuous hedgerow, could be a suitable solution for pome and stone fruit. In mild climates with long growing seasons, potted trees of small dimensions could be used to rapidly obtain medium-high density 'fruiting walls' where no regular axis distribution is required as long as canopy continuity is satisfied.

Figure 32 Celeste grafted onto Gisela 5. Left: before deblading. Right: plants after deblading. Apical leaflets removal (apical defoliation) was carried out when primary shoots were 50 cm long and with a high growth rate).

Figure 33 Left: Sunburst grafted onto Mazzard seedlings few days after deblading. Right: Summit grafted onto Colt few weeks after deblading.

Figure 34 Kordia budded onto Gisela 5 in August. Left: control plants. Right: plants after deblading and BA application.

Both of these approaches need specific research in the nursery and orchard to regulate the physiological status for efficiently achieving both early bearing and rapid lateral growth. Research should start with breeding to select new varieties suitable for the optimized approach.

For stone fruits, the challenge is to promote natural lateral (sylleptic) branching and a high growth rate during nursery scion formation, then shift to predominately proleptic shoot growth and a reduced growth rate after the second year. The target would be to obtain all the laterals well-distributed along the trunk and with wide crotch angles during the first growing season. Each lateral would then become a short branch with high flower differentiation, resulting in a uniform fruiting wall with a high potential for efficient mechanical fruit thinning, harvest and pruning (possibly in the summer).

For pome fruits, the challenge is to produce suitable trees on more vigorous rootstocks to overcome some replant problems. The use of well-feathered, pre-formed scions with one or two axes, with training as a hedgerow system, remains the most convenient solution to reduce tree vigour and induce early bearing.

In the nursery, the challenge is to propagate the scion genotype in the minimal time, ideally only a few months. This could be accomplished with small potted plants (either grafted or self-rooted). Such plants can be produced in a protected environment (e.g. greenhouse or screen house) to reduce the probability of infection by viruses or phytoplasmas, compared to normal 2-year nursery production cycles.

7 References

Ade, G. and Musacchi, S. 2003. La meccanizzazione della operazioni colturali nei cicli di produzione vivaistica dei fruttiferi e della fragola. *Rivista di Frutticoltura E di Ortofloricoltura* 65(1), 13–22.

Alpi, A., Pupillo, P. and Riganò, C. 1992. *Fisiologia delle piante*. Napoli: EdiSES, pp. 1-505.

Amiri, M. E. 2006. *In vitro* techniques to study the shoot-tip grafting of *Prunus avium* L. (cherry) var. Seeyahe Mashad. *International Journal of Food, Agriculture and Environment* 4(1), 151–4.

Andrews, P. K. and Serrano Marquez, C. 1993. Graft incompatibility. *Horticultural Reviews* 15, 183–232.

Argles, G. K. 1937. A review of the literature on stock-scion incompatibility in fruit trees, with particular reference to pome and fruits. In: *Imperial Bureau of Fruit Production*, Technical communication, no. 9.

Arndt, F., Rusch, R. and Stillfreid, H. V. 1976. SN49537, A new cotton defoliant. *Plant Physiology* 57, 99.

Baldini, E. 1986. *Arboricoltura generale*. Bologna: Clueb, pp. 1–402.

Bhattacharya, S. C. and Dutta, S. 1952. A preliminary observation on rootstocks for citrus in Assam. *Indian Journal of Horticulture* 9(3), 1–11.

Bonga, J. M. and Von Aderkas, P. 1992. *In Vitro Culture of Trees*. Dordrecht: Kluwer Academic Publishers, pp. 1–123.

Brand, M. H. and Lineberger, R. D. 1992. *In vitro* rejuvenation of *Betula* (*Betullaceae*): Morphological evaluation. *American Journal of Botany* 79(6), 618–25. doi:10.1002/j.1537-2197.1992.tb14603.x.

Breen, P. J. 1974. Cyanogenic glycosides and graft incompatibility between peach and plum. *Journal of the American Society for Horticultural Science* 99, 412–5.

Breen, P. J. 1975. Effect of peach/plum graft incompatibility on seasonal carbohydrate changes. *Journal of the American Society for Horticultural Science* 100, 253-9.

Breen, P. J. and Muraoka, T. 1975. Seasonal nutrient levels and peach/plum graft incompatibility. *Journal of the American Society for Horticultural Science* 100, 339–42.

Bruce, R. J. and West, C. A. 1989. Elicitation of lignin biosynthesis and isoperoxidase activity by pectic fragments in suspension-cultures of castor bean. *Plant Physiology* 91(3), 889-97. doi:10.1104/pp.91.3.889.

Buchloh, G. 1960. The lignification in stock-scion junctions and its relation to compatibility. In: Priham, J. B. (Ed.), *Phenolics in Plants in Health and Disease*. Oxford, UK: Pergamon Press, pp. 67–71.

Buchloh, G. 1962. Verwachsung und Verwachsungsstorungen als Ausdruck des Affinitatsgrades bei Pfropfungen von Birnenvarietaten auf Cydonia oblonga. *Beiträge zur Biologie der Pflanzen* 37, 183–240.

Cappellini, P. and Nicotra, A. 1973. Traslocazione di alcune sostanze rizogene in tale legnose. *Annal dell'Istituto Sperimentale per la Frutticoltura* 4, 109–17.

Catalano, L. and Martinelli, A. 2005. Il vivaismo si aggiorna fra nuove tecniche di propagazione e normative di processo. *Rivista di Frutticoltura E di Ortofloricoltura* 67(12), 28–36.

Cherubini, S. and Preka, P. 2001. Variazione stagionale della radicazione di alcuni portinnesti propagate per talea legnosa. *Rivista di Frutticoltura E di Ortofloricoltura* 63(5), 43–6.

Cioccolanti, T., Lodolini, E. M., Bastianelli, M., Perez, V., Zuccherelli, G. and Neri, D. 2012. Confonto fra piante di olivo micropropagate e da talea in un impianto ad alta densità. *Rivista di Frutticoltura E di Ortofloricoltura* 74(9), 66–71.

Colombo, R. 2003. Cotogno, franco o autoradicato. *Il divulgatore* 4, 24–37.

Copes, D. L. 1980a. Anatomical symptoms of graft incompatibility in *Pinus monticola* and *P. ponderosa*. *Silvae Genetica* 29, 77–82.

Copes, D. L. 1980b. Development of internal graft incompatibility symptoms in *Pinus radiata* D. Don. *New Zealand Journal of Forestry Science* 10, 367–80.

Cozzolino, E. and Neri, D. 2017. Esperienze, caratteristiche e prospettive delle piante di noce micropropagate. *Rivista di Frutticoltura E di Ortofloricoltura* 79(3), 44–9.

Cummins, J. N. and Gonsalves, D. 1982. Recovery of tomato ringspot virus from inoculated apple trees. *Journal of the American Society for Horticultural Science* 107, 798–800.

Curtis, A. S. 1978. Cell-cell recognition: Positioning and patterning systems. *Symposia of the Society for Experimental Biology. Cambridge University Press* 32, 51–82.

D'Oria, F. 2012. 'In Vitro' evaluation of new pear rootstock selections. (Master Thesis). Bologna: University of Bologna.

De Carlo, A., Benelli, C. and Lambardi, M. 2009. La micropropagazione per la salvaguardia della biodiversità vegetale: Esperienze condotte presso il CNR-IVALSA. *Italus Hortus* 16(2), 175–9.

De Paoli, G. 1996. Il Pero autoradicato: Situazione attuale e prospettive. *L'Informatore Agrario* 52(48), 56–8.

Deloire, A. and Hebant, C. 1983. Le greffage de combinaisons compatibles et incompatibles du genre *Prunus*: Une étude histophysiologique de la zone de jonction. *Agronomie. EDP Sciences* 3(3), 207–12. doi:10.1051/agro:19830303.

Dobránszki, J. and da Silva, J. A. T. 2010. Micropropagation of apple – A review. *Biotechnology Advances* 28(4), 462–88. doi:10.1016/j.biotechadv.2010.02.008.

Dolcet-Sanjuan, R., Claveria, E., Asín, L., Vilardell, P., Bonany, J. and Simard, M. H. 2008. Towards the selection of a new pear rootstock: In vitro and field evaluation for tolerance to iron chlorosis, low vigor and micropropagation of selected clones. *Acta Horticulturae* 800(800), 683–91. doi:10.17660/ActaHortic.2008.800.92.

Edwards, P. J. 1992. Resistance and defence: The role of secondary plant substances. In: Ayres, P. G. (Ed.), *Pest and Pathogens*. Milton Park, Didcot, UK: BIOS Scientific Publishers, pp. 69–84.

Eleftheriou, E. P. 1985. A decline-like disorder of pear in Greece: Pear decline or scion–Rootstock incompatibility? *Scientia Horticulturae* 27(3-4), 241–50. doi:10.1016/0304-4238(85)90028-7.

Errea, P. 1991. Compatibilidad de injertos en Albaricoquero - *P. armeniaca* L.- Anatomia y bioquimica de uniones compatibles e incompatibles. (Doctoral Dissertation). Pamplona, Spain: Universidad de Navarra.

Errea, P. 1998. Implications of phenolic compounds in graft incompatibility in fruit tree species. *Scientia Horticulturae* 74(3), 195–205. doi:10.1016/S0304-4238(98)00087-9.

Espen, L., Cocucci, M. and Sacchi, G. A. 2002. Vascular tissue regeneration *'in vitro'* graft of pear/quince. *Acta Horticulturae* 596, 431–5.

FAO. 2016. *The World of Organic Agriculture 2016*. FiBL – Research Institute of Organic Agriculture and IFOAM – Organics International Organization. ISBN: 978-3-03736-307-2.

Favre-Bonvin, J., Massias, M., Mentzer, C. and Massicot, J. 1968. Sites de biosynthèse des coumarines chez *Prunus mahaleb*: Étude par greffage et administration d'un précurseur marqué. *Phytochemistry* 7(9), 1555–60. doi:10.1016/S0031-9422(00)88605-7.

Fazio, G., Aldwinckle, H. S., Robinson, T. L. and Cummins, J. 2005. Geneva® 41: A new fire blight resistant, dwarf apple rootstock. *HortScience* 40(4), 1027.

Feucht, W. 1988. Graft incompatibility of tree crops: An overview of the present scientific status. *Acta Horticulturae* 227(227), 33–42. doi:10.17660/ActaHortic.1988.227.1.

Feucht, W. and Nachit, M. 1977. Flavolans and growth-promoting catechins in young shoot tips of *Prunus* species and hybrids. *Physiologia Plantarum* 40(4), 230–4. doi:10.1111/j.1399-3054.1977.tb04063.x.

Feucht, W. and Schmid, P. P. S. 1988. Incompatibility of grafted cherry trees (*P. avium/P. cerasus*): Loss of metabolic tolerance as a consequence of environmental stress. *Acta Horticulturae* 227(227), 87–9. doi:10.17660/ActaHortic.1988.227.9.

Feucht, W. and Treutter, D. 1989. *Phenolische Naturstoffe: ihre Bedeutung für Gartenbau, Land-und Forstwirtschaft; praktische Aspekte für die ngewandte Naturwissenschaft*. München: Obst- und Gartenbauverlag, 150pp.

Feucht, W. and Treutter, D. 1991. Phenol gradients in opposing cells of *Prunus heterografts*. *Advances in Horticultural Science* 5(3), 107–11.

Feucht, W., Schmid, P. P. S. and Christ, E. 1985. Flavanole und Wundheilung nach Hagelschäden beim Baumobst. *Obstbau Weinbau* 22, 249–50.

Finnegan, E. J. 2001. Is plant gene expression regulated globally? *Trends in Genetics: TIG* 17(7), 361–5. doi:10.1016/S0168-9525(01)02319-8.

Franclet, A. 1979. Rejuvenation of adult trees for vegetative propagation. *Estudes et Recherches, AFOCEL, France* (12), 3–18.

Freudenberg, K. and Weinges, K. 1962. *The Chemistry of Flavonoid Compounds*. Oxford, UK: Pergamon Press, pp. 1–11.

Garner, R. J. 1979. *The Grafter's Handbook 4th ed*. New York: Oxford University Press, pp. 49–67.

Gebhardt, K. and Goldbach, H. 1988. Establishment, graft union characteristics and growth of *Prunus* micrografts. *Physiologia Plantarum* 72(1), 153–9. doi:10.1111/j.1399-3054.1988.tb06637.x.

Goodwin, T. W. and Mercer, E. I. 1983. *Introduction to Plant Biochemistry*. Great Britain: Pergamon Press.

Grapin, A. 2009. La micropropagazione in Francia e nord Europa. *Italus Hortus* 16(2), 21–4.

Gross, G. G. 1979. Recent advances in the chemistry and biochemistry of lignin. In: Swain, T. (Ed.), *Biochemistry of Plant Phenolics*. Boston, MA: Springer, pp. 177–220.

Guerrjn, C. 1970. Contribution l'étude histologique et biochimique de l'incompatibilité, au greffage de l'abricotier sur le prunier. (Doctoral Dissertation). Bordeaux, Nouvelle-Aquitaine: Université de Bordeaux.

Gur, A. and Blum, A. 1973. The role of cyanogenic glycoside in incompatibility between peach scions and almond rootstocks. *Horticulture Research* 13, 1–10.

Gur, A., Samish, R. M. and Lifshitz, E. 1968. The role of cyanogenic glycoside of the quince in the incompatibility between pear cultivars and quince rootstocks. *Horticulture Research* 8, 113–34.

Hackett, W. P. 1985. Juvenility, maturation, and rejuvenation in woody plants. In: Janick, J. (Ed.), *Horticultural Reviews* (vol. 7). Hoboken, New Jersey: John Wiley & Sons, Inc.,, pp. 109–55.

Harborne, J. B. 1980. Phenolic compounds derived from shikimate. *Biosynthesis* 6, 40–75.

Harkin, J. M. and Obst, J. R. 1973. Lignification in trees: Indication of exclusive peroxidase participation. *Science* 180(4083), 296–8. doi:10.1126/science.180.4083.296.

Hartmann, H. T. and Kester, D. E. 2002. *Plant Propagation – Principles and Practices*. Upper Saddle River, NJ: Pearson Education.

Haslam, E. 1989. *Plant Polyphenols. Vegetable Tannins Revisited*. Cambridge, UK: Cambridge University Press.

Herrero, J. 1951. Studies of compatible and incompatible graft combinations with special reference to hardy fruit trees. *Journal of Horticultural Science* 26(3), 186–237. doi:10.1080/00221589.1951.11513736.

Heslop-Harrison, J. 1978. Cellular recognition systems in plants. In: *Studies in Biology* (vol. 100). London, UK: Edward Arnold Ltd., pp. 1–60.

Heuser, C. W. 1987. Graft incompatibility: Effect of cyanogenic glycosides on almond and plum callus growth. *Proceedings of the International Plant Propagators' Society* 36, 91–7.

Hibino, H., Kaloostian, G. H. and Schneider, H. 1971. Mycoplasma-like bodies in the pear psylla vector of pear decline. *Virology* 43(1), 34–40.

Higuchi, T. 1985. Biosynthesis of lignin. In: Higuchi, T. (Ed.), *Biosynthesis and Biodegradation of Wood Components*. New York, NY: Academic Press, pp. 141–60.

Howard, B. H. 1987. *Propagation. Rootstocks for Fruit Crops*. Hoboken, NJ: Wiley.

Howard, B. H., Skene, D. S. and Coles, J. S. 1974. The effect of different grafting methods upon the development of one-years-old nursery apple-trees. *Journal of Horticultural Science* 49(3), 287–95. doi:10.1080/00221589.1974.11514581.

Howard, B. H., Harrison-Murray, R. S. and Fenlon, C. A. 1983. Effective auxine treatment of leafless winter cuttings. Monograph 10. British Plant Growth Regulator Group, pp. 73–85.

Jeffree, C. E. and Yeoman, M. M. 1983. Development of intercellular connections between opposing cells in a graft union. *New Phytologist* 93(4), 491–509. doi:10.1111/j.1469-8137.1983.tb02701.x.

Jones, M. B. 1961. Seasonal trend of cyanide in peach leaves and flower buds and its possible relationship to the rest period. *Proceedings of the American Society for Horticultural Science* 71, 117–20.

Kefeli, V. I. and Kutacek, M. 1977. Phenolic substances and their possible role in plant growth regulation. In: Pilet, P. E. (Ed.), *Plant Growth Regulation*. Berlin, Heidelberg: Springer, pp. 181–8. doi:10.1007/978-3-642-66589-9_20.

Kester, D. E., Hansen, C. J. and Panetsos, C. 1965. Effect of scion and interstock variety on incompatibility of almond on Marianna 2624 rootstocks. *Proceedings of the American Society for Horticultural Science* 86, 169–77.

Kirby, E. and Granatstein, D. 2017. Certified organic acreage and sales in Washington State: 2007–2015 Trends in Washington organic crop production series. Available

at: http://pubs.cahnrs.wsu.edu/publications/wp-content/uploads/sites/2/publications/TB43E.pdf.
Lambardi, M. and Ozudogru, A. E. 2010. La propagazione *'in vitro'* dei portinnesti e delle specie da frutto. *Rivista di Frutticoltura E di Ortofloricoltura* 72(12), 18–25.
Lang, G. A. and Howell, W. 2001. Lethal sensitivity of some new cherry rootstocks to pollen borne viruses. *Acta Horticulturae* 557, 151–4.
Lang, G. A., Howell, W., Ophardt, D. and Mink, G. 1997. Biotic and abiotic stress responses of interspecific hybrid cherry rootstocks. *Acta Horticulturae* 451, 217–24.
Lapins, K. 1959. Some symptoms of stock-scion incompatibility of apricot varieties on peach seedling rootstock. *Canadian Journal of Plant Science* 39(2), 194–203.
Lewin, B. 1994. *Genes (V)*. Oxford, UK: Oxford University Press.
Ligonniere, G. 1981. Le chip-budding. *Le Fruit Belge* 396, 235–7.
Liu, B. and Wendel, J. F. 2003. Epigenetic phenomena and the evolution of plant allopolyploids. *Molecular Phylogenetics and Evolution* 29(3), 365–79. doi:10.1016/S1055-7903(03)00213-6.
Lodolini, E. M., Tarragoni, A., Cioccolanti, T., Pollastri, L. and Neri, D. 2017. Architectural characteristics of six olive cultivars with respect to their suitability for high density orchards. *Acta Horticulturae* 1160, 127–34.
Long, L., Lang, G., Musacchi, S. and Whiting, M. 2015. Cherry training systems. Pacific Northwest Ext. Publ. 667, pp. 1–63.
Lugli, S. and Musacchi, S. 1993. Tecniche e cicli produttivi delle pomacee in vivaio. *Rivista di Frutticoltura e Ortofloricoltura* 1, 25–33.
Martienssen, R. A. and Colot, V. 2001. DNA methylation and epigenetic inheritance in plants and filamentous fungi. *Science* 293(5532), 1070–4. doi:10.1126/science.293.5532.1070.
Massetani, F., Dalmonte, P., Giorgi, V. and Neri, D. 2008a. Innesto a mini chip-budding: Impiego di diversi materiali per la legatura. *Atti convegno nazionale sulla peschicoltura meridionale*, pp. 226–30.
Massetani, F., Neri, D., Savini, G. and Dalmonte, C. 2008b. Mini-innesto a chip-budding: Controllo della qualità delle gemme. *Atti convegno nazionale sulla peschicoltura meridionale*, pp. 188–93.
McCully, M. E. 1983. Structural aspects of graft development. In: *Vegetative Compatibility Responses in Plants*. Waco, TX: Baylor University Press, pp. 71–88.
Mencuccini, M. 2009. Impiego della coltura in vitro dell'olivo per screening di resistenza. *Italus Hortus* 16(2), 202–5.
Mitov, P. and Zocca, A. 1966. La diversa attitudine ad emettere rami anticipate nelle cultivar di melo. *Rivista di Frutticoltura E di Ortofloricoltura* 28, 71–9.
Moing, A. and Salesses, G. 1988. Peach/plum graft incompatibility: Structural, physiological and genetic aspects. *Acta Horticulturae* 227(227), 57–62. doi:10.17660/ActaHortic.1988.227.3.
Moing, A., Salesses, G. and Saglio, P. H. 1987. Growth and the composition and transport of carbohydrate in compatible and incompatible peach plum grafts. *Tree Physiology* 3(4), 345–54. doi:10.1093/treephys/3.4.345.
Monteuuis, O. 2004. *In vitro* rooting of juvenile and mature *Acacia mangium* microcuttings with reference to leaf morphology as a phase change marker. *Trees – Structure and Function* 18(1), 77–82. doi:10.1007/s00468-003-0283-x.

Moore, R. 1984a. A model for graft compatibility-incompatibility in higher plants. *American Journal of Botany* 71(5), 752–8. doi:10.1002/j.1537-2197.1984.tb14182.x.
Moore, R. 1984b. Graft formation in *Solanum pennellii* (*Solanaceae*). *Plant Cell Reports* 3(5), 172–5. doi:10.1007/BF00270192.
Moore, R. and Walker, D. B. 1981a. Studies of vegetative compatibility-incompatibility in higher plants. I. A structural study of a compatible autograft in *Sedum telephoides* (Crassulaceae). *American Journal of Botany* 68(6), 820–30. doi:10.1002/j.1537-2197.1981.tb12416.x.
Moore, R. and Walker, D. B. 1981b. Studies of vegetative compatibility-incompatibility in higher plants. II. A structural study of an incompatible heterograft between *Sedum telephoides* (Crassulaceae) and *Solanum pennellii* (Solanaceae). *American Journal of Botany* 68(6), 831–42. doi:10.1002/j.1537-2197.1981.tb12417.x.
Mosse, B. 1958. Further observations on growth and union structure of double-grafted pear on quince. *Journal of Horticultural Science* 33(3), 186–93. doi:10.1080/00221589.1958.11513925.
Mosse, B. 1960. Graft incompatibility in plums; observations on a ten-year-old field trial. *Journal of Horticultural Science* 35(4), 260–5. doi:10.1080/00221589.1960.11513990.
Mosse, B. 1962. Graft-incompatibility in fruit trees with particular reference to its underlying causes. In: *Commonwealth Bureau of Horticulture and Plantation Crops*. Technical communication no. 28, p. 36.
Murashige, T. and Skoog, F. 1962. A revised medium for rapid growth and bio assays with tobacco tissue cultures. *Physiologia Plantarum* 15(3), 473–97. doi:10.1111/j.1399-3054.1962.tb08052.x.
Musacchi, S. 1996. Studies on physiological and biochemical aspects of graft-incompatibility. (Doctoral Dissertation). Bologna: University of Bologna.
Musacchi, S. 2007a. Le scelte dell'agricoltore per l'impianto del pero. *L'informatore Agrario* 43, 37–40.
Musacchi, S. 2007b. Tecnica colturale. In: Fideghelli, C., and Barbera, G. (Ed.), *Il Pero*. Bayer CropScience, pp. 92–137.
Musacchi, S. 2008a. Training systems and soil management for Southern European pear orchards. *Acta Horticulturae* 772(772), 447–57. doi:10.17660/ActaHortic.2008.772.76.
Musacchi, S. 2008b. Bibaum®: A new training system for pear orchards. *Acta Horticulturae* 800(800), 763–9. doi:10.17660/ActaHortic.2008.800.104.
Musacchi, S. and Green, D. 2017. Innovations in apple tree cultivation to manage crop load and ripening. In: Evans, K. (Ed.), *Achieving Sustainable Cultivation of Apples*. Cambridge, UK: Burleigh Dodds Science Publishing Limited, pp. 195–237.
Musacchi, S., Pagliuca, G., Kindt, M., Piretti, M. V. and Sansavini, S. 2000. Flavonoids as markers for pear-quince graft incompatibility. *Journal of Applied Botany – Angewandte Botanik* 74, 206–11.
Musacchi, S., Ganino, T., Grassi, S. and Fabbri, A. 2004. Grafting in vitro shoots on acclimating plants: Morphological and anatomical aspects of compatible and incompatible pear/quince combinations. *Acta Horticulturae* 658(658), 559–63. doi:10.17660/ActaHortic.2004.658.85.
Musacchi, S., Gagliardi, F. and Serra, S. 2015. New training systems for high-density planting of sweet cherry. *HortScience* 50(1), 59–67. doi:10.21273/HORTSCI.50.1.59.

Navarro, L. 1988. Application of shoot-tip grafting in vitro to woody species. *Acta Horticulturae* 227(227), 43-56. doi:10.17660/ActaHortic.1988.227.2.

Nelson, S. H. 1968. Incompatibility survey among horticultural plants. *Proceedings of the International Plant Propagators' Society* 18, 343-407.

Neri, D. 1990a. Giovanilità e ringiovanimento nelle piante arboree. *Rivista di Frutticoltura E di Ortofloricoltura* 52(12), 69-73.

Neri, D. 1990b. Tecniche di innesto con materiale proveniente da colture *in vitro*. *Rivista di Frutticoltura E di Ortofloricoltura* 52(3), 69-72.

Neri, D. 1996. Giovanilità e ringiovanimento nelle piante arboree. Seminari, Dottorato di ricerca 'Colture Arboree', 1985-1995, Bologna, pp. 203-7.

Neri, D. 2004. La formazione delle piante da frutto in vivaio. *Rivista di Frutticoltura E di Ortofloricoltura* 66(12), 26-31.

Neri, D., Massetani, F. and Dalmonte, C. 2008. Più razionale la produzione degli astoni con il ciclo breve di vivaio. *Rivista di Frutticoltura E di Ortofloricoltura* 70(7), 30-34.

Neri, D. and Massetani, F. 2011. Spring and summer pruning in apricot and peach orchards. *Advances in Horticultural Science* 25(3), 170-8.

Neri, D., Sansavini, S. and Capobianco, M. 1989. Performance of micropropagated and budded pear trees: Interaction between scion and dwarf-quince clones. *Acta Horticulturae* 256(256), 69-76. doi:10.17660/ActaHortic.1989.256.6.

Neri, D., Papachatzis, G., Sabbatini, P. and Zucconi, F. 2001. La formazione dei rami anticipati. *Rivista di Frutticoltura E di Ortofloricoltura* 63(5), 31-6.

Neri, D., Mazzoni, M. and Dradi, G. 2004. Feathering induction in sweet cherry (*Prunus avium* L.) by leaf defoliation and cytokinin foliar application in the nursery. *Acta Horticulturae* 636, 119-27.

Neri, D., Giovannini, D., Massai, R., Di Vaio, C., Sansavini, S., Del Vecchio, G., Guarino, F., Mennone, C., Abeti, D. and Colombo, R. 2010. Efficienza produttiva e gestionale degli impianti di pesco in un confronto Nord-Sud. *Italus Hortus* 17(3), 46-62.

Neri, D., Massetani, F. and Murri, G. 2015. Pruning and training systems: What is next? *Acta Horticulturae* 1084(1084), 429-43. doi:10.17660/ActaHortic.2015.1084.60.

Neri, D., Lodolini, E. M. and Bastianelli, M. 2017. Acclimation *ex vitro*: How to improve the root system. *Acta Horticulturae* 1155(1155), 475-82. doi:10.17660/ActaHortic.2017.1155.70.

Nitsch, J. P. 1962. Composés phenoliques et croissance végétale. Annales de Physiologie Végétale 4, 211-25.

Nocito, F. F., Espen, L., Fedeli, C., Lancilli, C., Musacchi, S., Serra, S., Sansavini, S., Cocucci, M. and Sacchi, G. A. 2010. Oxidative stress and senescence-like status of pear calli co-cultured on suspensions of incompatible quince microcalli. *Tree Physiology* 30(4), 450-8. doi:10.1093/treephys/tpq006.

Ossani, V. 1991. Pesco autoradicato: Le problematiche agronomiche e l'influenza sulla qualità. *Rivista di Frutticoltura E di Ortofloricoltura* 53(6), 47-9.

Ozudogru, E. A. and Lambardi, M. 2016. Cryotechniques for the long-term conservation of embryogenic cultures from woody plants. In: Germana, M. A., and Lambardi, M. (Eds.), *In Vitro Embryogenesis in Higher Plants*. New York, NY: Humana Press, pp. 537-50. doi:10.1007/978-1-4939-3061-6_32.

Pasqual, M., Cavalcante-Alves, J. M., Chalfun, N. N. J., Silva, A. B., Dutra, L. F. and Bianchi, J. V. 2002. *In vitro* rooting ad shoot growth of *Pyrus betulaefolia* rootstock. *Acta Horticulturae* 596, 447-50.

Pastyrik, L. and Jvanicka, J. 1977. Influence of winter dormancy on the rooting of harwood cuttings of the wild apricots *Armoniaca vulgaris* and *Armoniaca manshurica*. *Physiologia Plantarum* 14, 87–95.

Pina, A. and Errea, P. 2005. A review of new advances in mechanism of graft compatibility-incompatibility. *Scientia Horticulturae* 106(1), 1–11. doi:10.1016/j.scienta.2005.04.003.

Poëssel, J. L. 1983. Composés phenoliques et peroxidases de l'abricotier: Etude comparative de deux varietés (Luizet et Canino) en relation avec l'incompatibilité, auf greffage. (Doctoral Dissertation). Montpellier: Université des Sciences et Technique du Languedoc.

Poëssel, J. L., Martinez, J., Macheix, J. and Jonard, R. 1980. Variations saisonnieres de l'aptitude au greffage *in vitro* d'apex de pecher (*Prunus persica*, Batsch). Relations avec les teneurs en composes phenoliques endogenes et les activites peroxydasique et polyphenoloxydasique. *Physiologie Vegetale* 18, 665–75.

Poething, R. S. 1990. Phase change and the regulation of shoot morphologenesis in plant. *Science* 250(4983), 923–30. doi:10.1126/science.250.4983.923.

Preka, P. and Cherubini, S. 2001. Tecniche di innesto erbaceo per la propagazione di piante arboree da frutto. *Rivista di Frutticoltura E di Ortofloricoltura* 63(5), 39–41.

Quessada, M. P. and Macheix, J. J. 1984. Caractérisation d'une peroxydase impliquée spécifiquement dans la lignification, en relation avec l'incompatibilité au greffage chez l'Abricotier. *Physiologie Végétale* 22(5), 533–40.

Randhawa, S. S. and Bajwa, B. S. 1958. Histology of bud-union in citrus. *Indian Journal of Horticulture* 15(1), 4–5.

Ravaglia, G. F. and Corelli Grappadelli, L. 1990. Le piante in vaso nel vivaismo frutticolo. *Rivista di Frutticoltura E di Ortofloricoltura* 52(5), 65–70.

Razdan, M. K. 2003. *Introduction to Plant Tissue Culture*. Enfield, Plymouth: Science Publishers, Inc.

Rhodes, M. J. C. 1985. The physiological significance of plant phenolic compounds. *Annual proceedings of the Phytochemical Society of Europe* 25, 99–117.

Rogers, W. S. and Beakbane, A. B. 1957. Stock and scion relations. *Annual Review of Plant Physiology* 8(1), 217–36. doi:10.1146/annurev.pp.08.060157.001245.

Rugini, E., Pesce, P. G. and Cristofori, V. 2009. La micropropagazione di specie da frutto: Stato attuale e rospettive. *Italus Hotus* 16(2), 15–20.

Salesses, G. and Al Kai, N. 1985. Simply inherited grafting incompatibility in peach. *Acta Horticulturae* 173(173), 57–62. doi:10.17660/ActaHortic.1985.173.7.

Sansavini, S. and Zocca, A. 1965. La diversa attitudine ad emettere rami anticipati nelle cultivar di pero. *Rivista di Frutticoltura E di Ortofloricoltura* 27(3), 233–41.

Sansavini, S., Ponti, I. and Ancarani, V. 2004. Variazioni degenerative in ceppaie micropropagate di meli M9 T337. *Rivista di Frutticoltura E di Ortofloricoltura* 66(12), 38–42.

Sansavini, S., Venturi, S. and Lugli, S. 2006. Innovazioni tecniche e sviluppo del vivaismo frutticolo. *Rivista di Frutticoltura E di Ortofloricoltura* 68(12), 18–32.

Sansavini, S., Neri, D. and Ancarani, V. 2008. Innovazioni tecnologiche per migliorare l'efficienza produttiva e la qualità di frutti nelle coltivazioni intensive. *Rivista di Frutticoltura e Ortofrutticoltura* 70(10), 10–26.

Schloupt, R. M., Splittstoesser, W. E. and Skirvin, R. M. 1994. Introduction and avoidance of vitrification of micropropagated onion (*Allium cepa* L.). *HortScience* 29(5), 465.

Schmid, P. P. S. and Feucht, W. 1984. Proteins, peroxidases and polyphenols in reciprocal grafts of *P. avium* L. and *P. cerasus* L. *Gartenbauwissenschaft* 49(4), 170–5.

Schmid, P. P. S. and Feucht, W. 1985. Compatibility in *Prunus avium/Prunus cerasus* graftings during the initial phase. III. Isoelectrofocusing of proteins, peroxidases and acid phosphatases during union formation. *Journal of Horticultural Science* 60(3), 311–8. doi:10.1080/14620316.1985.11515634.

Schmid, P. P. S. and Freitag, I. 1988. Differential substrate specificity of isoperoxidase from horseradish and cherry phloem after isoelectric focusing. *Angewandte Botanik* 62, 161–8.

Schmid, P. P. S., Schmitt, E. R. and Zorn, W. 1988. Water relations and some organic compounds in cherry leaves of grafting with delayed incompatibility. *Acta Horticulturae* 227(227), 90–2. doi:10.17660/ActaHortic.1988.227.10.

Smith, R. H. 2000. *Plant Tissue Culture – Techniques and Experiments*. Cambridge, MA: Academic Press.

Sriskandarajah, D., Mullins, M. G. and Nair, Y. 1982. Induction of adventious rooting in vitro in difficult-to-propagated cultivars of apple. *Plant Science* 24, 1–9.

Stafford, H. A. 1988. Proanthocyanidinidins (PAs) and the lignin connection (a review). *Phytochemistry* 27(1), 1–6. doi:10.1016/0031-9422(88)80583-1.

Stenlid, G. 1976. Effects of flavonoids on the polar transport of auxins. *Physiologia Plantarum* 38(4), 262–6. doi:10.1111/j.1399-3054.1976.tb04001.x.

Stoddard, F. L. and McCully, M. E. 1980. Effects of excision of stock and scion organs on the formation of the graft union in *Coleus*: A histological study. *Botanical Gazette* 141(4), 401–12. doi:10.1086/337174.

Takeno, K., Taylor, S. J., Sriskanderijah, S., Pharis, P. R. and Mullins, M. G. 1982/83. Enodogenous gibberellin- and cytokinin-like substance in cultured shoot tissues of apple, *Malus pumila* cv. Jonathan, in relation to adventitious root formation. *Plant Growth Regulation* 1, 261–8.

Tretter, D., Galensa, R., Feucht, W. and Schmid, P. P. S. 1985. Flavanone glucosides in callus and phloem of *Prunus avium*: Identification and stimulation of their synthesis. *Physiologia Plantarum* 65(1), 95–101. doi:10.1111/j.1399-3054.1985.tb02366.x.

Treutter, D. and Feucht, W. 1988. Accumulation of the flavonoid prunin in *Prunus avium/P. cerasus* grafts and its possible involvement in the process of incompatibility. *Acta Horticulturae* 227(227), 74–8. doi:10.17660/ActaHortic.1988.227.6.

Tromp, J. 1996. Sylleptic shoot formation in young apple trees exposed to various soil temperature and air humidity regimes in three successive periods of the growing season. *Annals of Botany* 77(1), 63–70. doi:10.1006/anbo.1996.0008.

Webster, C. A. and Jones, O. P. 1992. Performance of field hedge and stoolbed plant of micropropagated dwarfing apple rootstock clones with different degrees of apparent rejuvenation. *Journal of Horticultural Science* 67(4), 521–8. doi:10.1080/00221589.1992.11516278.

Westwood, M. N. 1988. Rootstocks: Their propagation, function and performance. In: *Temperate Zone Pomology*. San Francisco, CA: WH Freeman and Company, pp. 77–107.

Wolffe, A. P. and Matzke, M. A. 1999. Epigenetics: Regulation through repression. *Science* 286(5439), 481–6. doi:10.1126/science.286.5439.481.

Yeoman, M. M. and Brown, R. 1976. Implications of the formation of the graft union for organisation in the intact plant. *Annals of Botany* 40(6), 1265–76. doi:10.1093/oxfordjournals.aob.a085247.

Yeoman, M. M., Kilpatrick, D. C., Miedzybrodzka, M. B. and Gould, A. R. 1978. Cellular interactions during graft formation in plants, a recognition phenomenon? *Symposia of the Society for Experimental Biology* 32, 139–60.

Yu, K. S. and Carlson, R. F. 1975. Gas liquid chromatography determinations of phenolic acids and coumarins in Mazzard and Mahaleb cherry seedlings. *HortScience* 10, 301–403.

Zilka, S., Faingersh, E., Rotbaum, A., Tam, Y., Spiegel, S. and Malca, N. 2002. *In vitro* production of virus-free pear plants. *Acta Horticulturae* 596(596), 477–9. doi:10.17660/ActaHortic.2002.596.79.

Zimmerman, R. H. 1992. Micropropagation: past, present, future. In: S. Sansavini (Ed.), *Convegno su colture in vitro e micropropagazione in ortofrutticoltura*. Grafiche MDM Litografia s.r.l. Forli. Cesena, pp. 1–16.

Chapter 7

Optimizing precision in orchard irrigation and nutrient management

Denise Neilsen and Gerry Neilsen, Summerland Research and Development Centre, Agriculture and Agri-Food Canada, Canada

1 Introduction

Sustainable management of water and nutrients requires that they are applied with precision within production systems that are resilient to the rapidly changing global environment. In particular, the uncertainty associated with climate change and variability has potentially serious consequences for the availability and reliable supply of water for agriculture (Cisneros et al., 2014) which increases the need for precision in water management. Widespread, multi-year droughts have occurred within major fruit growing regions within the last decade, including California, central Chile, North China, South Europe and Australia. At the same time, world population growth and increasing disposable income in emerging economies is accelerating the demand for fruit as an important component of a healthy diet. Irrigated production is more efficient and reliable than rain-fed production and it is estimated that 40% of the world's food production comes from 17% of the land that is irrigated (Postel, 2000). Given the uncertainties in precipitation and increasing future demands for food, it is evident that there is a pressing need for highly efficient irrigated production systems.

http://dx.doi.org/10.19103/AS.2018.0040.08
Published by Burleigh Dodds Science Publishing Limited, 2019.

In irrigated production systems, soil-applied nutrients are greatly affected by water management which can determine the retention and movement of nutrients within the root zone and plant availability (Neilsen et al., 2017a,b). In addition, water management can also greatly influence the environmental footprint of nutrients added to horticultural production systems. Mismatched water supply and demand can lead to losses of mobile nutrients, for example nitrate-N to groundwater through leaching, or less mobile nutrients, for example phosphorus and potassium, to surface water through overland flow and sediment loss. There is also some evidence that water and nutrient management can affect greenhouse gas emissions from horticultural production systems and soil C dynamics and sequestration. Regulated deficit irrigation (RDI) reduced CO_2 emissions in peach (*Prunus persica*) and apricot (*Prunus armeniaca*) orchards (Zornoza et al., 2016). Automatically scheduled irrigation, in conjunction with high carbon mulches, reduced N_2O emissions from an apple (*Malus domestica*) orchard (Fentabil et al., 2016). However, there are also indications that irrigation with high bicarbonate water supplies can stimulate CO_2 emissions (Hannam et al., 2016).

Managing nutrients and water with precision also increases the opportunity to maximize fruit production and fruit quality and thus increases economic sustainability. Application of nutrients with irrigation water (fertigation) in low pressure systems allows a more precise delivery of nutrients into the plant root zone that can be timed to meet the phenological requirements of plant demand for growth (and fruit quality) (Neilsen et al., 2017a,b). Not all nutrient requirements are best met by delivery through the soil to the roots. A notable example is the requirement of certain apple cultivars for high Ca in the fruit and foliar supplements can improve the precision of nutrient management both temporally and spatially if there is variability within an orchard.

This chapter will examine management options for water and nutrient applications in the production of deciduous fruit trees that will potentially optimize production and fruit quality while minimizing the environmental footprint through the uses of efficient systems.

2 Optimizing precision in orchard irrigation management

Four components of irrigation management will be outlined that contribute to sustainable production of horticultural crops: these include the use of efficient irrigation systems, matching water supply to plant demand, reducing evaporation from the soil surface and responding to drought.

Published by Burleigh Dodds Science Publishing Limited, 2019.

2.1 Benefits of efficient irrigation systems

2.1.1 Comparisons with other systems

Irrigation system efficiency can be defined as the ratio of water available for crop uptake to the amount of water applied. Factors which affect irrigation system efficiency include those that are inherent in the mode of water delivery and those associated with irrigation system design and operation. Water losses from irrigation systems include evaporation before soil infiltration, wind drift, deep percolation beneath the root zone and surface runoff. Fereres et al. (2003) noted that in orchard systems, tree interception of water from high-pressure sprinkler irrigation also contributed to inefficiencies. Any particular system can have a range of efficiencies, depending on management and inherent characteristics (Howell, 2003), and these have been summarized for those systems used in tree fruit production (Table 1). Low pressure irrigation systems are usually more efficient than high-pressure systems (Bryla et al., 2003) and have been increasingly used in tree fruit production, particularly for those crops grown with size-controlling rootstocks (apple, sweet cherry [*Prunus avium*]) and in response to water shortages (Proebsting et al., 1981). However, Fereres et al. (2003) suggested that furrow irrigation may be quite efficient for some crops when using best management practices, but is limited by topographic and soil constraints.

2.1.2 Reduced water use

The efficiency of micro-irrigation systems reflects both the reduction in evaporative water losses compared to those from impact sprinklers and also the reduction in coverage of the orchard surface. Micro-sprinklers are inherently less efficient than drip irrigation if installed and operated to maintain full orchard floor coverage. 'Crimson Lady' peaches grown in California with

Table 1 Irrigation system efficiencies

Irrigation class	Efficiency	Components
Gravity	0.4	Flood, rill
High-pressure systems		
Gun	0.55-0.75	Gun, solid set gun, travelling gun
Sprinkler	0.70-0.75	Solid set over-tree, solid set under-tree
Low pressure systems		
Micro-irrigation	0.70-0.95	Micro-spray micro-sprinkler, over-tree micro
Drip	0.70-0.95	Drip, subsurface drip irrigation

subsurface drip or drip irrigation required 50% of the applied water compared to micro-sprinkler irrigated trees to achieve the same growth and yield (Bryla et al., 2003). Compared to micro-sprinklers, drip irrigation reduced water consumption in 'Lapins' sweet cherries by 47.6% (Yin et al., 2012) and in apples by 40% (Fallahi et al., 2010a), with no detrimental effects on yield and fruit quality in both cases. The amount of water used with micro-sprinklers can be reduced if the radius of throw is constrained to apply water just within the row. In an experiment with 'Braeburn' apple, growing season water inputs were similar for small radius micro-sprinkler and drip applied to meet 100% ET replacement with no effects on fruit productivity and quality (Table 2). However, for newly established 'Skeena' sweet cherry under drip irrigation, fruit yield was higher after the first 2 years of cropping with no concomitant effects on fruit size (Table 3).

2.1.3 Precise placement

Micro-irrigation, particularly drip, allows for precise placement of water in the root zone if applications are made to meet demand, taking into account the size of the root zone and soil characteristics. This reduces deep drainage and

Table 2 Effect of irrigation type on fruit yield and average individual fruit weight for Braeburn/M.26[a]

	Fruit yield (kg/tree)			Fruit weight (g)			Water inputs (L/tree)		
Treatment	2002	2003	2004	2002	2003	2004	2002	2003	2004
Drip	2.3	5.7	7.6	299	273	277	1111	1713	1765
Micro-sprinkler	1.6	4	6.2	300	274	263	1321	1911	1916
Significance	**	**	NS	NS	NS	NS			

[a] Neilsen D. unpublished data.
** Means within columns significantly different at P = 0.05, 0.001 respectively, or not significantly different (NS).

Table 3 Effect of irrigation type on fruit yield and average individual fruit weight in 2016 and 2017 for Skeena/Gi.6 sweet cherry[a]

	Fruit yield (kg/tree)		Fruit weight (g)		Water inputs (L/tree)	
Treatment	2016	2017	2016	2017	2016	2017
Drip	1.96	7.19	11.6	9.11	1734	1483
Micro-sprinkler	1.73	5.82	11.7	9.60	1759	1485
Significance	NS	*	NS	NS		

[a] Adapted from Watson (2018).
* Means within columns significantly different at P = 0.05 respectively, or not significantly different (NS).

also leaching of nutrients if fertigation is used. The combination of rootstock, water and nutrient distribution provides the potential spatial boundaries of the root zone, while temporal dynamics and root efficiency may be influenced by a number of additional factors including plant phenology, soil temperature and crop load (Eissenstat et al., 2001). Root distribution of McIntosh apple grown on three rootstocks with different size-controlling properties, dwarf (M.9), semi-dwarf (M.26) and semi-vigorous (M.7), with three irrigation frequencies (daily, 2x/week and weekly), was affected by both rootstock and irrigation frequency (Neilsen et al., 1997a). Plots of average root distribution perpendicular to the row adjacent to the drip emitter indicated that roots were shallower and closer to the emitter for dwarfing rootstocks M.9 and M.26 for daily, compared to weekly, irrigations (Fig. 1a,b). For M.7 trees, roots had greater lateral spread but the same depth for daily, compared to weekly, irrigation (Fig. 1c). Similar plots of root distribution were developed for 'Gala' on M.26 apple trees grown with either micro-jet or drip irrigation (Fig. 1d) where both irrigation systems resulted in shallow root systems but nearly 50% of root intersections occurred within 30 cm of the drip emitter, whereas there was a more even lateral distribution for micro-jet-grown trees (Neilsen et al., 2000).

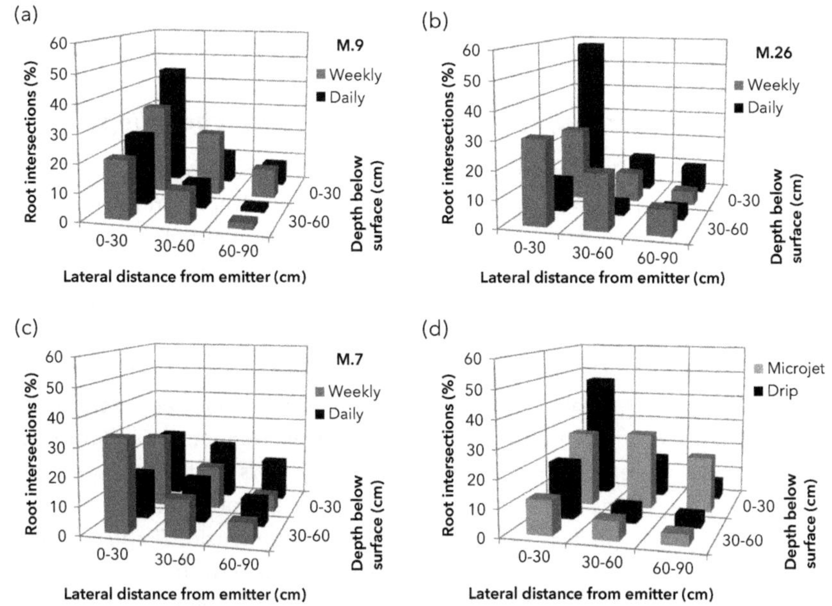

Figure 1 Spatial distribution of roots in response to weekly and daily drip irrigation for McIntosh apple on (a) M.9 rootstock, (b) M.26 rootstock and (c) M.7 rootstock (adapted from Neilsen et al., 1997a) and for Gala/M.9 apples grown with (d) drip or micro-jet irrigation (adapted from Neilsen et al., 2000).

In addition to supplying water for root uptake, micro-irrigation is frequently used to modify crop micro-climates. Hotter summers and changing production systems which leave fruit surfaces more exposed have led to an increase in the incidence of sunburn, particularly for apple (Racsko and Schrader, 2012; Darbyshire et al., 2015). Evaporative cooling with above-canopy micro-sprinklers has been used effectively (Parchomchuk and Meheriuk, 1996; Iglesias et al., 2005) as part of a range of tools to reduce sunburn (Racsko and Schrader, 2012), and may also have beneficial effects on fruit colour (Iglesias et al., 2005). It is important that water additions for the purpose of evaporative cooling be considered in the overall water budget. Total water use can be reduced if systems are cycled or pulsed to allow free water to be almost continually present to evaporate from the fruit surface (Evans and Van der Gulik, 2011). Micro-sprinklers can also be used for spring frost protection, particularly in crops which bloom early such as cherry and peaches. They are effective in off-setting radiation frosts when a temperature inversion traps cold air close to the surface, but not for advective frosts which occur when a cold air mass moves into a region (Perry, 1998). Both over-tree and under-tree systems can be used, and the same mechanism applies – the latent heat of fusion given off when water freezes increases the temperature of the above-ground parts of the tree. Wind speeds > 2 m/s can negate the beneficial effects of freezing if they result in evaporative cooling.

2.2 Matching water supply to demand (irrigation scheduling)

Scheduling for micro-irrigation systems has been reviewed extensively (Howell and Meron, 2007). In addition to methods of assessing and monitoring irrigation water demand, they reported on irrigation system design and its relationship with successful irrigation scheduling, which is beyond the scope of this chapter. Historically, two methods have been used to schedule irrigation applications to meet plant demand either (1) through estimates of evapotranspiration (ET) derived from Class 'A' evaporation pans or weather data (Allen et al., 1998) or (2) soil moisture measurements. In the past 20–25 years, there has been increasing interest in the use of plant-based water stress measurements, including leaf and stem water potential (Shackel et al., 1997), sensors such as gauges for trunk expansion (Goldhamer and Fereres, 2004; Livellaria et al., 2011) sap flow, (Dragoni et al., 2009; Juhász et al., 2013) and plant water tension (Pagay, 2014). More recently, field scale plant stress and water deficits have been measured using remote sensing, direct measurement of canopy temperatures, and spectral reflectance, with the aim of refining irrigation scheduling techniques (Jones et al., 2009). Where irrigation scheduling is used, often a range of methods and sensors are employed.

2.2.1 Estimates of ET

Irrigation scheduling for tree fruits that is based on ET requires ET estimates for a reference crop, usually well-watered grass from climate or Class 'A' evaporation pan data. Tree water requirements can be determined by modifying the ET_{ref} using crop coefficients (K_c) which take into account canopy development through the growing season according to FAO56 methodology (Allen et al., 1998).

$$\text{Crop water requirement}\left(ET_c\right) = ET_{ref} \times K_c \tag{1}$$

The crop coefficient has been subdivided into its transpiration and evaporation components

$$K_c = K_s K_{cb} + K_e \tag{2}$$

where K_c = crop coefficient; K_s = a dimensionless stress coefficient based on available soil water; K_{cb} = transpiration + small amount of evaporation from a dry soil (basal crop coefficient) and K_e = evaporation from a wet soil. There has been considerable effort to determine crop coefficients for tree fruits. Measurements of canopy area have been used to estimate both maximum K_c values and development through the growing season. Average canopy area estimates were used to modify Class A pan data in a cherry drought trial (Proebsting et al., 1981). Goodwin et al. (2004) used sap flow measurements in peach trees to estimate transpiration (T) and K_{cb} was calculated as the ratio between T and ET_{ref}. Gradual removal of branches and the effective canopy cover resulted in a reduction in K_{cb} and the strong relationship (K_{cb} = 1.4 × effective canopy cover) leading the authors to conclude that transpiration may be estimated from a simple measure of the shaded area and ET_{ref}. Marsal et al. (2009, 2010) used a ceptometer to measure PAR interception by sweet cherry canopies. At mid-season (peak canopy development), K_c was around 0.94, which is within the range 0.9–1.20 suggested by Allen et al. (1998) for sweet cherry grown with or without a ground cover. The challenge of accounting for the large variation in canopy cover and ground cover in orchard systems was addressed by Allen and Pereira (2009) who combined estimates of crop density and components of the crop coefficient to compute a K_{cb} for orchards with bare soil or an active ground cover that could also account for the ground shaded by the canopy. Detailed seasonal crop coefficients for a range of crop densities and ground covers are given in Table 3 in Allen and Pereira (2009).

More recently, remote sensing techniques have been used to assess K_c and the basal crop coefficient (K_{cb}) (Pereira et al., 2015). Vegetation indices such as the normalized vegetation difference index (NVDI) have been used to estimate the fraction of vegetation cover for application to ET estimates from either weather station variables and from other remotely sensed imagery used in energy, or water balance calculations. Few such studies have been reported for fruit trees, although there have been several for wine grapes (*Vitis vinifera*). Crop ET for high-density apple orchards with drip irrigation was evaluated using latent heat fluxes in an eddy covariance system and compared to a remote sensing-based soil water balance with reasonably good results (Odi-Lara et al., 2016). A soil adjusted vegetation index, which takes into account the fraction of unshaded ground, was used to calculate K_{cb} and a soil water balance used to estimate K_e. Mid-season K_{cb} values were lower than those reported in Allen and Pereira (2009), which was attributed to the lower percentage of orchard soil wetting under drip irrigation. Similar results were found for peach orchards when eddy covariance methods were compared with FAO56 methods to estimate crop coefficients (Paço et al., 2006, 2012). This emphasizes the need to take into account the wetted area when using standard crop coefficients with micro-irrigation. A useful rule of thumb in orchards with a herbicide strip is to consider that area as the potential wetted zone.

Weighing lysimeters have been extensively used to evaluate crop coefficients and crop water use (Allen and Pereira, 2009), although there are relatively few examples for tree fruits. Ayars et al. (2003) measured water use for late season 'O'Henry' peaches in large field lysimeters grown with drip irrigation. Measured water losses and calculated ET_{ref} were used to develop seasonal crop coefficients. The seasonal distribution of crop coefficients was best fitted by a cubic curve with a maximum K_c value of 1.10. There was a close relationship between K_c and canopy light interception. In a similar study, Girona et al. (2011) estimated seasonal K_c values for 'Golden Smoothie' apple and 'Conference' pear (*Pyrus communis*). Maximum K_c values increased over time for apple trees from 0.4 in year one to 1.0 in year four, while for pears K_c = 1.0 in all 4 years. Differences in behaviour were attributed to the taller, denser canopies in apple, which allowed less light penetration. In both studies, crop coefficients were within the ranges reported by Allen and Pereira (2009).

2.2.2 Soil moisture measurements

There is a number of techniques available for monitoring soil moisture which range from simple hand texturing to fully automated recording and reporting systems based on sensors which are either porous materials that equilibrate with soil moisture *in situ* (e.g. gypsum blocks, granular matrix sensors, tensiometers, psychrometers) or direct measurements of the soil dielectric constant (time domain reflectometry [TDR] or capacitance probes) or neutron dissipation (neutron probe). All have their limitations, which are described in Howell and Meron (2007). Not the least of these is the inherent variability in soil properties and the partially wetted zone under micro-irrigation which may be difficult to characterize with soil moisture measuring techniques that sample a small volume of soil. Porous material-based sensors, with the exception of gypsum blocks and psychrometers, measure soil moisture tension within the range recommended for optimum operation of micro-irrigation systems (−12 to −70 kPa), are relatively inexpensive and in some cases can now be electronically linked to irrigation controllers. Sensors based on measurements of the soil dielectric constant, TDR and capacitance probes measure volumetric soil moisture content over a wide range of conditions, have become less expensive and can be electronically linked to controllers. They have the disadvantage of measuring a small radial area, but can provide depth integrated measurements of soil moisture content. Neutron probes can measure a larger soil volume, but are expensive, challenged by the need for strict controls over radioactive sources, and are consequently used in commercial orchards through consulting services. The combination of manual measurements and expense make them less useful for micro-irrigation scheduling based on high-frequency irrigation and minimal soil dry down between irrigations. Soil moisture measurements are used in irrigation scheduling in several ways: (1) when using a water balance approach where both ET and soil moisture are used to determine irrigation requirements; (2) as a check on other scheduling techniques to ensure over- or under-watering does not occur; and (3) as a trigger for irrigation applications. In a water balance approach, Yin et al. (2012) used a neutron probe to determine weekly soil moisture content in a study on 'Lapins'/'Mazzard' sweet cherry. Irrigation amount was based on weekly soil moisture depletion and estimates of ET_c, precipitation and deep drainage. Implicit in this approach is the use of soil as a water storage reservoir often defined by the available water storage capacity (the soil water content between field capacity and permanent wilting point). The difficulties with this approach are well summarized in Howell and Meron (2007), who point out that the role of soil for water storage becomes less important as irrigation frequency increases – particularly if applied on a daily or sub-daily basis, as is possible with micro-irrigation (Hillel, 2004). Moreover, in dry, coarse-textured soils, hydraulic conductivity may become the limiting factor in water supply (Spinelli et al., 2017) at soil water potentials that are much higher than permanent wilting point. In comparison with traditional irrigation, micro-irrigation techniques are aimed at maintaining high soil water content in the crop root zone, thus preventing soil dry down between irrigation periods and avoiding detrimental effects on fruit growth. This is of particular importance for fruits grown on dwarfing rootstocks, as maximum root distribution and activity may be confined to as little as 30 cm depth, for example as in apple (Fig. 1) which contrasts with the reported apple rooting depth values of 1–2 m used for determining available soil water in irrigation scheduling manuals (Howell and Meron, 2007).

Soil moisture measurements are a useful tool to monitor irrigation management. Bryla et al. (2005) measured soil moisture using a neutron probe when comparing furrow, micro-spray, surface and subsurface drip irrigation and were able to demonstrate the large fluctuations in moisture content under furrow irrigation with a 7- or 14-day irrigation interval compared with the much more stable soil moisture levels under drip irrigation, which were reflected in decreased plant stress and larger fruit for the drip irrigated peach trees. Fully automated TDR systems were used to monitor soil moisture content in apple and sweet cherry experiments in response to drip and micro-sprinkler applications for which irrigation was controlled by a combination of reference ET estimated from an electronic atmometer (ETGage Co., Loveland, CO) calibrated against weather station estimates of ET_o and used in conjunction with K_c values (Fig. 2). Irrigation supplied at 100% ET replacement and a range of deficits were reflected in soil moisture content (Neilsen et al., 2016a).

With the advent of automated soil moisture monitoring, there is increasing interest in using soil moisture to control irrigation systems, although there are relatively few recent reports of soil moisture measurements being the sole controller of irrigation scheduling. Nevertheless, trigger values for soil moisture tension have been derived for micro-irrigation systems in extension publications, for example Van der Gulik (1999). More common, in addition to soil water monitoring for water balance and climate-based ET methodologies, is the combining of sensors for soil moisture, plant water status and stress.

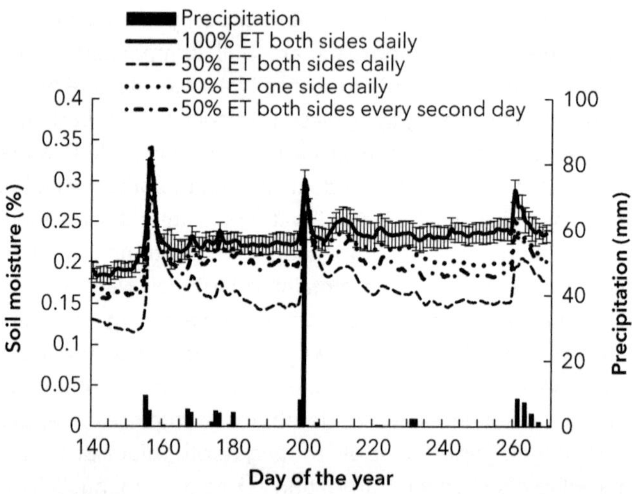

Figure 2 Average daily volumetric soil moisture content in response to 100% or 50% replacement ET irrigation and daily precipitation. Vertical bars represent ±SE. Adapted from Neilsen et al. (2016a).

2.2.3 Plant water status

The most commonly used method to assess plant water status is the pressure chamber which measures the water potential of leaves, stems or fruit, taken at different times during the day (Scholander et al., 1965; Shackel et al., 1997; Jones, 2004). A range of values exist as 'thresholds' of damaging stress for tree fruits, which are highly dependent on time of measurement as the diurnal time course of leaf and stem water potential is controlled by temperature and vapour pressure deficit. Large daily variations in leaf water potential, between >−1.0 MPa and up to −2.6 MPa, have been observed for 'Corum' and 'Napoleon' (syn. Royal Anne) (Tvergyak and Richardson, 1979) and 'Bing'/Gisela 6 sweet cherries (Oyarzún et al., 2010); 'Royal Glory' and 'Saturn' peaches (Bellvert et al., 2016); 'Big Top' nectarine (Bellvert et al., 2016); and 'Golden Delicious' apple and 'Spadona' pear (Naor et al., 2006). In general, highest leaf water potentials were found pre-dawn and lowest in mid-afternoon, but this was moderated for shaded leaves. Shackel et al. (1997) summarized research on the use of the mid-day stem water potential, assessed from covered leaves, as a more stable estimate of mid-day vapour pressure deficit and hence plant stress. Well-watered plum and cherry trees maintained mid-day stem water potential at >−1.0 MPa. Similar values were found for well-watered 'Skeena' sweet cherry (Fig. 3a) and 'Ambrosia' apple (Fig. 3b). Although threshold stem water potential values for detrimental plant stress and subsequent effects on fruit production and quality have been identified for many crops and used for assessing the effects of irrigation management practices on apples (Naor et al., 2006; Neilsen et al., 2016a), pears (Naor et al., 2006; Marsal et al., 2008), sweet cherries (Marsal et al., 2009, 2010), nectarines (Naor et al., 2006) and peaches (Bryla et al., 2005), the measurements are time-consuming and require manual intervention in irrigation systems for scheduling purposes. However, there are some very promising recent advances. As an example, continuous measurement of stem water potential was made with a prototype micro-tensiometer embedded in the stems of grape vines (Pagay, 2014). With wireless technology, this information could be transmitted to an irrigation controller.

Because of the difficulty in automating measurements of plant water status, a number of indirect measurement techniques have been developed. These include shrinkage and expansion of the trunk (e.g. Goldhamer and Fereres, 2001) and sap flow (e.g. Dragoni et al., 2009). Trunk diameter sensors or dendrometers (also called linear variable displacement transformer or transducers - LVDT) were assessed as indicators of plant water status by comparison to mid-day stem water potential for several crops, including almonds (*Prunus amygdalus*) and peach (Goldhamer and Fereres, 2001), plum (*Prunus salicina*) (Intrigliolo and Castel, 2004), apple, nectarine and pear (Naor et al., 2006). Considerable research has been undertaken to identify the efficacy and limitations of trunk diameter sensors for use in irrigation scheduling as reviewed by Fernández and Cuevas (2010) and Ortuño et al. (2010). The two most commonly used measures are maximum daily shrinkage (MDS) and stem or trunk growth rate (TGR). Goldhamer and Fereres (2001) found that TGR was a more sensitive measure than MDS for rapidly growing young peach trees and MDS for older trees exposed to periods of stress. These findings have been confirmed for other crops (Fernández and Cuevas, 2010; Ortuño et al., 2010). In addition to tree age, other factors affecting trunk diameter variation include crop load, seasonal growth patterns and phenology. Tree to tree variability may be large, which limits the use of absolute changes in diameter as a measure of stress for irrigation scheduling (Naor et al., 2006). There is also a need to remove the effect of daily fluctuations in evaporative demand on trunk diameter variation in order to determine underlying soil moisture limitations (Fernández and Cuevas, 2010). Reference equations relating MDS to ET_o, air temperature, VPD and mid-day stem water potential have been developed for plum (Intrigliolo

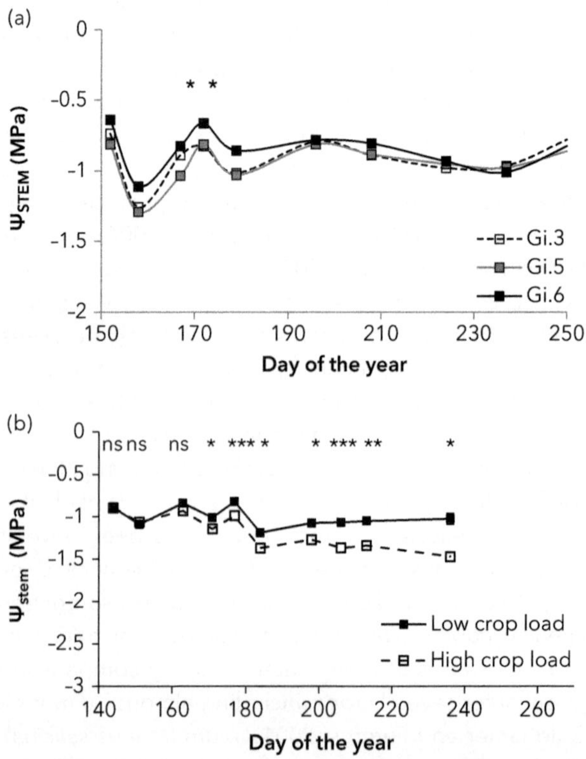

Figure 3 Effect on mid-day stem water potential for well-watered (a) 'Skeena' sweet cherry on three rootstocks of differing vigour and (b) 'Ambrosia'/M.9 apple with two crop loads. Adapted from Neilsen et al. (2016a,b).

and Castel, 2004, 2006), and relating MDS to mid-day stem water potential for peach (Conejero et al., 2010). Signal intensity, the ratio of a trunk displacement measurement to that of a well-watered control, has been successfully used to schedule irrigation for both full and deficit irrigation (DI) strategies (Fernández and Cuevas, 2010), but threshold values required to control irrigation systems need to be modified to account for other factors including crop load, seasonal growth patterns and phenology, as well as rainfall events (Ortuño et al., 2010).

Sap flow has been measured either by heat balance or heat pulse techniques. Use of this technique for irrigation scheduling has been pursued more in olive (*Olea europaea*), citrus (*Citrus* spp.) and grape vine production systems than for deciduous fruit trees, and often in conjunction with trunk diameter sensors (Fernández and Cuevas, 2010). Conejero et al. (2010) found that sap flow intensity measurements were less sensitive than MDS and concluded that MDS could be used alone as a scheduling technique for 5-year-old peach trees.

2.2.4 Remote sensing

In addition to estimates of canopy size through vegetation indices as discussed above, thermal imagery from both satellites and unmanned aerial vehicles (UAV) or drones has been assessed as an indicator of plant water stress (Jones et al., 2009). The seasonal evolution of crop water status was determined for peach and apricot trees subjected to different irrigation regimes using thermal imagery from a UAV (Gonzalez-Dugo et al., 2013). A crop water status index (CWSI) based on the difference between air and canopy temperature was identified and was closely related to tree water status defined by mid-day stem water potential.

$$CWSI = \frac{(T_c - T_a) - (T_c - T_a)_{LL}}{(T_c - T_a)_{UL} - (T_c - T_a)_{LL}} \tag{3}$$

where T_c = canopy temperature; T_a = air temperature; $(T_c - T_a)_{LL}$ = the lower limit and $(T_a - T_c)_{UL}$ = the upper limit for a well-watered control.

The authors considered that the observed differences in canopy temperature were sufficient to be of use both in spatial and temporal assessment of plant water stress and potentially irrigation requirements. Similar results were found for peach and nectarine trees in response to full irrigation and RDI, although in this case the measure of plant water stress was the mid-day leaf water potential (Bellvert et al., 2016). In a sensitivity analysis, it was determined that an image pixel size of 0.8 m or less was required to achieve precision in mapping leaf water potential. A number of indices were calculated, including $T_c - T_a$, NVDI and PRI (Photochemical Reflectance Index - a measure of Photosystem II efficiency and radiation use efficiency), from a combination of thermal, visible and near-infrared (NIR) imagery from a UAV to determine spatial variability in water stress in apple orchards (Gómez-Candón et al., 2016, 2017) and apricot and peach trees (Ballester et al., 2018).

2.2.5 Automation

Many of the techniques described above for estimating ET_c, canopy development, soil moisture and plant stress are frequently combined into expert advice systems which allow intervention in irrigation system management to achieve water savings and improve production and fruit quality. With the advent of wireless communication technologies (Majone et al., 2013) and improved soil and plant sensors, there is increased interest in automation of irrigation systems by using sensor data input into irrigation controllers to switch irrigation valves on and off (Romero et al., 2012). Micro-irrigation systems which are designed to provide frequent irrigations and maintain soil moisture close to field capacity are excellent candidates for fully automated irrigation. Casadesús et al. (2012) proposed a general algorithm for automated scheduling of drip irrigation for fruit tree orchards which would take into account seven elements: (1) estimation of irrigation needs (model); (2) adaptation to a specific irrigation system; (3) execution of the schedule; (4) soil/plant monitoring; (5) interpretation of sensor data; (6) reaction to occasional events; and (7) model tuning. This was successfully applied to different models of estimating crop needs and a DI scenario for peach. Osroosh et al. (2016) designed a fully automatic supervisory control and data acquisition system (SCADA), a wireless sensor network, and soil-, plant- and weather-based scheduling algorithms which could be assessed based on soil and plant water status, cost and simplicity of use when irrigating apple orchards with drip. They found that all scheduling methods reduced water use by at least two-thirds compared with the grower practice of applying water on set days of the week at fixed rates modified by two temperature thresholds. Soil moisture-based methods using granular matrix sensors (Watermark™) either with or without ET estimates resulted in under-watering. Soil moisture methods based on the neutron probe provided sufficient moisture for plants to remain unstressed, but were limited to weekly applications which may not be suitable for soils with coarser textures and freer drainage than this study's silt loam soil with a 1 m perched water table. Plant-based methods using infrared temperature sensors (CWSI), a time temperature threshold method (O'Shaughnessy and Evett, 2010) or ET_o estimates, maintained stem water potential above −1.0 MPa.

Automatic weather station data used to calculate reference ET and automated soil moisture monitoring was used to schedule supplemental irrigation based on maintenance of soil moisture between pre-defined trigger levels in sweet cherry (Juhász et al., 2013). A combination of reference ET estimated from an electronic atmometer (Etgage) calibrated against weather station estimates of ET_o has been used in conjunction with seasonal K_c values and precipitation measured with a tipping bucket rain gauge to automatically schedule and control daily or sub-daily irrigation through latching solenoid valves in high-density sweet cherry and apple orchards (Neilsen et al., 2010; 2014, 2016a,b; Parchomchuk et al., 1996). Sensor data were stored in a Campbell Scientific datalogger (e.g. CR10X) programmed with generalized crop coefficient curves designed to start at bloom. Irrigation applications were made to replace the amount of water used since the previous irrigation. Onset of irrigation was fixed and run time was controlled to meet water requirements. Feedback into the system was provided by automated measurements of soil moisture using TDR.

Automation appears to be the logical next step in irrigation management, but grower adoption is not a guaranteed response to the introduction of new technology (Boland et al., 2006) and even the use of many of the scheduling techniques outlined above are not fully adopted in important tree fruit growing areas, as indicated in Osroosh et al. (2016), resulting in over-irrigation. Limitations to adoption may include expense, system complexity, the difficulty of maintaining sensors in good operating condition and the flexibility of water supply systems which need to be compatible with commercially available components. There is also a need to design systems that make use of readily available technology, such as commercial irrigation controllers, and have user-friendly interfaces to allow easy access to back-up monitoring systems such as soil and plant sensors (Romero et al., 2012; Osroosh et al., 2016). However, in conjunction with an unrestricted water supply, fully automated systems provide the best option to meet tree fruit water requirements, including high-frequency pulsed irrigation as exemplified in an experiment in which plant growth (trunk cross-sectional area) and yield were greater for 'Skeena' and 'Cristalina' sweet cherry on Gisela 6 rootstock receiving high-frequency (four times daily) irrigation than for trees receiving irrigation every second day (Neilsen et al., 2010). High-frequency irrigation increased soil moisture content (Fig. 4a), tree growth (Fig. 4b) and improved yield and fruit quality (Table 4). While great strides have been made in the timing of irrigation supply to meet demand, spatial variability in requirements is a much more difficult proposition. Remote sensing of orchards has demonstrated the large tree to tree variation in plant water stress (Bellvert et al., 2016; Gómez-Candón et al., 2016; 2017; Ballester et al., 2018), but unless irrigation zones can be designed to reflect inherent variation in soil conditions prior to planting, there are very few strategies available to overcome spatial variability. With drip irrigation, emitters may be added to increase or decrease flow rates to equalize variation. There have been few advances in this field. Coates et al. (2006) designed an automated irrigation system with micro-sprinklers that were individually linked to an irrigation valve

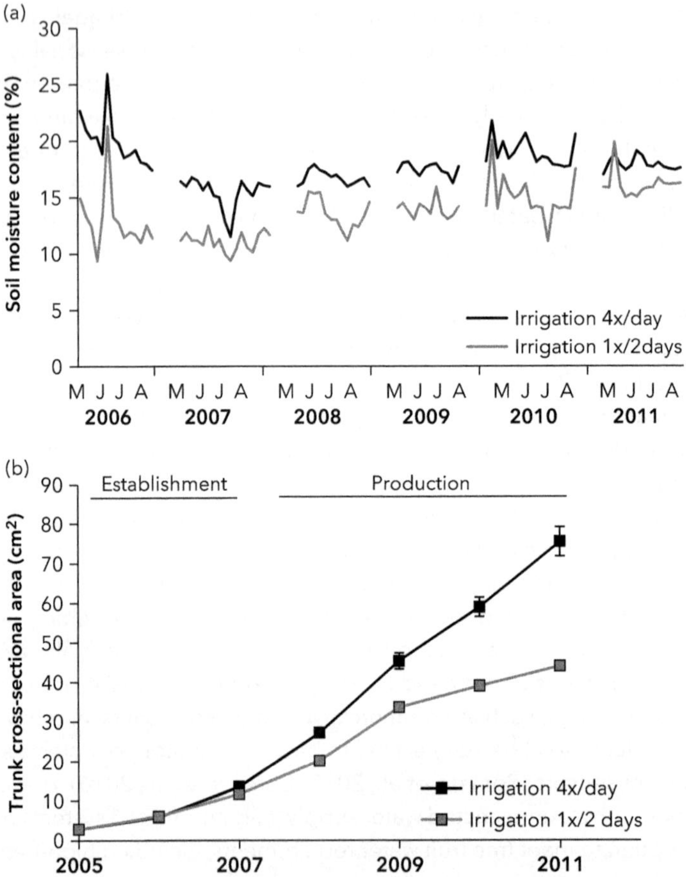

Figure 4 Effect of irrigation frequency on (a) soil moisture and (b) tree growth averaged for 'Skeena' and 'Cristalina' sweet cherry on Gisela 6 rootstock. Adapted from Neilsen et al. (2010, 2014).

Table 4 Effect of irrigation frequency on yield, fruit size and soluble solids for 'Skeena' and 'Cristalina' sweet cherry[a]

	Yield (kg/tree)				Fruit size (g)			Soluble solids (%)		
		2010								
Irrigation (L)	2009	Skeena	Cristalina	2011	2009	2010	2011	2009	2010	2011
1.0 (4x/day)	23.6	13.4	7.1	11.8	9.8	10.3	11.8	20.2	19.9	18.5
2.0 (1x/2 days)	15.5	8.8	6.6	6.4	9.6	9.6	11.5	21.4	20.8	18.7
Significance	**	*	NS	**	NS	*	NS	***	**	NS

[a] Adapted from Neilsen et al. (2014).

*,**,*** Means within columns significantly different at P = 0.05, 0.01, 0.001 respectively, or not significantly different (NS).

and could be programmed to switch on and off in response to external sensors. Thus, addressing spatial variability is one of the largest challenges in micro-irrigated systems.

2.3 Deficit irrigation strategies

There continues to be considerable interest in using DI as a strategy to cope with systemic water deficits, with possible benefits for improved fruit quality. Three types of DI have commonly been reported: (1) DI in response to drought conditions; (2) RDI with deficits applied at specific phenological stages; (3) partial root zone irrigation with deficits applied alternately to different parts of the root zone (PRI). Deficit irrigation in response to seasonal droughts can be catastrophic if prolonged, and successful DI depends on the ability to modify production practices (Lopez et al., 2012). These modifications may include reducing soil evaporation by reducing the wetted soil surface area using fewer drip emitters (Girona et al., 2010) or mulches if water is applied beneath the mulch (Long et al., 2014). Plant modifications include reducing leaf area by branch removal or summer pruning (Proebsting et al., 1981; Goodwin et al., 2006; Marsal et al., 2006), although care is needed that fruit yield is not affected if done pre-harvest (Lopez et al., 2006), and reducing crop load (Marsal et al., 2009, 2010; Naor et al., 2008; Neilsen et al., 2016a) which is most beneficial if applied early in the season. Other possibilities include the use of shade netting.

RDI has been applied to most deciduous fruit trees, including the pome fruits apple (Behboudian et al., 1998; Fallahi et al., 2010a; Neilsen et al., 2016a) and pear (Marsal et al., 2010, 2012), with deficits applied season long (Neilsen et al., 2016a), post cell division (Fallahi et al., 2010a; Marsal et al., 2010) or post-harvest (Marsal et al., 2012). Mild deficits, more than 50% of full irrigation, appear to have few detrimental effects on fruit production and quality. In general, pome fruits have long growing seasons and thus water savings with post-harvest deficits are small. Under RDI, shorter season 'Gala' apple production was found to conserve more water than longer season 'Fuji' (Lo Bianco and Francaviglia, 2012). The three-phase growth pattern of stone fruits and the shorter growing seasons of many cultivars have been important factors in the practice of RDI, particularly in regions with Mediterranean climates. Post-harvest RDI has been successfully applied to sweet cherry (Marsal et al., 2009, 2010), peach (Conejero et al., 2011; Marsal et al., 2016), plum (Intrigliolo and Castel, 2010) and apricot (Pérez-Sarmiento et al., 2010). The most successful outcomes occurred when RDI was avoided during the period of rapid fruit growth (Stage III – cell expansion). Long-term applications of pre-harvest RDI (Stages I, II) and post-harvest RDI reduced yield but not fruit quality or size in early peaches (Ruiz-Sánchez et al., 2018), although Marsal et al. (2016) indicated that yield reductions in mid-season peaches might indicate a need for a break from RDI after 3 years. Fruit

Published by Burleigh Dodds Science Publishing Limited, 2019.

maturity was advanced in the year after post-harvest RDI in sweet cherry (Marsal et al., 2009, 2010) and there was some indication that RDI helped break the cycle of biennial bearing when applications in the off-year reduced fruit set in the subsequent year, but there were no other consistent effects on fruit quality.

Partial root zone irrigation has been promoted as a strategy that both reduces water use and improves fruit quality. Part of the root zone is left dry by removing irrigation at some point in the growing season either from one side of the plant or by alternating wet and dry sides. Theoretically, it is expected that root to shoot signalling will cause partial closure of stomata to reduce transpiration while maintaining photosynthesis rates (Kang and Zhang, 2004). Jarvis (2011) suggested that there is also a compensatory mechanism by roots in the wet zone which increases uptake to maintain transpiration rates by increasing secondary growth and activity, as well as nutrient acquisition. At the same time, the area of the wetted soil surface and, consequently, evaporation decrease. These mechanisms have been well-reviewed in Chai et al. (2016). Yield was unaffected and water use efficiency increased for apple grown with PRI (Zegbe et al., 2016; Du et al., 2017) and for pear (Kang et al., 2002). Fruit colour and total soluble solids increased under PRI for 'Fuji' apple (Fallahi et al., 2010a) and for 'Fuji' and 'Gala' apple (Francaviglia et al., 2013), and PRI advanced maturity for 'Royal Gala' (Zegbe et al., 2016). There are few comparisons of alternating and fixed PRI. Fallahi et al. (2010a) found no additional benefits from alternating the irrigation for 'Fuji' apple. Applying a reduced water volume to one side (PRI) of the tree rather than both sides (DI) resulted in similar yields to full irrigation for 'Fuji' (Fallahi et al., 2010a) and 'Ambrosia' (Neilsen et al., 2016a) apples, which was attributed to the higher soil moisture content and soil moisture potential maintained in the 'wet' side of the PRI. This suggests that PRI might be a better option than DI for pome fruit, and DI is likely better suited to the growth pattern and early harvest associated with stone fruit.

3 Optimizing precision in orchard nutrient management

Two general approaches to improving the precision of nutrient application in orchards will be discussed in this section. These include application of nutrients dissolved in irrigation water, a process commonly referred to as ferti-irrigation or fertigation, and application of dissolved nutrients directly to the fruit tree in dilute water sprays, often described as foliar application, although spray coverage can include fruits, tree trunks and limbs with some leakage to the orchard floor. The efficiency of fertigation is strongly influenced by the degree of control of water application, particularly for soluble nutrients, such as N, which can readily be leached below the root zone. As a consequence, improving the precision of irrigation, as discussed previously, will improve fertigation precision. Foliar applications are made to overcome nutrient deficiencies or when high concentrations of nutrients at specific times improve crop quality.

4 Fertigation

Fertigation has been considered desirable due to its potential to improve the efficiency of crop nutrient uptake. Fluctuations in soil nutrient supply can be reduced as a consequence of the ability to make more frequent applications of smaller amounts of fertilizer with flexibility in timing of application allowing a closer match of nutrient supply to crop demand. Delivery of soluble nutrients directly to the root zone enhances root growth in affected soil volumes. General principles and practices of fertigation have previously been reviewed (Haynes, 1985; Bar-Yosef, 1999). In addition, specific research studies have been conducted on a wide range of perennial fruit crops (Table 5). Effective

Table 5 Selected summary of fertigation research studies, nutrients, soil texture and issues identified for a range of perennial fruit crops

Crop	References	Nutrient	Soil texture	Critical issues identified
Apple	Neilsen et al. (2009)	N	Loamy sand	Lower N rates possible. High N negatively affects fruit quality. Cultivar differences
	Klein and Spieler (1987a,b)	N	Sandy clay to clay loam	Usefulness of soil solution monitoring
	Neilsen et al. (2008a,b)	P	Loamy sand	Importance of P for tree establishment. Mobility of drip-fertigated P
	Neilsen et al. (1998b)	K	Loamy sand	Can correct and prevent deficiency induced by drip irrigation
	Neilsen et al. (2004c)	B	Loamy sand	B fertigation can be effective at low rates. Few cultivar differences
Blueberry (*Vaccinium corymbosum*)	Vargas and Bryla (2015) and Vargas et al. (2015)	N	Silty clay loam	Drip fertigation more effective than micro-jet. Lower N rates possible. Soil texture important
Cherry (sweet)	Neilsen et al. (2004a, 2007, 2010)	NPKB	Gravelly sandy loam	Drip more effective than micro-sprinkler fertigation, especially for P and K. Lower N rates best for fruit yield and size. P improves tree establishment. Crop load and irrigation frequency important
Cherry (sour)	Callan and Westcott (1996)	K	Sandy loam	Drip fertigation of K prevents K-deficiency. Balance required with Mg and Ca. K form important
Peach	Bussi et al. (1991, 1994)	NK	Sandy loam	Fertigated N improves vigour and yield relative to broadcast N. K ineffective. Annual response
Prune (*Prunus domestica*)	Southwick et al. (1999)	N	Clay loam	Moderate N application rates best to minimize leaching. Yield (but not fruit size) responded to fertigated N

application strategies have been developed, especially for the nutrients N, P and K, for which there is a high demand in the plant. Successful application of nutrients contained at lower concentrations in the plant, including Mg and B, have also been reported (Table 5). Fertigation has been applied over a wide range of soil types, although the majority of studies have been focused on coarse-textured soils ranging from sands to loamy sands, many with inherently limited nutrient and water holding capacity and therefore judged to be more likely to respond to fertigation. Important crop- and nutrient-specific issues have been identified in the various studies, suggesting a need to adjust fertigation strategies to site-specific conditions (Table 5). Soil fertility, the age of the planting, cultivar and rootstock differences, variations in crop load and the need to balance crop quality and yield may need to be considered. Issues such as these will subsequently be considered in detail for individual nutrients using research on apple and sweet cherry as representative orchard crops.

5 Fertigation methods

Fertigation requires soluble fertilizer sources which will not precipitate with co-applied fertilizers or other soluble constituents such as Ca and Mg which can occur naturally in irrigation waters. Potential chemical precipitation problems when using local irrigation water can be assessed by observing any precipitation after mixing fertilizer sources and the water in the proportions that will occur in irrigation lines. If lines become plugged, acidic injections (for chemical precipitates) and chlorination (for biofilm growth) are possible solutions (Bar-Yosef, 1999). A method of injection of fertilizer directly into the irrigation system also is required. Engineering guidelines and equipment are widely available in irrigated regions, with equipment ranging from expensive injector pump systems to low-cost venturi systems that rely on the pressure drop created by the increase in water velocity as it passes through a constriction. A low-cost venturi system developed for applying several different fertigation treatments in research plots in the same orchard illustrates one of the range of possibilities (Fig. 5). Additional details of fertilizer sources, injection equipment and methods, and safety considerations (e.g. to prevent backflow contamination of water sources) are frequently contained in technical publications from most irrigation districts (e.g. Van der Gulik, 1993), including some oriented specifically to drip irrigation systems (e.g. Van der Gulik, 1999).

6 Nutrient requirements

An often-cited advantage of fertigation is the ability to improve nutrient efficiency by more closely matching nutrient application to plant requirements. It has been possible to calculate daily biomass increase and N, P and K nutrient

Figure 5 Components of an automated fertigation system. Adapted from Neilsen et al. (2017b).

requirements throughout the growing season by successive destructive harvest and nutrient analyses for annual crops, such as greenhouse- and field-grown tomatoes and peppers (Bar-Yosef, 1991). It has been more challenging to develop similar numbers for perennial apple and cherry orchards due to the potential for storage and recycling within the perennial framework and the logistical difficulties of whole tree harvest. Nevertheless, Weinbaum et al. (2001) stressed the importance of sequential whole tree excavations to elucidate the patterns and magnitude of nutrient uptake by mature fruit trees. A study by Cheng and Raba (2009), involving destructive harvests at six different developmental stages of 6-year-old 'Gala' apple on M.26 rootstock grown in sand culture, is a rare example in which this has been undertaken for fruit trees. Total uptake of a range of nutrients from budburst to harvest has been reported per tree, as well as expressed in kg/ha at the planting density used (Table 6). The greatest growing season requirements were for K and N, but there were differences in their within-season acquisition (Fig. 6). More N is stored within the tree framework at bud break, and the high requirement does not commence until after bloom during the period of rapid shoot and canopy development; this decreases later in the season as growth declines. In contrast, the requirement for K is high post-bloom during shoot growth and especially during the period of rapid fruit growth expansion just prior to harvest. With fertigation, the timing of nutrient applications can be adjusted to coincide with periods of high demand. For N, applications can be made during a 6- to 8-week period post-bloom, and for K, during an 8-week period prior to harvest. Annually nutrients are required for development of roots, shoots, leaves and fruit. High nutrient demand with large crops can be anticipated for K, and also

Table 6 Total tree nutrient uptake from bud break to fruit harvest, and proportion of total nutrients associated with harvested fruit, for mature (6-year-old) 'Gala' apple trees on M.26 rootstock grown under non-limiting nutrient conditions in sand culture at a density equivalent to 2790 trees/ha and yield of 52.5 t/ha[a]

	Total tree uptake		Harvested fruit uptake
Major nutrients	g/tree	kg/ha	% of total
N	19.8	55.2	38.9
P	3.3	9.2	60.6
K	36.0	100.4	68.8
Ca	14.2	39.6	11.3
Mg	4.4	12.3	27.2
S	1.6	4.5	37.5
Minor nutrients	mg/tree	kg/ha	% of total
B	93.6	0.26	72.4
Zn	60.9	0.17	18.1
Cu	46.5	0.13	25.3
Mn	184.8	0.52	18.0
Fe	148.7	0.42	52.7

[a] Adapted from Cheng and Raba (2009).

for P and B since more than 60% of the annual requirements for those nutrients was associated with the fruit (Table 6). For these nutrients, annual application rates may need to be increased in years of high yield. Increases in nutrient requirements as yield increases is calculated for K, P and B for both apple and sweet cherry at typical low and high whole fruit nutrient concentrations

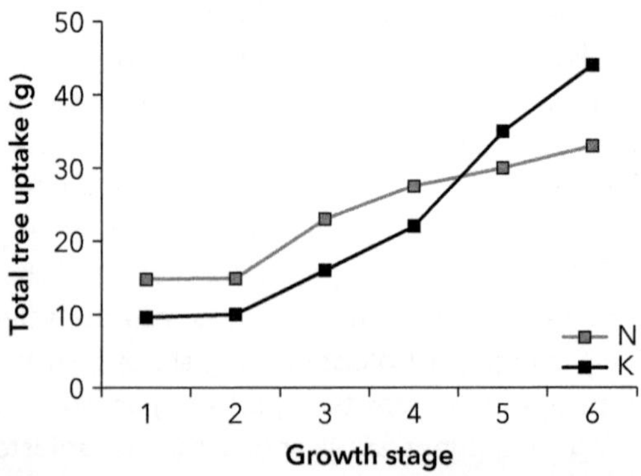

Figure 6 Whole tree N and K accumulation at various development stages for 'Gala'/M.26 apple destructively harvested after growth in sand culture. Growth stages correspond to (1) bud break, (2) bloom, (3) end of spur leaf growth, (4) end of shoot growth, (5) rapid fruit expansion, (6) fruit harvest. Adapted from Cheng and Raba (2009).

(Table 7; Neilsen et al., 2004a, 2007; Wolk et al., 1998). For both crops, unit area nutrient requirements become particularly large for K as a result of its high concentration in harvested fruit. This suggests that fertigated K rates need to be increased most in anticipated high crop years. Furthermore, measurement of leaf K, P and B concentrations in high crop years can be useful indicators of whether fertigation rates for these nutrients are sufficient for specific orchards. Similar studies have been reported for 'Chenin Blanc' wine grapes in South Africa (Conradie 1980, 1981).

In addition to considerations of the amount and timing of nutrient requirements, other factors affect the development of fertigation strategies. These include effectiveness at improving crop performance, especially marketable fruit quality, as well as efforts to improve nutrient use efficiency and minimize environmental impact. These issues will subsequently be discussed for key plant nutrients using apple and sweet cherry as representative perennial orchard crops and based upon results from published field experiments.

6.1 Nitrogen (N)

Nitrogen is the most commonly applied fertilizer in orchards and frequently at the highest rates. This is consistent with its importance to support annual development of the woody framework, leaf canopy and fruit crop. Many of the processes affecting the movement of N within orchards have been identified (Fig. 7), but much less is known concerning their magnitude. One site-specific study undertaken in a mature 'Golden Delicious' apple orchard on a silt loam soil in New Zealand measured a surplus of 32.4 kg N/ha in addition to leaching

Table 7 Range of typical whole fruit concentrations for K, P and B, and effect of yield increases on fruit nutrient requirements, in apple and sweet cherry[a]

	Apple				Sweet cherry		
	K	P	B		K	P	B
	Harvest nutrient concentration (mg/100 g F.W.)				Harvest nutrient concentration (mg/100 g F.W.)		
	100-135	8-10	0.2-0.5		120-220	15-25	0.2-0.5
Yield (t/ha)	Fruit nutrient requirements			Yield (t/ha)	Fruit nutrient requirements		
	kg/ha		g/ha		kg/ha		g/ha
25	25-34	2-2.5	50-125	10	12-22	1.5-2.5	20-50
50	50-68	4-5	100-250	20	24-44	3.0-5.0	40-100
75	75-101	6-7.5	150-375	30	36-56	4.5-7.5	60-150
100	100-135	8-10	200-500	40	48-88	6.0-10.0	80-200
				50	60-110	7.5-12.5	100-250

[a]Adapted from Neilsen et al. (2004a, 2007) and Wolk et al. (1998).

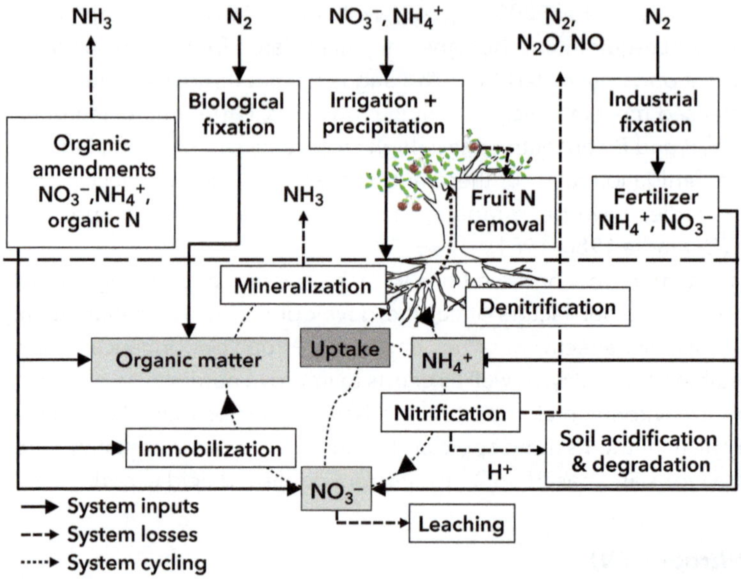

Figure 7 Potential inputs, losses and cycling of N within perennial fruit orchards.

losses of 23.4 kg N/ha (Table 8). The largest N inputs were associated with two annual broadcast applications of N-fertilizer (Haynes and Goh, 1980). Fertigated N applications have the potential to increase the frequency, but reduce the magnitude, of fertilizer N applications.

6.1.1 Soil N

Monitoring of the soil solution via suction lysimeters placed in the soil below and near drip emitters has indicated the predominance of the NO_3-N form and its ready mobility when ammonium nitrate is fertigated, whether on a clay loam (Klein and Spieler, 1987b) or silt loam or sandy loam soil (Neilsen et al., 1998a).

Table 8 Annual nitrogen inputs and losses for a 14-year-old 'Golden Delicious' orchard at a 4 m × 5 m spacing in Canterbury, New Zealand[a]

N inputs		N losses	
Source	kg/ha	Source	kg/ha
Precipitation (792 mm)	0	Fruit (5950 kg/ha)	21.3
Irrigation	9.6	Pruning wood (507 kg/ha)	3.9
Fertilizer	71.6	Leaching (361 mm)	23.4
Total	81.2	Total	48.6

[a] Adapted from Haynes and Goh (1980).

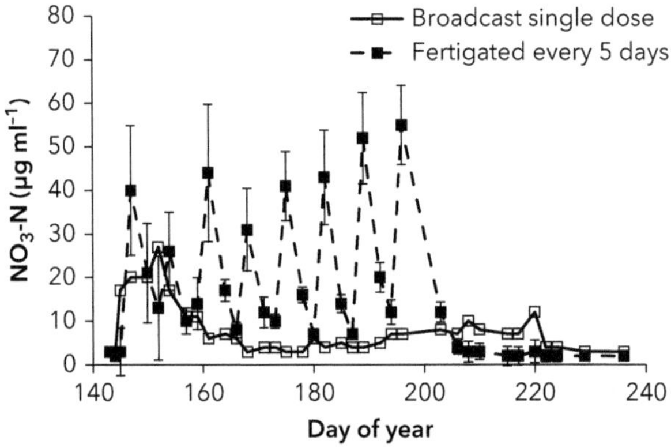

Figure 8 Soil solution NO_3-N concentration during a growing season in response to N applied at 25 g/tree as either a single dose of broadcast ammonium nitrate fertilizer with micro-sprinkler irrigation or periodic fertigation with drip irrigation. Vertical bars represent ±SE. Adapted from Neilsen et al. (1998a).

Much higher root zone NO_3-N concentrations over a longer period of time are achievable via eight weekly fertigation applications compared to a single surface broadcast application of ammonium nitrate fertilizer (Fig. 8). Similarly, root zone NO_3-N concentration responded to both amount of fertilizer and volume of irrigation water applied on both silt loam and loamy sand soils (Fig. 9), suggesting the potential for a high degree of control of root zone N concentration, particularly in regions with minimal summer precipitation. Subsequent studies in the same orchards indicated it might be possible to reduce N leaching loss below the root zone (0.75 m depth in these orchards) by restricting irrigation on the silt loam soil (Neilsen et al., 1998a). However, limiting leaching in coarse-textured soils without creating excessive water stress was more problematic, but might include pulse irrigation with smaller, more frequent applications of the same amount of water. Soil solution monitoring has generally been undertaken on soils with low organic matter content. As indicated by Haynes and Goh (1980), large amounts of N are mineralized from grass cuttings and potentially also from increased organic applications which are likely to be associated with increasing numbers of integrated and organic orchards (Neilsen et al., 2017a). The effect of N mineralized from organic sources on soil N is unlikely to be as pronounced as fertigated N as a result of its gradual increase during the growing season as soil temperature increases.

6.1.2 Plant N

In the sand culture study by Cheng and Raba (2009), the amount of N stored within the framework of 6-year-old 'Gala' apple trees was large, averaging

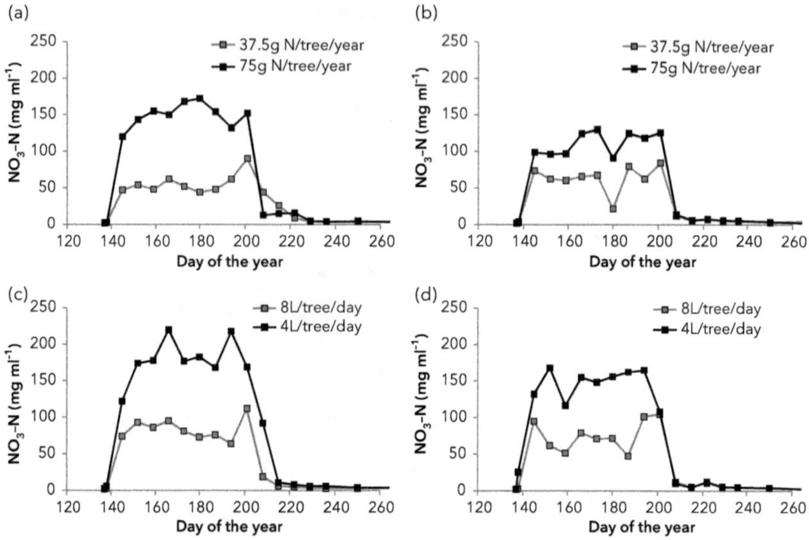

Figure 9 Soil solution NO_3-N concentration in response to rate of calcium nitrate applied to (a) a silt loam soil, (b) a loamy sand soil, and rate of irrigation applied to (c) a silt loam soil and (d) a loamy sand soil in a 'Gala'/M.9 apple planting. Adapted from Neilsen et al. (1998a).

75% of the total N taken up during the growing season (Fig. 6). Thus, stored tree N reserves potentially represent a significant source of annual N needs. The importance of N reabsorbed from senescing leaves in the autumn and recycled for early spring growth has been long recognized for apple (Oland, 1959; Titus and Kang, 1982) and more recently for sweet cherry (Millard et al., 2006; Ouzounis and Lang, 2011). Using ^{15}N tracer studies, the importance of remobilization of stored N for early spring growth has been identified for a range of fruit trees including greenhouse-grown apple trees (Millard and Neilsen, 1989) and field-grown cherry (Millard et al., 2006). This has practical implications for the timing of fertigating N, as indicated by the minimal uptake of fertigated ^{15}N until the period of rapid shoot, leaf and fruit growth for two-year-old 'Fuji' apple trees on M.9 rootstock (Fig. 10). For this reason, many N-fertigation applications are delayed until 6–8 weeks post-bloom in apple.

A common feature of many N-fertigation experiments is that maximum orchard performance is not associated with highest N-fertigation rates. Yield of fertigated 'Delicious' apple orchards did not increase above 150 kg N/ha and fruit colour was reduced above 250 kg N/ha (Klein et al., 1989). Declining fruit quality is an incentive to use lower N-fertigation rates, as has been reported for both apple and sweet cherry. For apples, the performance of five cultivars, all on M.9 rootstock and planted at high density (1 m × 3 m), was monitored for daily fertigation regimes which included high (125 kg N/ha) and low (25 kg N/ha) N

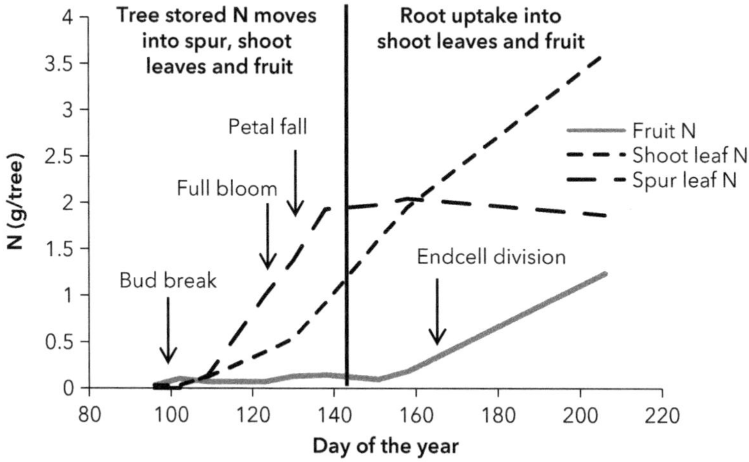

Figure 10 Uptake of fertigated ^{15}N-fertilizer into sequentially harvested 2-year-old 'Fuji'/M.9 apple trees relative to phenology. Adapted from Guak et al. (2003) and Neilsen et al. (2006).

rates applied during three different 4-week periods post-full bloom (Neilsen et al., 2009). The crop was generally unaffected by variation in the fertigation period. There was no difference in various measures of yield between N rates (Table 9), but fruit firmness and red colour (%) of red-skinned cultivars decreased at high N (Table 10). It was concluded from this study that cumulative annual application of 40 kg N/ha applied daily for 4 weeks post-bloom would have sufficed to maximize apple tree performance. Similarly, field trials with sweet cherry indicated best results from moderate N application rates (Neilsen et al., 2004a, 2007). Over the first six growing seasons, the maximum cumulative yield resulted from the lowest sprinkler-fertigated N application rate of 63 kg N/ha for 'Lapins' sweet cherry on Gisela 5 rootstock (Table 11). Average fruit size, an economically important fruit quality attribute for sweet cherry, was largest for the lowest rate of fertigated N.

6.1.3 N leaching

One of the advantages of fertigation is the ability to improve the efficiency of nutrient use by achieving similar crop performance at lower fertilizer application rates. The previously cited field experiments confirm such a possibility. The high solubility of NO_3-N means that avoiding excess water application when fertigating will be important in limiting N leaching. Passive capillary wicks, installed at 0.75 m depth and below the main rooting zone in a high-density apple orchard on M.26 rootstock, were used to monitor drainage losses of N under various fertigation regimes (Neilsen et al., 2008a). Drip irrigation systems

Table 9 Effect of fertigated N rate on cumulative yield, fruit number, average fruit weight and cumulative yield efficiency for 'Ambrosia', 'Cameo', 'Fuji', 'Gala' and 'Silken' apple cultivars over the first six fruiting seasons[a]

N rate[b]	Cumulative yield (kg/tree)	Fruit number (n/tree)	Average fruit weight (g)	Cumulative yield efficiency (kg/cm² TCSA[c])
Low (L)	52.6	277	191	3.48
High (H)	55.8	292	193	3.93
Significance	NS	NS	NS	NS

[a] Adapted from Neilsen et al. (2009).
[b] Fertigated at either 28 mg N/L (Low) or 168 mg N/L (High) as calcium nitrate (15.5-0-0).
[c] Trunk cross-sectional area (TCSA) measured at 0.3 m above the graft union.
NS = Means within columns not significantly different.

Table 10 Annual effects of fertigated N rate on harvest firmness of 'Ambrosia', 'Cameo', 'Fuji', 'Gala' and 'Silken' apples and red colour at harvest of 'Ambrosia', 'Cameo', 'Fuji' and 'Gala' over the first six fruiting seasons[a]

N Rate[b]	1999	2000	2001	2002	2003	2004
Fruit firmness (N)						
Low (L)	85.7	82.4	83.8	80.9	89.9	84.7
High (H)	84.7	80.9	80.6	78.7	87.3	81.4
Significance	NS	*	*[c]	***	***	***
Red colour (% solid red)						
Low (L)	85	80	80	85	85	80
High (H)	80	80	75	80	80	75
Significance	NS	NS	*	*	NS	*

[a] Adapted from Neilsen et al. (2009).
[b] Fertigated at either 28 mg N/L (Low) or 168 mg N/L (High) as calcium nitrate (15.5-0-0).
[c] Cultivar interaction, not significant for 'Ambrosia'.
*,*** Means within columns significantly different at $P = 0.05, 0.001$ respectively, or not significantly different (NS).

were found to be more sensitive than micro-sprinklers to the effects of water over-supply on drainage losses of N, particularly near the emitters. Lowering N concentrations in fertigating solutions and restricting fertigation periods to periods of low evaporative demand and high plant N requirements helped reduce N leaching.

6.1.4 Soil acidification

Acidification of soil as a consequence of fertigation is primarily associated with use of NH_4-N-containing fertilizers and their microbiologically-mediated conversion to the NO_3-N form (Haynes and Swift, 1987). Soil acidification can result in adverse orchard soil chemical changes, including accelerated leaching of essential nutrient bases such as K, Mg and Ca and excessive solubility

Table 11 Yield and average fruit size of 'Lapins' sweet cherry on Gisela 5 rootstock as affected by low and high rates of N over six fruiting seasons[a]

N Rate	2000	2001	2002	2003	2004	2005
Average yield (kg/tree)						
63 kg N/ha	1.8	13.5	8.6	37.8	46.5	12.5
254 kg N/ha	3.0	13.3	12.3	21.5	23.0	10.1
Significance	NS	NS	NS	****	***	NS
Average fruit size (g/fruit)						
63 kg N/ha	12.6	11.0	10.0	11.2	9.7	14.9
254 kg N/ha	12.5	9.6	9.6	8.5	9.4	13.8
Significance	NS	*	*	**	NS	**

[a] Adapted from Neilsen et al. (2004a, 2007).
*,**,***,**** Means within columns significantly different at P = 0.05, 0.01, 0.001, 0.0001 respectively, or not significantly different (NS).

of toxic elements including Mn and Al (Ross et al., 1985). The problem is particularly serious for soils with a low buffering capacity and limited resistance to pH change. An acidification resistance index (ARI) was developed to identify soils prior to fertigation with low ARI-values. For these soils, fertigating non-acidifying N fertilizers such as calcium nitrate would be recommended (Neilsen at al., 1995b). Also, routine soil pH monitoring should occur when undertaking long-term N-fertigation.

6.2 Phosphorus (P)

6.2.1 Soil P

Fertigating P using soluble P fertilizer sources such as phosphoric acid (O'Neil et al., 1979) elevates soil solution P concentration in both fine-textured, sandy clay loam (Klein and Spieler, 1987b) and coarse-textured sand soils (Neilsen et al., 1997b) resulting in improved P mobility relative to broadcast surface applications of P fertilizer (Neilsen et al., 1999). Post-fertigation soil sampling can indicate increased extractable soil P at considerable depth below the point of application over time, especially for drip-relative to micro-fertigation on sandy loam soils (Table 12). Phosphorus availability depends upon a soil's P sorption capacity with soil solution P concentrations increased more for a range of soil textures via a single large, relative to multiple smaller, P applications (Neilsen et al., 1993a).

6.2.2 Plant P

Many early field studies on fruit trees failed to observe responses to broadcast surface applications of P (e.g. Childers, 1966) which is consistent with low annual

Table 12 Average annual extractable (0.25M HOAc + 0.015M NH_4F) soil P concentration at various soil depths and locations as affected by drip or micro-jet irrigation averaged over three irrigation frequencies in a 3 year study[a]

Depth (cm)	Vertical distance from emitter (cm)	Irrigation	Extractable P (mg/g) Year 1	Year 2	Year 3
0–10	0	Drip	106	151	221
		Micro-jet	89	56	59
			**[b]	*	***[b]
10–20	0	Drip	100	122	128
		Micro-jet	51	46	43
			*	*[b]	***
20–40	0	Drip	48	76	76
		Micro-jet	24	36	36
			*	*	*
40–60	0	Drip	40	40	63
		Micro-jet	33	54	33
			NS	NS	*
20–40	60	Drip	23	15	17
		Micro-jet	24	24	27
			NS	*	*

[a] Adapted from Neilsen et al. (1997b).

[b] Values averaged over-irrigation frequency but affected by significant frequency × irrigation interaction.

*,**,*** Means within columns significantly different at $P = 0.05, 0.01, 0.001$ respectively, or not significantly different (NS).

requirements relative to other major nutrients (Table 6). Furthermore, research with some species, such as apple, indicated that P-deficiency is characterized by reduced leaf size and shoot growth but no obvious visual signs of deficiency (Benson and Covey, 1979). Nevertheless, subsequent research on apple provided evidence of a role for P-fertilization in initiating fruiting (Neilsen et al., 1990), improving establishment and initial growth of orchards being replanted on fumigated, old orchard soils (Neilsen and Yorston, 1991) and on soils with low available P (Cripps, 1987).

The greater mobility of fertigated P in soil can improve chances of a positive response of fruit trees to P application. Tree performance of five apple cultivars, planted at high density, was compared on trees fertigated with NPKB relative to those receiving NKB (Neilsen et al., 2008b). P-fertigation consistently increased leaf and fruit P concentration and was associated with a 20% increase in cumulative yield over the first five fruiting years, with increased fruit numbers and no decrease in average fruit size (Table 13). In a 2-year-old study in the

Table 13 Effect of fertigated P applied at bloom as ammonium polyphosphate on cumulative yield, fruit number and average fruit weight for 'Ambrosia', 'Cameo', 'Fuji', 'Gala' and 'Silken' apples over the first five growing seasons[a]

Treatment	Cumulative yield (kg/tree)	Total fruit number (n/tree)	Average fruit size (g/fruit)
+P	47.2	239	199
−P	39.3	203	194
Significance	*	*	NS

[a] Adapted from Neilsen et al. (2008b).
* Means within columns significantly different at $P = 0.05$ or not significantly different (NS).

same experiment, enhanced fruit P concentration had a role in maintaining apple fruit membrane stability by reducing the incidence of water core in susceptible cultivars and elevating antioxidant concentration of fruit after cold storage (Neilsen et al., 2008b). This research suggested that P-fertigation can provide long-term benefits to apple tree performance. Earlier research had indicated that P-fertigation was more effective at increasing P-uptake and tree growth in the first 2 years than equivalent P applications of solid mono-ammonium phosphate in the planting hole in replant orchards (Neilsen et al., 1993b). P-fertigation research on sweet cherry indicated that irrigation method can affect plant response. In a field experiment on a gravelly sandy loam with low soil P and 'Lapins' sweet cherry on Gisela 5 rootstock, applications of 20 g P/tree in the spring via sprinklers did not affect cherry growth or yield in the first four growing seasons (Neilsen et al., 2004a). No effects on tree performance and minimal effects on fruit quality occurred in the subsequent three seasons (Neilsen et al., 2007). In contrast, application of the same quantities of P at bloom via drip emitters on the same soil series consistently increased leaf and fruit P concentrations and yield over the first three growing seasons (Neilsen et al., 2010) and could delay cherry maturity in low crop years (Neilsen et al., 2014). The sensitivity of cherry fruit size to high crop load suggests it will be important to explore potential benefits of P-fertigation when crop load and fruit P demands are high (Table 7).

Commonly used soluble fertilizers such as ammonium polyphosphate or phosphoric acid, like NH_4-N based N fertilizers, acidify the orchard soil. Thus, long-term monitoring of soil pH is recommended, particularly below and near drip emitter locations used for P-fertigation. Concern has been expressed that high rates of P-fertilization via fertigation might also accelerate Zn-deficiency, as has been observed for some crops grown on sandy soils or in nutrient solution culture (Loneragan et al., 1979). Despite the susceptibility of both apple and sweet cherry to Zn-deficiency, there have been no field reports of P-fertigation increasing Zn-deficiency symptoms.

6.3 Potassium (K)

6.3.1 Soil K

Soil solution K concentrations are readily increased to 0.3 m depth beneath drip emitters in soils of a range of textures (Klein and Spieler, 1987b; Neilsen et al., 2004b). Soil solution K concentrations peak during fertigation, often declining to pre-fertigation levels over multiple years (Fig. 11). Nevertheless, five years of annual K-fertigation in a sandy loam soil elevated extractable soil K relative to no K-treatments at 0.6 m depth and lateral distance from drip emitters. This coincided with the most important rooting depth for 'Gala' apples on M.9 rootstocks in this orchard (Fig. 12). The movement of fertigated K to depth will be reduced in soils with high exchange capacity. However, the increased mobility of drip-fertigated K was used to move K to depth in an orchard soil with a high capacity to irreversibly adsorb (fix) K (Uriu et al., 1980).

6.3.2 Plant K

Drip fertigation has effectively corrected K-deficiency for both sour cherry (*Prunus cerasus*) (Callan and Westcott, 1996) and apple (Neilsen et al., 1998b). Annual applications of 15–30 g K/tree for apple and 290 g K for cherry trees were sufficient to increase leaf K concentrations from deficiency values, increase tree yield and, for apple, to increase red colour. The benefits of K-fertigation are less likely when leaf K concentrations of apple and cherry trees are adequate, exceeding 1.5%. High crop load can significantly increase apple and cherry K-requirements, as discussed previously (Table 7). Thus, fertigation application rates previously found to be adequate can become inadequate

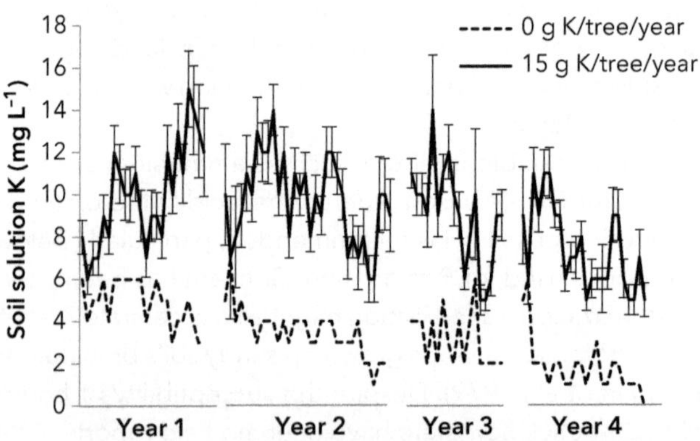

Figure 11 Soil solution potassium (K) concentration at 30 cm depth directly beneath the drip emitter as affected by daily fertigation on a sandy loam orchard soil. Vertical bars represent ±SE. Adapted from Neilsen et al. (2004b).

Published by Burleigh Dodds Science Publishing Limited, 2019.

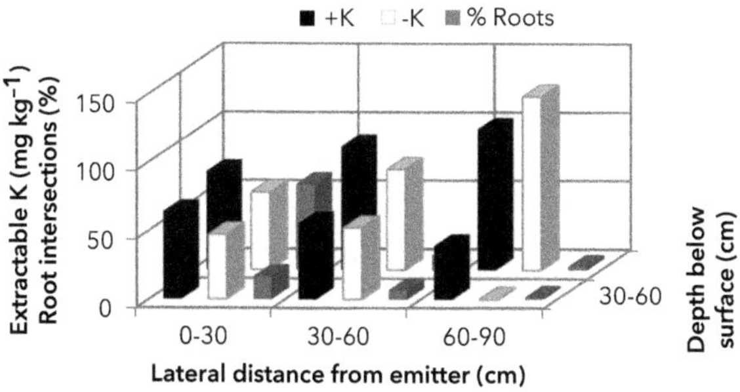

Figure 12 Average percentage of all root intersections compared to soil-extractable K for treatments with (+K) and without (−K) potassium fertigation as located in 30 cm × 30 cm zones in a plane originating at the emitter location and extending perpendicular to the tree row for 'Gala'/M.9 apple trees irrigated daily and fertigated for 5 years. Adapted from Neilsen et al. (2000).

with high crops (Neilsen et al., 2015). Also, within-season K-fertigation is often delayed until later in the growing season in order to more closely match the K-requirements of rapidly growing fruit. Callan and Westcott (1996) suggested KCl was not a suitable K source for sour cherry because of the risk of Cl-toxicity. In contrast, chloride, sulphate and thiosulphate forms consistently increased apple leaf and fruit K concentrations (Neilsen and Neilsen, 2006). However, no treatments improved tree vigour, yield or fruit quality since tree K status was already adequate. The extent of the wetted zone, as influenced by irrigation method, can affect K nutrition and hence fertigation strategies. Both apple and cherry leaf K concentrations decline over time, approaching deficiency, when fertigated with N-only through drip- rather than micro-jet- or sprinkler-irrigation systems (Table 14). The existence of a smaller wetted soil volume, especially on coarse-textured soils, appears to adversely affect the acquisition of K. In the long-term, K-fertigation may be a necessity when using drip emitters.

6.3.3 Nutrient balance

Most concern about the need for balanced nutrition has been expressed concerning K-fertigation, since high K-availability can be antagonistic to the uptake of other cations such as Ca and Mg. A high fruit Ca concentration at harvest, relative to fruit size and fruit K concentration, is critical to eliminate harvest and storage disorders in apple (Wolk et al., 1998). Fertigating K, when leaf and fruit K concentrations were adequate, did not decrease harvest fruit Ca concentrations nor increase bitter pit in a 'Jonagold' apple orchard over a 3-year period (Neilsen and Neilsen, 2006). K-fertigation can decrease leaf Mg concentration across a range of apple leaf Mg and K concentrations (Neilsen et al., 2004b; Neilsen and Neilsen, 2006). There have been few reports of the

Table 14 Average annual leaf K concentration as affected by irrigation, starting the year of planting (Year 1) for 'Lapins' sweet cherry[a] and 'Gala' apple[b] not receiving K fertilization

	Year 1	Year 2	Year 3	Year 4	Year 5	Year 6	Year 7	Year 8
Irrigation	Leaf K concentration (% DW)							
Sweet cherry								
Micro-sprinkler	2.78	1.91	2.08	2.0	nm[c]	1.47	1.35	1.91
Drip	2.47	1.72	1.75	1.45		1.19	0.89	1.41
Significance	**	**	**	****		***	****	***
Apple								
Micro-jet	1.84	2.17	1.89	1.92	2.09			
Drip	1.73	1.76	1.37	1.14	1.43			
Significance	NS	****	****	****	****			

[a]Adapted from Neilsen et al. (2004a, 2007).
[b]Adapted from Neilsen et al. (1995a).
[c]Not measured.
*,***,**** Means within columns significantly different at $P = 0.05, 0.001, 0.0001$ respectively, or not significantly different (NS).

effects of K-fertigation on cherry Mg and Ca status, although it is anticipated the same antagonisms would occur. It would, therefore, be prudent to routinely monitor fruit Ca and leaf Mg concentrations in fruit orchards where long-term K-fertigation is being undertaken.

6.4 Other nutrients

Other essential plant nutrients have been successfully fertigated. These include B applied via drip emitters, daily, during an 8-week period over two growing seasons (Neilsen et al., 2004c). Soil solution B concentrations at 0.3 m depth increased during the fertigation period, declining afterwards in a similar manner to NO_3-N concentrations following N-fertigation (Fig. 13). B fertigation was as effective as foliar applications in correcting B-deficiency symptoms which had developed on four apple cultivars grown in the orchard (Fig. 14). Fruit trees have a relatively narrow range between deficiency and toxicity for B, and fertigation allows effective application of low B unit area application rates (0.6 kg B/ha/year in this study). The ready response of leaf B concentrations to fertigated B indicates caution is required to avoid over-application of B. Similar effectiveness when fertigating 0.28 kg B/ha/year has been reported for high-density sweet cherry orchards on dwarfing rootstocks (Neilsen et al., 2016b).

Magnesium has also been successfully fertigated for wine grapes (Treeby, 2008), although there are few reports for apple or sweet cherry. Fertigation of soluble K and Mg sulphate in apple orchards increased soil Mg, but had few effects on leaf and fruit Mg concentrations or apple tree performance

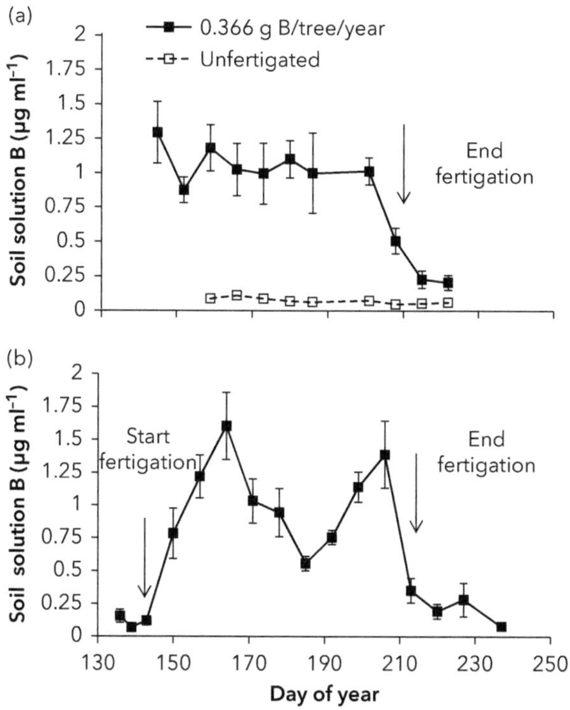

Figure 13 Soil solution concentration of B as influenced by fertigation of B in (a) 1995 and (b) 1996. Vertical bars represent ±SE. Adapted from Neilsen et al. (2004c).

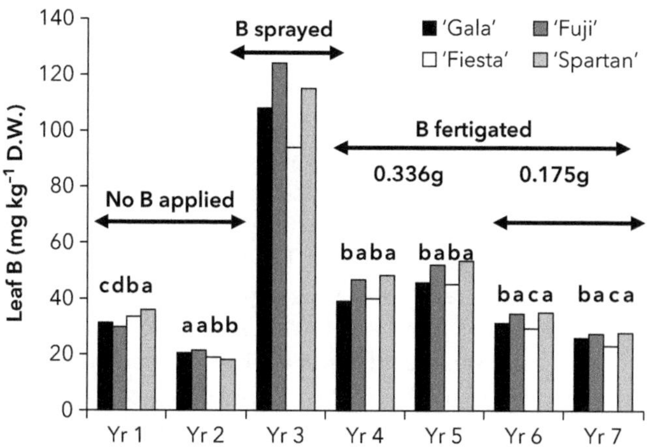

Figure 14 Leaf boron (B) concentration of 'Gala', 'Fuji', 'Fiesta' and 'Spartan' apple trees in response to B application method over the first seven growing seasons. Adapted from Neilsen et al. (2004c).

(Neilsen and Neilsen, 2006). It may be difficult to augment tree Mg status when co-applying K due to K-Mg antagonism.

Application of Ca via the soil, including by fertigation, has generally been considered ineffective relative to foliar applications. Similarly, foliar application has been the preferred method for most micronutrients, as discussed in the following section.

7 Foliar application

Foliar application has previously been reviewed for fruit crops and the reader is referred to these excellent and comprehensive reviews (Swietlik and Faust, 1984; Weinbaum, 1988). In these reviews, much information is provided concerning factors affecting the efficiency of nutrient uptake from foliar applications and the effect of foliar sprays on horticultural performance. This discussion will focus on foliar applications as a method of improving the precision of nutrient applications in orchards based on field experience with apple and cherry.

7.1 Nitrogen

The process of natural withdrawal of N from apple leaves prior to their senescence has been utilized as a strategy to augment dormant tree N status by spraying foliar urea solutions at concentrations ranging from 2% to 5% (w/v) post-harvest (Table 15). Stored N can be particularly important for spring growth of young apple trees when environmental conditions for root growth are suboptimal (Millard and Neilsen, 1989). Subsequent research using ^{15}N-tracers identified the limited amount of soil N uptake which occurs prior to bloom for both apple (Neilsen et al., 2001) and cherry (Grassi et al., 2003), indicating their early growth is dependent upon remobilization of stored N. Post-harvest foliar urea sprays have increased apple yield the subsequent year in some studies, while other research has failed to demonstrate benefits (Weinbaum, 1988). It is likely that enhanced plant performance occurs when strategically timed urea sprays overcome N-deficiency conditions. This may explain contradictory results when pre-bloom urea sprays made to apples have sometimes increased yield (Yogaratnam and Greenham, 1982). It is likely that cherry response to N is also dependent upon tree N status, as suggested by Wojcik and Morgas (2015). Recently, Ouzounis and Lang (2011) were able to increase dormant cherry tree N status and shoot cold hardiness by fall foliar urea applications at concentrations ranging from 3% to 5% (w/v). Application of urea sprays to other *Prunus* species, including nectarine and peach was initially considered to be ineffective (Norton and Childers, 1954). Subsequent research using ^{15}N labelled urea, intensive leaf sampling and tree partitioning demonstrated that urea was rapidly absorbed and retranslocated in nectarine (Rosecrance et al., 1998a,b;

Published by Burleigh Dodds Science Publishing Limited, 2019.

Tagliavini et al., 1998). Earlier research had also shown the effectiveness of urea sprays for improving the N status of prune (*Prunus domestica* L.) (Leece and Dirou, 1979). Four weekly late season applications of 0.5% and 1.0% urea, with the last spray applied 7–10 days prior to commercial harvest, were effective in promoting greenness in four green apple cultivars (Meheriuk et al., 1996). The sprays enhanced green pigmentation in 'Granny Smith' and 'Newtown' at harvest and retarded yellowing in all four cultivars tested during air storage at 0°C for 4 years. Targeted foliar sprays were more effective than doubling soil fertilizer applications in enhancing this quality attribute of green apples.

Since N is the fertilizer applied in the largest quantities in most orchards, it has been proposed that the higher efficiency of foliar N applications might reduce excessive N leaching from orchards by substituting for soil applications (Weinbaum, 1988). Dong et al. (2005) compared biweekly foliar to biweekly soil N applications in a mature 'Gala'/M.9 apple orchard at the same rate (50 g N/tree, 62.5 kg N/ha) over two growing seasons. Tree N status, yield and fruit quality were the same for both treatments and soil NH_4-N and NO_3-N concentrations near the irrigation emitters at 0.6 m depth were reduced for the foliar treatment. This study provided promising evidence that soil N leaching could be reduced while maintaining short-term apple tree performance in an orchard established under a soil N fertilization regime. The long-term consequences of relying only on foliar N needs further assessment for a wider range of fruit crops including *Prunus* species (Johnson et al., 2001).

7.2 Phosphorus and potassium

Foliar potassium dihydrogen phosphate sprays have been used successfully to increase apple fruit P concentration when applied 4–6 weeks post-bloom and to reduce the incidence of low temperature breakdown in 'Bramley' and 'Cox Orange Pippin' (Yogaratnam and Sharples, 1982) and 'McIntosh' apples (Webster and Lidster, 1986) (Table 15). There have been no reports of foliar P application benefitting fruit quality of cherries. K-deficiency can inhibit red colour development in apple, but soil applications are more effective than foliar applications in correcting the deficiency and improving fruit colour as a result of the high and continuous late season demand by fruit for K.

7.3 Calcium

Calcium is the nutrient which is most commonly applied by foliar applications. Cumulative research has identified inadequate Ca supply as being implicated in many physiological disorders in fruits and vegetables (Shear, 1975), including apple bitter pit, cork spot and internal breakdown. The development of localized Ca deficiency within apple fruit has been associated with the

Table 15 Formulation, concentration, timing and purpose of nutrients successfully applied to foliage of apple or cherry in randomized, replicated field trials

	Crop	Formulation (concentration w/v)	Timing	Purpose	Reference
N	Apple	Urea (0.36%)	Pre-bloom	Overcome winter damage	Stover et al. (1999)
		Urea (0.5–1.0%)	Multiple pre-harvest	Enhance fruit greenness	Meheriuk et al. (1996)
		Urea (1–4%)	Post-harvest	Enhance tree N reserves and correct deficiencies	Weinbaum (1988)
	Cherry	Urea (3–5%)	Pre-senescence	Enhance tree N reserves and cold acclimation	Ouzounis and Lang (2011)
P	Apple	Potassium dihydrogen phosphate (1%)	Weekly mid-June to mid-July	Reduce low temperature disorder	Yogaratam and Sharples (1982)
Ca	Apple	Calcium chloride (0.3–0.5%)	Multiple pre-harvest in growing season	Reduce bitter pit	Peryea et al. (2007)
		Calcium nitrate (0.6%)	Multiple late pre-harvest	Reduce post-harvest breakdown	Mason (1979)
		Calcium chloride (2.5, 5%)	Single spray just before harvest	Reduce fruit Ca disorders	Looney (1977)
	Cherry	Calcium chloride (0.35%)	Multiple late pre-harvest	Reduce cracking	Meheriuk et al. (1991)
		Calcium chloride (0.5–1.0%)	Multiple pre-harvest during rain	Reduce cracking	Lang et al. (1998)
		Calcium (3%) + copper (0.2%) hydroxides	6 and 3 weeks pre-bloom	Reduce cracking	Brown et al. (1995)
Mg	Apple	Magnesium sulphate (1.2–2.0%)	Multiple 2–6 weeks post bloom	Overcome deficiency Maintain tree levels	Ford et al. (1965)
		Magnesium nitrate (0.5–0.6%)	Multiple 2–6 weeks post bloom	Overcome deficiency Maintain tree levels	Webster (1973)

Zn	Apple, cherry	Zinc sulphate [36% Zn] (0.12–2.0%)	Late dormant silvertip to bud swell	Maintain tree levels Overcome deficiency	Swietlik and Faust (1984)
		Liquid zinc sulphate (2.5–3%)	Late dormant silvertip to bud swell	Maintain tree levels Overcome deficiency	Neilsen et al. (2017b)
		Chelated, organically complexed Zn	Late dormant silvertip to bud swell	Maintain tree levels Overcome deficiency	Neilsen et al. (2017b)
	Apple	Organically complexed Zn (0.2–0.4 g Zn/ha)	1–4 weeks post-petal fall	Maintain tree levels Overcome deficiency	Peryea (2006)
B	Apple	Sodium borate [Solubor] (0.2–0.5%)	Season long and post-harvest	Maintain tree levels Overcome deficiency	Neilsen and Neilsen (2003)
	Cherry	Sodium borate [Solubor] (0.25%)	Post-harvest	Overcome deficiency	Hanson (1991)
Mn	Apple, cherry	Manganese sulphate (0.2%)	0–4 weeks post-petal fall	Overcome deficiency	Neilsen et al. (2017b)
Fe	Apple, cherry	Ferrous sulphate	As soon as leaves develop	Overcome deficiency (temporary)	Neilsen et al. (2017b)
Cu	Apple, cherry	Copper sulphate (0.5%)	Season long unless bearing then pre-bloom; post-harvest	Overcome deficiency	Neilsen et al. (2017b)
Mn, Fe, Cu	Apple, cherry	Chelated forms (label rates)	Season long	Maintain tree levels Overcome deficiency	Neilsen et al. (2017b)

deterioration of xylem functionality later in the growing season (Drazeta et al., 2004) and inhibition of Ca translocation to fruit during periods of high tree vigour and gibberellin activity (Saure, 2005). Regardless of the mechanism, foliar spray applications directly to the fruit surface at dilute concentrations are more effective than soil applications for augmenting fruit Ca (Vang-Petersen, 1980). Various soluble Ca salts have been tested in apple orchards, with dilute sprays of calcium nitrate and calcium chloride recommended most often (Table 15). Food-grade calcium chloride is often favoured due to its effectiveness, low cost, and lack of N (Raese and Drake, 2000). Other Ca formulations have been tested over time for effectiveness in apple orchards in different production regions. These include calcium acetate (Lotze at al., 2008), amino acid-complexed Ca salts (Yuri et al., 2002), various synthetic or natural organic chelated forms (Neilsen and Neilsen, 2002) and powdered calcium carbonate slurries (Neilsen et al., 2005a). These materials can be effective at raising fruit Ca concentrations, providing they are soluble, but often they have to be applied at equal Ca application rates to standard materials such as calcium chloride. This can be economically unfeasible due to the high cost per unit of applied Ca for some materials. Dilute concentrations and early morning applications are favoured to avoid leaf damage, particularly in regions where high air temperatures and low humidity result in rapid drying of applied solutions. Single high concentration calcium chloride applications (2.5–5.0% w/v) have proven effective when applied immediately prior to harvest (Looney, 1977). The ensuing leaf damage, occurring late in the growing season, was judged to have no detrimental effects on fruit quality or subsequent tree performance. Application of Ca sprays later in the growing season more effectively increases Ca concentration of apples at harvest than early Ca sprays since apples have a much larger surface area for absorption at this time. Hence, Ca spray recommendations in many apple growing regions suggest multiple applications in the last half of the growing season, ending just prior to harvest. Recent research using ^{44}Ca isotopes as tracers indicates that the apple fruit is capable of Ca adsorption throughout its development (Kalcsits et al., 2017). Early season sprays, commencing within 40 days of full bloom, were effective at reducing incidence of bitter pit in 'Braeburn' and 'Honeycrisp', which are highly susceptible, despite resulting in lower fruit Ca concentration at harvest (Neilsen et al., 2005a; Peryea et al., 2007).

Foliar Ca applications have also been applied during the growing season to improve sweet cherry quality at harvest by reducing surface pitting and skin cracking (Table 15). Success has been achieved using calcium chloride and calcium hydroxide alone (Meheriuk et al., 1991) or in combination with copper hydroxide (Brown et al., 1995). Sprays are usually applied later in the growing season as cherries change colour and become more susceptible to splitting. A novel approach has been to apply dilute calcium chloride via over-tree sprinklers, with applications computer-activated during rainfall events to reduce the osmotic differential between rainwater on the surface of the fruit and the juice (Lang et al., 1998). However, rain-cracking of cherry fruit can be driven by root-absorbed water in addition to fruit surface-absorbed water, so the effects of Ca applications to the canopy can be inconsistent and fail to reduce fruit cracking (Knoche and Winkler, 2017).

7.4 Magnesium

For apple, multiple foliar applications of dilute magnesium sulphate solutions have been sufficient to correct Mg-deficiency symptoms and maintain adequate leaf Mg concentrations (Table 15). Magnesium nitrate (Webster, 1973) and magnesium chloride (Ford et al., 1965) also have been used successfully. Chelated Mg forms are less effective unless applied at similar Mg rates to the Mg-salts (Neilsen and Hoyt, 1984). In regions with low soil pH and low available soil Mg, annual foliar Mg applications offer rapid Mg availability to supplement dolomitic liming (Sadowski et al., 1976). Mg-deficiency symptoms are less commonly reported for cherry, and little work has thus been reported on correcting deficiencies. It is, however, likely that foliar spray recommendations suggested for apple would be effective for cherry.

7.5 Micronutrients

Fruit trees are susceptible to micronutrient deficiencies associated with zinc, boron, iron, manganese and, more rarely, copper. These deficiencies are associated with unfavourable soil conditions that are incapable of supplying tree micronutrient requirements despite their modest levels of <1 kg/ha (Table 6). Foliar sprays were developed as the preferred method of ameliorating deficiency symptoms and maintaining adequate nutritional levels.

7.5.1 Zinc

Zn is a long-recognized micronutrient deficiency of perennial fruit orchards with distinct deficiency symptoms including little leaf, rosetting, interveinal leaf chlorosis and in the extreme, shoot tip dieback (Neilsen and Neilsen, 2003). Factors affecting the development of Zn-deficiency, including high soil pH, have been reviewed comprehensively (Swietlik, 1999). Historically, zinc sulphate sprays applied at late dormancy from silver tip to bud swell have ameliorated Zn-deficiency symptoms and, when applied annually, have prevented their reoccurrence (Table 15; Woodbridge, 1954). Application of zinc sulphate post-bloom has generally been avoided because it can damage fruit finish. Recently, post-bloom sprays of inorganic, chelated, and humic- and fulvic-complexed Zn products have been identified which are capable of increasing tree Zn status without toxicity (Neilsen et al., 2005b; Peryea, 2006). The requirement for annual Zn sprays is necessitated since residual increases in leaf Zn are often minimal following dormant, post-bloom and post-harvest Zn applications (Neilsen and Hoyt, 1990). The inability of chelated Zn sprays (applied post-harvest) to improve Zn nutrition of sour cherry has been attributed to the inability of Zn to be withdrawn for subsequent storage prior to leaf senescence (Wojcik and Morgas, 2015). The low efficiency of foliar Zn sprays for *Prunus* species has recently been highlighted (Saa et al., 2018).

7.5.2 Boron

B-deficiency is widely reported for horticultural crops, with symptoms that can include abnormal development and dieback of shoot tips, misshapen and chlorotic young leaves, blossom death and fruit cracking for both apple (Neilsen and Neilsen, 2003) and sweet cherry (Neilsen et al., 2017b). Successful correction of B-deficiency and maintenance of plant B status for both apple and cherry has been achieved by both soil applications (see fertigation discussion) and via foliar applications of dilute concentrations of sodium borate (Solubor) (Table 15). Foliar applications have been applied successfully throughout the growing season for apple, ranging from early at pre-pink to improve fruit set, to post-bloom (Peryea et al., 2003) and post-harvest to benefit next year growth (Table 15). Stover et al. (1999) also found a pre-bloom foliar B application combined with Zn and urea enhanced apple cropping after cold damage. The flexibility in application timing is a reflection of the ready phloem transportation of B with sorbitol in *Prunus, Malus and Pyrus* species (Brown and Shelp, 1997). The effectiveness of foliar sprays is obvious when deficiency symptoms such as blossom blast are eliminated (Neilsen et al., 2004b). B sprays readily increased sour cherry bud, flower and often fruit B concentration whether applied post-harvest or around flowering, but increased yield only on trees of low B status (Hanson, 1991). In contrast, Wojcik and Wojcik (2006) increased soluble solids and anthocyanin colouration of sweet cherry fruit on trees with adequate leaf B concentrations with B sprays applied via the soil and foliar applications in the spring.

Caution is necessary to avoid indiscriminate application of B in orchards due to the narrow range between boron deficiency and toxicity for many crops. In apple, excess B can manifest itself as elevated fruit B concentrations and advanced fruit maturity with greater breakdown in storage (Fallahi et al., 2010b). Peryea and Neilsen (2013) reported a significant linear relationship between splitting and fruit B concentrations for 'Gala' apple. Thus, in regions with coarse-textured, low organic matter soil prone to B-deficiency, annual maintenance applications of dilute B sprays can provide the moderation required of B fertilizer management.

7.5.3 Iron, manganese and copper

The solubility of Fe, Mn and Cu is reduced in soil where pH exceeds seven. Therefore, deficiencies of these micronutrients are more likely to occur on calcareous or alkaline soils with Fe-deficiency frequently referred to as lime-induced chlorosis. The development of deficiency symptoms for each of these micronutrients has been documented for fruit trees, including apple and sweet cherry. Iron deficiency commences with shoot tip and interveinal chlorosis of the youngest leaves, which can result in complete loss of chlorophyll and photosynthetic capacity of these leaves. Mn deficiency is also expressed on the youngest leaves as symmetrical interveinal chlorosis, but it rarely reduces shoot growth. Cu deficiency is not common, especially for cherry, and when it occurs, symptoms are described as wither tip, characterized by summer dieback of chlorotic shoot tip leaves. In high pH soils, simultaneous deficiencies of multiple micronutrient deficiencies (such as Fe, Mn and Zn) can occur. Foliar applications of their respective chelated or sulphate salts can eliminate their respective deficiency symptoms, but correction is often temporary, requiring annual applications (Table 15). The challenges and alternative methods for correcting Fe deficiency on calcareous soils in important production regions, such as southern Europe, has been reviewed (Tagliavini et al., 2000).

7.5.4 Multiple nutrient deficiencies

Many commercial formulations have been developed to simultaneously apply several nutrients (Swietlik and Faust, 1984). With the exception of previously discussed micronutrient deficiencies, multiple nutrient deficiencies can be rare. Foliar applications have provided economic benefits by overcoming deficiencies and thereby increasing tree vigour, yield or fruit quality. Unfortunately, there have been few documented, randomized and replicated field trials to confirm the benefits of sprays containing multiple nutrients or sprays applied when nutrient levels are apparently sufficient. Given the high economic value of many perennial fruit crops relative to spray costs, there will always be a temptation to make foliar applications as a presumed 'insurance policy'.

8 Where to look for further information

Future trends in research for irrigation and nutrient management will include improved techniques for sensing water and nutrient deficits at the plant and landscape scale; automation of irrigation and nutrient management tailored to meet requirements at the individual tree and orchard scales; improved understanding of cultivar and rootstock influences on water and nutrient use through genomics and adaptation of water and nutrient management to changing climate and extreme weather.

Further information is available from a number of sources. The references section provides a comprehensive list of studies and reviews as cited in this chapter which can provide additional information. International scientific societies including the International Society for Horticultural Science (ISHS) (https://www.ishs.org) and the International Commission on Irrigation and Drainage (ICID) (http://www.icid.org) host regular conferences and specialized working group meetings and provide research reports and journals which showcase new developments in precision water and nutrient management. Many countries also have national horticultural, soil and water management scientific societies which organize annual meetings that provide venues for updating new developments in nutrient and water research. In addition many universities and national research institutes with active horticultural/plant and soil science departments can be accessed through their web pages.

9 References

Allen, R. G. and Pereira, L. S. (2009). Estimating crop coefficients from fraction of ground cover and height. *Irrig. Sci.* 28:17–34.

Allen, R. G., Pereira, L. S., Raes, D. and Smith, M. (1998). Crop evapotranspiration guidelines for computing crop water requirements. FAO Irrigation and Drainage Paper 56. United Nations Food and Agriculture Organization, Rome, 100p.

Ayars, J. E., Johnson, R. S., Phene, C. J., Trout, T. J., Clark, D. A. and Mead, R. M. (2003). Water use by drip-irrigated late-season peaches. *Irrig. Sci.* 22:187.

Ballester, C., Zarco-Tejada, P. J., Nicolás, E. Alarcón, J. J., Fereres, E., Intrigliolo, D. S. and Gonzalez-Dugo, V. (2018). Evaluating the performance of xanthophyll, chlorophyll and structure-sensitive spectral indices to detect water stress in five fruit tree species. *Precision Agric.* 19:178–93.

Bar-Yosef, B. (1991). Fertilization under drip irrigation. In: Palgrave, D. A. (Ed.), *Fluid Fertilizer Science and Technology*. Marcel Dekker, New York, NY, pp. 285–329.

Bar-Yosef, B. (1999). Advances in fertigation. *Adv. Agron.* 65:1–76.

Behboudian, M. H., Dixon, J. and Pothamshetty, K. (1998). Plant and fruit responses of lysimeter-grown 'Braeburn' apple to reduced irrigation. *J. Hortic. Sci. Biotech.* 73:781–5.

Bellvert, J., Marsal, J., Girona, J., Gonzalez-Dugo, V., Fereres, E., Ustin, S. L. and Zarco-Tejada, P. J. (2016). Airborne thermal imagery to detect the seasonal evolution of crop water status in peach, nectarine and saturn peach orchards. *Remote Sens.* 8:39–56.

Benson, N. R. and Covey Jr., R. P. (1979). Phosphorus nutrition of young 'Golden Delicious' apple trees growing in gravel culture. *J. Am. Soc. Hortic. Sci.* 104:682–5.

Boland, A. M., Bewsell, D. and Kaine, G. (2006). Adoption of sustainable irrigation management practices by stone and pome fruit growers in the Goulbourn/Murray Valleys, Australia. *Irrig. Sci.* 24:137–45.

Brown, P. H. and Shelp, B. J. (1997). Boron mobility in plants. *Plant Soil* 193:85–101.

Brown, G., Wilson, S., Boucher, W., Graham, B. and McGlasson, B. (1995). Effects of copper-calcium sprays on fruit cracking in sweet cherry. *Sci. Hortic.* 62:75–80.

Published by Burleigh Dodds Science Publishing Limited, 2019.

Bryla, D. R., Trout, T. J. Ayars, J. and Johnson S. R. (2003). Growth and production of young peach trees Irrigated by furrow, microjet, surface drip, or subsurface drip systems. *HortSci*. 38:1112-16.

Bryla, D. R., Dickson, E., Shenk, R., Johnson, R. S., Crisosto, C. H. and Trout, T. J. (2005). Influence of irrigation method and scheduling on patterns of soil and tree water status and its relationship to yield and fruit quality in peach. *HortSci*. 40:2118-24.

Bussi, C., Huguet, J.-G., Besset, J. and Girard, T. (1991). Fertilization scheduling in peach orchards under trickle irrigation. *J. Hortic. Sci.* 66:487-3.

Bussi, C., Huguet, J.-G. and Defrance, H. (1994). Effects of nitrogen fertilization during trickle irrigation on the growth and fruit yield of peach. *Eur. J. Agron.* 3:243-8.

Callan, N. W. and Westcott. M. P. (1996). Drip irrigation of potassium to tart cherry. *J. Plant Nutr.* 19:163-72.

Casadesús, J., Mata, M., Marsal, J. and Girona, J. (2012). A general algorithm for automated scheduling of drip irrigation in tree crops. *Comput. Electron. Agric.* 83:11-20.

Chai, Q., Gan, Y., Zhao, C., Xu H.-L., Waskom, R. M., Niu, Y. and Siddique, K. H. M. (2016). Regulated deficit irrigation for crop production under drought stress: A review. *Agron. Sustain. Dev.* 36:3-24.

Cheng, L. and Raba, R. (2009). Accumulation of macro- and micronutrients and nitrogen demand-supply relationship of 'Gala'/'Malling 26' apple trees grown in sand culture. *J. Am. Hortic. Sci.* 134, 3-13.

Childers, N. F. (1966). *Temperate to Tropical Fruit Nutrition*. Horticultural Publications, Rutgers, NJ.

Cisneros, J. B. E., Oki, T., Arnell, N. W., Benito, G., Cogley, J .G., Döll, P., Jiang, T. and Mwakalila, S. S. (2014). Freshwater resources. In: Field, C. B., Barros, V. R., Dokken, D. J., Mach, K. J., Mastrandrea, M. D., Bilir, T. E., Chatterjee, M., Ebi, K. L., Estrada, Y. O., Genova, R. C., Girma, B., Kissel, E. S., Levy, A. N., MacCracken, S., Mastrandrea, P. R. and White, L. L. (Eds), *Climate Change 2014: Impacts, Adaptation, and Vulnerability. Part A: Global and Sectoral Aspects. Contribution of Working Group II to the Fifth Assessment Report of the Intergovernmental Panel on Climate Change*. Cambridge University Press, Cambridge, UK, pp. 229-69.

Coates, R. W., Delwiche, M. J. and Brown, P. H. (2006). Control of individual microsprinklers and fault detection strategies. *Precision Agric.* 7:85-99.

Conejero, W., Ortuño, M. F., Mellisho, C. D. and Torrecillas, A. (2010). Influence of crop load on maximum daily trunk shrinkage reference equations for irrigation scheduling of early maturing peach trees. *Agric. Water Manage.* 97:333-8.

Conejero, W., Mellisho, C. D., Ortuño, M. F., Moriana, A., Moreno, F. and Torrecillas, A. (2011). Using trunk diameter sensors for regulated deficit irrigation scheduling in early maturing peach trees. *Environ. Exp. Bot.* 71:409-15.

Conradie, W. J. (1980). Seasonal uptake of nutrients by Chenin blanc in sand culture. I. Nitrogen. *S. Afr. J. Enol. Vitic.* 1:59-65.

Conradie, W. J. (1981). Seasonal uptake of nutrients by Chenin blanc in sand culture. II. Phosphorus, calcium, and magnesium. *S. Afr. J. Enol. Vitic.* 2:7-13.

Cripps, J. E. L. (1987). Response of apple trees to soil applications of phosphorus, nitrogen and potassium. *Aust. J. Exp. Agric.* 27:909-14.

Darbyshire, R., McClymont, L. and Goodwin, I. (2015). Sun damage risk of Royal Gala apple in fruit-growing districts in Australia. *N. Z. J. Crop Hortic. Sci.* 43:222-32.

Dong, S., Neilsen, D., Neilsen, G. H. and Fuchigami, L. H. (2005). Foliar N application reduces soil NO_3-N leaching loss in apple orchards. *Plant Soil* 268:357-66.

Dragoni, D., Lakso, A. N. and Piccioni, R. M. (2009). Transpiration of apple trees in a humid climate using heat pulse sap flow gauges calibrated with whole-canopy gas exchange chambers. *Agric. For. Meteorol.* 130:85–94.

Drazeta, L., Lang, A., Hall A. J., Volz, R. K. and Jameson, P. E. (2004). Causes and effects of changes in xylem functionality in apple fruit. *Ann. Bot.* 93:275–82.

Du, S., Kang, S., Li, F. and Du, T. (2017). Water use efficiency is improved by alternate partial root-zone irrigation of apple in arid northwest China. *Agric. Water Manag.* 179:184–92.

Eissenstat, D. M., Wells, C. E. and Wang, L. (2001). Root efficiency and mineral nutrition in apple. *Acta Hortic.* 564:165–84.

Evans, R. G. and Van der Gulik, T. W. (2011). Chapter 29: Irrigation for microclimate control. In: Stetson, L. and Dedrick, A. (Eds), *Irrigation*. 6th edition, The Irrigation Association, Falls Church, VA, pp. 1015–36.

Fallahi, E., Neilsen, D., Neilsen, G. H., Fallahi, B. and Shafii, B. (2010a). Efficient irrigation for optimum fruit quality and yield in apples. *Hortic. Sci.* 45:1616–25.

Fallahi, E., Fallahi, B., Neilsen, G. H., Neilsen, D. and Peryea, F. J. (2010b). Effect of mineral nutrition on fruit quality and nutritional disorders in apples. *Acta Hortic.* 868:49–59.

Fentabil, M. M., Nichol C. F., Jones M. D., Neilsen, G. H., Neilsen, D. and Hannam, K. D. (2016). Effect of drip irrigation frequency, nitrogen rate and mulching on nitrous oxide emissions in a semi-arid climate: An assessment across two years in an apple orchard. *Agric. Ecosyst. Environ.* 235:242–52.

Fereres, E., Goldhamer, D. A. and Parsons, L. R. (2003). Irrigation water management of horticultural crops. *HortSci.* 38:1036–42.

Fernández, J. E. and Cuevas, M. V. (2010). Irrigation scheduling from stem diameter variations: A review. *Agric. For. Meteorol.* 150:135–51.

Ford, E. M., White, G. C. and Allen, M. (1965). The response of magnesium-deficient Edward VII apple trees to variations in the timing and composition of foliar sprays. *J. Hortic. Sci.* 40:351–60.

Francaviglia, D., Farina, V., Avellone, G. and Lo Bianco, R. (2013). Fruit yield and quality responses of apple cultivars Gala and Fuji to partial rootzone drying under Mediterranean conditions. *J. Agric. Sci.* 151:556–69.

Girona, J., Behboudian, M. H., Mata, M., del Campo, J. and Marsal, J. (2010). Exploring six reduced irrigation options under water shortage for 'Golden Smoothee' apple: Responses of yield components over three years. *Agric. Water Manag.* 98:370–5.

Girona, J., Del Campo, J., Mata, M., Lopez, G. and Marsal, J. (2011). A comparative study of apple and pear tree water consumption measured with two weighing lysimeters. *Irrig. Sci.* 29:55–63.

Goldhamer, D. A. and Fereres, E. (2001). Irrigation scheduling protocols using continuously recorded trunk diameter measurements. *Irrig. Sci.* 20:115–25.

Goldhamer, D. A. and Fereres, E. (2004). Irrigation scheduling of almond trees with trunk diameter sensors. *Irrig. Sci.* 23:11–19.

Gómez-Candón, D., Virlet, N., Labbé, S., Jolivot, A. and Regnard J. L. (2016). Field phenotyping of water stress at tree scale by UAV-sensed imagery: New insights for thermal acquisition and calibration. *Precision Agric.* 17:786–800.

Gómez-Candón, D., Delalande, M., Vincourt, S., Mathieu, V., Crété, X., Labbé, S. and Regnard, J. L. (2017). Contribution of high-resolution multispectral and thermal-infrared airborne imagery to assess the behavior of fruit trees facing water stress: Proof of the concept and first results in an apple variety field trial. *EFITA WCCA 2017*

Conference, Montpellier SupAgro, Montpellier, France, 2nd - 6th July 2017. Paper number: 110. http://CIGRProceedings.org.

Gonzalez-Dugo, V., Zarco-Tejada, P. J., Nicolás, E., Nortes, P. A., Alarcón, J. J., Intrigliolo, D. S. and Fereres, E. (2013). Using high resolution UAV thermal imagery to assess the variability in the water status of five fruit tree species within a commercial orchard. *Precision Agric.* 14:660–78.

Goodwin, I., Whitfield, D. M. and Conner, D. J. (2004). The relationships between peach tree transpiration and effective canopy cover. *Acta Hortic.* 664:283–9.

Goodwin, I., Whitfield, D. M. and Connor, D. J. (2006). Effects of tree size on water use of peach (*Prunus persica* L. Batsch). *Irrig. Sci.* 24:59–68.

Grassi, G., Millard, P., Gioaccchini, P. and Tagliavini, M. (2003). Recycling of nitrogen in the xylem of Prunus avium trees starts when spring remobilization of internal reserves declines. *Tree Phys.* 23:1061–8.

Guak, S., Neilsen, D., Millard, P., Wendler, R. and Neilsen, G. H. (2003). Determining the role of N remobilization for growth of apple by measuring xylem-sap N flux. *J. Exp. Bot.* 54:2121–31.

Hannam, K. D., Kehila, D., Millard, P., Midwood, A., Neilsen, D., Neilsen, G. H., Forge, T. A., Nichol, C. and Jones, M. D. (2016). Bicarbonates in irrigation water contribute to carbonate formation and CO_2 production in orchard soils under drip irrigation. *Geoderma* 266:120–6.

Hanson, E. J. (1991). Sour cherry trees respond to foliar boron applications. *HortSci.* 26: 1142-5.

Haynes, R. J. (1985). Principles of fertilizer use for trickle irrigated crops. *Fert. Res.* 6:235–55.

Haynes, R. J. and Goh, K. M. (1980). Distribution and budget of nutrients in a commercial apple orchard. *Plant Soil* 56:445–57.

Haynes, R .J. and Swift, R .S. (1987). Effect of trickle irrigation with three forms of nitrogen on soil pH, levels of extractable nutrients below the emitter and plant growth. *Plant Soil* 102:211–21.

Hillel, D. (2004). *Introduction to Environmental Soil Physics*. Elsevier Academic Press, New York, NY.

Howell, T. A. (2003). *Irrigation Efficiency. Encyclopedia of Water Science*. Marcel Dekker, Inc., New York, NY. doi:10.1081/E-EWS 120010252.

Howell, T. A. and Meron, M. (2007). Chapter 3: Irrigation scheduling. In: Lamm, F. R., Ayars, J. E. and Nakayama, F. S. (Eds), *Microirrigation for Crop Production*. Elsevier B.V., Amsterdam and London.

Iglesias, I., Salvia, J., Torguet, L. and Montserrat, R. (2005). The evaporative cooling effects of overtree microsprinkler irrigation on 'Mondial Gala' apples. *Sci. Hortic.* 103:267–87.

Intrigliolo, D. S. and Castel, J. R. (2004). Continuous measurement of plant and soil water status for irrigation scheduling in plum. *Irrig. Sci.* 23:93–102.

Intrigliolo, D. S. and Castel, J. R. (2006). Performance of various water stress indicators for prediction of fruit size response to deficit irrigation in plum. *Agric. Water Manag.* 83:173–80.

Intrigliolo, D. S. and Castel, J. R. (2010). Response of plum trees to deficit irrigation under two crop levels: Tree growth, yield and fruit quality. *Irrig. Sci.* 28:525–34.

Jarvis, N. J. (2011). Simple physics-based models of compensatory plant water uptake: Concepts and eco-hydrological consequences. *Hydrol. Earth Syst. Sci.* 15:3431–46.

Johnson, R. S., Rosecrance, R., Weinbaum, S., Andris, H. and Wang, J. (2001). Can we approach complete dependence on foliar-applied urea nitrogen in an early-maturing peach. *J. Am. Soc. Hortic. Sci.* 126:364–70.

Jones, H. G. (2004). Irrigation scheduling: Advantages and pitfalls of plant-based methods. *J. Exp. Bot.* 55:2427–36.

Jones, H. G., Serraj, R., Loveys, B. R., Xiong, L. Z., Wheaton, A. and Price, A. H. (2009). Thermal infrared imaging of crop canopies for the remote diagnosis and quantification of plant responses to water stress in the field. *Funct. Plant Biol.* 36:978–89.

Juhász, A., Sepsi, P., Nagy, Z., Tokei, L. and Hrotkóba, K. (2013). Water consumption of sweet cherry trees estimated by sap flow measurement. *Sci. Hortic.* 164:41-9.

Kalcsits, L., van der Heijden, G., Reid, M. and Mullin, K. (2017). Calcium absorption during fruit development in 'Honeycrisp' apple measured using ^{44}Ca as a stable isotope tracer. *HortSci.* 52:1804–9.

Kang, S. and Zhang, J. (2004). Controlled alternate partial root-zone irrigation: Its physiological consequences and impact on water use efficiency. *J. Exp. Bot.* 55:2437–46.

Kang, S., Hu, X., Goodwin, I. and Jerie, P. (2002). Soil water distribution, water use, and yield response to partial root zone drying under a shallow groundwater table condition in a pear orchard. *Sci. Hortic.* 92:277–91.

Klein, I. and Spieler, G. (1987a). Irrigation of apples with nitrate or ammonium nitrogen under drip irrigation. I, Tree performance. *Commun. Soil Sci. Plant Anal.* 18:311–22.

Klein, I. and Spieler, G. (1987b). Irrigation of apples with nitrate or ammonium nitrogen under drip irrigation. II, Nutrient distribution in the soil. *Commun. Soil Sci. Plant Anal.* 18:323-39.

Klein, I., Levin, I., Bar-Yosef, B., Assaf, R. and Berkovitz, A. (1989). Drip nitrogen fertigation of 'Starking Delicious' apple. *Plant Soil* 119:305–14.

Knoche, M. and Winkler, A. (2017). Rain-induced cracking of sweet cherries. In: Quero-Garcia, J., Iezzoni, A., Pulawska, J. and Lang, G. (Eds), *Cherries, Botany, Production and Uses*. CABI, Wallingford, UK, pp. 140–65.

Lang, G., Guimond, C., Flore, J. A., Southwick, S., Facteau, T., Kappel, F. and Azarenko, A. (1998). Performance of calcium/sprinkler-based strategies to reduce sweet cherry rain-cracking. *Acta Hortic.* 468:649–56.

Leece, D. R. and Dirou, J. F. (1979). Comparison of urea foliar sprays containing hydrocarbon or silicone surfactants with soil-applied nitrogen in maintaining the leaf nitrogen concentration of prune trees. *J. Am. Soc. Hortic. Sci.* 104:644–8.

Livellaria, N., Saavedra, E. and Salgado, E. (2011). Plant based indicators for irrigation scheduling in young cherry trees. *Agric. Water Manag.* 98:684–90.

Lo Bianco, R. and Francaviglia, D. (2012). Comparative responses of 'Gala' and 'Fuji' apple trees to deficit irrigation: Placement versus volume effects. *Plant Soil* 357:41–58.

Loneragan, J. F., Grove, T. S., Robson, A. D. and Snowball, K. (1979). Phosphorus toxicity as a factor in zinc-phosphorus interactions in plants. *Soil Sci. Soc. Am. J.* 43: 966–72.

Long, L. E., Yin, X., Huang, X. L. and Jaja, N. (2014). Responses of sweet cherry water use and productivity and soil quality to alternate groundcover and irrigation systems. *Acta Hortic.* 1020:131–8.

Looney, N. E. (1977). A four-year study of single calcium chloride and growth regulator tree sprays to control storage breakdown of 'Spartan' apples. *J. Am. Soc. Hortic. Sci.* 102:85–8.

Lopez, G., Mata, M., Arbones, A., Solans, J. R., Girona, J and Marsal, J. (2006). Mitigation of effects of extreme drought during stage III of peach fruit development by summer pruning and fruit thinning. *Tree Physiol.* 26:469–77.

Lopez, G., Behboudian, M. H., Girona, J. and Marsal, J. (2012). Chapter 17: Drought in deciduous fruit trees: Implications for yield and fruit quality. In: Aroca, R. (Ed.), *Plant Responses to Drought Stress*. Springer-Verlag, Berlin Heidelberg.

Lotze, E., Joubert, J. and Theron, K. I. (2008). Evaluating pre-harvest foliar calcium applications to increase fruit calcium and reduce bitter pit in 'Golden Delicious' apples. *Sci. Hortic.* 116:299–304.

Majone, B., Viani, F., Filippi, E., Bellin, A., Massa, A., Toller, G., Robol, F. and Salucci, M. (2013). Wireless sensor network deployment for monitoring soil moisture dynamics at the field scale. *Proc. Environ. Sci.* 19:426–35.

Marsal, J., Lopez, G., Mata, M. and Girona, J. (2006). Branch removal and defruiting for the amelioration of water stress effects on fruit growth during Stage III of peach fruit development. *Sci. Hort.* 108:55–60.

Marsal, J., Mata, M., Arbones, A., del Campo, J., Girona, J. and Lopez, G. (2008). Factors involved in alleviating water stress by partial crop removal in pear trees. *Tree Physiol.* 28:1375–82.

Marsal, J., Lopez, G., Arbones, A., Mata, M., Vallverdu, X. and Girona, J. (2009). Influence of post-harvest deficit irrigation and pre-harvest fruit thinning on sweet cherry (cv. New Star) fruit firmness and quality. *J. Hortic. Sci. Biotech.* 84:273–8.

Marsal, J., Lopez, G., del Campo, J., Mata, M., Arbones, A. and Girona, J. (2010). Postharvest regulated deficit irrigation in 'Summit' sweet cherry: Fruit yield and quality in the following season. *Irrig. Sci.* 28:181-9.

Marsal, J., Lopez, G., Mata, M. and Girona, J. (2012). Postharvest deficit irrigation in 'Conference' pear: Effects on subsequent yield and fruit quality. *Agric. Water Manag.* 103:1-7.

Marsal, J., Casadesus, J., Lopez, G., Mata, M., Bellvert, J. and Girona, J. (2016). Sustainability of regulated deficit irrigation in a mid-maturing peach cultivar. *Irrig. Sci* 34: 201–8.

Mason, J. L. (1979). Increasing calcium content of calcium-sensitive tissues. *Commun. Soil Sci. Plant Anal.* 10:349–71.

Meheriuk, M., Neilsen, G. H. and McKenzie, D.-L. (1991). Incidence of rain splitting in sweet cherries treated with calcium or coating materials. *Can. J. Plant Sci.* 71:231–4.

Meheriuk, M., McKenzie, D.-L., Neilsen, G. H. and Hall, J. (1996). Fruit pigmentation of four green apple cultivars responds to urea sprays but not to nitrogen fertilization. *HortSci.* 31:992–3.

Millard, P. and Neilsen, G. H. (1989). The influence of nitrogen supply on the uptake and mobilization of stored N for the seasonal growth of apple trees. *Ann. Bot.* 63:301-9.

Millard, P., Wendler, R., Grassi, G., Grelet, G-A. and Tagliavini, M. (2006). Translocation of nitrogen in the xylem of field-grown cherry and poplar trees during remobilization. *Tree Physiol.* 26:527–36.

Naor, A., Gal, Y. and Peres, M. (2006). The inherent variability of a few water stress indicators in apple, nectarine and pear orchards, and the validity of a commercial leaf-selection procedure for water potential measurements. *Irrig. Sci.* 24:129–35.

Naor, A., Naschitz, S., Peres, M. and Gal, Y. (2008). Responses of apple fruit size to tree water status and crop load. *Tree Physiol.* 28:1255–61.

Neilsen, G. H. and Hoyt, P. B. (1984). Field comparison of chelated and Epsom salt magnesium foliar sprays on apple trees. *HortSci.* 19: 431–2.

Neilsen, G. H. and Hoyt, P. B. (1990). A comparison of methods to raise zinc concentrations of apple leaves. *Can. J. Plant Sci.* 90:599-603.

Neilsen, G. H. and Neilsen, D. (2002). Effects of foliar Zn, form and timing of Ca sprays on fruit Ca concentration in new apple cultivars. *Acta Hortic.* 594:435–43.

Neilsen, G. H. and Neilsen, D. (2003). Nutritional requirements for apple. In: Ferree, D. and Warrington, I. J. (Eds), *Apples, Botany, Production and Uses*. CABI publishing, Wallingford, UK, pp. 267-302.

Neilsen, G. H. and Neilsen, D. (2006). Response of high density apple orchards on coarse-textured soil to form of potassium applied by fertigation. *Can. J. Soil Sci.* 86:749–55.

Neilsen, G. H. and Yorston, J. (1991). Soil disinfection and monoammonium phosphate fertilization increase precocity of apples on replant problem soils. *J. Am. Soc. Hortic. Sci.* 116:651–4.

Neilsen, G. H., Hogue, E. J. and Parchomchuk, P. (1990). Flowering of apple trees in the second year is increased by first-year P fertilization. *HortSci.* 25:1247–50.

Neilsen, D., Parchomchuk, P. and Hogue, E. J. (1993a). Soil and peach seedling responses to soluble phosphorus applied in single or multiple doses. *Commun. Soil Sci. Plant Anal.* 24:881–98.

Neilsen, G. H., Parchomchuk, P., Wolk, W. D. and Lau, O. L. (1993b). Growth and mineral composition of newly planted apple trees following fertigation with N and P. *J. Am. Soc. Hortic. Sci.* 118:50–3.

Neilsen, G. H., Parchomchuk, P., Neilsen, D., Berard, R. and Hogue, E. J. (1995a). Leaf nutrition and soil nutrients are affected by irrigation frequency and method for NP-fertigated 'Gala' apple. *J. Am. Soc. Hortic. Sci.* 120: 971–6.

Neilsen, D., Hoyt, P. B., Parchomchuk, P., Neilsen, G. H. and Hogue, E. J. (1995b). Measurement of the sensitivity of orchard soils to acidification. *Can. J. Soil Sci.* 75:391–5.

Neilsen, G. H., Parchomchuk, P., Berard, R. and Neilsen, D. (1997a). Irrigation frequency and quantity affect root and top growth of fertigated 'McIntosh' apple on M9, M26 and M7 rootstock. *Can. J. Plant Sci.* 77:133-9.

Neilsen, G. H., Parchomchuk, P. and Neilsen, D. (1997b). Distribution of soil P and K as affected by NP-fertigation in high density apple orchards. *Acta Hortic.* 448:439–47.

Neilsen, D., Parchomchuk, P., Neilsen, G. H. and Hogue, E. J. (1998a). Use of soil solution monitoring to determine the effects of irrigation management and fertigation on nitrogen availability in high-density apples. *J. Am. Soc. Hortic. Sci.* 123:706–13.

Neilsen, G., Parchomchuk, P., Meheriuk, M. and Neilsen, D. (1998b). Development and correction of K-deficiency in drip-irrigated apple. *HortSci.* 33:258–61.

Neilsen, G. H., Neilsen, D. and Peryea, F. (1999). Response of soil and irrigated fruit trees to fertigation or broadcast application of nitrogen, phosphorus, and potassium. *HortTech.* 9:393–401.

Neilsen, G. H., Parchomchuk, P., Neilsen, D. and Zebarth, B. J. (2000). Drip fertigation of apple trees affects root distribution and development of K deficiency. *Can. J. Soil Sci.* 80:353–61.

Neilsen, D., Millard, P., Herbert, L. C., Neilsen, G. H., Hogue, E. J., Parchomchuk, P. and Zebarth, B. J. (2001). Remobilization and uptake of N by newly planted apple (*Malus domestica*) trees in response to irrigation method and timing of N application. *Tree Physiol.* 21:513–21.

Neilsen, G., Kappel, F. and Neilsen, D. (2004a). Fertigation method affects performance of 'Lapins' sweet cherry on Gisela 5 rootstock. *HortSci.* 39:1716–21.

Neilsen, G. H., Neilsen, D., Herbert, L. C. and Hogue, E. J. (2004b). Response of apple to fertigation of N and K under conditions susceptible to the development of K deficiency. *J. Am. Soc. Hortic. Sci.* 129: 26–31.

Neilsen, G. H., Neilsen, D., Hogue, E. J. and Herbert, L. C. (2004c). Zinc and boron nutrition management in fertigated high density apple orchards. *Can. J. Plant. Sci.* 84:823–8.

Neilsen, G., Neilsen, D., Dong, S. and Toivonen, P. (2005a) Application of $CaCl_2$ sprays earlier in the season may reduce bitter pit incidence in 'Braeburn' apples. *HortSci.* 40:1850–3.

Neilsen, G. H., Hogue, E. J., Neilsen, D. and Bowen, P. (2005b). Postbloom humic- and fulvic-based zinc sprays can improve apple zinc nutrition. *HortSci.* 40:205–8.

Neilsen, D., Neilsen, G. H., Herbert, L., Millard, P. and Guak, S. (2006). Allocation of dry matter and N to fruit and shoots in dwarf apple in response to sink size and N availability. *Acta Hort.* 721:33–40.

Neilsen, G., Kappel, F. and Neilsen. D. (2007). Fertigation and crop load affect yield, nutrition and fruit quality of 'Lapins' sweet cherry on Gisela 5 rootstock. *HortSci.* 42:1456–62.

Neilsen, D., Neilsen, G. H., Gregory, D., Forge, T. and Zebarth, B. (2008a). Drainage losses of water, N and P from micro-irrigation systems in a young, high density apple planting. *Acta Hortic.* 792:483–90.

Neilsen, G. H., Neilsen, D., Toivonen, P. and Herbert. L. (2008b). Annual bloom-time phosphorus fertigation affects soil phosphorus, apple tree phosphorus nutrition, yield and fruit quality. *HortSci.* 43:885–90.

Neilsen, G. H., Neilsen, D. and Herbert, L. (2009). Nitrogen fertigation concentration and timing of application affect nitrogen nutrition, yield, firmness and color of apples grown at high density. *J. Am. Soc. Hortic. Sci.* 44:1425–31.

Neilsen, G. H., Neilsen, D., Kappel, F., Toivonen, P. and Herbert, L. (2010). Factors affecting establishment of sweet cherry on Gisela 6 rootstock. *HortSci.* 45:939–45.

Neilsen, G. H., Neilsen, D., Kappel, F. and Forge, T. (2014). Interaction of irrigation and soil management on sweet cherry productivity and fruit quality at different crop loads that simulate those occurring by environmental extremes. *HortSci.* 49:215–20.

Neilsen, G. H., Neilsen, D., Guak, S. and Forge, T. (2015). The effect of deficit irrigation and crop load on leaf and fruit nutrition of fertigated 'Ambrosia'/'M.9' apple. *HortSci.* 50:1387–93.

Neilsen, D., Neilsen, G. H., Guak, S. and Forge, T. (2016a). Consequences of deficit irrigation and crop load reduction on plant water relations, yield and quality of 'Ambrosia' apple. *HortSci.* 51:98–106.

Neilsen, D., Neilsen, G. H., Forge, T. and Lang, G. (2016b). Dwarfing rootstocks and training systems affect initial growth, cropping and nutrition in 'Skeena' sweet cherry. *Acta Hortic.* 1130:199–206.

Neilsen, G. H., Neilsen, D., Forge, T. and Hannam, K. (2017a). Advances in soil and nutrient management in apple cultivation. In: Evans, K. (Ed.), *Achieving Sustainable Cultivation of Apples*. Burleigh Dodds Science Publishing, Cambridge, UK, pp. 239–78.

Neilsen, G. H., Neilsen, D. and Forge, T. (2017b). Environmental limiting factors for cherry production. In: Quero-Garcia, J., Iezzoni, A., Pulawska, J. and Lang, G. (Eds), *Cherries, Botany, Production and Uses*. CABI, Wallingford, UK, pp. 189–222.

Norton, R. A. and Childers, N. F. (1954). Experiments with urea sprays on the peach. *Proc Am. Soc. Hortic. Sci.* 63:23–31.

Odi-Lara, M., Campos, I., Neale, C. M. U., Ortega-Farias, S., Poblete-Echeverria, C., Balbontin, C. and Calera, A. (2016). Estimating evapotranspiration of an apple orchard using a remote sensing-based soil water balance. *Remote Sens.* 8:253–73.

Oland, K. (1959). Nitrogenous reserves of apple trees. *Physiol. Plant.* 12:594–648.

O'Neil, M. K., Gardner, B. R. and Roth, R. L. (1979). Orthophosphoric acid as a phosphorus fertilizer in trickle irrigation. *Soil Sci. Soc. Am. J.* 43:283–6.

Ortuño, M. F., Conejero, W., Moreno F., Moriana, A., Intrigliolo, D. S., Biel, C., Mellisho, C. D., Pérez-Pastor, A., Domingo, R., Ruiz-Sánchez, M. C., Casadesus, J., Bonany, J. and Torrecillas, A. (2010). Could trunk diameter sensors be used in woody crops for irrigation scheduling? A review of current knowledge and future perspectives. *Agric. Water Manag.* 97:1–11.

O'Shaughnessy, S. A. and Evett, S. R. (2010). Canopy temperature based system effectively schedules and controls center pivot irrigation of cotton. *Agric. Water Manag.* 97:1310–16.

Osroosh, Y., Peters, R. T., Campbell, C. S. and Zhang, Q. (2016). Comparison of irrigation automation algorithms for drip-irrigated apple trees. *Comput. Electron. Agric.* 128:87–99.

Ouzounis, T. and Lang, G. A. (2011) Foliar applications of urea affect nitrogen reserves and cold acclimation of sweet cherries (*Prunus avium* L.) on dwarfing rootstocks. *HortSci.* 46:1015–21.

Oyarzún, R., Stöckle, C. and Whiting, M. (2010). Analysis of hydraulic conductance components in field grown, mature sweet cherry trees. *Chil. J. Agric. Res.* 70:58–66.

Paço, T. A., Ferreira, M. I. and Conceição, N. (2006). Peach orchard evapotranspiration in a sandy soil: Comparison between eddy covariance measurements and estimates by the FAO 56 approach. *Agric. Water Manag.* 85:305–13.

Paço, T. A., Ferreira, M. I., Rosa, R. D., Paredes, P., Rodrigues, G. C., Conceição, N., Pacheco, C. A. and Pereira, L. S. (2012). The dual crop coefficient approach using a density factor to simulate the evapotranspiration of a peach orchard: SIMDualKc model vs. eddy covariance measurements. *Irrig. Sci.* 30:115–26.

Pagay, V. (2014). Physiological responses of grapevine shoots to water stress and the development of a microtensiometer to continuously measure water potential. PhD dissertation, Cornell University.

Parchomchuk, P. and Meheriuk, M. (1996). Orchard cooling with pulsed overtree irrigation to prevent solar injury and improve fruit quality of 'Jonagold' apples. *HortSci.* 31:802–4.

Parchomchuk, P., Berard, R. C. and Van der Gulik, T. W. (1996). Automatic irrigation scheduling using an electronic atmometer. In: Camp, C. R., Sadler, E. J. and Yoder, R. E. (Eds), *Evapotranspiration and Irrigation Scheduling*. CABI, Wallingford, UK, pp. 1099–104.

Pereira, L. S., Allen, R. G., Smith, M. and Raes, D. (2015). Crop evapotranspiration with FAO 56: Past and future. *Agric. Water Manag.* 147:4–20.

Pérez-Sarmiento, F., Alcobendas, R., Mounzer, O., Alarcón, J. and Nicolás, E. (2010). Effects of regulated deficit irrigation on physiology and fruit quality in apricot trees. *Span. J. Agric. Res.* 8:86–94.

Perry, K. B. (1998). Basics of frost and freeze protection for horticultural crops. *HortTech.* 8:10–15.

Peryea, F. J. (2006). Phytoavailability of zinc in postbloom zinc sprays applied to 'Golden Delicious' apple trees. *HortTech.* 16:60–5.

Peryea, F. J. and Neilsen, G. H. (2013). Influence of 6-benzyladenine tank mixed with boron, followed by mid-summer calcium plus copper sprays, on 'Gala' apple splitting and firmness. *Acta Hortic*. 984:393–9.

Peryea, F. J., Neilsen, D. and Neilsen, G. (2003). Boron maintenance sprays for apple: Early-season applications and tank-mixing with calcium chloride. *HortSci*. 38:542–6.

Peryea, F. J., Neilsen, G. H. and Faubion, D. (2007). Start-timing for calcium chloride spray programs influences fruit Ca and bitter pit in 'Braeburn' and 'Honeycrisp' apples. *J. Plant Nutr.* 30:1213–27.

Postel, S. L. (2000). Entering an era of water scarcity: The challenges ahead. *Ecol. Appl.* 10:941–8.

Proebsting Jr., E. L., Middleton, J. E. and Mahon, M. O. (1981). Performance of bearing cherry and prune trees under very low irrigation rates. *J. Am. Soc. Hortic. Sci.* 106:243–346.

Racsko, J. and Schrader, L. E. (2012). Sunburn of apple fruit: Historical background, recent advances and future perspectives. *Crit. Rev. Plant Sci.* 31:455–504.

Raese, J. T. and Drake, S. (2000). Effect of calcium spray materials, rate, time of spray application, and rootstocks on fruit quality of 'Red' and 'Golden Delicious' apples. *J. Plant Nutr.* 23:1435–47.

Romero, R., Muriel, J. L., García, I. and Munoz de la Pena, D. (2012). Research on automatic irrigation control: State of the art and recent results. *Agric. Water Manag.* 114:59–66.

Rosecrance, R. C., Johnson, R. S. and Weinbaum, S. A. (1998a). Foliar uptake of urea-N by nectarine leaves: A reassessment. *HortSci*. 33(1):158.

Rosecrance, R. C., Johnson, R. S. and Weinbaum, S. A. (1998b). The effect of timing of post-harvest foliar urea sprays on nitrogen absorption and partitioning in peach and nectarine trees. *J. Hort. Sci. and Biotech.* 73:856–61.

Ross, G. J., Hoyt, P. B. and Neilsen, G. H. (1985). Soil chemical and mineralogical changes due to acidification in Okanagan apple orchards. *Can. J. Soil Sci.* 65:347–55.

Ruiz-Sánchez, M. C., Abrisqueta, I., Conejero, W. and Vera, J. (2018). Chapter 6: Deficit irrigation management in early-maturing peach crop. In: García Tejero, I. F. and Zuazo, V. H. D. (Eds), *Water Scarcity and Sustainable Agriculture in Semiarid Environment*. Academic Press.

Saa, S., Negron, C. and Brown, P. (2018). Foliar zinc applications in Prunus: From lab experience to orchard management. *Sci. Hort.* 233:233–7.

Sadowski, A., Jadczuk, E., Pekacki, J. and Scibisz, K. (1976). Magnesium nutrition of apple trees I Effect of K and Mg fertilization and comparison of soil and foliar treatments with magnesium sulphate. *Acta Agrobot.* XXIX:179–99.

Saure, M. C. (2005). Calcium translocation to fleshy fruit: Its mechanism and control. *Sci. Hort.* 105:65–89.

Scholander, P. F., Bradstreet, E. D., Hemmingsen, E. A. and Hammel, H. T. 1965. Sap pressure in vascular plants: Negative hydrostatic pressure can be measured in plants. *Science* 148:339–46.

Shackel, K. A., Ahmadi, H., Biasi, W., Buchner, R., Goldhamer, D., Gurusinghe, S., Hasey, J., Kester, D., Krueger, B. and Lampinen, B. (1997). Plant water status as an index of irrigation need in deciduous fruit trees. *HortTech*. 7:23–9.

Shear, C. B. (1975). Calcium-related disorders of fruit and vegetables. *HortSci*. 10:361–6.

Southwick, S. M., Rupert, M. E., Yeager, J. T., Lampinen, B. D., Dejong, T. M. and Weis, K. G. (1999). Effects of nitrogen fertigation on fruit yield and quality of young 'French' prune trees. *J. Hortic. Sci. Biotech.* 74:187–95.

Spinelli, G. M., Shackel, K. A. and Gilbert, M. E. (2017). A model exploring whether the coupled effects of plant water supply and demand affect the interpretation of water potentials and irrigation management. *Agric. Water Manag.* 192:271–80.

Stover, E., Fargione, M., Risio, R., Stiles, W. and Lungerman, K. (1999). Prebloom foliar boron, zinc, and urea applications enhance cropping of some 'Empire' and 'McIntosh' apple orchards in New York. *HortSci*. 34:210–14.

Swietlik, D. (1999). Zinc nutrition in horticultural crops. *Hort. Rev.* 23:109–78.

Swietlik, D. and Faust, M. (1984). Foliar nutrition of fruit crops. *Hort. Rev.* 6:287–355.

Tagliavini, M., Millard, P. and Quartieri, M. (1998). Storage of foliar-absorbed nitrogen and remobilization for spring growth in young nectarine *(Prunus persica var. nectarina)* trees. *Tree Physiol*. 18: 203–7.

Tagliavini, M., Abadia, J., Rombola, A. D., Abadia, A., Tsipouridis, C. and Marangoni, B. (2000). Agronomic means for the control of iron deficiency chlorosis in deciduous fruit trees. *J. Plant Nutr.* 23:2007–22.

Titus, J. S. and Kang, S. (1982). Nitrogen metabolism, translocation and recycling in apple trees. *Hort. Rev.* 4:204–46.

Treeby, M. (2008). Manipulating grapevine annual shoot growth, yield and composition of grapes using fertigation. In: Imus, P. and Price, M. R. (Eds), *Fertigation: Optimizing the Utilization of Water and Nutrition*. International Potash Institute, Horgen, Switzerland, pp. 89–102.

Tvergyak, P. J. and Richardson, D. G. (1979). Diurnal changes of leaf and fruit water potentials of sweet cherries during the harvest period. *HortSci*. 14:520–1.

Uriu, K., Carlson, R. W., Henderson, D. W. and Shulbach, H. (1980). Potassium fertilization of prune trees under drip irrigation. *J. Am. Soc. Hort. Sci.* 105:508–10.

Van der Gulik, T. W. (1993). *Chemigation Guidelines for British Columbia BC Ministry of Agriculture, Fisheries and Food*. Soils and Engineering Branch, Abbotsford, BC.

Van der Gulik, T. W. (1999). *B.C. Trickle Irrigation Manual. British Columbia Ministry of Agriculture and Food.* Resource Management Branch, Abbotsford, BC.

Watson, T. T. (2018). Non-fumigant alternative soil management practices for mitigating replant disease of fruit trees: Mechanisms contributing to *Pratylenchus penetrans* suppression. PhD Thesis, University of British Columbia.

Vang-Petersen, O. (1980). Calcium nutrition of apple trees: A review. *Sci. Hort.* 12:1-9.

Vargas, O. L. and Bryla, D. R. (2015). Growth and fruit production of highbush blueberry fertilized with ammonium sulfate and urea applied by fertigation or as a granular fertilizer. *HortSci*. 50:479–85.

Vargas, O. L., Bryla, D. R., Weiland, J. R., Strik, B. C. and Sun, L. (2015). Irrigation and fertigation with drip and alternative micro irrigation systems in northern highbush blueberry. *HortSci*. 50:897–903.

Webster, D. H. (1973). Absorption of magnesium by McIntosh apple leaves as influenced by spray composition and weather. *Can. J. Plant Sci.* 53: 579–84.

Webster, D. H. and Lidster, P. D. (1986). Effects of phosphate sprays on McIntosh apple fruit and leaf composition, flesh firmness and susceptibility to low-temperature disorders. *Can. J. Plant Sci.* 66: 617–26.

Weinbaum, S. A. (1988). Foliar nutrition of fruit crops. In: Newman, P. E. (Ed.), *Plant Growth and Leaf-Applied Chemicals*. CRC Press, Boca Raton, FL, pp. 81–100.

Weinbaum, S. A., Brown, P. H., Rosecrance, R .C., Picchioni, G. A., Niederholzer, F. J. A., Youseffi, F. and Muraoka, T. T. (2001). Necessity for whole tree excavations in determining patterns and magnitude of macronutrient uptake by mature deciduous fruit trees. *Acta Hort.* 564:41–9.

Wojcik, P. and Morgas, H. (2015). Impact of postharvest sprays of nitrogen, boron and zinc on nutrition, reproductive response and fruit quality of 'Shattenmorelle' tart cherries. *J. Plant Nutr.* 38:1456–68.

Wojcik, P. and Wojcik, M. (2006). Effect of boron fertilization on sweet cherry tree yield and fruit quality. *J. Plant Nutr.* 29:1755–66.

Wolk, W. D., Lau, O. L., Neilsen, G. H. and Drought, B. G. (1998). Factors and time of sample collection for correlating storage potential of 'McIntosh', 'Spartan', and 'Golden Delicious' apples. *J. Am. Soc. Hort. Sci.* 123:104–9.

Woodbridge, C. G. (1954). Zinc deficiency in fruit trees in the Okanagan valley in British Columbia. *Can. J. Agr. Sci.* 34:545–51.

Yin, X., Long, L. E., Huang, X.-L., Jaja, N., Bai, J., Seavet, C. F. and le Roux, J. (2012). Transitional effects of double-lateral drip irrigation and straw mulch on irrigation water consumption, mineral nutrition, yield, and storability of sweet cherry. *HortTech.* 22:484–92.

Yogaratnam, N. and Greenham, D. W. R. (1982). Supplementing the nutrition of Bramley's seedling apple with phosphorus sprays. I. Effects on leaf composition. *J. Hort. Sci.* 57:45–52.

Yogaratnam, N. and Sharples, R. O. (1982). Supplementing the nutrition of Bramley's seedling apple with phosphorus sprays. II. Effects on fruit composition and storage quality. *J. Hort. Sci.* 57:53–9.

Yuri, J. A., Retamales, J. B., Mogia, C. and Vasquez, J. L. (2002). Bitter pit in apples cv Braeburn through foliar sprays of different calcium sources. *Acta Hort.* 594:453–60.

Zegbe, J. A., Behboudian, M. H. and Lang, A. (2016). Partial rootzone drying advances fruit maturity of Royal Gala apple. *Rev. Fitotec. Mex.* 39:187–92.

Zornoza, R., Acosta, J. A., Martínez-Martínez, S., De la Rosa, J. M., Faz, A. and Pérez-Pastor A. (2016). Regulated deficit irrigation can decrease soil CO_2 emissions in fruit orchards. Geophysical Research Abstracts Vol. 18, EGU2016-9099-1, 2016 EGU General Assembly 2016.

Chapter 8

Optimizing plant growth, yield and fruit quality with plant bioregulators

Duane Greene, University of Massachusetts, USA

1 Introduction

Plant bioregulators (PBRs) influence many processes in a plant including shoot growth, branch angle, fruit abscission, fruit ripening, fruit shape and fruit finish, bud break and flower bud formation. They function in this capacity as signal givers that lead to gene expression, gene activation or repression. One could view PBRs as the first domino in a series of events that ultimately results in a physiological response. They don't play a direct role or act as a component in a chemical reaction. Rather, they initiate responses to signal gene expression, activation or downregulation and these actually bring about the responses that we attribute to PBR control of processes within the plant.

PBRs are a group of organic compounds that, in small amounts, promote, inhibit or otherwise affect biological and biochemical processes in plants. There are three subcategories of compounds that are generally considered PBRs. First, there are endogenous plant hormones that are naturally occurring, synthesized by the plant. They are intimately involved in the regulations and control of all physiological processes important for regulating growth and development in a plant. There are also endogenous compounds, produced by some, but not all plants, that can regulate specific physiological functions.

http://dx.doi.org/10.19103/AS.2018.0040.09

These compounds are not found in all plants and have limited influence on growth and development. PBRs in this category may influence a plant even if they are not produced by that plant. Aminoethoxyvinylglycine (AVG) is an important PBR that falls into this category. There is a third group of organic compounds that are not produced by the plants, but they are active in regulating some function(s) within a plant. They are frequently referred to as synthetic PBRs. All plant hormones are considered PBRs but not all PBRs can be considered as plant hormones.

All aspects of fruit production have undergone many changes in the past 50 years. Among the components that have undergone the greatest change are development and use of new rootstocks, change in cultivars grown, pest management strategies, tree training and pruning systems. A prerequisite for all successful fruit growers is to have the ability to understand and embrace the appropriate innovations that give growers the best chance for success. Even if a grower does adopt all of these innovative practices, growers will not be able to achieve the greatest yield, highest fruit quality, the best pack-out and the highest return for harvested fruit if PBRs are not part of and incorporated into their production and management system. PBRs are unlikely to measurably improve the production of high-quality fruit unless sound management practices are applied to a well-designed orchard.

PBRs are used more extensively in tree fruit production than in any other horticultural or agricultural commodity, and they are essential for effective and profitable production of high-value horticultural crops. They are expensive so they must add significant value to the crop just so that a grower can retrieve the cost of material. The increased value is usually not measured in higher yield but rather in increased value of the harvested crop. This enhancement of value is often manifest in larger fruit size, improved red colour, improved fruit appearance or enhanced fruit quality at harvest or following storage.

The first group of plant hormones to be discovered was the auxins. The structure of indole-3-acetic acid (IAA) was confirmed in the 1930s and similar synthetic compounds that had biological activity that could significantly improve and aid the production of tree fruit were identified. Over the next several decades new PBRs were identified and they became commercially important. There was a lack of consistency, especially with the chemical thinning PBRs that were characterized by an inconsistent response even when used under very similar conditions. In the 1980s more effort was focussed on determining the mode of action of many of these growth regulating compounds. As a result, it was possible to more precisely and consistently use these compounds because of the greater understanding of how these PBRs actually participated in specific physiological events. A better understanding of how PBRs work allows more consistent and targeted use of these compounds in the orchard.

PBRs generally require approval and registration for use, and not all are universally available in every country. Although they may require governmental approval for use they should not be considered as pesticides. In some cases PBRs do serve a dual function as PBR and pesticide or at least function to reduce the amount of pesticide that may be required. For example, prohexadione-calcium (Pro-Ca) is a compound that was registered to inhibit vegetative growth on some tree fruit. It limits vegetative growth by inhibiting gibberellin biosynthesis. In the process of growth retardation it alters the structure of the growing shoot which in turn makes the stem less susceptible to invasion of the highly devastating fire blight disease (*Erwinia amylovora*). It functions as one of the most effective methods to control shoot blight rather than having to depend on antibiotic spray, where resistance can be developed if excessively used. Growth retardants can reduce shoot growth in a tree leading to less interior shading and faster drying, thus reducing disease pressure. Reduced shoot growth also allows for better penetration of pesticide sprays into the centre of the tree and reduced use of pesticides.

In this chapter the most important commercially used PBRs will be discussed. Further, how they are used to influence many important orchard activities and how they contribute to and play an important role in the sustainable production of apples will be elaborated.

2 Classification of PBRs

The growth of plants is directed and synchronized by plant hormones (Davies, 1995). These molecules are signal givers that essentially regulate every function of a plant. Not all plant cells respond to hormones and often they respond at only specific stages in the life of a cell. In the absence of plant hormones, a cell would probably remain undifferentiated. Plants can regulate their own synthesis, longevity and destruction as well as affect levels of their hormones in a plant. There are five major classes of hormones that are universally recognized as plant hormones.

2.1 Endogenous plant hormones

2.1.1 Auxins

The auxins were the first group of hormones to be discovered. Since early investigators were unaware of the existence of compounds within a plant that could direct growth and development, the discovery of the first hormone, indole-3-acetic acid (IAA), was a long process that spanned nearly 50 years (Jacobs, 1979). The structure of IAA was identified in the 1930s by Japanese investigators and was recognized as the dominant auxin present within plants. Other auxins were soon identified that were not naturally occurring including naphthaleneacetic acid (NAA) and indole-3-butyric acid (IBA). Research soon revealed that auxins played an important role in many plant processes including apical dominance, phototropism, fruit growth, fruit set, root initiation, fruit ripening and leaf abscission. Synthetic auxins, not the naturally occurring IAA, were more active in influencing physiological process and thus they were adopted and developed for commercial use.

2.1.2 Gibberellins

Unlike auxins which were dominated primarily by one compound, the gibberellins include a very large group that numbers in the dozens. Although there are many endogenous gibberellins identified, within a specific type of plant (e.g. apple) there are just a few that are dominant in the gibberellin pool (Cleland, 1969). For example, in apple, gibberellins A_4 and A_7 are present in the highest concentration (Phinney, 1983). The physiological activity of gibberellins differs widely. A gibberellin may be highly active in one type of plant yet inactive in another plant. Gibberellins are recognized for being very important in regulating terminal growth in plants. However, stimulation of terminal growth is not one of the commercial uses of gibberellins. More frequently, plants are treated with plant bioregulators that specifically inhibit gibberellin biosynthesis. Paclobutrazol, uniconazole and prohexadione-calcium (Pro-Ca) are examples of gibberellin biosynthesis inhibitors used to reduce plant growth. Each differ in the point along the gibberellin biosynthetic pathway where they inhibit the synthesis. Gibberellins can inhibit flowering in trees. In pome fruit, gibberellins produced by the seeds diffuse out of the seeds, migrate to growing points where flower bud formation occurs and inhibit the meristem from developing into a flower cluster. Gibberellins are an important hormone group that facilitate and help direct seed germination.

2.1.3 Cytokinins

In the 1940s studies were initiated to identify compound(s) that were responsible for increased growth of leaf disks. It was determined that an adenine-based compound present in yeast extract was involved. Carlos Miller in Folk Skoog's lab in Wisconsin fractionated yeast extract and herring sperm DNA and found cell division activity using a tobacco pith assay that was perfected there (Miller et al., 1955). The cell division factor was identified as an amino purine and it was called kinetin, because it stimulated strong cell division (cytokinesis) activity. Kinetin was not a naturally occurring compound but rather a breakdown product of the herring sperm DNA. Many analogs of kinetin were synthesized and activity confirmed using the tobacco callus assay. Likely structures were identified and efforts were then directed to identify a similar compound that may be present in plants. Letham (1963) isolated a very active compound from immature corn kernels and called this compound zeatin. The primary cytokinins found in plants are zeatin and zeatin riboside, although there are a few other cytokinins, with similar structures present in lower concentrations, that do possess cytokinin activity.

Cytokinins are associated with tissue undergoing rapid cell division. They interact with auxins to determine the morphological development of undifferentiated cells. A high ratio where auxin is favoured results in tissue developing into roots whereas changing the ratio by increasing the cytokinins will favour the growth of the cells into shoots. There is strong evidence that a major site of cytokinin synthesis in plants is the roots and the primary direction of movement in the plant is upward through the xylem.

2.1.4 Abscisic acid

Experiments that led to the discovery of abscisic acid (ABA) were conducted independently in two different laboratories in the 1960s (Moore, 1989). Addicott and his group were studying hormonal relationship in cotton. They identified a compound that had strong abscission promoting properties and they called it abscisin II. Wareing's group in Wales was studying dormancy in trees and identified a compound that promoted dormancy in woody plants. Later it was determined that these two compounds were identical and had the same structure leading to the adoption of the name abscisic acid (Milborrow, 1984). While it is true that ABA can cause leaf abscission, it is more prominently associated with the onset of dormancy in plants, and for playing a dominant role in the regulation of water relations in a plant by regulating the stomatal functions. When a plant is stressed, ABA level increases and this signals the pumping of cations out of the guard cells which results in osmatic movement of water from the guard cell, closing the stomates. When the leaves rehydrate ABA decreases in the leaf, cations are pumped back into the guard cells, causing them to become turgid and thus opening the stomates.

2.1.5 Ethylene

The hormonal properties of ethylene were first recognized by the Russian scientist Nejebulov in 1901 (Beyer et al., 1984). He showed that illuminating gas could cause leaf abscission and epinasty. Several scientists over the next three decades made discoveries that showed that ethylene had hormone-like activity. However, it was Crocker, a post-harvest physiologist, who first suggested that ethylene was an endogenous plant hormone (Crocker et al., 1935). Few scientists agreed with Crocker since most had a difficult time assigning ethylene a status as a legitimate endogenous hormone with important regulatory activity, since it was a gas and could move freely in the air and in the plant (Reid and Howell, 1995). It was not until the 1960s when the improved analytical properties of the gas chromatograph were perfected that scientists were able to easily measure small quantities of ethylene which then allowed them to recognize and acknowledge the hormonal properties of ethylene. Ethylene is unique among the plant hormones in that it can move within the intracellular space of a plant as a gas, dissolve in the cytoplasm and act as the precursor of ethylene, 1-amino-cyclo-propane-1-carboxylic acid (ACC). Ethylene is a dominant hormone involved in the ripening process. Ripening of fruit and the post-harvest life of many fruit and vegetables can be extended by regulating the production or the action of ethylene. Ethylene plays a central role in the abscission process and in some crops can reduce the fruit removal force to facilitate mechanical harvest of some fruit. It can act as a growth retardant and enhance flower bud formation in pome fruit, even in the absence of reducing crop load.

2.2 Minor (lesser known) plant hormones

There are a few other naturally occurring compounds that are considered by many to be plant hormones. While these compounds may not be involved in many physiological processes, they are compatible with the definition of a plant hormone and therefore they are included in the plant hormone category.

2.2.1 Jasmonates

Jasmonic acid (JA) and methyl jasmonate (MeJA) are a class of oxylipins composed of a cyclopentane ring with two substituent groups. JAs and their derivatives activate a succession of signalling pathways that can influence aspects of fruit ripening, production of viable pollen, root growth and plant resistance to insect and pathogen invasion (Creelman and Mullet, 1997). Following mechanical wounding JA biosynthesis is increased which in turn signals the expression of the appropriate gene response. Jasmonates appear to be involved in the early steps of the ripening of climacteric fruit in conjunction with ethylene (Fan et al., 1998). JA may also enhance the production of anthocyanins (Rudell et al., 2005). There is a JA proprietary formulation that is currently being used to enhance red colour development of apples when applied within 4 weeks prior to harvest.

2.2.2 Brassinosteroids

The brassinosteroids were first discovered in the 1970s in pollen grains of rape (*Brassica napus*). Since that time they have been found in most other structures in a plant, but in very low concentrations. Over 70 brassinosteroids have been discovered. They have the basic steroid nucleus of four fused rings and an alkyl side chain, but differing isomeric configuration and substituent groups. They are involved in cell expansion (Clouse, 1996), vascular differentiation (Fukuda, 1997), cell elongation (Katsumi, 1991) and others (Savaldi-Goldstein, 2006). Brassinosteroids are required for growth and reproduction (Clouse and Sasse, 1998). Brassinosteroids have shown promise for agricultural use, but the extremely low levels found within a plant have slowed progress.

2.3 Plant growth regulators

Perhaps the most important PBRs in general use in tree fruit production are compounds that are not produced by the plant or if they are, their occurrence is not widespread. Most are chemically synthesized similar to many biochemicals while others are produced by fermentation.

2.3.1 Aminoethoxyvinylglycine (ReTain®)

AVG is a compound that was identified as an ethylene biosynthesis inhibitor (Boller et al., 1979). Hoffman-La Roche was the first company to attempt to commercialize this compound. There was a flurry of research activity that confirmed that it had strong biological activity especially when used for delaying senescence. However, development of AVG was abandoned because of the high cost of production. While other companies looked at the compound, all lost interest and AVG remained undeveloped. In the late 1980s, registration of an important pre-harvest drop control compound for apples (daminozide) was dropped, leaving a void in the market. Bangerth (1978) previously showed that AVG could control pre-harvest drop of apples. Abbott Laboratories started to evaluate this compound for commercial use as a drop control compound on apples as well as for other uses. Following several years of development and testing AVG received full label registration in 1997 as ReTain® to retard pre-harvest drop, aid in harvest management and to improve fruit quality. It took nearly 20 years from the time the pre-harvest drop control activity of AVG was recognized in late 1970s until it was registered as a commercial product.

2.3.2 6-Benzyladenine

The road from discovery, through evaluation, field testing and marketing of 6-benzyladenine (6-BA) occurred over a 35-year period. There were three reasons for this extremely long development period (Greene et al., 2016). First, 6-BA is a cytokinin and abscission was not one of the physiological characteristics attributed to cytokinins. Second, the apple industry is constantly undergoing change. As the industry evolved (training systems, rootstock, cultivars etc.) new needs emerged, which in turn created the opportunities for new uses for existing PBRs. Third, there were real challenges posed in formulating and packaging BA as a registrable product that required time and formulation research to resolve. The ability of 6-BA and the gibberellins A_{4+7} to increase the length of 'Delicious' apples and to cause bud to break on apple stems was recognized by Williams and Stahly (1969). Abbott laboratories combined these two compounds into a proprietary product (Promalin®) that could both elongate 'Delicious' and cause bud break on young apple shoots (Williams and Billingsley, 1970). When this product was applied to young trees in the orchard to improve branching structure, it was found that, at the rates used it cause fruit thinning. It was determined that it was the 6-BA that was the main component that was causing the thinning (Greene et al., 2016). Rather than develop a new 6-BA-alone product for chemical thinning, Abbott Laboratories altered their Promalin® product by significantly reducing the GA component. All indications from the literature suggested that the small amount of GA in the formulation should have no effect on the abscission capability of the 6-BA. However, this newly formulated product did not thin trees as well and on treated trees there often remained clusters of small fruit that had reduced seed numbers. A new 6-BA product was then formulated, tested and registered for use as the product MaxCel® in 2004.

2.3.3 2-Chloroethylphosphonic acid (ethephon, Ethrel®)

In the 1960s ethylene was recognized as an important hormone that could influence many physiological processes. Since it is a gas in its natural state, it is difficult to administer it to plants unless they are confined in an enclosed space. Activities such as ripening bananas, avocado and tomatoes and degreening of citrus are activities compatible with treatment in enclosed rooms (Barry and Giovannoni, 2007). However, ethylene is very active in causing ripening, fruit abscission and other physiological responses where only application in the orchard or field is practical. In the 1960s a compound was identified by scientists at Amchem Products Inc. that would produce ethylene gas when exposed to alkaline conditions. This product was identified as Amchem 66-329 and eventually was given the common name of ethephon (Ethrel®) (Yang, 1969). It was stable under acid conditions and when sprayed on a plant it could be taken up by a plant. When it was absorbed into the cell where the cytoplasm was alkaline or had a neutral pH, it broke down and in the process ethylene gas was liberated in the plant. Over the past five decades ethephon has been the compound of choice for treating trees in the orchards and fields with ethylene. There are many commercial uses of this product in fruit production including vegetative growth control, enhanced flower bud formation, advanced ripening of fruit, reduce fruit removal force to aid mechanical harvesting and chemically thinning of apples.

2.3.4 Naphthaleneacetic acid

The auxins were the first group of hormones to be discovered and identified and NAA is one of the few auxins that still retains registration for use. It is a very important fruitlet thinner used on apples and it is also an effective pre-harvest drop control compound. When applied to plants it can stimulate ethylene production. Therefore, it is not uncommon for plants treated with NAA to display symptoms of ethylene production such as epinasty of leaves and shoots and advancement of fruit ripening.

2.3.5 Prohexadione-calcium (Pro-Ca, Apogee®, Kudos®)

The spacing between trees in contemporary orchards is being reduced resulting in orchards with less space between trees. Regulation and control of the vegetative growth in these orchards is becoming more important. The use of growth retardants is a strategy used by many growers to reduce vegetative growth and minimize shading within the orchard. It is a gibberellin biosynthesis inhibitor that was introduced in 1997. It can effectively control growth while having few negative side effects. This compound is the product of choice and the primary growth retardant used to manage vegetative growth in apples.

3 Application of PBRs

3.1 Foliar sprays

The majority of PBRs are applied as foliar sprays. The effectiveness of the application is generally gauged by observing or quantifying a physiological response as a result of the application. However, the spray application details and the environmental factors associated with the application may influence the uptake of the PBR and thus play an important role in determining the ultimate response to the PBR.

Traditionally, foliar sprays of PBRs were made as a dilute application where sprays were applied to the drip point. This approach is less common now due in part to loss of spray by runoff, compaction of the soil as a result of many trips through the orchard and increased fuel consumption which are not sustainable practices. Further, valuable time is lost by spending unproductive time driving through the orchard and having to fill the sprayer tank frequently. Further, the makeup and design of orchards has changed dramatically in recent years with the widespread adoption and use of dwarfing rootstocks and more efficient training systems which provided more attractive options for efficient application of plant growth regulators (PGRs).

The trend of the past few decades is to plant trees at much higher densities. The increase in orchard density has led to a reduction in the size of application equipment needed and a reduction in the volume of spray required to deliver pesticides and PBRs. Byers et al. (1971) recognized the problem and they were the first to introduce the concept of tree-row volume (TRV) to determine the volume of spray necessary to achieve adequate coverage of trees planted at different tree densities. An example of a dilute TRV calculation for an apple orchard is illustrated in Table 1. The concept has evolved and has been redefined over time. Effective and consistent use of PBRs must take TRV into account if spray applications of PBRs are to be applied at reduced spray volumes (Bukovac, 1980). For example, label recommendations for the commonly used PBRs, such as 6-BA, NAA or AVG, initially made label recommendations based upon grams active ingredient per ha. Wider acceptance of the TRV concept has led to recent label modifications which result in recommendations that use mLs or grams of product per volume of spray per ha.

The response to PBRs is generally rate sensitive. If too little is applied there is either no response or an insufficient response. If excess is applied there could be detrimental responses such as phytotoxicity or over responses. In developing protocols for grower use, it is important to assure that the correct amount of material is applied to achieve the desired response. This is relatively

Table 1 Calculations of dilute volume necessary to wet apple foliage in an orchard to saturation

Parameters to measure
Canopy height (m) - distance from first scaffold to the top of the canopy
Tree width (m) - average maximum width of a tree from branch tip to branch tip

$$\text{Row length} = \frac{\text{Area of a ha}(\text{m})}{\text{Row width}(\text{m})}$$

$$\text{Tree-row volume}(\text{TRV}) \text{ or canopy}(\text{m}^3) = \text{Canopy height} \times \text{Tree width} \times \text{Row length}$$

Dilute volume requirement per ha equals about 1 L of spray for each 10 m^3 of foliage

Example calculation of dilute spray requirement
Trees in a sample orchard have a canopy height of 4 m, a tree width of 3 m and a row width of 5 m

$$\text{TRV} = 3.4 \times 3 \times \frac{10000(\text{m}^2\ \text{ha}^{-1})}{5} = 24000$$

$$\text{Dilute volume} = \frac{\text{TRV}}{10} = 2400\ \text{L ha}^{-1}$$

easily accomplished by making a dilute application since all parts of the tree receive the desired dose and any excess will drip off.

However, application of pesticides and PBRs is generally done now using concentrate spray application. Concentrate spraying involves applying the desired and calculated amount of active ingredient on an orchard using a reduced amount of water. In order to do this correctly, it is first necessary to calculate the TRV of the orchard to be sprayed. From this you can calculate the amount of material suggested on the label for the area to be sprayed. For example if a block has a TRV of 3740 L ha^{-1} (400 gal $acre^{-1}$) and the grower decides to apply a PBR at a 2X concentration or using half the amount of water. The grower would then put in the sprayer tank the amount of material suggested to make the dilute application on that block. The block will then be sprayed with 1870 L of the water but it would also receive the same amount of active ingredient as the same block sprayed with a larger volume of water.

There are several problems associated with the application of PBRs in a low-volume spray that may result in variable responses (Bukovac, 1985). The possibility of over- or under-application is increased because the response to PBRs is generally linear. For example, if an error of 10% is made in calculation or measurement when making a dilute application there is likely to be a small or minimal influence on the final result since the error is only 10%. However, if the same 10% error is made in 6X concentration application this error will then be increased to 600% which will most likely have a significant and important influence on the final result. When a lower volume of water is used there is generally a faster drying time which is associated with reduced uptake. Finally, small spray particles are associated with concentrate spraying and these are the particles that are most likely to drift and the ones that are most difficult to reach the tops of trees.

Translocation of PBRs is generally limited. Therefore, it is important that the spray application be made directly to the target tree part and does not depend on translocation for more uniform distribution. For example, when AVG is applied to a tree to control pre-harvest drop maximum control is achieved when both the leaves and fruit receive the spray (Fig. 1). If only the fruit received the spray, little drop control was achieved. For maximum response to a PBR precise and directed application is necessary.

3.2 Environmental factors influencing foliar penetration

The plant response to foliar applied growth substances is frequently variable, even when that application is made under, what appears to be, similar or identical application conditions. Environmental factors can cause this variability in several ways.

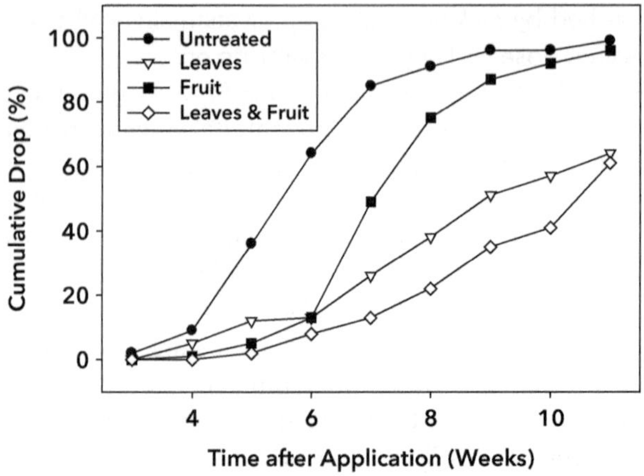

Figure 1 Influence of location of aminoethoxyvinylglycine (AVG) application on apple spurs on the extent of pre-harvest drop on 'McIntosh' apples. Application of AVG made directly to the fruit had minimal influence on pre-harvest drop whereas if the application was made to leaves or leaves plus fruit, pre-harvest drop was controlled for an extended period of time. Source: Greene (2006).

3.2.1 Environmental conditions prior to application

Uptake into the leaf must occur through the cuticle which is present on both the upper and lower leaf surfaces. During leaf development, epicuticular waxes are exuded from underlying cells onto the leaf surfaces and these influence uptakes into the leaf. It is generally accepted that the amount, structure and composition of epicuticular waxes influence leaf wettability and penetration of foliar applied substances (Bukovac, 1973). During cool, moist, cloudy periods leaf cuticle development is slowed whereas during hot, dry sunny weather the cuticle development is more robust. It has been observed in the field that response to a foliar applied substance is greater when leaves develop under cool, cloudy conditions and a reduced response is seen when leaves develop under sunny, dry and low humidity conditions. The difference in response may be attributed to the amount of PBR taken up by the trees whose leaves were exposed to these different weather conditions while developing.

3.2.2 Environmental conditions during and immediately following application

There are two major factors that influence the uptake of a growth substance into a leaf: temperature and droplet drying time (Greene and Bukovac, 1971). As temperature increases there is a linear increase in uptake of foliar applied substances as illustrated by penetration of NAA through the upper surface of 'Bartlett' pear leaves (Fig. 2). At 25–27°C there is an inflection point where the

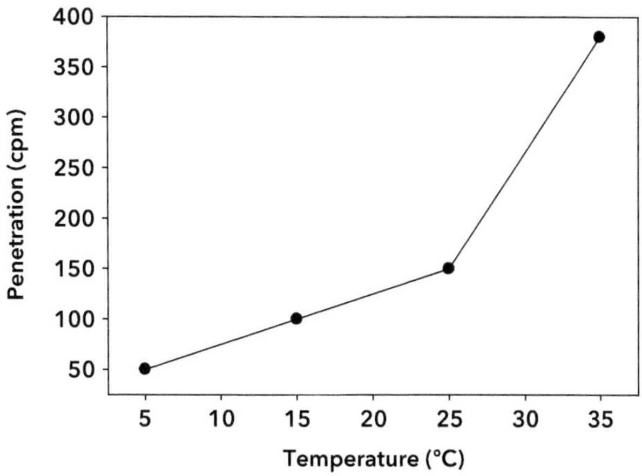

Figure 2 Influence of the temperature at which naphthaleneacetic acid (NAA) was applied on the penetration through the upper surface of 'Bartlett' pear leaves. Penetration of NAA was lower as the temperature increased to 25°C. There was an increase in the slope of update above 25°C that is attributed to a phase change in the cuticle. Source: Greene and Bukovac (1971).

rate of uptake substantially increases. This is attributed to a phase change that is occurring in the cuticle (Price, 1982). This rate in change of uptake makes it somewhat difficult to predict accurately the amount of substance that is being taken up by a leaf. Growers are cautioned to not apply PBRs or to apply these with caution when the temperature approaches 30°C because of the uncertainty associated with the amount of growth substances that are taken up at and above this temperature and consequently to the response of the applied PBR.

The time required for a spray droplet to dry can influence the amount of growth substance that is taken up. This can be illustrated by the penetration of NAA into the 'Bartlett' pear leaf discs when applied to the upper surface (Fig. 3). In general, when a spray is first applied there is a linear increase in penetration of a substance into a leaf that is attributed to diffusion. As a droplet starts to dry the concentration of the growth substance within a drop increases leading to an acceleration in penetration of the growth substance into a leaf. Once the droplet dries there is little or reduced additional penetration. As long as the spray droplet stays intact and does not dry, penetration may occur somewhat continuously (Greene and Bukovac, 1971). The enhancement in uptake will always occur as the droplet dries due largely to the increased concentration of the PBR in the droplet. Therefore, if droplet drying can be slowed, there will be greater uptake of the growth substance into the leaf. Frequently, PBR labels suggest application under high humidity conditions with the express purpose of increasing foliar penetration. In experiments in the more humid Eastern US, when application was made during the day when the temperature was warm

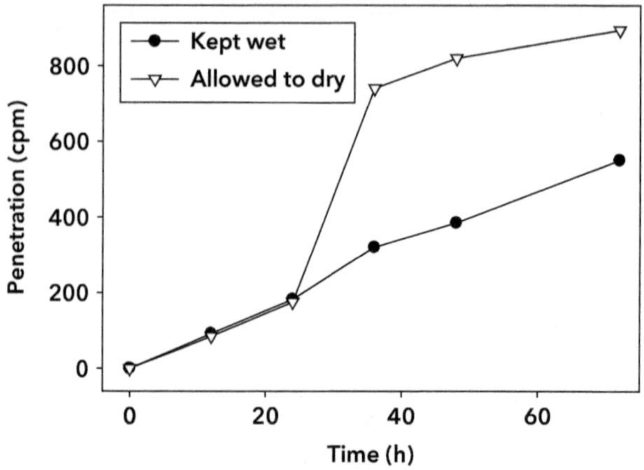

Figure 3 Influence of droplet drying on penetration of naphthaleneacetic acid (NAA) on penetration through the upper surface of 'Bartlett' pear leaves. Penetration is generally linear as long as the spray droplet remains hydrated. However, when the spray droplet starts to dry there is a rapid increase in penetration which tends to stop after the drop dries. Source: Greene and Bukovac (1971).

and drying time rapid it was similar to the uptake when the same amount of growth substance was applied at night when the temperature was cool and drying time was longer (Stover and Greene, 2005). It was concluded that the enhancement in penetration under warm conditions where drying time was shorter was offset by uptake over a longer period of time when the temperature was cooler.

3.3 Influence of rain on PBR response

Frequently, there is a narrow window of opportunity to apply PBRs to be effective. The weather is an uncontrollable factor and predictions are tenuous, especially when it comes to forecasting the probability of rain. Orchardists are often forced, out of necessity, to take chances and apply sprays even if there is a possibility of rain. Sometimes rain does occur following application. The concern then is, just how much activity was lost due to wash-off from the rain? The answer to this question is determined by the time after application (droplet drying) the rain occurred, how much rain occurred, what the solubility properties of the PBR are, the PBR formulation and whether a surfactant was included in the spray. In studies with NAA application on apples, Westwood and Batjer (1960) concluded that some activity was lost if rain occurred within 2 h of application but not if rain occurred after 6–8 h. It appears that this observation may depend on the compound applied and the formulation of that compound. Greene et al. (2000) reported that some loss of pre-harvest drop control activity

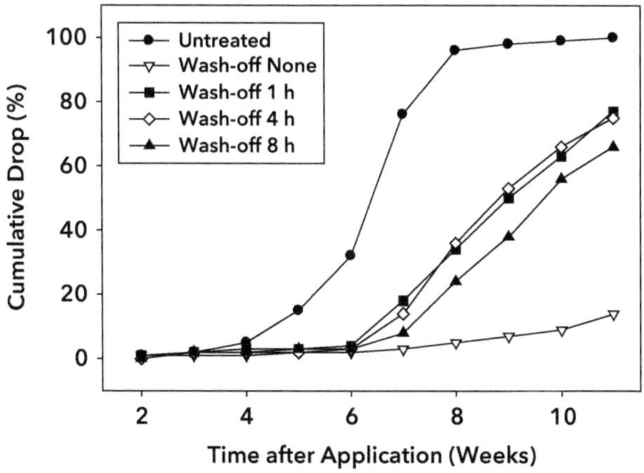

Figure 4 'McIntosh' apple trees were sprayed with aminoethoxyvinylglycine (AVG) that contained no surfactants. Illustrated is the influence of the time after application that trees were thoroughly washed by a high volume of water using a handgun on pre-harvest drop. When trees were sprayed 1–8 h after AVG application, a significant loss in pre-harvest drop control was noted. Source: Greene et al. (2000).

occurred on 'McIntosh' apple trees if AVG was washed off within 24 h after application (Fig. 4) but if the organosilicone surfactant L-77 was included, no activity was lost even if wash-off was done within 1 h after the spray had dried (Fig. 5). Results from field observations made over the years suggest that if the

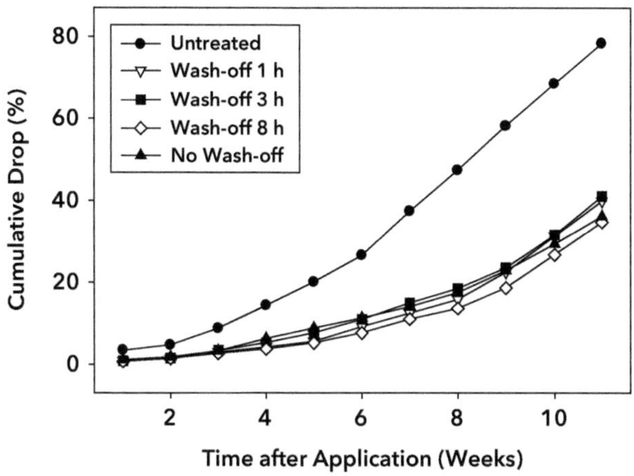

Figure 5 Mature 'McIntosh' apple trees were sprayed with aminoethoxyvinylglycine (AVG) containing 0.01% Silwet L-77 silicone surfactant. Illustrated is the influence of the time after application that trees were thoroughly washed by a high volume of water from a handgun sprayer. Regardless of the time after application pre-harvest drop control was not reduced by thoroughly washing the trees. Source: Greene et al. (2000).

spray droplets have completely dried before rain starts, a substantial amount of PBR activity is retained. A rule-of-thumb, which has not been substantiated by research but arrived at through experience and observation, is that if the spray has been dry for at least 30 min prior to the rain event, 80% of the PBR activity will be retained. Sometimes a small amount of rain occurs following application. In this case there is a possibility of enhancing activity rather than inhibition of activity. Bukovac (1965) applied the auxin chlorophenoxyacetic acid to peach leaves. If the leaves were rewetted twice there was a 17% increase in penetration and if the leaves were rewetted five times there was a 47% increase in penetration into the leaf. Field observation following rewetting of the blossom thinners such as dinitrocresol and ammonium thiosulphate (ATS) revealed increased phytotoxicity may be associated with the rewetting of these compounds, which is attributed to increased penetration.

4 Development and maintenance of tree structure

The makeup of apple orchards has changed dramatically over the past half century, especially the tree structure, the rootstocks that the trees are propagated on and the planting systems used. Regardless of the system or the component parts, a common concern expressed by all orchardist is that trees should be encouraged to develop a structure that will allow early and high production. In the 1970s and 1980s the predominant tree structure was a central leader tree propagated on a semi-dwarfing rootstock. The goal in training young trees was to encourage the development of lateral branches that were robust enough to serve as permanent or semi-permanent scaffold branches or develop into sturdy lateral shoots on scaffold branches. Heading cuts (cutting into 1-year-old wood) and the use of a proprietary mixture of 6-BA and gibberellins $A_4 + A_7$ (GA_{4+7}) at rates of 400–500 mg L^{-1} were common. In recent years the tree architecture in the apple industry has evolved rapidly and trees are now propagated primarily on dwarf rootstocks that are spaced very close together. The type of lateral branches that are desirable and are most productive are shorter and weaker shoots.

Most orchardists now order and hope to receive well-feathered trees from the nursery. These trees are the most productive, and come into full production earlier because these branches can produce a small crop as early as the next year. Frequently, trees arrive from the nursery having few if any lateral shoots, or if a few laterals are present, they are often not useful and are cut off. An option available to growers is to apply branch promoting PBR treatments to improve branch development.

Generally, attempts to increase branching on 1-year-old trees the first year in the ground have not been very successful (Forshey, 1982; Miller, 1988). The primary reason for the lack of success is that these trees do not have a well-established and active root system to support the lateral shoot growth desired. Branching successes reported on these types of trees has generally been on very healthy trees that have been planted early and trickle irrigation was installed and used early in the season. Lack of success at increasing branching on first-year trees is common on trees whose growth is compromised by late planting, lack of water, poor soil preparation, poor weed control or inadequate cultural management. After planting it is not uncommon to have sections of the leader that have blind wood. This is especially true with some cultivars that are particularly prone to blind wood: Granny Smith, Fuji, Jonagold and others (Miller, 1988). In general, it is easier and more predictable to stimulate lateral buds to grow when practised on trees the year after planting or on buds located on 2-year-old wood. These trees have an established and growing root system and the attempts to increase lateral branching can be initiated very early in the season when greater success is generally achieved.

4.1 Branching using physical techniques

There are physical techniques to increase branching on young trees including notching, heading, bending the central leader and pinching out very young leaves in the apex. All have disadvantages that have led to lack of extensive use. Of the physical techniques available, notching appears to be the most frequently used technique to improve branching on young apple trees (Greene and Autio, 1994). Notching involves removing a thin piece of bark immediately above an inhibited bud (Greene and Autio, 1994). The effectiveness of this technique is attributed to blocking or preventing the downward movement of auxin from the shoot tip. Auxin is the hormone that is primarily responsible for preventing lateral buds from growing. Experience has shown that at least a 2 mm, and, preferably a 4 mm (McArtney and Obermiller, 2015), strip of bark must be removed from immediately above the bud and this bark removal must extend around at least one-third of the stem. Use of a hacksaw blade, either single or double, is a rapid and useful tool to remove this strip of bark in the orchard. Callus will start to form at the cut, and once the callus grows together and fills the gap, auxin again can migrate down to the lateral bud, re-establishing apical dominance thus slowing or stopping the growth of the lateral shoot. Therefore, severing the bark with a knife blade or other sharp instrument that removes a small amount of bark has a temporal influence on stimulating lateral bud growth. The time at which notching is done is important since it influences the number of buds that can be induced to grow. Greene and Autio (1994) reported that notching 'Delicious' buds between 2 and 4 weeks prior to bloom resulted in about 60% of the notched bugs growing, whereas notching done 6 weeks before bloom or at bloom reduced bud break. A larger percentage of the bud broke if buds were located at the tip of a shoot. There was a linear reduction in the percentage of bud that broke the further away the notched bud was from the shoot apex. McArtney and Obermiller (2015) reported that the highest proportion of bud stimulated to grow on 'Granny Smith' apples was when notching was done between bud break and two weeks after bud break. The difference in the optimum time to notch may be related to the early warm temperatures experienced in North Carolina compared with cool temperatures generally experienced in the Northeast US in the spring.

4.2 Branching using PBRs

A number of PBR branching agents have been tested. However, the two primary proprietary products in use today contain 6-BA alone or 6-BA plus GA_{4+7} present in equal amounts. Both products have other uses on apple trees. The 6-BA product has the advantage over the 6-BA plus GA_{4+7} product since the gibberellins in the combined product can inhibit flower bud formation for the next year. Consequently, any tree sprayed with this product may have reduced flower bud formation thus delaying or reducing cropping potential on treated trees the following year (Musacchi and Greene, 2017). On trees where cropping is not planned or not expected for 2 years, a reduction in flower bud formation would not be a problem. 6-BA has little or no direct influence on flower bud formation, although it is used as a proprietary product to thin apples on trees that carry an excessively heavy crop load.

The 6-BA formulation may be applied to young trees in the orchard as either a spray application or painted directly on a bud in latex paint. The latex paint application is made at rates of 5000–7500 mg L^{-1} before bud break. Later application of paint when buds are starting to grow will lead to phytotoxicity and a reduced response. This method is infrequently used. The most common and frequently used method to increase branching on trees is with the use of PBR sprays.

Spray applications of 6-BA or 6-BA + GA_{4+7} when buds start to break at concentrations from 500 to 2500 mg L^{-1} have resulted in variable responses (McArtney and Obermiller, 2015). On similar trees when application is delayed until buds have broken, greater bud break is observed. The variable response is attributed to several factors. In a 2-year study using second leaf 'Granny Smith' and 'Fuji' apples, McArtney and Obermiller (2015) showed that notched buds could cause buds to grow in some instances but if notching was followed by 6-BA application the buds not only started to break but they continued to grow and develop into functional lateral shoots. Apparently, the buds created a somewhat impervious barrier to entrance of the 6-BA spray and notching opened an avenue for the spray to enter and allow the continued growth of the bud.

Application of PBRs to stimulate lateral branch development on second and third leaf trees is not a routine management activity. It has been observed in many orchards that using cultural and management techniques that assure good tree growth may be sufficient to encourage the development of an appropriate number of lateral branches, thus not requiring intervention with PBRs.

5 Control of vegetative growth

Management of the growth of apple trees is an ongoing activity. The goal is to maintain a balance between vegetative growth and cropping but management is only partially under the control of the orchardist. Frost, or freeze damage, during the winter may kill flowers thus reducing the crop. Under these circumstances, terminal growth is frequently excessive and requires supplemental control. The use of PGRs is the most effective, convenient and cost-effective tool available to accomplish this and they can be used occasionally or yearly depending on the growth control required. There are two primary growth retardants in general use on apples today: prohexadione-calcium (Pro-Ca) and ethephon (Ethrel).

5.1 Growth control using prohexadione-calcium

Pro-Ca is a gibberellin biosynthesis inhibitor. Because it blocks GA synthesis late in the biosynthetic pathway (between GA_{20} and GA_1) (Evans et al., 1999) it has relatively few effects that can't be attributed directly to inhibition of GA synthesis. The primary reason to apply Pro-Ca is to reduce terminal growth by reducing internode length. A consequence of this is that there is less crowding between trees due to shoot competition, light penetration into the tree is increased which then leads to improved fruit colour and fruit quality and improved flower bud formation. The amount of pruning is also reduced.

There are three Pro-Ca formulations available. Apogee® and Kudos® contain 27.5% a.i., whereas Regalis® is a 10% a.i. formulation. Preparation for spraying is different for the 27.5% and 10% a.i. products. Pro-Ca may precipitate out of solution if 'hard' water is used. To remedy this situation when using the 27.5% product you should add an equivalent weight of ammonium sulphate to Apogee® or Kudos® to the spray tank. Alternatively, a commercial water conditioner may be used following label directions. Regalis® is the 10% a.i. formulation that does not require water adjustment since it is formulated with the proper amount of ammonium sulphate. A non-ionic or organosilicate-containing surfactant is recommended to use with all formulations. Pro-Ca should not be applied with any calcium sprays since it will reduce the effectiveness of the Pro-Ca.

It requires 10–14 days after application for growth control to become effective (Fig. 6). Regardless of the rate applied, the reduction in growth was initiated at approximately the same time. Apple shoot growth is most rapid in early near bloom. Therefore, to achieve the greatest growth control the initial Pro-Ca application should be made as soon as there is sufficient leaf area to absorb Pro-Ca so that growth control is in place when rapid shoot growth begins. Measureable and commercially significant growth retardation can be achieved with an initial application of 3 oz/100 gal (62.5 mg L^{-1}). Slightly better growth control is possible with rates of 6–12 oz/100 gal (125–250 mg L^{-1}). A higher rate may be selected for the first application which may help compensate for the reduced leaf absorptive area present in the tree. An additional application is required 2–4 weeks after the first and follow-up applications are frequently needed. As the season progresses acceptable growth control can be achieved with lower rates. The amount applied and the number of applications required for season-long control depends on several factors including the cultivar, vigour of the tree, the rootstock, soil, the weather, moisture during the growing season and the geographical location. Usually more applications are required in orchards in warm climates where the growing season is longer and terminal growth continues later into the season. Typically 2-3 applications of Pro-Ca would be used for season-long control in the major fruit growing areas. The highest rate should be applied in the first application and lower rates may be used later.

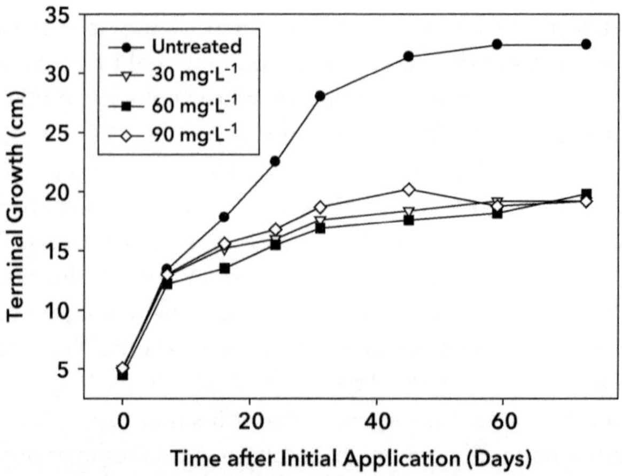

Figure 6 Influence of the rate of prohexadione-calcium (Pro-Ca) applied starting at petal fall on terminal growth of 'Macoun' apples. Regardless of the initial rate of Pro-Ca applied, the time at which growth control was initiated and the extent of growth control was similar. Source: Greene (1999).

Pro-Ca often results in increased fruit set (Table 2). It does this by reducing the intensity or extent of 'June drop'. This is the same physiological period that chemical thinners operate in. Pro-Ca inhibits drop of some fruit that would normally be encouraged to drop by post-bloom chemical thinners. Therefore, when Pro-Ca is used in blocks where thinners are applied, it is recommended that more aggressive thinning should be done to achieve an appropriate final fruit set.

Pro-Ca provides excellent control of shoot fire blight. It does not have any antibacterial activity to directly inhibit the fire blight bacteria. It is believed that it acts indirectly by thickening the cell walls in the shoot thus making them less permeable and susceptible to infection. The initial application should be made early, at petal fall of the king flower. Since effective control depends on growth retardation, the spray will only confer resistance to fire blight when growth retardation has been initiated, 10–14 days after application. Rates of 6–12 oz/100 gal (125–250 mg L^{-1}) are generally used. Pro-Ca does not afford any direct protection from blossom blight. Pro-Ca is a very good alternative to the use of antibiotics for control of shoot fire blight bacteria. Since Pro-Ca works by altering the cell structure of the inhibited shoot, bacteria can't build up resistance to shoot fire blight. Antibiotics can be reserved for blossom blight and control of shoot blight following a traumatic event in the orchard such as hail.

5.2 Control of vegetative growth with ethephon

Ethephon is a very effective growth retardant especially when applied from bloom to a week after bloom. It is also a good fruit thinner when applied over the same time period. Consequently, when ethephon is used to control growth it may also cause significant crop reduction. It is generally accepted that ethephon should not be used routinely for growth control on bearing trees due to its thinning potential. Ethephon may be useful in orchards that have lost a crop due to frost. Application of 500 mg L^{-1} when shoots are 10–15 cm in

Table 2 Influence of prohexadione-calcium on fruit set and return bloom when applied at petal fall on 'McIntosh'/M.7 apples to control vegetative growth

Treatment (mg L^{-1})	Fruit per cm limb cross-sectional area	Return bloom per cm limb cross-sectional area
0	5.8	8.4
125	6.2	6.1
250	8.2	4.1
375	9.8	3.3
Significance	l***	l***

Source: Greene (1999).

length will retard growth and help the trees to re-establish the delicate balance between vegetative growth and cropping. It should be noted that ethephon promotes flower bud initiation on apples and when applied on bearing trees the potential exists that a 'snowball bloom' may be promoted the following year. This may lead to the trees slipping into a biennial bearing cycle because of the increased difficulty in chemically thinning, especially on difficult to thin cultivars.

Ethephon may be used to control growth of non-bearing trees but this use is far less common now because of the dominant use of precocious dwarfing rootstocks in new plantings. On non-bearing trees, rates of 300–500 mg L^{-1} are commonly applied when the terminal growth reaches 7–12 cm. Growth control is more difficult on trees propagated on vigorous rootstock. Ethephon will enhance flower bud formation on these more vigorous trees.

5.3 Promotion of flowering on bearing trees

Apple trees tend to be biennial, producing a crop one year followed by a year where production is low or there is no crop, due primarily to the reduction in flower bud formation. Cultivars differ in their biennial bearing tendencies. NAA and ethephon are used to thin fruit and they also have the capability to increase flower bud formation, independent of their ability to thin. If these PGRs are applied enough times, at low rates and not during the normal thinning 'window of opportunity' (bloom to 25 mm) they have the ability under some circumstances to increase flower bud formation without causing thinning. There is a general consensus, based on experience, that either 5 mg L^{-1} NAA or 100 mg L^{-1} ethephon may be effective when fruit reach 25–30 mm in diameter. The spray should be repeated three or four more times at 7–10 day intervals. The effectiveness of these sprays differs from year to year and there is no guarantee that they will actually increase flowering. However, this is the best current option available to influence flowering on bearing trees once a crop has been set. Ethephon should not be used on 'McIntosh', 'Macoun' or 'Honeycrisp' or other 'McIntosh' types because if applied there is the possibility of advancing ripening if ethephon is used even at this late timing.

6 Crop load management

6.1 Chemical thinning

Chemical thinning has been practised in orchards since it was first recognized to help alleviate biennial bearing in the 1930s. Initial efforts were made to learn how to most effectively use the thinners available. Many empirical thinning studies were done but variability and lack of consistency of the thinning responses persisted. More recently, progress has been made and more emphasis has been placed on attempting to determine how individual thinners work, gaining a greater understanding of how environmental conditions influence fruit set and thinner response and recognizing the changing susceptibility of fruit to thinners at different physiological stages of fruit development. Interest in thinning has continued to increase because of the diminishing pool of workers available to hand thin orchards and because of the economic consequences fruit growers suffer from reduced crop value due to insufficient thinning.

A number of studies have been done to determine the mode of action of specific thinners. While there is no consensus on a specific mode of action, it appears that they may act by either influencing the amount of carbohydrate available to developing fruit or they may influence the auxin production or movement from the fruit. Bangerth (2004) reported that young fruit that persist to harvest have a continuous supply of auxin moving from the developing fruit, down the pedicels and through the abscission zone. Abscising fruit have a reduced supply of auxin moving from the fruit through the abscission layer which then triggers the production of enzymes that result in the destruction of these cells in the abscission zone leading to fruit abscission. When carbon supply available to young developing fruit drops to a critical low level, abscission of fruitlets is irreversibly initiated. When the carbohydrate available to young developing fruit is reduced either by shading (Byers et al., 1990a) or by applying photosynthetic inhibitors (Byers et al., 1990b), fruit could be induced to abscise. Lakso et al. (2006) showed that conditions that result in low availability of carbohydrates during peak carbohydrate demand periods result in fruit abscission. It may be that the carbohydrate deficit may serve as the triggering event which then signals the reduction and movement of auxins through the abscission zone.

Variability of the weather has been recognized as a major contributing factor in the lack of thinning consistency, even in the same block using the same thinners and the same rates in different years. Many researchers have made contributions to help quantify and clarify the weather influence on thinning. Williams and Edgerton (1981) identified 15 factors that could influence thinning response but temperature and solar radiation before, during and the 3–4 days following thinner application have emerged as the factors that appear to have the greatest influence on the response to thinners.

Apple may be chemically thinned over a wide range of times starting as early as bloom and extending until fruit reach about the 25 mm fruit growth stage. However, as a fruit grows it undergoes physiological changes that of necessity require use of different thinning strategies and different thinning chemicals. During bloom the objective of applied thinners is to damage the stigma or interfere with pollen germination, pollen tube growth or fertilization. Once the king flower has been successfully pollinated and the pollen tube has grown down to the ovary most commonly used blossom thinners inflict some type of damage to the remaining unpollinated flowers. Blossom thinners are routinely used in arid growing regions. If they are applied in humid growing climates, these compounds may cause some phytotoxicity, which in some cases may be severe. In addition to potential damage, growers in these humid growing regions appear to be reluctant to thin before the pollination period ends and when the danger of frost has passed so that they can assess the potential for initial fruit set. Thinning decisions in humid regions should be re-evaluated since thinning at bloom provides the best opportunity to increase fruit size and ensure return bloom (Batjer and Hoffman, 1951) and the potential danger of thinning at bloom may be more perceived than real or that can actually be documented.

Petal fall, immediately after bloom, affords an extremely important thinning opportunity. The bloom period has passed and the danger of frost has diminished. Meaningful thinning can occur but over thinning rarely occurs if thinners are applied at this time (Greene, 2002). Physiologically, cell division is occurring and seed development is starting but in the receptacle (fruit) visible changes are small and subtle and difficult to measure.

Fruit are most vulnerable to applied thinners during the 7–14 mm growth stage when typically there are 3–5 fruit present on each spur and they are in the log growth phase. Storage carbohydrate has been nearly depleted and the primary source of carbohydrate for growing fruit is now supplied by the leaves associated with each spur. Impact of photosynthate from other portions of the tree does not generally occur until later when shoot growth reaches 10 inches (25 cm). Chemical thinners applied at this time create a stress which may amplify the differences between photosynthate required for all fruit to persist and the ability of the spurs to supply carbohydrate. When a carbohydrate deficit occurs, an internal signal is given resulting in the abscission of the weakest fruit.

As fruit grow larger (>15 mm) they become progressively more difficult to thin (Greene, 2002; Schwallier, 1996). At this development stage, ethylene production declines, there is less intra-fruit competition due to previous fruit abscission within the spur and fruit are starting to accumulate starch. Carbohydrate deficits are less likely to occur unless unusual weather conditions occur. As fruit approach 17–20 mm diameter they regain more sensitivity to ethylene and ethephon or 1-amino-cyclopropane-1-carboxylic acid (ACC) application appear to be more effective. Once fruit have grown to about 25 mm, further reduction in crop load is generally very difficult.

6.2 Thinning chemicals

Thinning chemicals may be roughly placed in two categories: caustic thinners and hormone thinners. Caustic thinners are generally applied during bloom with the objective of damaging the flower, flower parts, or damaging or preventing the pollen germination or inhibiting pollen tube growth and thus preventing fertilization of the ovule. The king flower is generally the largest flower in the cluster and the first to open so the goal is to assure that flower set fruit while inhibiting set of three remaining flowers. Attempts are made to time the blossom thinning application to allow pollination of the king then apply a blossom thinner to inhibit set of the remaining flowers in the cluster. Sometimes the window of opportunity to apply blossom thinners is narrow and can be measured in hours.

A concerted efforts was made to find a replacement for the popular blossom thinner Elgetol (DNOC, sodium 4, 6-dinitro-*ortho*-cresylate) when its registration was cancelled in 1990, but few replacement thinners are in general use or remain registered for this purpose. Lime sulphur is a very commonly used blossom thinner. It is frequently combined with fish oil or other oils to enhance activity. ATS is a foliar nitrogen fertilizer. Although not specifically registered for use as a thinner, if applied to trees at bloom at a rate of 2–4 gal/100 gal (20–40 L/1000 L) it has been shown to be a viable thinner.

Blossom thinners are routinely used in dry, low humidity climates. If blossom thinners are applied before wet or high humidity periods it is not uncommon to have a significant amount of foliage damage. Spur leaves are very important to support fruit cell division and early fruit growth. The observation has been made that application of blossom thinners in humid-wet conditions may lead to thinning but it may not be manifest in increased fruit size at harvest due to foliage damage. Blossom thinning removes fruit early, thus providing the best opportunity to increase return bloom. It therefore remains a viable option to be used, especially on cultivars that are noted for biennial bearing and in areas where weather does not routinely create a carbohydrate stress within the tree during the 7–14 mm fruit size period.

The majority of thinning is done using hormone-type thinners. Unlike caustic thinners that act by causing some type of damage, hormonal thinners act by creating some form of stress and interacting with some physiological events in the plant, which are frequently modified by the weather. Fruit physiological changes occur as the fruit continues to grow. Therefore, selection of thinner(s) may change over time as fruit grows and undergoes developmental changes. The mode of action of specific thinners is not definitively known, although many investigators have identified physiological events that may be affected such as a reduction in photosynthesis, an increase in respiration, limitation of photosynthate transport or enhanced production of ethylene (Dennis, 2002). As a fruit increases in diameter above 15 mm it becomes progressively more difficult for thinners to cause fruit abscission. In general, when fruit diameter exceeds 25 mm it no longer responds to hormonal thinners (Greene, 2002).

The time applied and the circumstances under which individual thinners are used has been previously reviewed: Byers (2003), Greene (2002), Greene and Costa (2013), Schwallier (1996), Wertheim (2000) and Williams and Edgerton (1981).

6.2.1 Naphthaleneacetic acid (NAA)

NAA is the oldest and perhaps the most potent of the thinners now in general use. It may be used in concentrations from 3 to 20 mg L^{-1} but the rates used are generally between 5 and 12 mg L^{-1}. It may be applied between bloom and about 15 mm fruit size. When higher rates are applied on fruit larger than 15 mm especially in very warm weather, reduced fruit size or no increase in fruit size may be noted even though crop load is reduced. It is frequently chosen as the thinner to use on very difficult to thin cultivars. Because NAA has the reputation for being a strong thinner, growers use it with caution. It is commonly combined at lower rates with other thinners, especially carbaryl.

6.2.2 Naphthaleneacetamide (NAAm, NAD)

NAD was tested and evaluated during the same time that NAA was evaluated. NAA became the thinner of choice because it was more potent. It is generally used at a rate of 40–50 mg L^{-1} and the time of application is generally at bloom or petal fall. It can cause pygmy fruit formation when applied late and on 'Delicious' (and perhaps other cultivars). It can cause leaf epinasty but it is far less severe than that caused by NAA. NAD is frequently used in combination with carbaryl and then followed by another thinner at the 7–14 mm stage.

6.2.3 Carbaryl

Carbaryl is perhaps the most versatile thinner in general use. It is an insecticide and a mild thinner that can be used over a wide range of developmental stages from petal fall until fruit reach 18–20 mm. It is a popular thinner since it rarely over thins. Unfortunately, it is very toxic to bees so the earliest time it should be used is after bloom, when beehives have been removed from the orchard. Unlike most thinners, the thinning is concentration independent.

6.2.4 Benzyladenine (BA)

BA is the newest thinner to be registered for use on apples. Effective thinning concentrations are 50–150 mg L^{-1}, although the most commonly used rates are between 75 and 100 mg L^{-1}. BA when used alone is a mild thinner but when combined with carbaryl it is considered a potent thinner that may be similar in strength to NAA. BA can increase fruit size beyond that attributed to a reduction in crop load due to its ability to increase cell division (Wismer et al., 1995). It is most effective when temperatures in the 3–4 days following application rise to 22–25°C.

6.2.5 Ethephon

Ethephon is generally not considered a mainstream thinner in most growing regions. Flower and fruit susceptibility to ethephon varies depending on the stage of development. Ethephon is an effective blossom thinner (Jones et al., 1990) but it loses some of its effectiveness as fruit develops. Fruit redevelop thinning sensitivity between 16 and 22 mm (Marini, 1996). It is the thinner that most effectively thins larger fruit (18–25 mm), although thinning may be inconsistent when applied during this period. It is sometimes referred to as a 'last chance thinner'. The primary use of ethephon now appears to be in orchards where growers failed to achieve acceptable thinning during the normal thinning time period.

6.2.6 Metamitron (Brevis®)

The newest thinner, to be evaluated and in some countries registered for use as a thinner, is metamitron (McArtney and Obermiller, 2012). It was originally registered for use as an herbicide to be used on sugar beets. It is a photosynthetic inhibitor that blocks electron transfer in photosystem II (Abbaspoor et al., 2006). When used at rates of 300–400 mg L^{-1} it thins apples, presumably by reducing the carbon balance within a tree resulting in reduced fruit set. It appears to have the greatest activity when applied during the 7–14 mm stage when apples are generally most sensitive to a reduction in photosynthesis and when the weather conditions following application indicate that the tree will be under a carbon deficit. Experience has shown that having prevailing weather conditions 4–5 days following application that create a negative carbon balance are necessary for metamitron to thin effectively. When applied at rates above 300 mg L^{-1} some minor leaf chlorosis may be possible, but it does not appear to be commercially important.

6.3 Precision thinning

Thinning is the most important management activity a grower is required to do. Insufficient thinning will result in the harvest of small, poor quality fruit and reduced return bloom which leads to reduced income for 2 years. Establishment and growing costs have escalated to a point where large fluctuations in income are no longer economically sustainable. Precision thinning is a strategy developed to assure that consistent thinning is achieved each year thus assuring a somewhat steady income. When fully implemented, Precision Crop

Load Management involves three steps. These include (1) precision dormant pruning adjustment, use of computer models to (2) predict thinning prior to thinner application and (3) early prediction of effectiveness of the thinning sprays (Robinson et al., 2016).

6.3.1 Precision dormant thinning

Precision dormant thinning will be initiated by growers who determine during the dormant season the number of fruit that they would like to harvest from a tree or unit. This is a deviation from earlier orchard practices where crop loads per tree were generally not preset. Setting fruit crop goals is quite easy on new high-density plantings where tree numbers and total desired yield per acre are known. It is much more difficult on larger trees where branch subunits are relied on.

Once the target number of fruit per tree is established, the number of flower buds on a tree needs to be determined. This may be challenging at first but with a little practice it can be done somewhat accurately, even during the dormant season. Select five uniform and representative trees in a block and count all flower buds. Next, calculate the blossom cluster averages for the measured trees. Multiply the predetermined and desired number of fruit on each tree by 1.5 and this will for most cultivars be the number of flower buds that should remain after pruning.

Assume an orchard is planted at a 12 × 3 ft. (0.9 × 3.7 m) spacing it would contain 1210 trees (2990 trees ha^{-1}). If the desired yield is 1200 bu $acre^{-1}$ (59 300 kg ha^{-1}) then 100 fruit per tree would be the desired final crop load. If you multiply 100 fruit by 1.5 the target number of flower clusters to remain should be 150. Trees would be pruned in an attempt to reduce the flower cluster number down to the calculated number. Initially, pruning followed by counting would be required to fine-tune the process.

Thinning at pruning time offers several advantages. It reduces early competition among flowers. Weaker flowers may be pruned off thus allowing resources and new growth on a tree to grow the large flower buds that have a greater potential to produce large fruit.

6.3.2 Apple carbon balance model

Precision thinning done during the bloom and post-bloom period utilizes two recently developed computer models, the apple carbon balance model and the apple fruit growth model. These two models work together to allow methodical reduction in crop load.

The apple tree carbon balance model was developed by Alan Lakso at Cornell using a mathematical model to estimate apple tree photosynthesis, respiration and growth of fruit, leaves, roots and woody structure (Lakso et al., 2006, 2007). The model uses daily maximum and minimum temperatures and solar radiation to calculate the daily production of carbohydrate and allocates the available carbohydrate to the organs in the tree including developing fruit. Temperature and light are primarily responsible for determining the carbon balance in a tree. In low light intensity and high temperature conditions there would be a carbon deficit and thinning would be easy. Conversely, high light and lower temperatures favour carbohydrate accumulation and thinning would be difficult. The carbon balance model is able to quantify the carbon balance using these inputs. The data can then be used as a guide to determine if thinning at a given time is appropriate and if so it provides guidance about the strength of the thinner to use.

6.3.3 Apple fruit growth model

This model is based on the observation that fruit that abscise following thinner application slow growth well in advance of the time that they actually abscise (Greene et al., 2013). Generally 3–4 days is required for a thinner to affect fruit growth. In this model, spurs are tagged prior to thinner application. At 3–4 days after thinner application all fruit on the tagged spurs are measured. Generally, at 3–4 days after application of the thinner, fruit that are induced to abscise will start to show a reduction in growth rate. Fruit must be at least 6 mm in diameter to start measuring. A second measurement is made 4–5 days after the initial measurement (Fig. 7). At 7–8 days after thinner application a significant

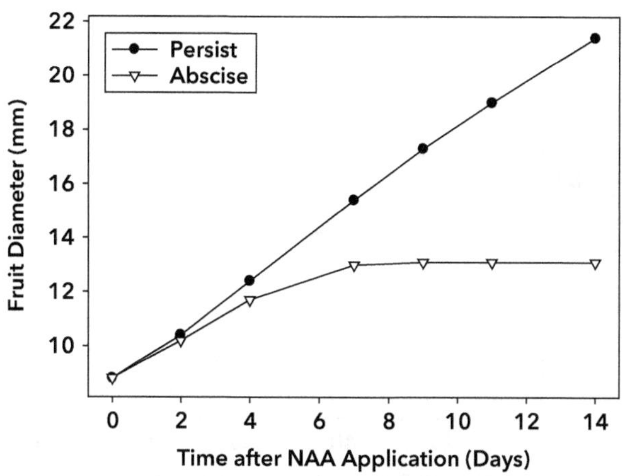

Figure 7 Fruit growth of an apple that will persist to harvest and one that was induced to abscise as a result of a naphthaleneacetic acid (NAA) thinning spray. Fruit that are induced to abscise exhibit slow growth 3–4 days after application and by 8 days after application fruit growth has essentially stopped.

reduction in fruit growth can be measured. These data are entered into an Excel spreadsheet programme and the growth rate of each fruit is calculated. Fruit growth rate is a good indication of the fate of young fruit (Fig. 8). This figure illustrates that slow growing fruit will abscise and fast growing fruit will persist. The fruit growth model considers fruit that grow at less than 50% of the rate of the fastest growing fruit will abscise whereas those growing faster than 50% will persist. This model requires establishing a desired end crop load. The spreadsheet predicts the crop load if no further thinning is done. If the predicted crop load is insufficient then a follow-up thinner spray is applied. Fruit will then be measured again to determine if the crop load has been lowered to the target level. Implementation and details for grower use are outlined by Robinson et al. (2013, 2014) and Schwallier and Irish-Brown (2015).

6.3.4 Pollen tube growth model

Blossom thinning is done routinely in areas that are typically arid. It has the advantage over post-bloom thinning since it can have the greatest effect on fruit size and return bloom (Greene, 2002). Thinners act by preventing pollination and fertilization by damaging the anthers, stigma and style of flowers or inhibiting pollen tube growth (McArtney et al., 2006). Blossom thinners can be very erratic since timing is critical for its success. Ideally, there should be a sufficient number of king flowers open and pollinated to achieve a full crop. Once this is accomplished a blossom thinner should be applied to prevent pollination of the remaining flowers. Timing of the spray is important to the success of blossom thinners. The pollen tube growth model

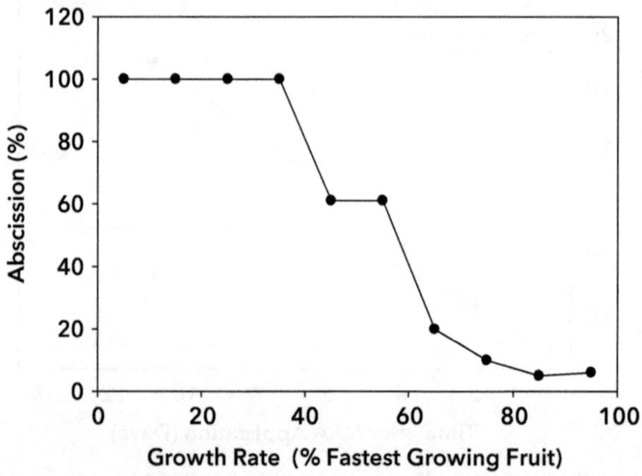

Figure 8 The fruit growth rate 4–8 days after thinner application is an excellent indicator/predictor of which fruit will persist and which will abscise. Source: Greene et al. (2013).

was developed to provide guidance to accurately time the application of the blossom thinners (Peck et al., 2016; Yoder et al., 2013). Temperature influences the rate of pollen tube growth. The length of the style is initially measured, and taking temperature into account, the model indicates the time required for the pollen tube to grow through the style to fertilize the ovule. As soon as the model predicts the pollen tube has reached the ovules in the ovary, the blossom thinner should be applied to discourage pollination and fertilization of all lateral flowers. Since pollen tubes grow at variable rates of different cultivars, a model for each cultivar must be established.

7 Influencing flowering and fruit set

Developing and executing a management strategy to assure adequate bloom each year is a significant challenge. A key component to achieve this goal is to successfully manage a chemical thinning programme. However, in situations where less precocious rootstocks are used, PBR treatments may be useful to help control vegetative growth and increase flowering bud formation. The rapid shift to the use of dwarfing rootstock that is occurring has changed the way orchardists approach the problem of regulating flowering and fruiting in an orchard. The most commonly used rootstocks now are very precocious, and with some varieties, excessive flowering and fruit set may occur before the tree has had the opportunity to fill its allotted space. In these situations the major problem is to establish an appropriate balance between allowing fruit set to occur on young trees while limiting the crop load to allow adequate return bloom and growth of young trees. The need for increased flower bud formation in orchards has changed over time. The need remains but the methods used are different.

7.1 Promotion of flowering on non-bearing trees

The age that flower bud formation and fruit set occur on young apple trees is frequently delayed when trees are propagated on less precocious semi-dwarfing rootstocks. This delay in age at which flowering starts limits early production and extends the amount of time required to achieve full return on the investment made in an orchard. There are cultural methods that growers can use to promote flower bud formation. Scoring or girdling are methods of severing the phloem thus reducing or preventing the movement of photosynthate from the top of the tree down to the root system. This starving of the root system encourages flower bud formation while reducing vegetative growth. Scoring generally is the least aggressive method since a knife is often used to make the cut around the trunk. Girdling is more aggressive since it involves the removal of a band of bark on the trunk of a tree. The wider the cut or the larger amount of bark removed the more severe the girdling effect will be. A hacksaw blade may be used to remove a narrow strip whereas a chainsaw can remove a much wider strip of bark. The timing that girdling is generally done is shortly after bloom when shoots are 5–10 mm in length. Even on trees where the use of this method is appropriate, growers are frequently reluctant to use this method of growth control now because it creates an avenue for the fire blight bacteria to enter the tree at a time of year when this devastating disease is frequently active. This is a greater concern now that fire blight has become a more prevalent and widespread problem.

Branch spreading and bending are cultural practices that have been used on trees to reduce vegetative growth and increase flower bud formation. In the past, wooden limb spreaders with sharpened nails in the ends were used on permanent and semi-permanent scaffold branches. This was a very common practice used on trees on vigorous semi-dwarfing rootstock. Many methods have been used to spread limbs and flatten branches on a tree including toothpicks, numerous types of weights and clothespins to mention just a few. More recently branch bending has been used as a method to control growth and increase flower bud formation on large limbs on newly planted trees in a tall spindle planting (Robinson et al., 2008).

The primary PBR that is used to enhance flowering on non-bearing trees is ethephon, although the frequency of use has diminished as the number of trees propagated on semi-dwarfing rootstocks has declined. In general, ethephon is initially applied on trees that are large enough to bear at least a partial crop of apples and this is usually on trees 2–4 years old. The recommended rate is 300–450 mg L^{-1} and the application is usually made 10–14 days after bloom. A reduction in both terminal and lateral growth can be expected. Sometimes application of ethephon at this time does not result in increased bloom or subsequent fruit set and application for a second year may be appropriate (Greene and Lord, 1978).

7.2 Promotion of flowering on bearing trees

Flowering on bearing trees is regulated primarily by crop load adjustment using chemical thinners. However there are several reasons why orchardists may wish to employ supplemental methods to encourage flower bud formation. First, as many trees start to come into production they may be very easily nudged into biennial bearing when 3 or 4 years old if crop load on these young trees is not regulated carefully. Some popular varieties have a natural tendency to become biennial. Chemical thinning is not a precise science since success is often dependent upon the weather conditions working in conjunction with thinners. It is not always possible to time thinner applications with favourable weather since weather forecasts are subject to change after a thinner has been applied. In some years favourable weather may never occur. Where conditions are not favourable, trees may be overset at the end of 'June drop'. There is a period of time after the end of June drop that flowering can be affected by PBRs, although thinning may not be possible. There are two PBRs that may be used on bearing trees. NAA is the most common PBR used on bearing trees for this purpose. NAA at 5-7.5 mg L^{-1} can be applied 3-4 times at 7-10 day intervals, starting 4-6 weeks after bloom when fruit size is generally greater than 25 mm. Alternatively, ethephon at 100-150 mg L^{-1} may be used in a similar manner to NAA (Byers, 1993). These treatments do not guarantee increased flower bud formation, especially on difficult varieties to increase bloom on such as 'Honeycrisp', but they are the best 'insurance policy' available to affect bloom on bearing trees.

7.3 Inhibition of flowering

Regulation of flower bud formation in apples is generally accomplished by causing fruit abscission during the first 3 weeks after bloom. Fruit or seeds in fruit are primarily responsible for the inhibition of flower bud formation in spurs (Chan and Cain, 1967). It is widely acknowledged that gibberellins produced in the seeds of developing fruit diffuse out of the seed, migrate to the bourse bud associated with that spur and inhibit flower bud formation in the spur. By simply removing the fruit from some spurs, flower bud formation on those spurs is then allowed to proceed. Therefore, the removal of developing fruit does not increase flower bud formation per se, but rather removal of fruit removes the flower bud inhibitor, gibberellin present in the seeds, allowing flower bud formation to proceed unimpeded in the bourse bud associated with the spur.

Apples and pears have a biennial bearing habit which results in a heavy bloom and fruit set in the 'on year', followed by a year where there are few flowers present which result in a few fruit setting, the 'off year'. The normal way to break or modify this cycle is to thin trees in the 'on year' which will then leave spurs with no fruit allowing for flower bud formation to proceed. However, some cultivars are difficult to thin making this strategy difficult and trees such as peach can't be thinned with hormonal sprays effective on apples and pears. It is possible to apply gibberellins to trees in the 'off year' and even out flowering in the 'on year' by inhibiting flower bud formation. To be most effective the gibberellins must be applied early in the season to achieve maximum effect, although some inhibition of flowering can occur up to 2 months after bloom (Tromp, 1982). McArtney (1994) reported that a single spray of gibberellin A_3 (GA_3) or GA_{4+7} at full bloom in the 'off year' reduced the subsequent season's flowering in the 'on year' and the severity of biennial bearing in 'Braeburn' apple.

7.4 Commercial PBRs that inhibit flower bud formation

There are proprietary gibberellin products that are registered for use on apples, although inhibition of flowering is not one of the registered label uses. GA_{4+7} (Provide®, Novagib®, TypRus®) is used to reduce fruit russetting and to reduce cracking on some apple varieties. Although applications for russet control start at petal fall, inhibition of flowering probably does not occur as the result of this treatment due primarily to the low rate used. A 50% 6-BA plus 50% GA_{4+7} formulation (Promalin®, Perlan®, Typy®) has several uses in tree fruit but the use that has the greatest influence on flowering is when employed to increase branching on trees in the nursery and to enhance branching on newly planted trees. Application is made early in the spring. The high rates used at this time can inhibit flower bud formation. This may be considered a positive response since it is generally undesirable to have fruit development too early on young trees which in turn could delay tree development and filling of the allotted space.

7.5 Increase fruit set

In tree fruit production emphasis is frequently placed on strategies to reduce fruit set as a way to improve fruit quality at harvest and to encourage good return bloom. However, there are a number of situations where it is appropriate to increase fruit set. Situations where increased fruit would be appropriate include frost damage to flowers, poor weather during the pollination period, light blooming trees, winter cold damage to buds and on cultivars where natural fruit set is typically low.

Foliar sprays of gibberellins can increase fruit set and parthenocarpic fruit development on pears when applied near bloom and this strategy can be used commercially to increase cropping on blocks of pear trees that have poor natural fruit set. Yarushnykov and Blanke (2005) reported that gibberellins could increase fruit set of pears after frost damage at bloom. Gibberellins, particularly GA_{4+7} can cause parthenocarpic fruit set on apples (Bukovac, 1963) but the response is quite variety dependent. McArtney et al. (2014) reported that the proprietary mixture of 6-BA plus GA_{4+7} increased parthenocarpic fruit set on frost damaged apple trees. While final fruit set was reduced in all trees by the frost, treated trees had a significantly higher yield than non-treated trees. It appears that the treatments must be applied within 4 days of the frost event to be effective.

It was established as early as 1980 that AVG could increase the fruit set on apples (Williams, 1980). The commercial formulation of AVG (ReTain®) was not registered for commercial use until 1997 and addition of increasing fruit set on the label did not occur until later. The suggested time of application is from the pink stage until full bloom. Use for this purpose is not widespread, in part because the need is frequently not great and application at bloom reduces the amount that can be applied nearer harvest where the use is common and widespread. Some cherry cultivars (e.g. Regina) routinely have low fruit set and AVG application between the balloon stage and first bloom can increase set (Bound et al., 2014).

Early in the development of prohexadione-calcium (Pro-Ca) it was recognized that it had the potential to increase the fruit set of apples when applied to control vegetative growth. Application timing is from pink to when the new shoot growth is 2–5 cm in length and it requires 10–14 days for the compound to start to exhibit signs of growth retardation. This time coincides somewhat with the initiation of 'June drop'. Pro-Ca generally does not increase fruit set per se but it may inhibit June drop which frequently does lead to an increase in final fruit set. Pro-Ca can negate some of the abscission-promoting effects of chemical thinners. Therefore, it is routinely suggested for orchardists to be more aggressive in their thinning programmes on blocks treated with Pro-Ca to help negate the increased fruit set tendencies induced by Pro-Ca. However, the primary reason to apply Pro-Ca remains to achieve growth retardation and to aid in the control of shoot fire blight rather than using it to specifically increase fruit set.

8 Pre-harvest application of plant bioregulators

PBRs are frequently used during the 4 weeks prior to normal harvest to influence the ripening process and to retard pre-harvest drop. The PBRs used and the time of application will influence the time of ripening, fruit quality, storage potential and fruit abscission.

8.1 Delay pre-harvest drop and retard ripening

The apple is a climacteric fruit and as such the onset of the climacteric signals a definitive stage in the development of maturation and ripening (Watkins, 2003). In climacteric fruit, ethylene advances the time of the onset of the climacteric. Ethylene is an endogenous hormone that plays a dominant role in the ripening process in apples and an initiator of pre-harvest drop. Stressful conditions (drought, high temperature) can trigger and exacerbate abscission prior to fruit reaching an acceptable level of maturity for harvest. Additionally, if orchardists do not have sufficient labour to harvest fruit in a timely manner, drop control measures would be required to avoid substantial economic loss especially on drop-prone cultivars.

Many commercially important apple cultivars suffer from the pre-harvest drop malady before they have reached maturity that would indicate optimum commercial harvest date (Greene et al., 2014). Drop-prone cultivars such as 'McIntosh', 'Macoun' and 'Honeycrisp' may suffer substantial loss due to drop.

With these cultivars, it is often normal to have 20% of the fruit drop prior to harvest and in some years, it is not uncommon to have losses up to 50%.

8.1.1 Pre-harvest drop control PBRs

There are three commercially available drop control compounds: NAA (many formulations), AVG (ReTain®) and 1-methylcyclopropene (1-MCP, Harvista™).

8.1.1.1 Aminoethoxyvinylglycine

AVG is a naturally occurring PBR that was discovered in the early 1970s. Its primary mode of action is to inhibit ethylene biosynthesis (Boller et al., 1979). Bangerth (1978) was the first to demonstrate that it could inhibit pre-harvest drop of apples. Development of this compound for drop control and other uses was not continued because of the high cost of production. In response to the withdrawal of registration of daminozide in 1989, development of AVG was resumed. Abbott Laboratories registered a proprietary formulation of AVG in 1997, ReTain®, to control pre-harvest drop and delay ripening of apples.

The extent of and the strategy for pre-harvest drop control is affected by the cultivar, the time of application, the objective of the application and the amount applied (Greene and Schupp, 2004; Schupp and Greene, 2004). For most cultivars where pre-harvest drop is a significant problem, the initial application is generally made 3–4 weeks before the anticipated start of normal harvest. Following application, it generally requires 10–14 days for drop control to become effective (Greene, 2006). If an application is made later, there is a chance that early drop may occur before AVG can become effective. Earlier application may result in early drop control but loss of drop control later during the harvest season when pre-harvest drop control is characteristically more severe. Split applications are a partial solution.

For most drop-prone cultivars, an initial application of one pouch (333 g of 15% ReTain® formulation per acre, 825 g ha^{-1}) is used. This rate will frequently control and restrict drop loss to about 20% between 5 and 6 weeks after application. It has long been known that higher rates of AVG will increase drop control. In 2015, the ReTain® label was updated to allow application of two 333 g packages of ReTain® per acre (1650 g ha^{-1}) in a season. This additional AVG further extended the effective period of drop control (Fig. 9). A lower (half) rate is usually used on low ethylene producing cultivars such as 'Gala' and 'Honeycrisp'. While 'Gala' does not suffer from pre-harvest drop, ReTain® is used to delay ripening which helps reduce fruit cracking and developing an oily 'greasy' feel which is frequently associated with ripe fruit. If full rates are used on these cultivars, ripening and red colour may be delayed too

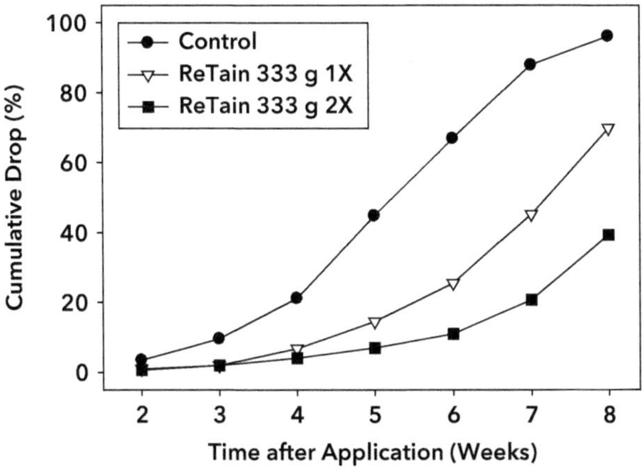

Figure 9 Influence of one full rate application of ReTain® (333 g acre^{-1}, 825 g ha^{-1}) or two full rates on pre-harvest drop of 'McIntosh' apples. The double application slows drop and allows greater drop control for a longer period of time. If one uses 20% drop as an acceptable loss, then the double application extends the period of drop control for 7–10 additional days.

long (3–4 weeks). The maximum amount of ReTain® that can be applied in an orchard is two pouches (666 g acre^{-1} (1659 g ha^{-1} year^{-1})). This amount may be applied at one time or it can be applied in two or more instalments. The timing and the amounts applied at each time depend on the objectives of the grower. The earlier in the season ReTain® is applied, the greater the delay in ripening. Maximum delay in ripening would occur if the highest rate of ReTain® was applied 4 weeks before harvest. Frequently growers use a split application strategy to have less reduction in ripening and use later applications to extend the period of drop control.

A 100% organosilicone surfactant accompanying ReTain® will enhance foliar uptake and improve performance but not all surfactants are equally effective. Currently, Silwet L-77 and Sylguard 309 are recommended for the United States at the rate of 0.1% v/v. Other organosilicate surfactants may be effective, but they would require testing. The organosilicone surfactants also provide some rain fastness if the spray has had a sufficient amount of time to dry (Greene, 2006).

8.1.1.2 1-methylcyclopropene (1-MCP, Harvista™)

1-MCP is an inhibitor of ethylene action that acts by binding irreversibly to ethylene receptors in the treated tissue. If binding sites are occupied by 1-MCP a plant is unable to respond to ethylene, even if present (Fan et al., 1999). It is applied to harvested fruit in sealed rooms. Because 1-MCP is a gas, application to trees in the orchard provides logistical challenges. An initial approach was to mix 1-MCP with oil and water immediately prior to application to ensure a more targeted application. Harvista™ can be applied using a proprietary in-line injections system developed by AgroFresh. However, AgroFresh is developing its own sprayer adaptation to allow the application of Harvista™.

The pre-harvest drop that occurs on trees treated with ReTain® differs from that on trees treated with Harvista® (Fig. 10). On ReTain®-treated trees drop occurs slowly at first and then after 5–6 weeks the rate of drop accelerates as more fruit become climacteric. On Harvista™-treated trees, initial drop control is nearly complete but new ethylene binding sites are generated in the fruit during the ripening process. Ethylene produced can then attach to those new binding sites, thus initiating the abscission process. This results in a rather rapid increase in pre-harvest drop that can be substantial.

Harvista™ like ReTain® may also delay red colour development, starch degradation, flesh firmness loss, degradation of starch and water core development. Following application, drop control may be achieved within 3–5 days. Since it can be applied closer to normal harvest, fruit have a longer time to mature on the tree normally and to develop red colour compared with ReTain® which must be applied earlier. Loss of drop control with Harvista™ is associated with ripening fruit starting to generate new ethylene binding sites which can then respond to ethylene being produced by the fruit. A follow-up application of Harvista™ may saturate newly generated binding sites thus significantly extending the period of drop control (Fig. 11).

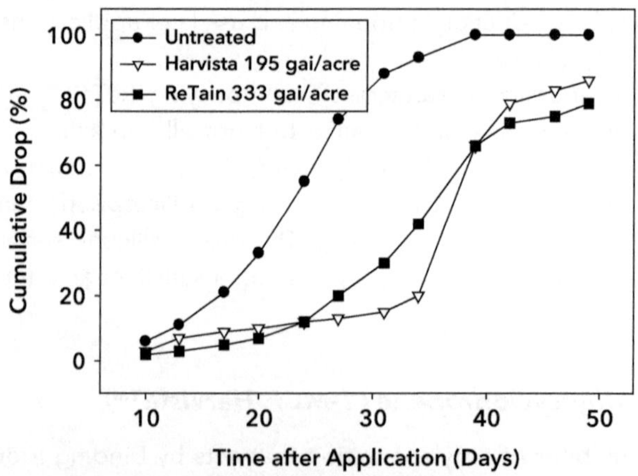

Figure 10 Comparison of pre-harvest drop control of ReTain® and Harvista™ on 'McIntosh' apples. Both compounds retarded drop comparably early in the season. Drop control on Harvista™-treated trees was lost rapidly but drop from ReTain®-treated trees was more gradual.

8.1.1.3 Naphthaleneacetic acid (NAA)

It was discovered in the 1930s that auxins can inhibit pre-harvest drop of apples. Over the past 75 years several auxins have been registered for pre-harvest drop control, but NAA remains the only auxin registered for use for this purpose.

The standard recommendation for drop control is to make the initial application before the first sound fruit start to drop. NAA at 10 mg L^{-1} is frequently used and drop control typically lasts for 7–9 days. A second supplemental application can be made to extend the period of drop control up to 14 days. NAA may advance ripening on apples when used to control drop. The amount of stimulation depends on the cultivar used, the weather following application, time interval between application and harvest and the amount NAA applied. Higher rates of NAA, very warm weather (mid-80°F (30°C)) and a delay in the harvest of treated fruit are likely to advance ripening and lead to reduced storage life. Drop control following NAA application can be seen within 2–3 days of application unless significant drop was occurring at the time of application.

NAA is sometimes used in conjunction with ReTain®. Since ReTain® requires 10–14 days for drop control to be initiated, NAA may be included in the application for near-term drop control. A minimum of a half rate of ReTain® is required to counteract the tendency for NAA to advance ripening. Recently, Robinson et al. (2010) and Yuan and Carbaugh (2007) reported addition of NAA in the second part of a ReTain® split application that was reported to enhance

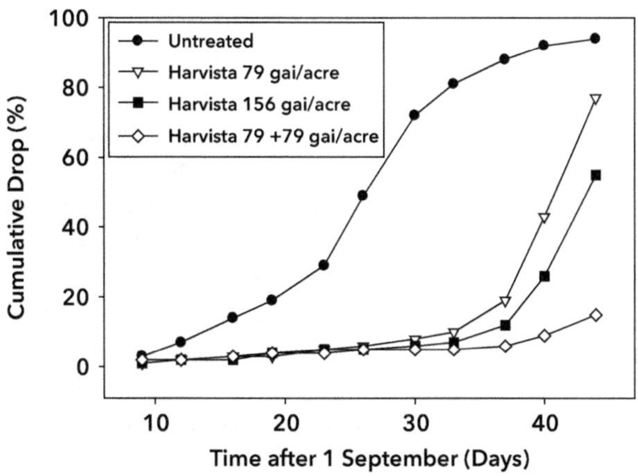

Figure 11 Influence of single and double application of Harvista™ on pre-harvest drop of 'McIntosh' apples. The higher rate of Harvista™ controls drop more effectively than a lower rate. However, a second application of the lower rate of Harvista™ made 2 weeks after the first resulted in drop control better than the higher rate applied only once.

the drop control of a ReTain®-alone application. An alternative approach to drop control by NAA preloading was reported by Marini et al. (1993). Application of several low rates of NAA (5 mg L^{-1}) starting 4 weeks before anticipated harvest resulted in excellent drop control of 'Delicious' apples that was equal to or better than a higher rate made at the traditional time nearer harvest. This approach has been useful on 'Delicious' but it has not been successful on more ethylene-sensitive cultivars such as 'McIntosh'.

8.2 Advance fruit ripening and increase red colour

As normal harvest approaches and seasons start to change, there is usually heightened interest by consumers to purchase first-of-the-season apples. However, this interest usually precedes the time that the fruit are ripe, palatable and ready to harvest. There is the temptation by orchardists to harvest fruit early when it lacks adequate colour, is starchy, acidic and tannic in order to satisfy this demand. However, selling unripe and inferior quality fruit would depress the market and discourage customers from making return purchases. Fortunately, there are PBR combinations available to allow the harvest of good quality fruit that have the varietal characteristics of fruit harvested later during the normal harvest season.

8.2.1 Drop control when using ethephon

Whenever ethephon is applied to apple trees a drop control PBR must be used as well. Ethephon causes ethylene evolution when absorbed by the tree and increased endogenous ethylene production is stimulated which advances ripening resulting in increased pre-harvest drop, especially on drop-prone cultivars. There are three drop control compounds available but only ReTain® and NAA have been extensively evaluated to be used with ethephon. In situations where ethephon is being applied primarily to advance ripening, NAA is the drop control most often selected since the advanced ripening caused by NAA is not necessarily detrimental and the cost is less. One application of NAA at 10 mg L^{-1} will generally control drop for about 7 days and a second application may extend drop control to 12–14 days. Some orchardists choose to delay application of NAA until 4 days after ethephon application thus extending the period of drop control without having to make a second application. This strategy comes with some peril since poor weather may prevent a grower from making the application in a timely manner and drop starts before the NAA can be applied or become effective.

In situations where ethephon is applied primarily to increase red colour but storage of harvested fruit is planned then the use of ReTain® may be the appropriate choice to control drop. Since ReTain® delays ripening as well as red colour development, a reduced rate may be appropriate in this situation. It is suggested that reduced rate be used (½ pouch, 166 gal $acre^{-1}$) and it should be applied 2–2.5 weeks before ethephon is applied.

8.2.2 Ethephon to advance ripening and increase red colour

Ethephon is the primary PBR used to advance fruit ripening. It should be applied between 1 and 3 weeks before normal harvest at rates between 62.5 and 300 mg L^{-1}. Increased red colour will be noted 5–7 days after application. Flesh softening may accompany red colour development. The closer the application is made to the time of normal harvest, the shorter the time required to observe a red colour and maturity response. If warm cloudy weather follows application, advanced ripening may occur with little or no increase in red colour development.

Ethephon-treated fruit should be monitored carefully following harvest since a reduction in storage life should be expected. Factors that may shorten the storage potential of fruit include use of high rates of ethephon, a long interval between the time of application and harvest, high temperatures while fruit is on the tree, the drop control compound selected and a delay in the time to cool the fruit down so that the internal temperature is reduced to 0°C. Therefore, treating apples with ethephon will mean early harvest of fruit, which will result in reduced yield due to smaller fruit size. A rule-of-thumb is that a fruit increases in size about 1% per day provided it remains on the tree.

8.2.3 Increase red colour with ethephon

There are some geographical locations where fruit do not develop good red colour and there are years when warm and cloudy weather persists resulting in poorly coloured fruit. In situations such as this, some growers wish to enhance red colour without appreciable advancing of ripening. In situations such as this a rate of ethephon of between ¼ and ½ pint of ethephon per 100 gal (75–150 mg L^{-1}) should be applied 7–10 days prior to the anticipated time to harvest. Care should be taken to monitor fruit maturity to make sure that fruit do not become too ripe on the tree. In recent years more emphasis has been placed on using starch testing as guide to help determine maturity and when to harvest fruit.

9 Improving fruit appearance and shape

Appearance of fruit often plays a dominant role in purchasing decisions made by customers and greatly impacts the price that growers receive from the packing house. Good cultural techniques and management of the tree are essential prerequisites but frequently PBRs may be required to improve fruit appearance even when management is excellent.

9.1 Fruit russet control

Russet is a physiological disorder that occurs during the early stages of fruit development. Rapidly dividing epidermal cells increase in size faster than the cuticle can expand. This leads to the cuticle cracking, exposing and damaging some of the epidermal cells below. As a result, cambium that is formed produces corky periderm russetted areas (Faust and Shear, 1972). Although not the only cause of russetting, water plays a critical and central role in its formation, as it tends to be more pronounced under conditions of high humidity or frequent rain or dew. These conditions lead to an apple developing a thin cuticle that is prone to cracking under high turgor conditions (Faust and Shear, 1972). Some varieties are also genetically pre-disposed to develop russet.

Foliar sprays of GA_{4+7} can either eliminate fruit russetting or reduce its severity (Taylor, 1975). When applied near petal fall at rates of 25–250 mg L^{-1} GAs are known to inhibit bud formation in flowers even when used at moderate rates (Tromp, 1982). By making multiple sprays (generally 4) of GA_{4+7} at a rate of 10 mg L^{-1}, fruit russetting can be reduced while having minimal effect on return bloom (Miller, 1988). High rates may improve russet control but the increased control may not be justified because of its adverse effect on return bloom (Eccher and Boffelli, 1981).

Scarf skin is a disorder of apples that is characterized as fruit having a grey flecking and/or a milky appearance on the surface. Its presence on the fruit leads to a reduction in grade and therefore lower returns to the grower. Treatment of susceptible cultivars with multiple applications of GA_{4+7} (as described for russet) will reduce the incidence of scarf skin but it may not reduce the incidence to commercially acceptable levels (Ferree et al., 1984).

9.2 Fruit shape

Fruit shape can be a significant component when customers choose an apple to purchase. Cultivars that are blocky and have prominent calyx lobes are perceived by customers as being preferred. Apple elongation and calyx lobe development are promoted by low temperatures during the very early fruit growth stage (Shaw, 1914). Cool nights during early fruit development characterize the growing conditions in Washington. The Washington apple industry has been able to market shape, especially for 'Delicious' as a quality parameter. Cool nights cause increased production of endogenous GAs. Abbott Laboratories developed a proprietary product, Promalin®, to elongate the shape of 'Delicious'. It contains equal quantities of GA_{4+7} and 6-benzyladenine

(BA) and this was registered in 1979. The application of this product near bloom allows growers in other regions that do not have cool temperatures at night to produce elongated 'Delicious' with prominent calyx lobes. Unrath (1974) demonstrated that when Promalin® was applied at 25 mg L^{-1} at bloom time, fruit length/diameter ratio and weight increased. Using GA rates above 50 mg L^{-1} could result in a reduction in return bloom. GA_4 and GA_7 increase fruit elongation but GA_7 is known to reduce flower bud formation more than GA_4 (Tromp, 1982).

10 References

Abbaspoor, M., Teicher, H. P. and Streibig, J. C. 2006. Effect of root-absorbed inhibitors on Kautsky curve parameters in sugar beet. *Weeds Res.* 46:226–35.

Bangerth, F. 1978. The effect of a substituted amino acid on ethylene biosynthesis, respiration, ripening and preharvest drop of apple fruit. *J. Am. Soc. Hort. Sci.* 103:401–4.

Bangerth, F. 2004. Internal regulation of fruit growth and abscission. *Acta Hort.* 636:235–48.

Barry, C. S. and Giovannoni, J. J. 2007. Ethylene and fruit ripening. *J. Plant Growth Reg.* 26:143–59.

Batjer, L. P. and Hoffman, M. B. 1951. Fruit thinning with chemical sprays. U.S. Dept. Agr. Circ. 867. US Department of Agriculture, Washington, DC.

Beyer Jr., E., Morgan, P. W. and Yang, S. F. 1984. Ethylene, pp. 111–26. In: M. B. Wilkins (Ed.), *Advanced Plant Physiology*. Pitman Publishing Ltd., London, UK.

Boller, T., Herner, R. C. and Kende, H. 1979. Assay for the enzymatic formation of an ethylene precursor 1-aminocylopropene-1-carboxylic acid. *Planta* 145:293–303.

Bound, S. A., Close, D. C., Jones, J. E. and Whiting, M. D. 2014. Improving fruit set of 'Kordia' and 'Regina' sweet cherry with AVG. *Acta. Hort.* 1042:285-92.

Bukovac, M. J. 1963. Induction of parthenocarpic growth of apple fruit with gibberellins A_3 and A_4. *Bot. Gaz.* 124:191–5.

Bukovac, M. J. 1965. Some factors affecting the absorption of 3-chlorophenoxy-α-propionic acid by leaves of peach. *Proc. Am. Soc. Hort. Sci.* 87:131-8.

Bukovac, M. J. 1973. Foliar penetration of plant growth substances with special reference to tree fruit. *Acta Hort.* 34:69–78.

Bukovac, M. J. 1980. The performance of growth regulators when applied in low volume sprays. *Proc. Michigan State Hort. Soc.* 93:63–7.

Bukovac, M. J. 1985. Spray application technology: A limiting factor in growth regulator performance. *Proc. Wash. State Hort. Soc.* 81:170–4.

Byers, R. E. 1993. Controlling of bearing apple trees with ethephon. *HortScience* 28:1103–5.

Byers, R. E. 2003. Flower and fruit thinning and vegetative: fruiting balance, pp. 409-36. In: D. D. Ferree and I. J. Warrington (Eds), *Apples: Botany, Production and Uses*. CABI Publishing, Cambridge, MA.

Byers, R. E., Lyons, C. G. and Hill, C. H. 1971. Base gallonage per acre. *Virginia Fruit* 60:19–23.

Byers, R. E. Barden, J. A. and Carbaugh, D. H. 1990a. Thinning of spur 'Delicious' apples by shade, terbacil, carbaryl, and ethephon. *J. Am. Soc. Hort. Sci.* 115:9–13.

Byers, R. E., Barden, J. A., Young, R. E. and Carbaugh, D. H. 1990b. Apple thinning by photosynthetic inhibition. *J. Am. Soc. Hort. Sci.* 115:14–19.

Chan, D. and Cain, J. 1967. The effect of seed formation on subsequent flowering in apple. *Proc. Am. Soc. Hort. Sci.* 91:63–8.

Cleland, R. E. 1969. The gibberellins, pp. 49–81. In: Wilkins, M. B. (Ed.), *Physiology of Plant Growth and Development*. McGraw-Hill, Maidenhead, UK.

Clouse, S. D. 1996. Molecular genetic studies confirm the role of brassinosteroids in plant growth and development. *Plant J.* 10:1–8.

Clouse, S. D. and Sasse, J. M. 1998. Brassinosteroids: Essential regulators of plant growth and development. *Annu. Rev Plant Physiol. Plant Mol. Biol.* 49:427–51.

Creelman, R. A. and Mullet, J. E. 1997. Biosynthesis and action of jasmonates in plants. *Annu. Rev. Plant Physiol. Plant Mol. Biol.* 48:355–81.

Crocker, W., Hitchcock, A. E. and Zimmerman, P. W. 1935. Similarities in the effect of ethylene and the plant auxins. *Contrib. Boyce Thompson Inst.* 7:231–48.

Davies, P. J. 1995. *Plant Hormones. Physiology, Biochemistry and Molecular Biology*. Kluwer Academic Publishers, Dordrecht, The Netherlands, 833pp.

Dennis, F. G. 2002. Mechanisms of action of apple thinning chemicals. *HortScience* 37:471–4.

Eccher, T. and Boffelli, G. 1981. Effects of application of GA_{4+7} on russeting, fruit set, and shape of 'Golden Delicious' apples. *Sci. Hort.* 14:307–14.

Evans, J. R., Evans, R. R., Rogusci, C. L. and Rademacher, W. 1999. Mode of action, metabolism and uptake of BAS 125W prohexadione-calcium. *HortScience* 34:1200–1.

Fan, X., Mattheis, J. P. and Fellman, J. K. 1998. A role for jasmonates in climacteric fruit ripening. *Planta* 204:444–9.

Fan, X., Blakenship, S. M. and Mattheis, J. P. 1999. 1-methylcyclopropene inhibits apple ripening. *J. Am. Soc. Hort. Sci.* 124:690–5.

Faust, M. and Sherer, C. B. 1972. Russeting of apples, in interpretive review. *HortScience* 7:233–5.

Ferree, D. C., Ellis, M. A. and Bishop, B. L. 1984. Scarf skin on 'Rome Beauty': time of origin and influence of fungicides and GA_{4+7}. *J. Am. Soc. Hort. Sci.* 109:422–7.

Forshey, C. G. 1982. Branching responses of young apple trees to applications of 6-benzylamino purine and gibberellins A_{4+7}. *J. Am. Soc. Hort. Sci.* 107:128–32.

Fukuda, H. 1997. Tracheary element differentiation. *Plant Cell* 9:1147–6.

Greene, D. W. 1999. Tree growth management and fruit quality of apple trees treated with prohexadione-calcium. *HortScience* 34:1209–12.

Greene, D. W. 2002. Chemicals, timing, and environmental factors involved in thinner efficacy on apple. *HortScience* 37:9–13.

Greene, D. W. 2006. An update of preharvest drop control of apples with aminoethoxyvinylglycine (ReTain). *Acta Hort.* 727:311–19.

Greene, D. W. and Autio, W. R. 1994. Notching techniques increase branching of young apple trees. *J. Am. Soc. Hort. Sci.* 119:678–82.

Greene, D. W. and Bukovac, M. J. 1971. Factors influencing the penetration of naphthaleneacetamide into leaves of pear. *J. Am. Soc. Hort. Sci.* 96:240–6.

Greene, D. W. and Costa, G. 2013. Fruit thinning in pome and stone-fruit: State of the art. *Acta. Hort.* 998:93:101.

Greene, D. W. and Lord, W. J. 1978. Evaluation of scoring, limb spreading, and growth regulators for increasing flower bud formation and fruit set on young 'Delicious' apple trees. *J. Am. Soc. Hort Sci.* 103:208–10.

Greene, D. W. and Schupp, J. R. 2004. Effect of aminoethoxyvinylglycine (AVG) on preharvest drop, fruit quality and maturation of 'McIntosh' apples. II. Effect of timing and concentration relationships and spray volume. *HortScience* 39:1036–41.

Greene, D. W., Krupa, J. and Clements, J. 2000. Effects of simulate rain following ReTain application on preharvest drop and fruit quality of McIntosh apples. *Fruit Notes* 65:57–60.

Greene, D. W., Lakso, A. N., Robinson, T. L. and Schwallier, P. 2013. Development of the fruit growth model to predict thinner responses to apple. *HortScience* 48:584–7.

Greene, D. W., Krupa, J. and Autio, W. 2014. Factors influencing preharvest drop of apples. *Acta Hort.* 1042:231–5.

Greene, D. W., Crovetti, A. J. and Pienaar. J. 2016. Development of 6-benzyladenine as an apple thinner. *HortScience* 51:1448–51.

Jacobs, W. 1979. *Plant Hormones and Plant Development*. Cambridge University Press, New York, NY.

Jones, K. M., Koen, T. B., Oakford, M. J. and Bound, S. A. 1990. Thinning 'Red Fuji' apples using ethephon at two timings. *J. Hort. Sci.* 65:381–4.

Katsumi, M. 1991. Physiological modes of brassinolide action in cucumber hypocotyl growth, pp. 246–54. In: H. G. Cutler, J. Yokota and G. Adams (Eds), *Brassinosteroids Chemistry, Bioactivity and Applications*. American Chemical Society, Washington, DC.

Lakso, A. N., Robinson, T. L. and Greene, D. W. 2006. Integration of environment, physiology, and fruit abscission via carbon balance modeling – Implications for understanding growth regulator response. *Acta Hort.* 727:321–6.

Lakso, A. N., Robinson, T. L. and Greene, D. W. 2007. Using an apple tree carbohydrate model to understand thinning responses to weather and chemical thinners. *N. Fruit Quart.* 15(3):17–20.

Letham, D. 1963. Zeatin, a factor inducing cell division from *Zea Mays*. *Life Sci.* 2:569–73.

Marini, R. P. 1996. Chemically thinning spur 'Delicious' apples with carbaryl, NAA, and ethephon at various stages of fruit development. *HortTechnology* 6:241–6.

Marini, R. P., Byers, R. E. and Sowers, D. L. 1993. Repeat applications of NAA control preharvest drop of 'Delicious' apples. *J. Hort. Sci.* 68:247–53.

McArtney, S. J. 1994. Exogenous gibberellin affects biennial bearing and the fruit shape of 'Braeburn' apple. *New Zeal. J. Crop Hort. Sci.* 22:343–6.

McArtney, S. J. and Obermiller, J. D. 2012. Comparison of the effects of metamitron on chlorophyll fluorescence and fruit set in apple and peach. *HortScience* 47:509–14.

McArtney, S. J. and Obermiller, J. D. 2015. Effect of notching 6-benzyladenine and 6-benzyladenine plus gibberellin $A_4 + A_7$ on budbreak and shoot development from paradormant buds on the leader of young apple trees. *HortTechnology* 25:233–7.

McArtney, S. J., Palmer, J., Davies, S. and Seymour, S. 2006. Effects of lime sulfur and fish oil on pollen tube growth, leaf photosynthesis and fruit set in apple. *HortScience* 41:357–60.

McArtney, S. J., Greene, D., Robinson, T. and Wargo, J. 2014. Evaluation of GA_{4+7} plus 6-benzyladenine as a frost-rescue treatment for apple. *HortTechnology* 24:171–6.

Milborrow, B. V. 1984. Inhibitors, pp. 76–110. In: M. B. Wilkins (Ed.), *Advanced Plant Physiology*. Pitman Publishing Ltd., London.

Miller, S. S. 1988. Plant bioregulators in apple and pear culture. *Hort. Rev.* 10:309–401.

Miller, C., Skoog, F., Okumura, F., Von Saltza, M. and Strong, F. 1955. Kinetin, a cell division factor from deoxyribonucleic acid. *J. Am. Chem. Soc.* 77:1392.

Moore, T. C. 1989. *Biochemistry and Physiology of Plant Hormones*. 2nd Edition. Springer-Verlag, New York, NY, 300pp.

Musacchi, S. and Greene, D. 2017. Innovations in apple tree cultivation to manage crop load and ripening. In: K. Evans (Ed.), *Achieving Sustainable Cultivation of Apples*. Burleigh Dodds Science Publishing, Cambridge, UK.

Peck, G. M., Combs, L. D., DeLong, C. and Yoder, K. 2016. Precision apple flower thinning using organically approved chemicals. *Acta Hort.* (*In press*).

Phinney, B. O. 1983. The history of gibberellins, pp. 19-52. In: A. Crozier (Ed.), *The Biochemistry and Physiology of Gibberellins Vol. 1*. Proeger Publishing, New York, NY.

Price, C. E. 1982. A review of factors influencing the penetration of pesticides through plant leaves, pp. 237-52. In: D. F. Cutler and C. E. Price (Eds), *The Plant Cuticle*. Linnaeus Soc. Symp. Ser. No. 10, Academic Press, London, UK.

Reid, J. B. and Howell, S. H. 1995. Hormone mutants and plant development, pp. 448-85. In: P. J. Davies (Ed.), *Plant Hormones: Physiology, Biochemistry and Molecular Biology*, 2nd ed. Kluwer Academic Publishing, Boston, MA.

Robinson, T., Hoying, S. and Reginato, G. 2008. The tall spindle planting system: principles and performance. *Acta Hort.* 903:571-80.

Robinson, T. L., Hoying, S., Ungerman, K. and Kviklys, D. 2010. ReTain combined with NAA controls preharvest drop of 'McIntosh' apples better than either chemical alone. *N. Y. Fruit Quart.* 18(3):9-13.

Robinson, T., Lakso, A., Greene, D. and Hoying, S. 2013. Precision crop load management. *N. Y. Fruit Quart.* 21(2):3-9.

Robinson, T. L., Hoying, S., Miranda Sazo, M. and Rufato, A. 2014. Precision crop load management Part 2. *N. Y. Fruit Quart.* 22(1):9-13.

Robinson, T. L., Lakso, A. N., Greene, D., Reginato, C. and De R. Rufato, A. 2016. Managing fruit abscission in apple. *Acta Hort.* 1119:1-14.

Rudell, D. R., Fellman, J. K. and Mattheis, J. P. 2005. Preharvest application of methyl jasmonate to 'Fuji' apples enhances red coloration and affects fruit size, splitting, and bitter pit incidence. *HortScience* 40:1760-2.

Savaldi-Goldstein, S. 2006. Brassinosteroids, pp. 617-34. In: L. Taiz and E. Zeiger (Eds), *Plant Physiology*. Sinauer Associates Inc., Sunderland, MA.

Schwallier, P. G. 1996. *Apple Thinning Guide*. Great Lakes Publishing Co., Sparta, MI.

Schwallier, P. G. and Irish-Brown, A. 2015. Predicting apple fruit set model. *N. Y. Fruit Quart.* 23(1):15-18.

Schupp, J. R. and Greene, D. W. 2004. Effect of aminoethoxyvinylglycine (AVG) on preharvest drop, fruit quality and maturation of 'McIntosh' apples. I. Concentration and timing of dilute application of AVG. *HortScience* 39:1030-5.

Shaw, J. K. 1914. A study of variations in apples. *Mass. Agr. Expt. Sta. Bull.* 149:29-36.

Stover, E. W. and Greene, D. W. 2005. Environmental effects on the performance of foliar applied plant growth regulators: A review focusing on tree fruit. *HortTechnology* 15:214-21.

Taylor, B. K. 1975. Reduction of apple skin russeting by gibberellin A_{4+7}. *J. Hort. Sci.* 50:169-72.

Tromp, J. 1982. Flower-bud formation in apple as affected by various gibberellins. *J. Hort. Sci.* 57:277-82.

Unrath, C. R. 1974. The commercial implication of gibberellin A_4 A_7 plus benzyladenine for improving shape and yield of 'Delicious' apples. *J. Am. Soc. Hort. Sci.* 99:381-4.

Watkins, C. B. 2003. Principles and practices of postharvest handling and stress, pp. 585–615. In: D. C. Ferree and I. J. Warrington (Eds), *Apples, Botany, Production and Uses*. CABI Publishing, Cambridge, MA.

Wertheim, S. J. 2000. Developments in chemical thinning of apple and pear. *Plant Growth Reg.* 31:85–100.

Westwood, M. N. and Batjer, L. P. 1960. Effects of environment and chemical additives on absorption of naphthaleneacetic acid by apples leaves. *Proc. Am. Soc. Hort. Sci.* 76:16–29.

Williams, M. W. 1980. Retention of fruit firmness and increase in vegetative growth and fruit set of apples with aminoethoxyvinylglycine. *HortScience* 15:76–7.

Williams, M. W. and Billingsley, H. D. 1970. Increasing the number and crotch angles of primary branches of apple trees with cytokinins and gibberellic acid. *J. Am. Soc. Hort. Sci.* 95:649–51.

Williams, M. W. and Edgerton, L. J. 1981. Fruit thinning of apples and pears with chemicals. U.S. Dept. Agr. Bul. 289. US Department of Agriculture, Washington, DC.

Williams, M. W. and Stahly, E. A. 1969. Effect of cytokinins and gibberellins on shape of 'Delicious' apple fruit. *J. Am. Soc. Hort. Sci.* 94:17–19.

Wismer, P. T., Proctor, J. T. A. and Elfving, D. C. 1995. Benzyladenine affects cell division and cell size during apple fruit thinning. *J. Am. Soc. Hort. Sci.* 120:802–7.

Yang, S. F. 1969. Ethylene evolution from 2-chloroethylphosphonic acid. *Plant Physiol.* 44:1203–4.

Yarushnykov, V. V. and Blanke, M. W. 2005. Allocation of frost damage to pear flowers by application of gibberellin. *Plant Growth Reg.* 45:21–7.

Yoder, K. S., Peck, G. M., Combs, L. D. and Byers, R. E. 2013. Using the pollen tube growth model to improve apple bloom thinning for organic production. *Acta Hort.* 1001:207–14.

Yuan, R. and Carbaugh, D. H. 2007. Effect of NAA, AVG, and 1-MCP on ethylene biosynthesis, preharvest drop, fruit maturity, and quality of 'Golden Supreme' and 'Golden Delicious' apples. *HortScience* 42:101–5.

[illegible] 2003 [illegible]
CABI Publishing, Cambridge, MA [illegible]

[illegible]

Chapter 9

Optimizing fruit production efficiencies through mechanization

Manoj Karkee, Center for Precision and Automated Agricultural Systems, Washington State University, USA; Jacqueline Gordón, Washington State Tree Fruit Association, USA; Bernardita Sallato, Washington State University, USA; and Matthew Whiting, Center for Precision and Automated Agricultural Systems, Washington State University, USA

1 Introduction: challenges in tree fruit harvesting

Many technological advances and new mechanization technologies in farming have been adopted over the last century. Some of the technologies that have transformed various farming operations include row crop-harvesting machines (e.g. combine harvesters for corn and wheat), field preparation and planting machines, chemical application technologies, and auto-guidance and steering technologies. However, various operations in fruit and vegetable production, such as tree fruit and berry harvesting (particularly for fresh market consumption), have not changed for centuries (Fig. 1). Currently, tree fruit and berry crops such as apples (*Malus domestica*), cherries (*Prunus cerasus* and *Prunus avium*), peaches (*Prunus persica*), blueberries (*Vaccinium* spp. and hybrids), and strawberries (*Fragaria* x *ananassa*) are harvested manually around the world. This operation is highly laborious and labor intensive, demanding a huge seasonal workforce in rural areas around the world where fruit crops are concentrated. As the availability of skilled labor and its associated cost is

http://dx.doi.org/10.19103/AS.2018.0040.10

Figure 1 (a) A worker emptying a bag of apples into a bin in the Shenandoah Valley, Virginia, in 1939 (by Peter Killian, USDA); (b) a worker emptying a bag of "Jazz" apples into a bin at an orchard near Prosser, Washington, in 2016.

becoming an increasingly critical challenge (Calvin and Martin, 2010; Hertz and Zahniser, 2013) in the United States and around the world, development of mechanization and automation solutions for fresh market fruit harvesting remains an important priority for these industries.

As shown in Fig. 2, research and development efforts to build fruit-harvesting machines began in the 1970s (Schupp et al., 2011). Over the last half-century, there has been a wide range of research, development, and commercialization efforts around the world to develop technologies that could improve productivity and efficiency, and/or reduce use of manual labor in fruit harvesting. These efforts can, in general, be categorized into three different levels of mechanization that are as follows: (1) harvest aids; (2) mass harvesting systems; and (3) automated or robotic picking systems. Harvest-aid (or labor-assist) machines (or platforms), as the name suggests, are designed to assist human fruit pickers. Mass and robotic harvesting systems, on the other hand, are designed to eliminate the need of manual picking. These technologies will be discussed with specific examples in the following subsections of this chapter.

2 Harvest-aid systems

Harvest-aid machines generally include multiple worker platforms with flexibility to change the height of each platform so that workers can reach any part of the fruit tree canopy without the use of ladders (Fig. 3). A number of researchers and private companies have developed and evaluated various types of harvest-aid machines with different features (Peterson, 2005; Schupp

Figure 2 A canopy-shaking harvester evaluated in a citrus grove in Florida used a number of force transmission fingers that were inserted into the canopy and vibrated to detach fruit.

(a) (b)

Figure 3 (a) Argiles Harvest Assist Platform (Model: AF10-Evolution, Spain) with a singulating conveyor; and (b) DBR Conveyor Concepts vacuum-based harvest-assist machine (Michigan).

et al., 2011). A harvest-aid machine can carry two to eight or more workers at a time, about half of them at a lower level (sometimes workers would be on the ground harvesting onto a lower conveyor) and another half at a higher level to reach upper portions of the canopy. These machines generally also include a fruit conveying and bin-filling system that allows workers to pick fruit directly onto the conveying system (avoiding the filling and dumping of picking bags or buckets). These machines, therefore, improve the working conditions by avoiding the need for moving ladders around and the need for walking around (up and down the ladder, and to the bin to empty the bags) with a heavy load of harvested fruit. The machines also improve, in general, the productivity of individual workers. Research has shown that productivity may be improved from 5% to 30%, depending on the type of orchards (modern narrower orchards with uniform fruit distribution facilitates higher productivity) and team of workers on the machine (Peterson, 2005; Lewis and Hanrahan, 2014). Furthermore, harvest-aid machines have the potential to reduce fruit damage from transfers between full bags/buckets and the bin.

There have been a number of manufacturers from the United States, Europe, and other parts of the world (e.g. *Revo Pioneers*, Italy; *Picker Technologies,* Seattle, Washington; *Van Doren,* East Wenatchee, Washington; *Argiles,* Lleida, Spain; *DBR Conveyor Concepts*, Conklin, Michigan) offering different types of harvest-aid machines with or without self-propelling capability. Some of these machines also have self-steering capability. However, for a number of reasons, adoption of the harvest-assist systems is still quite limited (particularly in the United States). First, when pickers must work together in a nonuniform orchard environment (e.g. with a variable number of fruit in different parts of the canopy), each worker may not have an equal opportunity to reach his/her full potential. Farmers also have cited current worker-compensation practices as a barrier to the adoption of harvest-aid systems. Workers are generally paid by piece rate to encourage higher productivity and longer working hours. It is possible that slower workers in a harvest-aid team might slow down everybody, thus reducing the team's productivity and potential earnings for some of the workers with higher productivity.

An obvious strategy, in principle, to minimize this challenge would be to identify and assign workers with similar productivity to a single platform. However, the variation in work load would still impact the overall productivity of the team. Alternatively, there has been some research in dynamically adjusting platform speed and worker locations (platform height and location) in the local neighborhood of the canopies based on estimated variability of the work load (Vougioukas et al., 2016). However, adding such capability into harvest-aid machines will increase the complexity and cost of the machines, and therefore an even higher level of overall productivity will be necessary to offset the cost of the technology.

Hand harvest activities pose a high risk of back strain and musculoskeletal problems for workers, due to repetitive hand motions, awkward postures from picking at heights with loads, high forces and awkward postures while ascending and descending ladders with loads, and carrying heavy ladders from tree to tree. Harvest-aid technologies have the additional advantage of reducing these injuries in the orchards by avoiding ladders and bags.

3 Mass harvesting techniques: use of handheld tools

Mass harvesting of tree fruit and berry crops involves some transfer of energy to the canopies so that a high proportion of the fruit will detach from the canopies very quickly for collection using some sort of catching mechanism. Fruit detachment for mass harvesting is commonly achieved by applying vibration to the canopy via transfer of shaking energy that is sufficient to produce a force on the fruit that is greater than the pedicel-fruit-retention force, which can vary by cultivar. Energy can be applied to the canopies using different types of shaking methods, including (1) branch-shaking; (2) trunk-shaking (e.g. *SMF Technologies* tart cherry harvester); and (3) canopy-shaking (e.g. *Oxbo Internatinal* citrus harvester, *Littau* blueberry harvester). These techniques, also called nonselective harvesting, do not have the ability to differentiate fruit ripeness stages for selective harvest unless a specific known energy optimization value could be applied for specific cultivars.

As discussed before, manual harvesting of fruit crops such as apples and cherries involves laborious and risky tasks such as ladder movement and fruit picking while climbing up and down the ladder. Handheld harvesting tools is one approach to utilize mechanization technologies to reduce the physical burden of workers, while improving worker productivity and safety. Such devices are generally used to shake portions of the tree (i.e. branches). For fruit trees with complex, three-dimensional (3D) canopy structures that prevent machine harvesting (mass or robotic), handheld harvest tools could provide a compromise solution. For fruit trees with simplified canopies, handheld harvest tools may provide an interim solution while machine-harvest technologies are being improved for commercial adoption for fresh market crops.

Advantages of these tools include avoidance of the use of ladders to reach fruit in the higher parts of the canopy. As an example, one start-up company in India (*Sickle Innovations*) has commercialized a fruit-picking tool that workers can use to harvest apples, mangoes, or other similar fruit in tall, conventional canopies that are common in the country (Fig. 4a). Second, these tools can help workers improve productivity by several magnitudes. A team of researchers at Washington State University investigated various types of handheld shaking-and-catching mechanisms for sweet cherry harvest. One type utilized two workers, one to operate the handheld mechanical shaker while the other

Figure 4 Handheld (labor-assisted) fruit harvesting techniques: (a) apple-harvesting tool (Sickle Innovations, India) and (b) sweet cherry-harvesting technique (Washington State University, Prosser, Washington).

managed fruit catching and transport (Fig. 4b); results from a trial in a modern sweet cherry orchard found that this system removed up to 97% of the fruit (Zhou et al., 2014) and individual worker productivity increased up to seven times in some orchards (Zhou et al., 2016a). However, more research and development is necessary to improve harvested fruit quality and further validate the harvest efficiency and productivity results.

Over the past few years, a team of scientists led by University of Georgia has been developing a handheld blueberry harvesting system (Takeda et al., 2017). In this system, shaking units can be operated by multiple workers on a large platform that would have a catching surface that goes around the canopies, similar to the catching system of current canopy-shaking machines (e.g. *Oxbo International* blueberry harvester).

4 Shake-and-catch machines

Shaking fruit off trees or bushes and catching them underneath is a typical method of mass harvesting called shake-and-catch. Contrary to handheld devices, shake-and-catch machines perform the harvesting operation with little or no human involvement. Commercial variations of mass harvester systems include both trunk and canopy-shaking machines for various crops primarily destined for processing, including tart cherries, citrus for juice, red raspberries for IQF (individual quick freezing), and wine grapes (particularly for low to medium quality wine).

Canopy-shakers use a number of fingers that can be inserted into the canopy and vibrated to transmit shaking energy to the fruit-bearing branches

(a) (b)

Figure 5 Shake-and-catch harvesting machines: (a) a canopy-shaker for citrus harvest (Oxbo 3220, http://www.oxbocorp.com/Portals/0/Oxbo/Brochures/citrus2013.pdf) and (b) a prototype trunk-shaking machine (Peterson, 2005) being evaluated for sweet cherry harvest at Washington State University.

(Fig. 5a). Alternatively, a number of horizontal bars can be used to contact the canopy from either side, with vibrational energy applied to the bars to transfer the energy into the canopy to the points of fruit attachment.

Trunk-shaking machines are designed to firmly grab the tree trunk close to the ground using a powerful shaking arm to apply energy that is transferred into the canopy (Fig. 5b). USDA scientist Don Peterson and his team worked on this concept for many years, developing different versions of trunk-shaking machines for evaluation with apples and cherries (e.g., Peterson et al., 1999; Peterson and Wolford, 2003). These machines were primarily designed for inclined (V or Y) canopy architectures; fruit falling from the trees landed on an inclined padded surface that rolled to convey the fruit into a container. The team reported up to 95% average fruit removal, with catching (recovery) percentage varying from 60% to 85%.

Alternatively, some shaking machines use a wire cable and a reinforced nylon sling placed around the tree trunk to apply shaking energy using tractor hydraulics (e.g. a mechanical tree shaker by SFM Technology Limited). Depending on branch geometric characteristics, different parts of the canopy might get different amounts of vibration, potentially leading to nonuniform and less effective fruit detachment unless a great deal of force is applied to shake the tree really hard. The benefit of this approach is that the machine-tree engagement is only at the trunk, which avoids the potential for direct mechanical contact with fruit that can be possible when the shaking force is transmitted via mechanical fingers or bars.

With shake-and-catch harvesting systems (whether branch-, trunk- or canopy-shaking), fruit damage remains an important concern for commercial viability (Fig. 6; Zhou et al., 2014; De Kleine, 2014). There have been efforts to optimize shaking frequency, amplitude, and/or shaking duration to maximize fruit removal with a low level of fruit damage. In recent years, new innovations

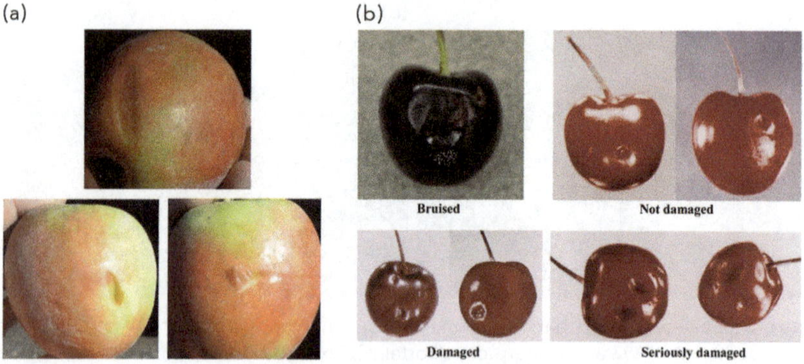

Figure 6 (a) An apple fruit with a bruise, cut, and puncture caused by a shake-and-catch harvesting system (from De Kleine, 2014); (b) different types of cherry fruit damage caused by a shake-and-catch harvesting system (from Zhou et al., 2014).

have been made in shaking, fruit collection, and handling materials and methods that show increasing potential for commercial use of mass harvesting techniques for certain types of fresh market tree fruit and berry crops.

With modern fruiting wall orchards increasingly being used in cherry, apple, and other fruit orchards (e.g. Fig. 7), shake-and-catch harvesting systems can be designed such that only a small canopy area is targeted for shaking at a time. Smaller, targeted canopy-shaking can minimize the optimal energy

Figure 7 A modern fruiting wall orchard of "Aurora" apple in Vantage, WA (Auvil Fruit Company, Orondo, Washington).

transmitted to the fruit to improve shaking uniformity for detachment from the branches. Such a shaking method will reduce the initial acceleration (amount and direction) of the detached fruit, creating the possibility of minimizing the distance for catching it and thus minimizing fruit damage potential. Using this concept, research teams at Washington State University have developed and evaluated targeted shake-and-catch machines for shaking individual branches of cherry trees trained in upright fruiting offshoots (UFO) canopy architecture and apple trees trained in a fruiting wall architecture (Fig. 7; Larbi et al., 2015; De Kleine and Karkee, 2016; Zhou et al., 2016b,c; He et al., 2017a,b, 2018). In addition, an orchard architecture like that shown in Fig. 7 provides an opportunity to catch fruit using multilayers of catching surfaces (Fig. 8a; De Kleine and Karkee, 2016), which has the potential to further reduce fruit damage (Fig. 8b).

Besides fruit damage, it is necessary to avoid premature fruit drop caused by the harvesting system. With shake-and-catch harvesting, this issue is highly challenging. For some crops and varieties with a uniform ripening window, it is not that critical to have the ability for multiple selective harvests. However, for crops and varieties that ripen over an extended period of time, such as mangoes and some apple varieties, it is crucial to have the capability to selectively harvest mature fruit multiple times. When pedicel-fruit retention force is substantially different between ripe and immature fruit, it could be possible to optimize shaking frequency, stroke (amplitude), and duration so that only mature fruit are harvested. When there are only subtle differences in ripe and unripe fruit, though, multiple selective harvests will be difficult, if not impossible. Future research is essential to further explore, understand, and develop solutions to address this issue.

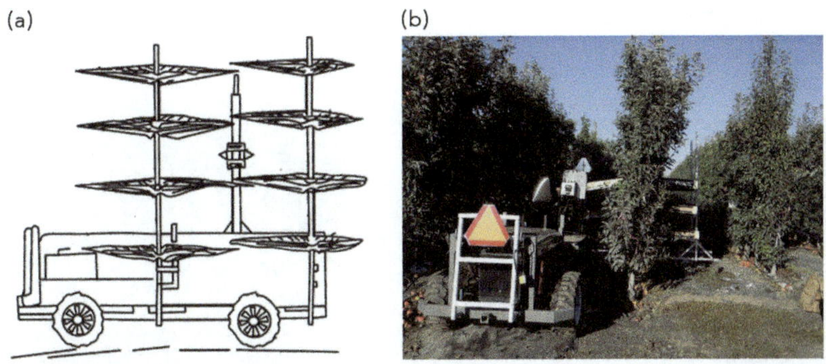

Figure 8 A multilayer system for targeted shake-and-catch harvesting of apples: (a) drawing of a patented multilayer catch-and-shake concept (De Kleine and Karkee, 2016); (b) a prototype multilayer machine in a commercial orchard in Prosser, Washington.

5 Robotic harvesting

Robotic harvesting machines are equipped with sensors, machine vision, robotic arms (also called manipulators), and robotic hands (also called end effectors) to detect, locate, reach, remove, and/or store fruit individually depending on the overall system design. Though mass harvesting methods described in Section 3 has the potential to achieve a high level of throughput with a relatively simpler machine, as discussed above, such a technique is limited by its potential to damage fruits and its limited ability to differentiate ripe and immature fruit for selective harvesting. Robotic harvesting is seen as having potential for various high value, relatively large-fruited crops such as apples, pears (*Pyrus communis*), peaches, and oranges (*Citrus* spp.), although robotic harvesters for strawberries is also under development. Robotic machines can be designed to handle ripe fruit gently to avoid bruising and other damage. In addition, with robotic harvesting, picking only ripe fruit is much more manageable since the fruit are being harvested individually. With current machine vision and sensing technologies, color, shape, size, and even internal parameters such as soluble solids content can be estimated using noncontact sensors. Such information then can be used by the robot to determine which fruit to pick and which to leave intact on the plant. Consequently, robotic harvesting is sometimes called a selective harvesting method.

Robotic fruit-harvesting techniques have been investigated extensively over the past several decades, with researchers around the world proposing different types of sensing, machine vision, control, and automation mechanisms. Scientists have developed and evaluated fruit-picking robots in apples (Fig. 9) (Rabatel et al., 1995; Ceres et al., 1998; Kataoka et al., 1999), citrus (D'Esnon et al., 1987; Harrel et al., 1989, 1990; Muscato et al., 2005), kiwifruit (*Actinidia deliciosa*), cherries, and other crops. These systems, however, were limited by speed, accuracy, robustness, and/or cost, and therefore the technology has not yet been commercialized.

(a) (b)

Figure 9 Examples of fruit-harvesting robots: (a) a French apple-harvest robot; (b) an apple-harvest robot being developed at Washington State University (Silwal et al., 2017).

In recent years, with increasing interest from the industry and increasing production challenges due to hand labor costs and availability, governments, stakeholders, researchers, and private companies have increased investment in the development of robotic harvesting systems for fruit crops. Research institutions actively working on novel robotic concepts and technologies for apple harvesting include Washington State University (Davidson et al., 2016; Silwal et al., 2016, 2017), University of Florida (Mehta et al., 2006), a multicountry team in Europe (e.g. Sweeper Robot, http://www.sweeper-robot.eu/), University of Sydney (Australian Center for Field Robotics), and University of Auckland (http://robotics.auckland.ac.nz/current-work/robots-in-agriculture/). In addition, start-up companies such as Robotic Harvesting LLC (California), FFRobotics (Israel), Abundant Robotics (California), Digital Harvest (Oregon), Robotics Plus (New Zealand), and agricultural automation groups of larger companies such as Panasonic (Japan) and Bosch (Germany) are working on, or have plans to develop, robotic apple-harvesting systems.

6 Fruit conveyance and handling

Irrespective of the harvesting technique used (harvest aid, mass harvesting, or robotic harvesting), harvested fruit needs to be conveyed from the picking point to a collection bin or container into which it is loaded. Different types of conveyance and bin-filling mechanisms have been developed, evaluated, and integrated with harvesting systems. Mechanical conveying is one of the most simple and common methods used. While designs may vary in terms of arrangement of conveying belts and bin-filling mechanisms, this type of system uses a series of mechanical conveyors to deliver fruit from the pickers to the bins. For example, some harvest-aid machines use flat conveyor belts (e.g. Puma Platform, Revo, Pioneers, Italy) whereas others use mechanical conveyance designs with the ability to individualize fruit (e.g. Argiles harvest-aid machine). Fruit individualization can be achieved using cups or fingers, which avoids contact between individual fruit once they are placed on the conveyor, thus reducing the potential for handling-induced fruit bruising.

Another technique uses vacuum tubes to convey fruit from the picker (manual or robotic) to the bin filler (e.g. DBR Conveyance Concepts, Conklin, Michigan). One unique need of vacuum conveying systems is the mechanism to decelerate the delivered fruit for gentle loading into the bin. The effectiveness and robustness of the fruit deceleration mechanism is critical in keeping the handling-induced bruising to a very low level.

Sorting apples into two or three quality grades during harvest has been another area of research and development (Lehnert, 2013; Zhang et al., 2017). A computer vision-based automated sorting apparatus can be integrated into the fruit harvesting system (harvest aid, mass harvesting, or robotic harvesting) with multiple mechanical fruit conveyors and two to three bin-filling mechanisms. Such machines can support a number of picking heads (manual or robotic), and can sort harvested apples according to color, size, shape, and weight for filling into corresponding bins. This technology will enable growers to separate out defective fruit in the orchard, reducing postharvest problems as well as transport, postharvest storage, and packing costs, thereby improving overall harvest efficiency and product safety.

Gravity flow or drop is another method to convey fruit. However, any drops (similar to shake-and-catch harvesting) will generate impact forces that can lead to fruit damage. Various approaches can be used to receive fruit safely on a surface after gravity drop. A gravity drop and a novel catching mechanism (using soft, roller wheels) has been investigated to move fruit from a platform into bins (Lu et al., 2018). Another approach could be to receive fruit in an air suspension system and then gradually reduce the air pressure to let the floating fruit settle down into the bin. Water drop and conveyance has also been used in harvest-aid machines (Picker Technologies, Bellevue, Washington).

Once the fruit gets to the end of conveyance system, different types of bin-filling mechanisms can be used to fill the container. An elephant ear-type bin filler has been utilized by companies like DBR Conveyor Concepts in their harvest-aid machines (Fig. 10). In each of these bin-filling mechanisms, depth sensors are used to determine the instantaneous fill height so that fruit-drop height is controlled precisely. With such a mechanism, the fruit-dropping unit can be lowered to the bottom of the empty bin at the beginning of harvest and gradually raised up as the bin is filled with fruit so that the drop height always remains at the minimum possible level.

Because the process of fruit conveyance and bin filling has the potential to cause substantial fruit damage (particularly bruising), fruit damage percentage

Figure 10 Elephant-ear distributor in a bin-filling system (from Schupp et al., 2011).

is used as an important measure to assess the performance of these systems. Some machines have been reported to cause minimal fruit damage (comparable to completely manual harvest, around 5% or less), but other platforms have been shown to perform unreliably in terms of fruit quality (Peterson, 2005; Lewis and Hanrahan, 2014).

7 Mechanical pruning of tree fruit

Left unpruned, orchard trees will form their natural shape, typically large globular or pyramidal canopies, depending on the genotype. Pruning is therefore a perennial necessity for tree fruit growers to manage canopy form and improve production efficiency and fruit quality. In general, pruning objectives are straightforward: remove dead or diseased wood, improve intra- and inter-canopy light distribution and interception, balance crop load by removing future fruiting sites, reshape a canopy for equipment accessibility, rejuvenate a tree with new fruiting wood, and so on. The specific pruning decisions and processes, including the timing and type of pruning, will vary significantly within a species, depending on genotype, vigor, fruiting habit, and training system. Even within a species, pruning strategies may vary significantly. For example, in the Tatura trellis system, trees are pruned and trained to have permanent vertical leaders with renewable, horizontal fruiting wood; in contrast, trees pruned and trellised as UFO are comprised of a horizontal leader and renewable, vertical fruiting wood. In traditional orchards with low tree density, specific pruning steps and decisions are esoteric, often artful, requiring considerable expertise and variable interpretation on a tree-by-tree basis. Indeed, great skill is required to prune trees in traditional, low density orchards (Fig. 11) to maintain annual bearing, mitigate excessive vigor, and regularly renew fruiting wood from the permanent scaffolds.

Traditionally, pruning has been accomplished manually, using a variety of tools including loppers, secateurs, or saws, and workers climbing ladders to reach the upper canopy (Fig. 12). More recently, these tools have become powered, either by hydraulic, pneumatic, or battery power. This dependence upon skilled labor for pruning typically represents the second greatest expense among annual orchard production costs (Gallardo et al., 2010). In addition, pruning also presents a high injury risk for workers due to the use of ladders, particularly in traditional orchard systems with low-density plantings of large trees (Forshey, 1976; Craig and Embree, 2006).

The gradual evolution of orchard tree architectures from large, 3D, chaotic structures to formally trained planar structures (Fig. 13) and the adoption of precision canopy management strategies offer the opportunity to adopt mechanization and reduce pruning costs (Forshey, 1976; Whiting, 2018). For example, in the UFO training system for sweet cherry, a vertical or angled

Figure 11 Traditional open-center, low-density sweet cherry orchard near Pasco, Washington, immediately following dormant pruning. Photo: M. Whiting.

Figure 12 A dormant pruning crew using long-handled loppers to prune a moderate-density steep-leader sweet cherry orchard in Washington. Photo: M. Whiting.

Figure 13 High-density, vertical fruiting wall apple orchard with seven horizontal fruiting tiers near Prosser, Washington. Photo: M. Whiting

fruiting wall is comprised of vertical fruiting wood from which all lateral branching is removed (Fig. 14). A key objective in the development of this system was the simplification of pruning processes, minimizing interpretation/decision-making, and the need for skilled workers.

Mechanical pruning systems are comprised of either a series of circular sawblades or a sickle bar that is mounted temporarily to a tractor, fork lift, other equipment (Sansavini, 1978) (Fig. 15). Sickle cutting bars are considered to be more effective than rotating blades because they can limit cuts and remove young wood, whereas the larger machines with rotating circular blades are better suited to more severe pruning and remove bearing surface (Forshey, 1976). These systems function similarly to cut limbs in a plane at a distance and

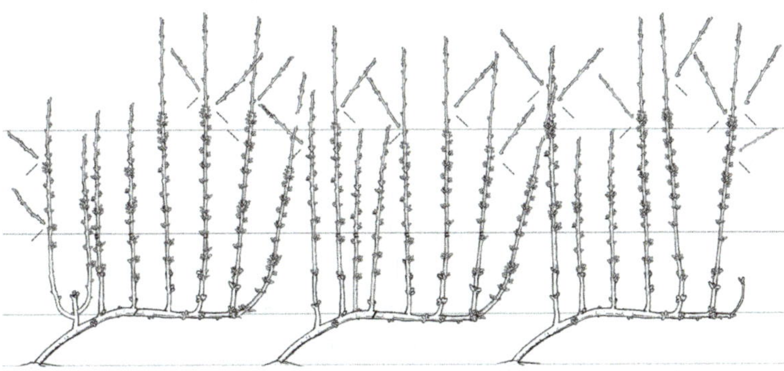

Figure 14 Illustration of pruning removal of lateral branching in three young sweet cherry trees trained to the UFO system.

Figure 15 Sickle bar-type mechanical pruning system mounted to tractor. This system is capable of pruning at any angle, from horizontal topping to vertical hedging. Photo: M. Whiting.

orientation set by the operator. Mechanized systems impose a 'surface' pruning that is described as either topping (i.e. cutting apparatus oriented horizontally to set the upper limit of the canopy) or hedging (vertical or angled orientation of the cutting apparatus to set the side limits of the canopy).

It is essential that growers plan the timing, intensity, and periodicity of the mechanized pruning depending on the cultivar and characteristics of the tree– otherwise the trees may response with undesirable regrowth, fruit quality, and/ or yield (Sansavini, 1978; Stebbins, 1992). Unlike hand pruning, mechanical pruning is not selective. Indiscriminate cuts can negatively influence the growth habit of the tree and stimulate undesirable and vigorous regrowth, which ultimately will modify the shape of the tree and produce unfruitful branches (Forshey, 1976; Sansavini, 1978). Complementary hand pruning following mechanized pruning can yield better results than mechanical pruning alone and costs can be greatly reduced, but it is important to consider factors such as tree size and canopy structure.

Research on the potential for mechanical pruning began in citrus in the 1950s in the United States, and the variability in results due to pruning type and severity, species, tree age, varieties, and other parameters has led to ongoing further investigation (Martin-Gorriz et al., 2014). Kallsen (2005) reported that mechanical pruning did not affect yield nor fruit size of mature 'Frostnucellar' navel orange (*Citrus* × *sinensis*) compared to hand pruning over 4 years. Manual pruning reduced fruit number compared with mechanical topping. Yildirim et al. (2010) reported that mechanical topping and hedging of 13-year-old "Star Ruby" grapefruit (*Citrus* x *paradisi*) trees in Turkey improved fruit yield significantly compared to unpruned trees. Over this 3-year study, trees that were hedged or topped mechanically showed no significant differences in fruit quality. Further, they found that yield from mechanically topped and hedged trees was more consistent compared to unpruned trees, which exhibited large year-to-year variations. This observation underscores the importance of conducting multiyear pruning trials because pruning effects are not manifested entirely during the pruning year, but need to be assessed in following years. Velazquez and Fernandez (2010) compared the effects of mechanical pruning, hand pruning, or the combination of both on yield and fruit quality on 'Valencia Late' orange and found that mechanical pruning alone resulted in low yields and excessive vegetative regrowth. In contrast, mechanical pruning combined with subsequent manual pruning improved fruit production and had similar costs to traditional, manual pruning, thereby increasing profits. Martin-Gorriz et al. (2014) reported a 22% reduction in yield of 20-year-old 'Fortune' mandarin (*Citrus reticulata*) trees pruned mechanically alone over 3 years compared to hand pruning. When exclusively mechanical pruning in 1 year was followed with mechanical + hand pruning in the subsequent year, there was no reduction in yield, and the mechanical pre-pruning reduced the time to perform hand pruning cleanup by 13%, compared to hand pruning alone. These authors concluded that this benefit was not sufficient to offset the expense of the mechanical pruning equipment.

Pruning is also a labor-intensive and costly task in olive production, and several studies have been conducted to find a less expensive and more efficient alternative that does not affect yield and crop quality (Dias et al., 2012). Giametta and Zimbalatti (1997) conducted field trials with 10-year-old 'Leccino' olive (*Olea europaea*) trees in Italy, testing mechanical pruning efficiency and the effect on yield compared to hand pruning and mechanical pruning complemented with traditional pruning. Mechanical pruning was 32 times more efficient than hand pruning, requiring 4 h to prune 100 trees, compared to 128 h necessary for hand pruning. The combination of both methods was also highly efficient, with a labor utilization of 21 h for the same number of trees. There was no effect of mechanical pruning on yield during the 3 years of research and no apparent permanent damage from mechanical pruning. Similarly, pruning efficiency and yield were evaluated by Dias et al. (2012) over 8 years comparing mechanical pruning using a cutting bar of circular saws, hand pruning, and a combination of the two in an olive grove in Portugal. They reported a ca. 20-fold increase in pruning speed (tree number/ha/person) from mechanical pruning compared to pruning by hand and the mechanical/manual combination, which were similar, and fruit yield was significantly higher from trees mechanically pruned which did not differ with mechanical and hand pruning combined. Cherbiy-Hoffmann et al. (2012), on the other hand, found that pruning methodology did not influence yield or fruit set in 'Arbequina' olive trees, but the vegetative regrowth produced by mechanical pruning possibly limited fruit weight and oil content due to low light interception in the interior of the canopy.

Mechanized apple pruning has been tested for decades concomitant with the transition from traditional to high-density systems and increasing labor shortages. Both sickle bar and circular saw machines have been used, but with limited success due to the excessive vegetative growth response in the periphery of the canopy from all of the heading cuts to large limbs in apple trees on vigorous rootstocks. Excessive shading from the mechanical pruning reduced fruit quality compared with hand pruning (Ferree and Short, 1972; Miranda Sazo and Robinson, 2013). Cain (1972) described slot pruning on 'McIntosh' apples in New York performed with a slotting saw during 4-year cycles; a continuous slot of about 2 ft wide was cut 2 ft into the side of the trees along the entire hedgerow, with a similar slot above that cut the next year with no additional pruning. The shoots resulting from cutting one-fourth of the tree each year were less vigorous than those produced by hedging the whole tree, which improved light penetration, fruit color, and the production of fruiting spurs. The slotting saw was compared with a cutter bar during 3 years; results showed that slot pruning performed better, producing three times more new spurs, six times more bearing spurs, and greater light penetration which improved fruit color and yields. More recently, a similar concept of mechanically pruning 'windows' or slots in high-density apple orchards has been evaluated and compared with traditional complete canopy hedging (Dorigoni, 2013). Mechanically pruned 'windows' with multiple-angled, short sickle-type cutting bars (ca. 36 cm) improves light distribution and better mimics hand pruning than indiscriminate hedging (Dorigoni, 2013).

Other work in apple found that late summer mechanical hedging did not influence tree height or average shoot length of four different varieties trained to central leader or trellis systems; however, it controlled tree volume better than root pruning (Ferree and Rhodus, 1993). Yield from hedged trees was significantly lower than from trees with standard pruning; however, their results were also influenced by training system, planting spacing, and cultivar growth habit. Miranda Sazo and Robinson (2013) investigated summer mechanical hedging on several apple cultivars trained to the tall spindle and super spindle systems. They reported that 70% of growing points remained on the trees after pruning, light distribution to the lower part of tree was improved by 10%, and shoot regrowth was minimal, with terminal buds that would likely become flower buds. They also reported better fruit color in mechanically pruned trees compared to control treatments. This study indicates that summer mechanical pruning could be a good approach to reduce apple production costs without affecting fruit quality and regrowth, but further investigation is required to determine long-term effects on tree growth and yield.

More recently, the nonselective outcomes of mechanical pruning, and developments in machine vision systems, have led to interest in the development of robotic pruning systems. Recent advances in precision orchard management have led to training systems that are prescriptive, uniform, and compact (Whiting, 2018) with simplified pruning rules. Karkee et al. (2014) developed a machine vision system using a 3D camera to identify branches in apple trees trained to a spindle system. Similarly, Akbar et al. (2016) reported on 3D apple canopy reconstruction using single-depth images, a system that correctly estimated branch diameter in nearly 90% of limbs. Karkee et al. (2014) reported using algorithms to establish simple pruning rules in the spindle training system to define optimum branch spacing and branch length; future work includes identification of dead branches and large branches, and localization of pruning points in the limbs. Likewise, Elfiky et al. (2015) were able to identify pruning points with 96% accuracy with skeleton-based geometric features in 3D reconstruction of apple trees in the first phase of a research project that is planned to automate dormant pruning of specialty crops.

Limited studies on mechanized pruning have been conducted on stone fruit and almost none on sour and sweet cherry. Summer hedging has been regarded as an invigorating process with negative tree responses if it is used repeatedly and uncoupled with dormant pruning (Webster and Looney, 1996). On the other hand, sweet cherry hedging during the summer can improve light penetration throughout the trees, and tipping can reduce shoot length and control tree size with little effect on floral bud numbers (Flore, 1992). Kesner et al. (1981) reported that hedging one-third to one-half of the current season shoots of 'Montmorency' sour cherry from 40 to 47 days after petal fall reduced shoot length, and increased flower buds, number of spurs, fruit set, yield, and fruit size, probably due to improved light interception or removal of competing sinks. Studies showed that tree size can also be controlled on sour cherry by summer hedging when it is combined with dormant hand pruning every 3–4 years to increase light distribution (Flore, 1992). Hedging timing is an important consideration; when performed too early, it can affect carbon acquisition and partitioning, and if done too late, it can delay cold acclimation due to growth stimulation. Hedging at 45 days after full bloom has been identified as a good timing since it promotes shoots and spur development, which influences yield and fruit quality (Kesner, 1990; Flore, 1992).

In sweet cherry, preliminary mechanical pruning research using rotary discs for hedging and topping showed good potential to improve pruning efficiency and reduce costs when compared to hand pruning of 6-year-old 'Selah'/'Gisela 6' trees trained to the UFO system (Ampatzidis and Whiting, unpublished). Mechanical pruning was accomplished in ca. 8% of the time required for hand pruning despite requiring three passes for each tree (topping plus hedging on both east and west sides of trees). The relative

Table 1 Effects of pruning treatments on sweet cherry efficiency and yield

Pruning technique	Pruning time/ tree (s)	Weight of pruned growth (kg/tree)	Yield/tree (kg)
Hand	158	3.7	11.8 ± 3.2
Full mechanical	13	1.5	12.8 ± 3.8
Mechanical + hand	127	3.6	10.4 ± 2.5

Treatments were imposed during the dormant season in a sixth leaf "Selah"/"Gisela 6" orchard trained to the vertical UFO system north of Prosser, Washington.

speed of mechanical pruning is related directly to tractor speed, which was 1.9 km/h (1.2 mph) in this trial. Traveling faster resulted in excessive limb damage (Whiting, unpublished). In addition, a mechanical pre-pruning reduced manual pruning time by 20% compared to full hand pruning. In this trial, mechanical pruning also removed 60% less wood than hand pruning (Table 1). This is due to the inability of the machine to remove (1) growth oriented parallel with the row, and (2) lateral wood closer than about 10 cm from the trunk. Most of this growth was comprised of current season shoots, thereby having little effect on yield in the subsequent season (Table 1) (i.e. no fruiting spur-bearing wood was pruned). This is expected in the UFO training system since all lateral wood is removed or pruned back to short stubs (Whiting, 2018). More recently, using a sickle bar mechanical pruning system (Fig. 15), postharvest summer pruning of 'Tieton' sweet cherry trained to the UFO system was nearly 30 times faster than hand pruning, with no negative effects on yield or fruit quality in the subsequent season (Whiting and Gordón, unpublished).

8 Conclusion: challenges and opportunities for mechanization in fruit harvesting and pruning

Over the past century, agricultural mechanization and automation technologies have transformed various farming operations in many types of crops, from field preparation to chemical applications and harvesting. Tree fruit and berry crop production has also benefited from technological advances to mechanize operations such as field preparation, precision planting in straight rows, chemical application, nonselective pruning and thinning, and mechanical harvesting (for crops destined for processed markets, e.g. sour cherries and raspberries sent to IQF). However, harvesting tree fruit and berry crops for fresh market consumption remains highly laborious and labor intensive. Research and development efforts in the past have not achieved adequate success for commercial adoption of automated harvesting and pruning systems. Primarily, these efforts have been challenged by (1) variable and uncertain outdoor conditions, (2) complex plant structures, and (3) variable shape and size of delicate tree fruits and berries. Due to these challenges, the technologies developed in the past have been limited by insufficient cycle time, lack of robustness, excessive fruit damage, and high cost.

There are a few areas of research and development that will drive fruit-harvesting technologies closer to practical adoption. First, model-based design should be used extensively to reduce the time and resources necessary to go through repeated cycles of prototype development, evaluation, and redesign to achieve an improved machine. Along with the utilization of model-based design and virtual reality simulation, it is advantageous for modular machines. A multifunction robotic system with modular tools (e.g. to crop-thin, harvest, and prune during various growth stages) will improve growers' returns-on-investment in, and therefore adoption of, these technologies.

Second, agricultural environments and operations are so complex that the development of solutions that optimize collaboration between human and machine are likely to be most practical. Agricultural control stations could be developed in which some human operators supervise (remotely or from the machine cab at the field) and provide input to machines to deal with the most difficult challenges, while the machines take care of most of the job independently. For example, a shake-and-catch harvesting system could be operated remotely using a virtual environment generated by a set of cameras (much like a remote station for piloting airplanes in dangerous areas) to guide placement of the shaking arms to each desired location for actuating the shaking operation. Growers might find an 80–20 solution useful, in which a robotic machine harvests 80% of the fruit in an orchard and the remaining 20% is harvested by a human crew. Such hybrid solutions will offer relatively simpler (and less expensive) machines in combination with a much smaller (but permanently employed) labor force to perform critical refinement of activities such as harvesting and pruning.

Last, but not least, it is important for engineering solutions to be developed through close collaboration with horticulturists. In general, a systematic approach is essential so that crop architectures, machine systems, and socio-economic parameters are optimized as a whole. Even with the latest fruiting wall tree architectures, a substantial percentage of fruit may still be inaccessible to robotic hands. Further research on novel training, thinning, and pruning strategies would be valuable to present fruit in such a way that they are fully accessible to robotic hands with the simplest maneuvering (e.g., linear motion). Closer collaboration between growers, horticulturalists, and engineers is vital to achieve these desirable parameters in the design of future orchards.

To summarize, the importance of fruit harvesting and pruning technology has been recognized around the world and many private and public institutions are involved in research and development in this area. In addition to a number of research units that are working in fruit harvesting and agricultural robotics, there are a lot of private companies making an effort in this area. New developments such as John Deere acquiring Blueriver Technologies (a robotic thinning and weeding company), Bosch investing significantly in their smart farming efforts, and Panasonic unveiling a tomato-harvesting robot, indicate that there will be a number of smart and robotic farming solutions available in the near future. With such extensive efforts from around the world, coupled with access to huge computational power at reasonable costs, availability of advanced sensing, machine vision, control and automation algorithms and technologies (e.g. 'deep learning' methods for fruit detection), and adoption of modern fruiting wall tree architectures, the feasibility of robotic harvesting and pruning in the orchard environment is ever increasing. Smart, robotic farming (also called 'Agriculture 4.0') is going to be a reality sooner than later.

9 References

Akbar, S. S., Elfiky, N. M. and Kak, A. 2016. A novel framework for modeling dormant apple trees using single depth image for robotic pruning application. In: *Proceedings of 2016 IEEE International Conference on Robotics and Automation (ICRA)*. doi:10.1109/ICRA.2016.7487718.

Cain, J. C. 1972. Slotting saw pruning of hedgerow apples improves production and quality. *New York's Food and Life Sciences Bulletin*, extension bulletin No. 15, pp. 1–4.

Calvin, L. and Martin, P. 2010. The U.S. produce industry and labor: facing the future in a global economy. Economic Research Report No. ERR-106. USDA-ERS, Washington DC.

Ceres, R., Pons, J. L., Jiménez, A. R., Martín, J. M. and Calderón, L. 1998. Design and implementation of an aided fruit-harvesting robot (Agribot). *Industrial Robot: an International Journal* 25(5), 337–46. doi:10.1108/01439919810232440.

Cherbiy-Hoffmann, S. U., Searles, P. S., Hall, A. J. and Rousseaux, M. C. 2012. Influence of light environment on yield determinants and components in large olive hedgerows following mechanical pruning in the subtropics of the Southern Hemisphere. *Scientia Horticulturae* 137, 36–42. doi:10.1016/j.scienta.2012.01.019.

Craig, W. E. and Embree, C. G. 2006. Pruning and training apple trees. *AgraPoint International Inc. Extension*. Publication ACC 1208, pp. 1–18.

Davidson, J., Silwal, A., Karkee, M., Mo, C. and Zhang, Q. 2016. Hand picking dynamic analysis for undersensed robotic apple harvesting. *Transactions of the ASABE* 59(4), 745–58. doi:10.13031/trans.59.11669.

De Kleine, M. 2014. Semi-automated end-effector concepts for localized removal and catching of fresh-market apples in fruiting wall orchards. *PhD Dissertation*. Washington State University.

De Kleine, M. E. and Karkee, M. 2016. A semi-automated harvesting prototype for shaking fruit tree limbs. *Transactions of the ASABE* 58(6), 1461–70.

D'Esnon, A., Rabater, G., Pellenc, R., Journean, A. and Aldon, M. J. 1987. Magali: a self propelled robot to pick apples. *American Society of Agricultural Engineers* 87, 1037.

Dias, A. B., Peca, J. O. and Pinheiro, A. 2012. Long-term evaluation of the influence of mechanical pruning on olive growing. *Agronomy Journal* 104(1), 22–5. doi:10.2134/agronj2011.0137.

Dorigoni, A. 2013. The fruiting wall: From hedging to window mechanical pruning. In: *Proceedings of Great Lake Fruit and Vegetable Expo*. Available at: http://glexpo.com/summaries/2013summaries/Applel.pdf (accessed 31 March 2018).

Elfiky, N. M., Akbar, S. A., Sun, J., Park, J. and Kak, A. 2015. Automation of dormant pruning in specialty crop production: an adaptative framework for automatic reconstruction and modeling of apple trees. Online. Available at: http://www.cv-foundation.org/openaccess/content_cvpr_workshops_2015/W05/papers/Elfiky_Automation_of_Dormant_2015_CVPR_paper.pdf (accessed on 17 January 2016).

Ferree, D. C. and Rhodus, W. T. 1993. Apple tree performance with mechanical hedging or root pruning in intensive orchards. *Journal of the American Society for Horticultural Science* 118(6), 707–13. doi:10.21273/JASHS.118.6.707.

Ferree, D. C. and Short, T. H. 1972. A slotting saw for mechanical pruning. *Fruit Crops Research*. Research summary 60, pp. 1-2.

Flore, J. A. 1992. The influence of summer pruning on the physiology and morphology of stone fruit trees. *Acta Horticulturae* 322(322), 257–64. doi:10.17660/ActaHortic.1992.322.28.

Forshey, C. 1976. Training and pruning apple trees. *Cornell Coop. Extension Info Bul. 112*.

Gallardo, R., Taylor, M. and Hinman, H. 2010. 2009 cost estimates of establishing and producing gala apples in Washington. Washington State University Extension Factsheet FS005E. Washington State University, Pullman, WA.

Giametta, G. and Zimbalatti, G. 1997. Mechanical pruning in new olive-groves. *Journal of Agricultural Engineering Research* 68(1), 15–20. doi:10.1006/jaer.1997.0175.

Harrell, R. C., Slaughter, D. C. and Adsit, P. D. 1989. A fruit-tracking system for robotic harvesting. *Machine Vision and Applications* 2(2), 69–80. doi:10.1007/BF01212369.

Harrell, R. C., Adsit, P. D., Munilla, R. D. and Slaughter, D. C. 1990. Robotic picking of citrus. *Robotica* 8(4), 269–78. doi:10.1017/S0263574700000308.

He, L., Fu, H., Sun, D., Karkee, M. and Zhang, Q. 2017a. Shake and catch harvesting for fresh market apples in trellis trained trees. *Transactions of the ASABE* 60(2), 353–60.

He, L., Fu, H., Karkee, M. and Zhang, Q. 2017b. An effect of fruit location on apple detachment with mechanical shaking. *Biosystems Engineering* 157, 63–71. doi:10.1016/j.biosystemseng.2017.02.009.

He, L., Karkee, M. and Zhang, Q. 2018. Evaluation of a localized shake-and-catch harvesting system for fresh market apples. *Agricultural Engineering International: CIGR Journal* 19(4), 36–44.

Hertz, T. and Zahniser, S. 2013. Is there a farm labor shortage? *American Journal of Agricultural Economics* 95(2), 476–81. doi:10.1093/ajae/aas090.

Kallsen, C. E. 2005. Topping and manual pruning effects on the production of commercially valuable fruit in a midseason navel orange variety. *HortTechnology* 15(2), 335–41. doi:10.21273/HORTTECH.15.2.0335.

Karkee, M., Adhikari, B., Amatya, S. and Zhang, Q. 2014. Identification of pruning branches in tall spindle apple trees for automated pruning. *Computers and Electronics in Agriculture* 103, 127–35.

Kataoka, T., Ishikawa, Y., Hiroma, T., Ota, Y., Motobayashi, K. and Yaji, Y. 1999. Hand mechanism for apple harvesting robot. *Journal of the Japanese Society of Agricultural Machinery* 61(1), 131–9.

Kesner, C. D. 1990. Mechanical summer tipping of tart cherry. Online. Available at: http://agbioresearch.msu.edu/uploads/files/Research_Center/NW_Mich_Hort/Training_Pruning_Varities/MechSummTippingTartCherry.pdf (accessed on 17 January 2016).

Kesner, C. D., Hansen, C. M. and Fouch, S. 1981. Tree training and mechanical pruning of tart cherry in high density plantings. *Compact Fruit Tree* 14, 135–9.

Larbi, P., Amatya, S., Karkee, M., Zhang, Q. and Whiting, M. 2015. Modification and field evaluation of an experimental mechanical sweet cherry harvester. *Applied Engineering in Agriculture* 31(3), 387–97.

Lehnert, R. 2013. New apple harvester shows promise. *Good Fruit Grower*, November 18, 2013. Available at: https://www.goodfruit.com/in-orchard-sorting/.

Lewis, K. and Hanrahan, I. 2014. Field testing of two harvest aids in Washington apple orchards. Available at: http://www.crec.ifas.ufl.edu/harvest/pdfs/presentations/Lewis_ISHS_Session_7.pdf (accessed on 20 March 2016).

Lu, R., Pothula, A. K., Mizushima, A., Vandyke, M. and Zhang, Z. 2018. U.S. Patent Application No. 15/256, p. 931.

Martin-Gorriz, B., Porras Castillo, I. and Torregrosa, A. 2014. Effect of mechanical pruning on the yield and quality of 'Fortune' mandarins. *Spanish Journal of Agricultural Research* 12(4), 952–9. doi:10.5424/sjar/2014124-5795.

Mehta, S., Ton, C., Asundi, S. and Burks, T. 2017. Multiple camera fruit localization using a particle filter. *Computers and Electronics in Agriculture* 142, 139–54.

Miranda Sazo, M. and Robinson, T. L. 2013. Recent advances of mechanization for the tall spindle orchard system in New York state – part 1. *New York Fruit Quarterly* 21(1), 15–20.

Muscato, G., Prestifilippo, M., Abbate, N. and Rizzuto, I. 2005. A prototype of an orange picking robot: past history, the new robot and experimental results. *Industrial Robot: an International Journal* 32(2), 128–38. doi:10.1108/01439910510582255.

Peterson, D. 2005. Development of a harvest aid for narrow-inclined-trellised tree-fruit canopies. *Applied Engineering in Agriculture* 21(5), 803–6. doi:10.13031/2013.19707.

Peterson, D. and Wolford, S. 2003. Fresh-market quality tree fruit harvester part II: apples. *Applied Engineering in Agriculture* 19(5), 545–8.

Peterson, D., Bennedsen, B., Anger, W.. and Wolford, S. 1999. A systems approach to robotic bulk harvesting of apples. *Transactions of the ASAE* 42(4), 871–6.

Rabatel, G., Bourely, A., Sevila, F. and Juste, F. 1995. Robotic harvesting of *Citrus*: state-of-art and development of the French Spanish EUREKA project. In: *Proceedings of the International Conference on Harvest and Postharvest Technology for Fresh Fruits and Vegetables*, Guanajuato, Mexico, pp. 232–9.

Sansavini, S. 1978. Mechanical pruning of fruit trees. *Acta Horticulturae* 65(65), 183–98. doi:10.17660/ActaHortic.1978.65.28.

Schupp, J., Baugher, T., Winzeler, E., Schupp, M. and Messner, W. 2011. Preliminary results with a vacuum assisted harvest system for apples. *Fruit Notes* 76(Fall), 1–5.

Silwal, A., Karkee, M. and Zhang, Q. 2016. A hierarchical approach of apple identification for robotic harvesting. *Transactions of the ASABE* 59(5), 1079–86. doi:10.13031/trans.59.11619.

Silwal, A., Davidson, J. R., Karkee, M., Mo, C., Zhang, Q. and Lewis, K. 2017. Design, integration, and field evaluation of a robotic apple harvester. *Journal of Field Robotics* 34(6), 1140–59. doi:10.1002/rob.21715.

Stebbins, R. L. 1992. Pruning apple trees in commercial orchards. *Pacific Northwest Extension*. Publication PNW 404, pp. 1–14.

Takeda, F., Yang, W., Li, C., Freivalds, A., Sung, K., Xu, R., Hu, B., Williamson, J. and Sargent, S. 2017. Applying new technologies to transform blueberry harvesting. *Agronomy* 7, 33–51.

Velazquez, B. and Fernandez, E. 2010. The influence of mechanical pruning in cost reduction, production of fruit, and biomass waste in citrus orchards. *Applied Engineering in Agriculture* 26(4), 531–40. doi:10.13031/2013.32056.

Vougioukas, S. G., He, L. and Arikapudi, R. 2016. Orchard worker localisation relative to a vehicle using radio ranging and trilateration. *Biosystems Engineering* 147, 1–16. doi:10.1016/j.biosystemseng.2016.03.006.

Webster, A. D. and Looney, N. E. 1996. *Cherries: Crop Physiology, Production and Uses* (1st edn.). CAB International, Wallingford.

Whiting, M. D. 2018. Precision orchard systems. In: Zhang, Q. (Ed.), *Automation in Tree Fruit Production: Principles and Practices*. CAB International, pp. 75–92.

Yildirim, B., Yesiloglu, T., Incesu, M., Kamiloglu, M., Ozguven, F., Tuzcu, O. and Aka, Y. 2010. The effects of mechanical pruning of fruit yield and quality in 'Star Ruby' grapefruit. *Journal of Food, Agriculture and Environment* 8(2), 834–8.

Zhang, Z., Pothula, A. K. and Lu, R. 2017. Economic evaluation of apple harvest and in-field sorting technology. *Transactions of the ASABE* 60(5), 1537–50. doi:10.13031/trans.12226.

Zhou, J., He, L., Zhang, Q. and Karkee, M. 2014. Effect of excitation position of a handheld shaker on fruit removal efficiency and damage in mechanical harvesting of sweet cherry. *Biosystems Engineering* 125, 36–44. doi:10.1016/j.biosystemseng.2014.06.016.

Zhou, J., He, L., Whiting, M., Amatya, S., Larbi, P. A., Karkee, M. and Zhang, Q. 2016a. Field evaluation of a mechanical-assist cherry harvesting system. *Engineering in Agriculture, Environment and Food* 9(4), 324–31. doi:10.1016/j.eaef.2016.05.003.

Zhou, J., He, L., Karkee, M. and Zhang, Q. 2016b. Analysis of shaking-induced cherry fruit motion and damage. *Biosystems Engineering* 144, 105–14. doi:10.1016/j.biosystemseng.2016.02.007.

Zhou, J., He, L., Karkee, M. and Zhang, Q. 2016c. Effect of catching surface and tilt angle on bruise damage of sweet cherry due to mechanical impact. *Computers and Electronics in Agriculture* 121, 282–9. doi:10.1016/j.compag.2016.01.004.

Chapter 10

Optimizing pest management in fruit cultivation

Arthur Agnello, Cornell University, USA

1 Introduction[1]

Sustainable pest management for fruit production must address numerous agricultural concerns. In many production areas, there are horticultural, economic, social and political pressures to reduce the environmental impact and use of pesticides. Public concerns with nutrient and sediment movement into ground and surface water and pressure against pesticide applications is growing. Agricultural producers are being asked to submit nutrient and soil management plans to address the offsite impacts of their practices. In addition, the development of pesticide resistance in key pests, registration of fewer and more expensive new chemicals for pest control, loss of existing products and increased competition from other regions continue to push the agricultural community to look for nonchemical alternatives.

Integrated crop management requires a combination of long- and short-term production strategies to maximize net profit while minimizing risks of undesirable environmental impacts of practices. Some of these practices include site selection, crop-specific production strategies, nutrient management and cover cropping. Integrated pest management (IPM) is a pest control strategy that promotes the use of a variety of tactics including pest-resistant cultivars

1 Adapted from Hoffmann et al. (1996).

http://dx.doi.org/10.19103/AS.2018.0040.11

and biological, cultural and physical controls (Rabb and Guthrie 1970; Brunner 1994; Bajwa and Kogan 2002). Pesticides are a control tactic employed in IPM, but they are used preferably only when needed. Pesticide use is thus minimized without jeopardizing crop quality or yield. Applying multiple control tactics minimizes the chance that pests will adapt to any one tactic and allows growers to choose the most environmentally sound, efficacious and economically efficient pest management programme for their situation.

1.1 Practising integrated pest management

In an IPM programme, it is important to accurately identify the pests (vertebrates, diseases, insects and weeds) and assess pest abundance. It is important to have knowledge of the biology and ecology of the pest(s) attacking the crop and the factors that can influence pest infestations. An understanding of the influence of factors such as weather and natural enemies on pest abundance will aid the choice of management tactics. IPM programmes stress suppression of insect and disease populations to levels that do not cause economic damage, rather than total eradication of a pest. In the case of insect pests, it may be important to have at least some pests present to ensure that natural enemies will remain in the crop to suppress subsequent infestations.

1.2 Components of integrated pest management

1.2.1 Monitoring (Scouting)

Scouting includes detecting, identifying and determining the level of pest populations on a timely basis. Insect traps can often be used to detect pests and identify times when scouting should be intensified or control measures should be taken. Monitoring individual orchard blocks throughout the season is the most effective way of assessing the insect, disease and weed situation and, therefore, the need for chemical treatment in that block. Scientifically based, accurate and efficient monitoring methods have been studied for many pests on fruit crops (Madsen and Vakenti 1973; Vakenti and Madsen 1976; Madsen and Madsen 1980; Riedl 1980; Nyrop et al. 1989; Agnello et al. 1990; Delisle 1992; Leskey and Wright 2004; Kovanci and Walgenbach 2005).

1.2.2 Forecasting

Weather data and other information help predict when specific pests will most likely occur (Olatinwo and Hoogenboom 2013). Weather-based pest forecast models for many diseases and insect pests of fruit crops have been developed (e.g. University of Massachusetts Amherst, https://ag.umass.edu/fruit/weather-pest-monitoring-forecasting; Cornell University, NEWA: http://newa.cornell.

edu/; Michigan State University, Enviroweather: https://enviroweather.msu.edu/; Washington State University, Decision Aid System: https://decisionaid.systems/; University of California, Weather, Models and Degree Days: http://ipm.ucanr.edu/WEATHER/index.html#PESTPLANTMODELS). Access to a computer network to obtain local weather, and regional insect and disease forecasts, is useful but not essential. Simple weather recording equipment such as thermometers, hygrometers and rain gauges placed in orchards will assist in the prediction of pest outbreaks. Information on the potential for pest outbreaks generally can also be obtained from local cooperative extension offices, newsletters and regional crop advisers.

1.2.3 Thresholds

Thresholds are used to determine when pest populations have reached a level that could cause economic damage. Thresholds have been scientifically determined by many fruit researchers (Hoyt and Burts 1974; Hoyt and Tanigoshi 1983). Following thresholds has been shown to reduce pesticide use by as much as 50%, saving significant money for growers. Good judgement should be exercised in applying the general advice to individual orchard situations, and potential problems should be anticipated to avoid pest outbreaks and pesticide waste. The suggested thresholds are not always exact; nonetheless, they represent the best guidelines available to the commercial grower concerned with effective and efficient management of tree-fruit pests.

1.2.4 Management tactics

Appropriate management tactics to control pests include cultural, biological and physical controls, as well as chemical controls when needed. Taking advantage of some of the simple and relatively inexpensive pest management tactics available for fruit pests can result in significant savings to growers in terms of both pesticide use and crop loss. Often a thoughtful preventive measure taken before the pest is a problem can result in significant savings of crop-rescue treatment expenses later in the season.

1.2.5 Recordkeeping

Records kept from year to year on pest occurrence in orchards can be valuable tools for avoiding pests in the future.

1.3 Integrated pest management tactics

An important aim of an IPM strategy is to integrate the available pest management options (Croft and Bode 1983). Some pests are endemic and usually require pesticide treatment, applied either preventively or in response to threshold numbers during the season. However, the incidence of these pests and the need for pesticides can often be reduced through a combination of control tactics:

1.3.1 Pest-resistant varieties

If available, insect- and disease-resistant or -tolerant varieties or rootstocks can reduce losses to pests. Using such varieties can be one of the simplest methods of reducing costly management procedures and negative environmental impacts during the growing season.

1.3.2 Cultural and physical controls

Removal of overwintering sites, such as cull piles, mummies, suckers, damaged branches or trees and alternate hosts, can minimize damage by insects and diseases. Techniques are recommended that expose pests to natural enemies or environmental stress, or that make the crop less susceptible to insects or diseases. Ensuring vigorous crop growth through proper nutrition and weed control will help avoid stress that may predispose crops to attack by insects, diseases or physiological disorders. Irrigation schedules should be managed to avoid long periods of high relative humidity, which encourage disease pests to develop. Avoiding planting of trees into areas of known, high pest pressure is a primary consideration that should be made at the outset, as well as orienting orchards to provide maximum air drainage and circulation.

1.3.3 Biological control

Natural enemies of insect and mite pests can be conserved by using fungicides and insecticides only when needed. As much as possible, pesticides should be used that are most selective to specific pests and least disruptive to beneficial organisms, and use of 'seeding' releases of predator mites can augment endemic populations (Breth et al. 1998).

1.3.4 Chemical control

Pesticides should be used only if pest pressure, monitoring, economic thresholds or disease forecasts indicate a need. Choice of pesticides should be based on efficacy, previous use patterns, the incidence of or potential for resistance, and the impact on the environment and natural enemies. Full and uniform spray coverage can be ensured by using recommended spray rates and accurately calibrated equipment that targets key crop locations that need to be protected. Pesticides should not be applied when wind velocity is high enough to cause drift to non-target sites.

2 Arthropod biology

A basic understanding of how arthropods develop, reproduce, feed and attack fruit crops is central to formulating a sound and sustainable management programme. Insects and mites are found almost everywhere and can survive the most extreme conditions found on earth. Their short lifespan provides a relatively greater opportunity for mutation and evolutionary change; however, their lifestyle has considerable drawbacks. Their small size allows relatively little space for the brain and complex organs, limiting behavioural complexity. Their physiology also prevents them from producing body heat, so they must cease activity in cold weather.

2.1 Types of injury inflicted

Fruit arthropod pests are classified according to the kind of injury they produce, which parts of the tree they infest, the timing and occurrence of their attack, and the severity level of their injury, which relates to the importance of controlling them (Hoyt et al. 1983). Insects with chewing mouthparts inflict great damage on foliage, causing leaves to be skeletonized, riddled with holes, eaten around the edges or entirely consumed (e.g. larvae of moths, sawflies and beetles). Other insects suck sap from leaves, stems, fruits or other plant parts, producing a characteristic spotting or browning, curling or wilting. Sap feeders also can transmit diseases such as viruses or mycoplasmas. Feeding on stems or twigs results in dwarfing or wilting. Damage is caused both by removal of the sap and by injury to the plant tissue (e.g. scale insects, aphids and true bugs).

Scale insects are usually minute, but if they are abundant enough to encrust barks, twigs or stems, they can kill orchard trees or predispose them to diseases. Aphids cause a curling of leaves and, when feeding on fruit, may cause it to be stunted or misshapen, and may change the sugar content, negatively impairing the flavour. Many fruit insect pests feed as miners in leaves or as borers in stems, roots or fruits. Feeding between the upper and lower surfaces of the leaf may cause as much defoliation as does external feeding. There are numerous leaf mining species that can affect tree fruits (e.g. spotted tentiform leaf miner and apple blotch leaf miner, *Phyllonorycter blancardella* (F.) and *P. crataegella* Clemens, respectively). Tunnelling causes serious damage; tunnelling insects include codling moth, *Cydia pomonella* (L.); oriental fruit moth, *Grapholita molesta* (Busck); apple maggot, *Rhagoletis pomonella* (Walsh); and plum curculio, *Conotrachelus nenuphar* (Herbst). Injection of an insect-produced chemical into plant tissues while the insect feeds causes abnormal growth (e.g. rosy apple aphid, *Dysaphis plantaginea* [Passerini]) or produces a gall (e.g. woolly apple aphid, *Eriosoma lanigerum* [Hausmann]) or necrotic tissue (e.g. brown marmorated stink bug, *Halyomorpha halys* [Stål]). Each species of gall insect produces a characteristic gall on a certain part of a particular plant. Insects at the larval and nymphal stages (e.g. woolly apple aphid) that live in the soil and attack the underground plant parts cause extensive damage. Shelters in plants are built by leaf rollers and leaf folders, which roll or fold the leaves and tie them together with silk, feeding in the shelter so formed. Leaf tiers and webworms tie several leaves or even branches together, producing large silken webs or tents. A few insects injure plants when they lay their eggs, particularly in stems or fruits, for example fruit flies (*Drosophila suzukii*), plum curculio, apple maggot, periodical cicada, tree crickets and leafhoppers.

2.2 Life history

2.2.1 Metamorphosis

Another trait contributing to insect and mite survival is *metamorphosis*, the process of changing from one form, or stage, into another. One stage can be specialized for hibernation, another for feeding and growing, and another for reproduction and dispersal. Insects can exploit different habitats in each stage; for example, a dragonfly spends its immature stages as a predacious underwater insect and its adult phase as a winged non-aquatic form that rarely feeds.

2.2.2 Physiology of metamorphosis

An understanding of the process of metamorphosis in insects and mites will help in the development of control strategies. Three different hormones are involved in metamorphosis. It is initiated and controlled by a *brain hormone*. *Ecdysone*, also called moulting hormone, promotes growth and induces the moulting sequence. *Juvenile hormone* (JH) promotes larval or nymphal development and prevents metamorphosis. JH is produced by glands that are active during the early instars and cease activity in the last pre-adult instar. Its absence in this instar results in metamorphosis. Several hormone analogs have been developed as insect growth regulator insecticides, which have several modes of action: for example inhibiting chitin synthesis (which forms the exoskeleton); mimicking ecdysone to induce a premature moult; JH analogs that interfere with normal metamorphosis; or inhibition of lipid biosynthesis (Chapman 2012).

2.2.3 Reproduction

Insects normally reproduce only once during their lifetime. Natural populations usually consist of individuals about the same age, with little or no generational overlap. *Parthenogenetic* reproduction, which does not require a combining of gametes from both sexes, is common in bees, wasps and roaches. The life cycle of some insects alternates between bisexual and parthenogenetic reproduction (e.g. many aphids). Most insects lay eggs. In some cases, such as in aphids and numerous flies, live young hatch just before or while eggs are being laid. The adults of such insects may differ in appearance and occur on different host plants from one generation to the next.

2.2.4 Generations

In temperate regions, adults of most insects appear for a limited time during a particular season, and overwinter as a particular life stage in a state of dormancy. Most insect pests in the United States have a single generation per year (e.g. plum curculio in the north, apple maggot). Many large insects in the northern US require two or more years to complete their life cycle (e.g. larger beetles, cicadas, some moths). Many insects have more than one generation per year. Occasionally, a species will have more generations in the southern part of its range, for example plum curculio and some moths and butterflies, such as oriental fruit moth, follow this trend. A few insects, usually rather small species that can complete their life cycle in a few weeks (e.g. aphids and mites) have many generations a year, as long as the weather is favourable.

2.2.5 Dormancy

A period of quiescence, or *dormancy*, occurs in the life cycle of many insects. This period may last from a few days to several months or even years, and may occur at any stage in the life cycle. Visible activity ceases, and physiological processes occur at greatly reduced rates. Winter dormancy is usually called *hibernation* (e.g. Mexican bean beetle, *Epilachna varivestis* Mulsant, will hibernate for several months if subjected to low temperatures). Dormancy during high temperature is called *aestivation* (e.g. plum curculio in late summer). Dormancy is controlled by environmental and genetic factors. Most insects enter dormancy when an environmental factor, such as temperature, becomes unfavourable; they resume activity when conditions are again favourable.

2.2.6 Diapause

Some species do not breed continuously throughout the year, even under favourable conditions, but enter dormancy before conditions become unfavourable. This period of *diapause* is normally broken only when the insect is subjected to low temperature and then returned to a temperature favourable for development. For example, second-generation codling moths remain as larvae in cells under apple bark unless subjected to at least three weeks of below-freezing weather; under normal developmental temperatures, they pupate and complete development. Diapause is controlled in part by *photoperiod* (day length). The day length that triggers diapause differs among species. All codling moth larvae that enter the soil for pupation before a certain date will complete their development and emerge as moths and reproduce, but any entering the soil after that date go into diapause and do not complete their development until the following spring.

2.3 Developmental rates and thresholds

Mammals are warm-blooded, developing at a constant rate regardless of the environmental temperature because they are able to maintain an internal temperature that allows their biochemical reactions to progress normally. Insects, which are poikilothermic, remain at the same temperature as their environment. They do not generate body heat and therefore depend on favourable external temperatures. At a certain temperature, which varies among species, an insect's biochemical reactions cannot proceed and development stops. This temperature is known as the insect's developmental *threshold* or developmental *base*. Charting the ambient temperature makes it possible to track insect development, which is directly proportional to the amount of time accumulated above the developmental threshold (up to some maximum not often reached during the season). This time is divided arbitrarily into heat units or *degree-days* (DD).

2.3.1 Degree-day calculation methods

There are different ways to determine the quantity of heat units accumulated, which is equivalent to the area under a temperature-versus-time graph on a given day (Baskerville and Emin 1969; Sevacherian et al. 1977; Carroll 2011). The following methods are listed in order of increasing precision in measuring small changes during the day or departures from idealized heating and cooling trends (Fig. 1).

- *Average or Max/Min Method* – This method is the simplest and least precise. It assumes that the daily temperature graph is linear and that the area beneath it is triangular.

$$\text{DD} = \left[\frac{\text{Daily max temp} + \text{Daily min temp}^*}{2}\right] - \text{Developmental Threshold}$$

(* - If Daily min temp < Developmental Threshold, substitute Developmental Threshold)

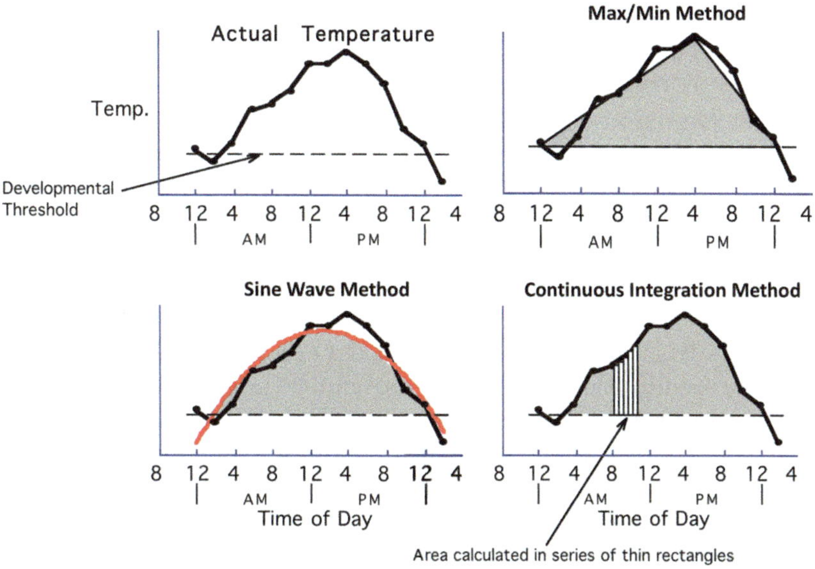

Figure 1 Different methods of calculating degree-days. Shaded areas are portions of the curve used to obtain an estimate of heat units accumulated.

- *Sine Wave (Baskerville-Emin) Method* - This method is more precise and assumes that the daily temperature cycle takes the form of a sine wave. The area beneath this curve is determined by integration, which requires calculus. This method makes the same use of daily maximum and minimum temperatures and developmental threshold as does the average method. Using the sine wave method tends to accumulate more DDs than the average method, particularly during the early part of the season.
- *Continuous Integration Method* - This method is the most precise and requires multiple temperature readings hourly or more frequently throughout the day to obtain a temperature-versus-time graph that is truly representative of a field situation. The area beneath the curve is again calculated using integration. The data collection is most efficient if done by a computer.

2.3.2 Relating degree-days to life cycle and development

The following methods are attempts to correlate a pest event or activity with another event that can be measured more precisely. Events in an insect's life cycle often occur after the same general number of heat units have accumulated each year, but many years' observations must be collected to measure this precisely. Degree-days can be used to predict insect (or plant) life cycle events wherever weather data are available.

Temperature – By monitoring temperature and pest activity simultaneously for many years, it is possible to build up a database of events and the range of accumulated DDs that correspond with them.

Phenology – Some events occur reliably at the same time as other, easily observed biological events in the field; for example, European red mite (*Panonychus ulmi* [Koch]) hatches from late tight cluster to pink bud stage of apple flower bud development; codling moth starts to fly during king bloom of 'Red Delicious' apple. European apple sawfly (*Hoplocampa testudina* [Klug]) lays eggs from late bloom to petal fall. These rules of thumb often draw on the evolved relationships between pests and their hosts.

Biofix – This is a distinct, easily monitored event in the life history of an organism, used to fine-tune our predictions of its activity; for example, first flight, first egg laid, first mine observed.

These principles have been used by fruit pest researchers to formulate pragmatic population prediction methods for a range of key insect pests (Welch and Croft 1983), including apple maggot, *Rhagoletis pomonella* (Walsh) (Reissig et al. 1979); codling moth, *Cydia pomonella* L. (Riedl et al. 1976); obliquebanded leafroller, *Choristoneura rosaceana* (Harris) (Onstad et al. 1985); oriental fruit moth, *Grapholita molesta* (Busck) (Hull et al. 2003); plum curculio, *Conotrachelus nenuphar* (Herbst) (Reissig et al. 1998); San Jose scale, *Quadraspidiotus perniciosus* (Comstock) (Mague and Reissig 1983); and spotted tentiform leaf miner, *Phyllonorycter blancardella* F. (Schmaedick and Nyrop 1995).

2.4 Influence of weather on pest activity

Of all the factors that can possibly affect the development of a given pest population, the weather is certainly one of the most critical. Nearly every discussion of how moderate or severe an insect or mite problem is, was or might be in a given season, starts with a general estimation of the temperature, wind, humidity and rainfall conditions to which that pest is subjected. Growers have numerous anecdotal examples of how the spring rains in one year reduced the usual increase in one insect population, or how the summer heat promoted the increase of another. The intent here is not to document specific effects, which are complex and abundant, but rather to indicate the need to take weather patterns into account when planning pest management programmes, both before (prevention) and after (rescue) the fact.

Some general principles can be stated. First of all, in many apple producing regions, early spring is considered to be the defining period; the growth of most prebloom arthropod populations is essentially determined for the first half of the season by what type of spring weather occurs. European red mite, rosy apple aphid, spotted tentiform leafminer, tarnished plant bug (*Lygus lineolaris* [Palisot de Beauvois], San Jose scale and mullein bug (*Campylomma verbasci* [Meyer]) are only the most obvious of the species that suffer from a cold, wet, rainy and windy spring. They may be slowed considerably until the summer generations, or they might fail to show up at all in some cases. Conversely, a warm, dry, quick spring can result in what appears to be spontaneous generation of most of these pests. After the petal fall period, the rate of heat unit (DD) accumulation is a primary factor in the duration of plum curculio oviposition (warmer = a shorter oviposition period) and the speed of summer mite population growth. This latter case is especially crucial, as the first summer mite eggs generally hatch in June in the Northern hemisphere, so the population is already primed to expand; additionally, the trees are particularly susceptible to foliar feeding stress, so a failure to act against a threshold-level infestation early will result in a long, difficult battle for the remainder of the summer.

Moving into midsummer, an abundance of rainfall stimulates foliar growth, which may be advantageous to the tree's development, but can also promote undesirable infestations of pests such as green aphids, leafhoppers and even leaf rollers. Hot and dry weather can be a mixed blessing, as it is associated on the one hand with localized outbreaks of twospotted spider mite (*Tetranychus urticae* Koch), but at the same time tends to discourage emergence of woolly apple aphid aerial colonies and apple maggot adults if the ground is hard and dried out. The IPM objective is to keep in mind which problems the prevailing conditions might require monitoring (and which to de-emphasize) as the season progresses. Much needless effort can be prevented in some cases, and an effective response to otherwise serious infestations made in other cases, by simply being aware of these basic trends.

2.5 Weather effects on pesticide activity

The effect of rainfall and humidity on pesticide efficacy is a topic that is much-debated, but about which few hard details exist. Certainly, everyone becomes apprehensive about the impact of a long, hard rainfall immediately following a pesticide spray. How much rain does it take to wash off a spray residue? Does it need to be reapplied? If so, how soon? In fact, the factors that determine the need for a re-spray are usually very specific to each case, and generalizations never give adequately specific answers. Research on this topic has shown that there are intrinsic differences between insecticides, and that advice on whether to re-spray following a rain event is mostly dependent on the insecticide and its formulation, and not so much on the intensity of the rainfall. The guidelines often used are heuristic and anecdotal - in other words, inexact - but they may help decide on the advisability of a re-spray. In general, it should be assumed that a spray deposit is fairly well established on the plant surface if allowed to dry for two hours after application; less than this may result in some cases where thorough drying has not occurred. After two hours, the potential loss in efficacy from a rain will generally vary with the duration and frequency of the rain, but not necessarily with rainfall intensity. Advancing specific numbers as guidelines are difficult, because each case is different; however, some notable studies have been conducted (e.g. Wise 2017).

3 Population monitoring

3.1 Trapping

Insect trapping is used to determine a biofix, to provide an index of population trends or to indicate population levels. Pheromone traps rely on chemical cues to attract insects (e.g. many moth pests of apples: obliquebanded leafroller, spotted tentiform leaf miner, codling moth, oriental fruit moth). Visual traps rely on visual cues to attract insects (e.g. red spheres for apple maggot, white cards for tarnished plant bug or European apple sawfly). Physical traps collect insects by obstructing them (e.g. double-sided tape placed around branches to trap San Jose scale crawlers).

3.2 Sampling

Insect sampling is conducted to learn about a population in a certain location (Binns et al. 2001). A *population* is made up of all the members of a specific group. It can be general or specific, such as all the mites, all the adult mites or all the adult European red mites in an orchard. A *sample* is a portion of the population that is examined and used to make a best guess about the rest of the population. For example, to estimate the number of mites per leaf for all the leaves on all the trees in an orchard, one could sample 50 representative leaves, examine all the mites on them, and from that sample make a best estimate.

To take a *representative* sample, the sample must first be truly representative of the population of interest. For instance, all the sampled leaves should not be taken from a single tree or only from border trees if the population of interest occurs on all the trees in the block. Second, samples should be selected *randomly* so that any one leaf has the same chance of being examined as any other, and that leaves chosen do not happen to be only those that look, for example, bronzed, healthy or are easiest to reach.

A sample is a *random variable*. Simple examples of random events or observations are the rolling of a pair of dice or flipping a coin. The labels used to describe the result of such an event (e.g. the number showing on the dice) are called random variables. One can describe random variables by setting a likelihood or *probability* for each possible outcome. For example, the probability of obtaining a six with any roll of a single die is one in six, and the probability of obtaining heads with a coin toss is one in two. The result of a sample of insects or mites is also a random variable. For example, if a sample of 50 leaves is used to estimate the average number of mites per leaf, this estimate is a random variable that will differ somewhat each time a sample is taken. If random variables are analysed statistically, they can help in understanding insect and mite populations in an orchard.

Random variables are important in making pest control decisions. Insect populations are sampled to estimate their average, or *mean*, size. Each sample of a given population will vary, but if they are representative samples they will be close to some average value. Because any estimate of pest population size using a sample is a random variable, there is always some uncertainty about the population's true size. Despite this uncertainty, the estimate made of population size using representative sampling is more likely to be close to the true size than far from it. When sampling a population that is greater than a given treatment threshold, there is a possibility that the sample will indicate that it is in fact below threshold; however, if the sample is truly representative, this problem is more likely to occur when the true population size is *very* close to (i.e. slightly above or below) the IPM threshold. This means that, although there is a chance the sample will lead to an incorrect management decision (such as no treatment when the population is actually above threshold), there is less likelihood of this happening as the true population size gets further from the threshold.

To illustrate, if the IPM threshold is 5.0 mites $leaf^{-1}$ and the true (but unknown) mite population in the orchard is exactly 5.0 mites $leaf^{-1}$, there is a 50% chance that the sample will indicate that the population is below threshold (Fig. 2). If the true population is slightly higher (e.g. 5.2 mites $leaf^{-1}$), the chance of a wrong conclusion drops to approximately 40%. If it is actually 6.0 mites $leaf^{-1}$, it drops to less than 10%, and at 7.0 mites $leaf^{-1}$ it drops to zero; that is, at some point when the actual population is above, but still relatively close to, the threshold, it will *never* be estimated to be below the threshold. Because thresholds are only general guidelines, the consequences of making a decision error with a population only marginally greater than threshold are not likely to be too serious. Also, the greater the number of samples taken, the greater the precision of the estimate.

Samples can be either fixed or sequential, which refers to their size; for example, the number of leaves that must be examined to reach a conclusion. A *fixed* sample is more basic and generally conservative. This means that, whereas more samples may be taken than are actually needed to make a decision, too few samples will never be taken. A *sequential* sample is a more advanced technique used when there is greater familiarity with the insect's biology, specifically, its distribution in particular field situations (Nyrop et al. 1989). The information from each sequential sample is used to determine whether more samples are needed. This is more time-efficient because it allows a rapid decision to be made in extreme cases, such as when populations are extremely high or low.

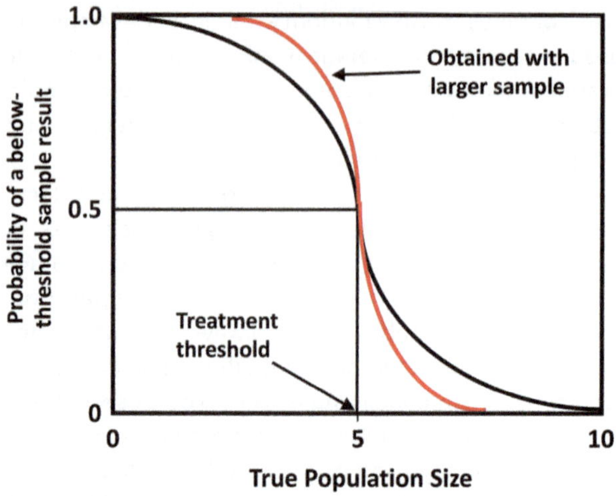

Figure 2 Likelihood of obtaining a below-threshold sample result as the true population size increases.

When taking a sequential sample, it should be kept in mind that nearness to the edge of the 'Continue Sampling' band has *nothing* to do with how close the population is to the threshold. It merely indicates how close *you* are to being able to make a decision based on the number of samples taken so far (Fig. 3).

The presence/absence method is a refinement of the sample and observation process. In special cases of mites and small insects, such as European red mites on apple leaves, the number of mites per leaf is mathematically related to the number of leaves on which one or more mites are present. By taking numerous samples of leaves and recording whether mites are present and the number of mites per leaf, a table of correlations can be constructed and used to develop a sequential sampling procedure. The grower or scout can use the number of leaves containing mites to make a decision about the need for treatment (Fig. 4).

3.3 Thresholds

The most common estimate pest managers make from samples of insects is whether enough insects are present in the true population to justify taking action to manage them (Hoyt and Tanigoshi 1983). This is called an *action threshold* and is based on a comparison of the costs of control versus the probable economic loss if no action is taken. An action threshold is the result of a complex process required to gather enough information about a pest-crop relationship to predict if and when to take control measures. Behind a threshold are generalizations about the pest's feeding and reproductive behaviour, rate of growth, most

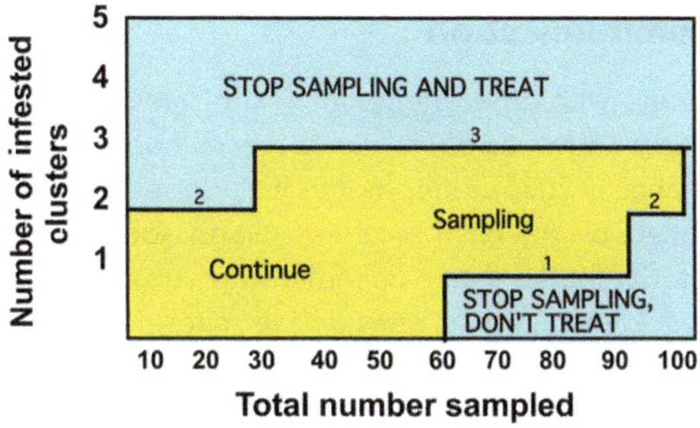

Figure 3 Example of a sequential sampling scheme for obliquebanded leafroller.

susceptible stage of growth, population trends, and the type of injury and costs and benefits of preventing it. As such, it can only serve as a guideline for action in most situations; thus, thresholds are usually constructed to be conservative, but many rely on information that is rarely given much thought in practice. For example, a grower will probably not try to estimate the market price of apples or the pack out expected from a specific block when deciding whether to spray for leaf rollers, although the action threshold may have been set with some attention to these factors. Action thresholds should always be regarded as management aids, not immovable laws and they do not replace common sense.

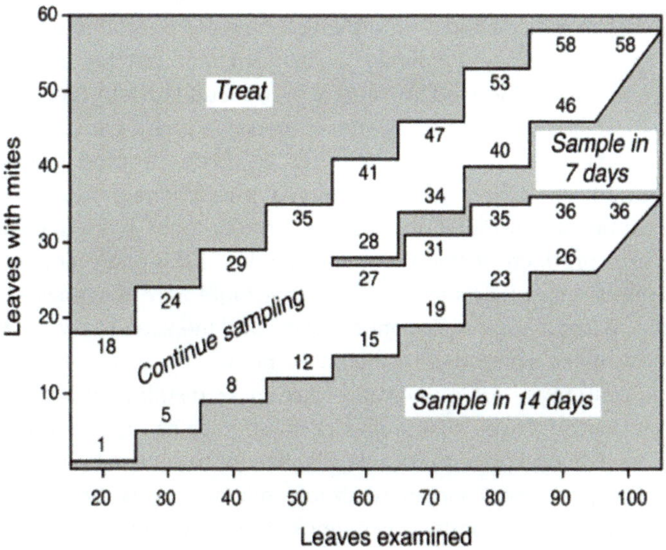

Figure 4 Example of a presence/absence sequential sampling scheme for European red mite.

4 Chemical communication

Insects communicate with other members of the same species by secreting chemical substances known as *pheromones* (social hormones) to the outside of the body (Welter et al. 2005). Pheromones include sex attractants, the queen substance produced by queen honeybees, alarm substances, aggregation substances, territorial markers and trail-following substances used by many ants. They are best known in ants, bees and termites. Pioneering work in the identification of pheromones of economically important insects has been done in connection with apple production since the late 1960s (Roelofs 2016). These advances have led to the manufacture of synthetic sex pheromones, which are used as lures in traps for monitoring insect presence and population magnitude, as developmental indicators and in controlling pests by tactics such as mass trapping and lure-and-kill (Cork 2016) by permeating the air to interfere with mate-finding behaviour. This approach is known as mating disruption (Evenden 2016).

4.1 Practical uses of mating disruption in fruit production

The overall objective of mating disruption is the suppression of sexual communication using synthetic pheromones. There are two general mechanisms of mating disruption (Miller and Gut 2015): (1) inhibition of the male's ability to respond to the sex pheromone produced by the female; this is a non-competitive mechanism, which can work by, for example desensitization, exhaustion or camouflage and (2) competition between pheromone dispensers and 'calling females', a competitive mechanism that can be described as competitive attraction, induced arrestment or induced allopatry (i.e. mimicking a situation in which the two sexes are not in the same physical location).

The evolution of mating disruption as a widely adopted practice has closely followed advances in the formulation of pheromone products available for use. A variety of considerations are taken into account in the development of pheromone products, including ability to release the pheromone over an extended period of time; protection of the active ingredient from environmental degradation; ease of application; affordability; and the extent to which finding of females by males is impeded by the pheromone.

Different strategies of pheromone dispenser distribution are used for different products. Densely distributed point sources can range from 120 to 400 acre^{-1}, as in hand-applied reservoir dispensers, to several hundred million release points per acre, used in sprayable microcapsules, flakes or fibres. Sparsely distributed point sources can range from only a few units per acre, as in mega dispensers such as puffers and misters, to approximately 20–30 devices per acre, used for meso dispensers such as plastic sachets and solid matrix products. Differences in application density are related to pheromone emission rate and distribution. Products deployed at a higher density have a lower pheromone emission rate from each individual point source, whereas sparsely distributed dispensers emit the pheromone at a higher rate. Densely applied formulations, such as sprayable and hand-applied products, are distributed via the application process, whereas sparsely distributed pheromone sources such as meso dispensers and puffers rely on the wind for ultimate distribution throughout the orchard.

4.2 Examples of pheromone formulations

Some of the most common products are *hand-applied reservoir dispensers*, in which the pheromone is enclosed in plastic or dispersed in synthetic polymers, as in twist-ties, twin-tubes, puzzle pieces and membrane-type products. The pheromone slowly diffuses from these reservoirs for up to 180 days. They are hand-applied at rates of 500–1000 ha^{-1} (200–400 $acre^{-1}$), and each dispenser releases up to several μg of pheromone per hour. A limitation of these products is their high labour requirement for application, with its associated higher costs. Disruption relies mainly on competitive attraction, and the effect is strongest when the competing sources (dispensers) greatly outnumber females in the orchard. Efficacy of this type of formulation is greatly affected by the number of point sources applied (direct relationship), and pest population density (inverse relationship). Despite the relatively high degree of orientation disruption attained with these dispensers, complete efficacy is elusive, and growers often feel it is most cost-efficient to use them at slightly lower application rates, which allows them to apply fewer dispensers and supplement with insecticides as needed.

Meso dispensers attempt to combine the best qualities of reservoir dispensers, yet reduce the application effort required, as they release substantially more pheromone than standard hand-applied reservoirs, and so can be applied at lower rates per unit of area. This approach can allow for much lower densities (as few as 20 ha^{-1} [8 $acre^{-1}$]), which therefore requires less labour. These dispensers also employ a competitive attraction mechanism, but possibly could also operate by desensitization.

In *sprayable dispensers*, the pheromone is encapsulated in microscopic polymer capsules (20 μm average size) that are sprayed onto the crop at 250 million+ per ha (100 million+ per acre) and cumulatively deliver 20–100 g pheromone per ha (8–40 g pheromone per acre) over 3–4 weeks. They are easy to apply and have greatly reduced labour costs over hand-applied products, but generally have a short field life, which can lead to inconsistent efficacy. Microcapsules exhibit a rapid decline in pheromone release rate, and are subject to being washed off by rain; therefore, frequent low-dose applications are recommended to maximize effectiveness. They employ a non-competitive mechanism, and operate via desensitization or camouflage.

Mechanically applied dispensers (e.g. flakes, fibres, flowable wax matrix droplets) are designed to release pheromone at about the same rate as calling females, so are referred to as 'female-equivalent' formulations. These products also employ a competitive disruption mechanism. One advantage is their ease of application, but they may divert males' attention away from calling females for only a short period, allowing more opportunities for them to search for actual females.

Mega dispensers are ultra-sparsely distributed pheromone sources, used at 2.5–5 ha^{-1} (1–2 $acre^{-1}$). These are aerosol devices that release large amounts of sex attractants; mg-size quantities of pheromone are released every 15–30 min over a 6–12 h period each day. The pheromone supply is maintained inside a pressurized canister, which provides a controlled, constant release rate, and a stable environment that protects the pheromone prior to its release. Male insect captures within the pheromone plume are inhibited for considerable distances downwind of the device, which has been described as a type of induced allopatry. The low deployment density of these dispensers leaves some areas with sparse pheromone coverage, where mate finding can occur. Orchard edges are a problem; therefore, supplemental treatment of the borders with hand-applied dispensers is advised, and this approach is recommended preferably for larger-size orchards (greater than 8 ha [20 acres]).

4.3 Candidate species for mating disruption

Control of some pests by IPM programmes that incorporate mating disruption is better than others (Cardé and Minks 1995). For instance, clearwing borer moth species (Sessiidae: peachtree borer, *Synanthedon exitiosa* [Say], lesser peachtree borer, *S. picitpes* [Grote & Robinson], and dogwood borer, *S. scitula* [Harris]) are extremely sensitive and susceptible to disruption by their species-specific pheromones (Pfeiffer et al. 1991; Leskey et al. 2009). In some cases, it is possible to use mating disruption against them as a stand-alone tactic; however, their pheromone blends can be more complex, which poses formulation challenges for manufacturers.

Oriental fruit moth and lesser appleworm (Tortricidae: *Grapholita molesta; G. prunivora* [Walsh]) are somewhat less sensitive to their sex pheromone, but still quite susceptible to mating disruption (Vickers et al. 1985). This tactic may be more suitable against these species in stone fruits, which have fewer additional direct pests that would require supplementary insecticide sprays. Both hand-applied and sprayable formulations are available, which are affordable and very effective; they are combined with codling moth pheromone in several products, for use in orchards where a complex of more than one species occurs.

Codling moth (Tortricidae: *Cydia pomonella*) is not as susceptible, as it is less sensitive to its sex pheromone, and therefore more difficult to disrupt;

however, mating disruption is still considered a valuable complement to insecticide programmes for management of this species (Howell et al. 1992; Knight 2008). There are a range of products available, with the hand-applied products being more effective than alternative dispensing mechanisms such as sprayables, aerosols and other flowables. Codling moth pheromone is more expensive than that for oriental fruit moth, with which it is often combined in several hand-applied products.

Obliquebanded leafroller (Tortricidae: *Choristoneura rosaceana*) is generally not considered a good candidate for mating disruption; although there is a synthetic pheromone available that can shut down trap captures in an orchard (Knight et al. 1998), mated females can fly long distances from non-disrupted sites to lay eggs in a mating disruption block (Lawson et al. 1996; Trimble and Appleby 2004).

4.4 Strategies for reducing mating disruption failures

Location likely has the greatest role in the potential success of a mating disruption strategy. Factors that come into play include orchard block size (the larger the better), block shape (square or rectangular is better than long and thin) and pest pressure (better against lower populations, and where less immigration occurs from untreated areas). All these considerations should be used in determining product selection, release rate and number of point sources desired.

It is prudent to monitor the target species by placing pheromone traps within disrupted orchards, as a way of verifying that there are no gaps in pheromone coverage that could allow moths to find and mate with each other. Trap design should be selected to facilitate efficient, regular checking. Trap placement in the orchard interior will provide more useful information about moth breakthroughs than those on the orchard edge, although if possible, some traps should also be placed in non-disrupted sites, to provide a comparison for assessment of population pressure. Lures with extended longevity will reduce the need for frequent replacement during the season.

Finally, the need for supplemental insecticide treatments should not be overlooked, particularly for species like codling moth that are more difficult to control. If using high-dose lures for monitoring, which can attract moths even in disrupted blocks, use trap numbers as a basis for making treatment decisions. Focus particularly on border infestations, as orchard edges show greater pheromone dissipation. Other considerations to keep in mind are that mated codling moths and oriental fruit moths can migrate from fruit bins (or non-disrupted trees) into the orchard; female oriental fruit moths fly 2–4 times as far as male moths. Bin piles (or other sources) infested with these pests can provide a source of adults to re-infest the orchard, make timing of insecticide sprays difficult, introduce insecticide-resistant populations from other areas and generally disrupt pheromone mating disruption programmes.

Is it possible to reduce insecticide applications when implementing mating disruption? For codling moth, it is not a stand-alone tactic, but sprays can be reduced, particularly after 2–3 years of use. Oriental fruit moth may not require as many sprays, especially in stone fruits, where it could be the primary species of concern. In the case of peachtree borers, mating disruption can completely replace sprays, and has been recognized as the tactic of choice. For dogwood borer, its potential value is dependent on the site; in larger plantings not subject to an influx of moths from unmanaged areas, this could become a primary tactic.

5 Organic tree fruit production

Traditionally, few growers have attempted to produce apples and other tree fruits organically because of the practical difficulties involved in controlling pests without using conventional, broad-spectrum pesticides, which are more effective than most organically acceptable products. The greatest amount of organic production tends to be where pest pressures are lowest, such as the arid fruit-growing regions of the Pacific Northwest in the United States However, during the last 10–15 years, studies have been conducted to develop management programmes that may be able to replace current strategies that rely primarily on these pesticide applications (Peck and Merwin 2010). For example, recent studies have shown that predaceous mites, such as *Typhlodromus pyri* (Scheuten) and *Neoseiulus* (=*Amblyseius*) *fallacis* (Garman), which are native to many apple production regions, can successfully control populations of the key mite pest, European red mite, in commercial apple orchards so that no miticide applications are required for seasonal control (Nyrop et al. 1998; Agnello et al. 2003). Recent research has also shown that pheromones can be deployed in orchards to disrupt mating of key lepidopteran species such as codling moth, oriental fruit moth and borer species, and substantially reduce fruit damage from this complex of pests. In addition to some of these newer types of organically compatible pest control technologies, traditional control methods such as selective fruit thinning, pruning, sanitation (frequent removal of dropped fruit infested or infected with pests), removal of wild hosts near commercial plantings and exclusion of pests with biological or physical barriers near or around trees have also been shown to reduce populations of many types of pests in fruit plantings in this region.

Ideally, organic fruit production is the synthesis of an entire suite of practices intended to take advantage of natural ecosystem interactions and minimize synthetic inputs. Such a system might start with the selection of disease-resistant cultivars, to circumvent the need for the majority of normal disease sprays. This one tactic could eliminate or substantially reduce the need to manage apple scab, powdery mildew, cedar apple rust and fire blight (Ellis et al. 1998). In lieu of resistance, a combined strategy of orchard sanitation and frequent applications of elemental sulphur and copper throughout most of the season would be the next practical alternative.

Although not as effective as their conventional counterparts, organically acceptable control options for arthropod pests have improved in recent years and more products are being developed as the number of growers interested in organic production increases (Phillips 2005, 2011; Agnello et al. 2017). Biopesticides may offer some useful alternatives for arthropod management in organic apple production in future years, even in regions with high pest population pressure. Traditional efforts rely on the most well-known representatives of this class of insecticides: *Bacillus thuringiensis*, azadirachtin and pyrethrin. However, others have been registered and are being evaluated for efficacy in various crop systems, and some may ultimately prove effective against selected orchard pests typically found in certain regions. Along with more familiar active ingredients such as *Beauveria bassiana* and *Metarhizium brunneum*, there are several new products being tested that contain bacterial extracts, for example *Chromobacterium subtsugae* and *Burkholderia* spp. More products are under development, but have yet to be thoroughly evaluated in apple production systems. Because there is much variability in organic certification standards from one market to another, every grower must verify that a given crop protectant or management practice is acceptable to their prospective buyer or consumer.

6 Case studies

To evaluate the practicality of implementing specifically timed sampling and monitoring procedures for making management decisions, a five-year, simplified pest-monitoring programme was conducted in New York apple orchards from 1987 to 1991, which entailed individual instruction sessions with grower cooperators throughout the growing season (Agnello et al. 1994). Programme participants saved as many as 3.4 pesticide sprays in one year of the programme, for a saving of up to \$77 ha^{-1} (\$31 $acre^{-1}$). For the remainder of the time, pesticide applications and costs were comparable with those in conventionally managed blocks. Except for 1987, when an unrefined prediction model for one insect pest was responsible for unacceptable fruit damage at harvest, IPM orchard blocks did not show significant differences in fruit quality from that obtained in conventionally managed blocks. Examination of participant compliance with recommended pest decisions indicated that growers were more risk-averse (in treating when not recommended) for direct fruit-feeding insects, but with successively greater reliance on scouting decisions for foliar pests, some tended to be slightly more lax than was recommended by programme guidelines. A follow-up survey to evaluate participant attitudes and permanent changes in their practices found that many now rely on someone else to do their scouting for them. Those who scout their own orchards often use modified procedures or treatment thresholds, and most do not keep written pest-sampling records. A computer-based decision support system was developed to aid in the implementation of whole-farm pest management guidelines, but a lack of adequate support curtailed efforts to maintain and upgrade its operating system and knowledge base.

From 2002 to 2005, studies in the United States were conducted to determine the effectiveness of reduced-risk (RR) tactics for managing key pests in 50 commercial apple orchards in Michigan, North Carolina, New York, Pennsylvania, Virginia and West Virginia and 20 peach orchards in Michigan, New Jersey, Pennsylvania and West Virginia, comprising 190 and 114 ha for apple and peach, respectively (Agnello et al. 2009). At each apple site, a block of up to 5 ha received a seasonal programme of selective RR and organophosphate (OP) replacement insecticides, with or without pheromones for mating disruption of key lepidopteran pests of apple (codling moth, oriental fruit moth and obliquebanded leafroller) and peach (oriental fruit moth, lesser peachtree borer and peachtree borer). A comparison block at each site with the same varieties and tree training was managed using each grower's standard programme of conventional insecticides (STD). Pheromone traps for lepidopteran species were hung in all plots and monitored weekly. Foliar samples were taken during the season to estimate phytophagous and predator mite densities. Red sphere traps baited with fruit volatiles were used to monitor apple maggot adults in apple orchards. Fruits were inspected for insect damage at harvest, and graded according to USDA standards. Partial budget analysis was used to assess the net profitability of RR programmes to produce apples and peaches for their intended market in each state. Fruit damage at harvest caused by direct fruit pests was generally low across all blocks and treatments. There were no statistically significant differences in fruit damage for either crop or mite populations between the RR blocks, with or without pheromones, and the grower standards. Insecticide use patterns in the RR plots represented up to 88% and 78% reduction in the amount of active ingredient applied per ha, and an 85% and 77% decrease in their Environmental Impact Quotient for apples and peaches, respectively (Kovach et al. 1992). However, RR programmes were more expensive and generally less profitable compared with growers' STD programmes. Regression analysis estimated that RR apple programmes with and without mating disruption were, on average, \$465 and \$144 ha^{-1} more expensive, and \$544 and \$159 ha^{-1} less profitable, respectively, compared with standard programmes. Reduced risk + mating disruption programmes for peaches cost an average \$314 ha^{-1} more and returned about \$284 ha^{-1} less than STD peach programmes.

7 Where to look for further information

7.1 Publications

Agnello, A., Chouinard, G., Firlej, A., Turechek, W., Vanoosthuyse, F. and Vincent, C. (2006). Tree fruit field guide to insect, mite, and disease pests and natural enemies of Eastern North America. *NRAES-169*. Ithaca, NY: Plant and Life Sciences Publishing, 238pp. http://palspublishing.cals.cornell.edu/, Online at: http://www.ipm.msu.edu/search

Blommers, L. H. (1994). Integrated pest management in European apple orchards. *Annu. Rev. Entomol.* 39: 213–41. doi:10.1146/annurev.en.39.010194.001241.

Beers, E. H., Brunner, J. F., Willett, M. J. and Warner, G. M. (Eds) (1993). *Orchard Pest Management: A Resource Book for the Pacific Northwest.* Yakima, WA: Good Fruit Grower, 276pp. http://jenny.tfrec.wsu.edu/opm/toc.php?h=3

Croft, B. A. and Hoyt, S. C. (Eds) (1983). *Integrated Management of Insect Pests of Pome and Stone Fruits.* New York, NY: Wiley, 454pp. ISBN:0471053341.

Gruys, P. (1982). Hits and misses: The ecological approach to pest control in orchards. *Entomol. Exp. & Appl.* 31: 70–87. https://onlinelibrary.wiley.com/doi/pdf/10.1111/j.1570-7458.1982.tb03120.x

Hogmire, H. W., Jr. (Ed.). (1995). Mid-Atlantic orchard monitoring guide. *NRAES-75*. Ithaca, NY: Plant and Life Sciences Publishing, 361pp. http://palspublishing.cals.cornell.edu/

Kovach, J. and Tette, J. P. (1988). A survey of the use of IPM by New York apple producers. *Agric. Ecosys. Envir.* 20: 101–8.

Sutton, T. B., Aldwinckle, H. S., Agnello, A. M. and Walgenbach, J. F. (Eds) (2013). *Compendium of Apple and Pear Diseases and Pests*, 2nd ed. St Paul, MN: American Phytopathological Society, 224pp. http://www.apsnet.org/apsstore/shopapspress/Pages/44303.aspx

University of California. (1999). *Integrated Pest Management for Apples and Pears, 2nd ed.* UC Division of Agriculture and Natural Resources Publication No. 3340. Oakland, CA, 231pp.

Whalon, M. E. and Croft, B. A. (1984). Apple IPM implementation in North America. *Annu. Rev. Entomol.* 29: 435–70. doi:10.1146/annurev.en.29.010184.002251.

Williams, K. (Ed.) (1991). *New Directions in Tree Fruit Pest Management.* Yakima, WA: Good Fruit Grower, 214pp. ISBN: 0-9630659-0-4.

7.2 Tree fruit and IPM-related web sites

http://www.fruit.cornell.edu – Cornell Fruit Resources homepage.

https://cropandpestguides.cce.cornell.edu/ – Cornell Tree Fruit Pest Management Guidelines (paid access only).

http://www.nysipm.cornell.edu – NYS IPM Program, containing Tree Fruit Pest Fact Sheets and other information.

https://blogs.cornell.edu/treefruit/pest-management/apple-ipm-for-beginners/ – Apple IPM for Beginners, a guide for new apple growers to protect apple orchards from pests, in fact sheet format.

http://virtualorchard.net – The Virtual Orchard – a site for the dissemination of information on all aspects of sustainable apple production.

http://hdl.handle.net/1813/42886 - A Grower's Guide to Organic Apples (2009).

http://hdl.handle.net/1813/5219 - New York Integrated Fruit Production Protocol for Apples.

http://newa.cornell.edu/index.php?page=apple-insects - NEWA Apple Insect Model website. Ref: Agnello, A. M. and Reissig, H. (2010). Development and validation of a 'Real-Time' Apple IPM Website for New York. *NY Fruit Quarterly* 18(2): 25–8.

http://www.ams.usda.gov/NOP – National Organic Program website.

http://www.agriculture.ny.gov/AP/Organic/docs/Organizations-Providing-Organic-Certification-Services.pdf – NYS Dept of Agriculture & Markets list of organic certifiers.

http://www.omri.org/omri-lists – Organic Materials Review Institute/OMRI.

7.3 Meetings and conferences

Orchard Pest and Disease Management Conference, Portland, Oregon (annual). Started in 1927, this conference is held each January. The meeting is devoted exclusively to the arthropod and disease pests of orchard crops, including deciduous tree fruits and nuts. Presentations on arthropod and disease pests are in separate (concurrent) sessions. While the majority of meeting attendees are from the western states, the meeting is national and international in scope. Most of the presentations are submitted from state and federal research and extension personnel, and from the agrochemical industry. http://www.tfrec.wsu.edu/pages/wopdmc/

Cumberland-Shenandoah Fruit Workers Conference, Winchester, Virginia (annual). Founded in 1924, the CSFWC is a meeting for university and government researchers representing fruit pathology, entomology and horticulture, fruit crop consultants, university extension agents and representatives from the agrochemical industry. The goal of the conference is to discuss and disseminate results from fruit research programmes. http://csfwc.grapepathology.org/

Great Lakes Fruit Workers Meeting (annual). The Great Lakes Fruit Workers (GLFW) is a group of researchers, extension agents/educators and consultants from public or private institutions, who have met annually since 1998 to share preliminary research and extension programming, and to discuss challenges common to fruit production in the Great Lakes Region. Rotating annual meetings are held in Michigan, New York or Ontario. Annual meetings often include a local tour of fruit operations. The group meets to (1) identify crop management priorities for fruit producers, (2) foster collaborations, (3) develop innovative crop management techniques and (4) share new IPM technologies, management guides, pest alert systems and outreach efforts that result from these collaborations. https://greatlakesfruitworkers.weebly.com/

The International Organisation for Biological and Integrated Control, West Palaearctic Regional Section (IOBC-WPRS) Fruit Working Groups (e.g. Integrated Production of Fruit Crops; Pheromones and Other Semiochemicals in Integrated Production) convenes bi- or tri-annual meetings in various European venues. http://www.iobc-wprs.org/

8 References

Agnello, A. M. Spangler, S. M. and Reissig, W. H. (1990). Development and evaluation of a more efficient monitoring system for apple maggot (Diptera: Tephritidae). *J. Econ. Entomol.* 83(2): 539-46. doi:10.1093/jee/83.2.539.

Agnello, A. M., Kovach, J., Nyrop, J. P., Reissig, W. H., Breth, D. I. and Wilcox, W. (1994). Extension and evaluation of a simplified monitoring program in New York apples. *Am. Entomol.* 40(1): 37–49. doi:10.1093/ae/40.1.37.

Agnello, A. M., Reissig, W. H., Kovach, J. and Nyrop J. P. (2003). Integrated apple pest management in New York State using predatory mites and selective pesticides. *Agric. Ecosyst. Environ.* 94: 183–95.

Agnello, A. M., Atanassov, A., Bergh, J. C., Biddinger, D. J., Gut, L. J., Haas, M. J., Harper, J. K., Hogmire, H. W., Hull, L. A., Kime, L. F., Krawczyk, G., McGhee, P. S., Nyrop, J. P., Reissig, W. H., Shearer, P. W., Straub, R. W., Villanueva, R. T. and Walgenbach, J. F. (2009). Reduced-risk pest management programs for eastern U.S. apple and peach orchards: A 4-year regional project. *Am. Entomol.* 55(3): 184–97. doi:10.1093/ae/55.3.184.

Agnello, A., Cox, K., Lordan, J., Francescatto, P. and Robinson, T. (2017). Comparative programs for arthropod, disease and weed management in New York organic apples. *Insects* 8(3): 96–116. doi:10.3390/insects8030096.

Bajwa, W. I. and Kogan, M. (2002). Compendium of IPM definitions (CID). *Integrated Plant Protection Center Publication No. 998*. Corvallis, OR: Oregon State University. http://ipmnet.org/IPMdefinitions/index.pdf

Baskerville, G. L. and Emin, P. (1969). Rapid estimation of heat accumulation from maximum and minimum temperatures. *Ecology* 50(3): 514–17. doi:10.2307/1933912.

Binns, M. R., Nyrop, J. P. and Van der Werf, W. (2001). *Sampling and Monitoring in Crop Protection: The Theoretical Basis for Developing Practical Decision Guides*. Wallingford, UK: CABI Press, 284pp. ISBN: 0-851999-347-8.

Breth, D. I., Nyrop, J. P. and Kovach, J. (1998). Achieving biological control of European red mite in northeast apples. *NYS IPM Publication No. 215*. Geneva: New York State IPM Program, 8pp. http://hdl.handle.net/1813/43090

Brunner, J. F. (1994). Integrated pest management in tree fruit crops. *Food Rev. Int.* 10(2): 135–57. doi:10.1080/87559129409540994.

Cardé, R. T. and Minks, A. K. (1995). Control of moth pests by mating disruption: Successes and constraints. *Annu. Rev. Entomol.* 40: 559–85. doi:10.1146/annurev.en.40.010195.003015.

Carroll, J. (2011). About degree days. *Network for Environment and Weather Applications (NEWA)*, NYS Integrated Pest Management Program, Cornell University. http://newa.cornell.edu/index.php?page=about-degree-days

Chapman, R. F. (2012). *The Insects: Structure and Function, 5th ed.* S. J. Simpson and A. E. Douglas (Eds). Cambridge, UK: Cambridge University Press. ISBN: 9780521113892.

Cork, A. (2016). Pheromones as management tools: Mass trapping and lure-and-kill. In J. D. Allison and R. T. Cardé (Eds), *Pheromone Communication in Moths: Evolution, Behavior, and Application* (pp. 349-63). Oakland, CA: University of California Press. ISBN: 9780520278561.

Croft, B. A. and Bode, W. M. (1983). Tactics for deciduous fruit IPM. In B. A. Croft and S. C. Hoyt (Eds), *Integrated Management of Insect Pests of Pome and Stone Fruits* (pp. 219–70). New York, NY: Wiley. ISBN: 0471053341.

Delisle, J. (1992). Monitoring the seasonal male flight activity of *Choristoneura rosaceana* (Lepidoptera: Tortricidae) in Eastern Canada using virgin females and several different pheromone blends. *Environ. Entomol.* 21(5): 1007–12. doi:10.1093/ee/21.5.1007.

Ellis, M. A., Ferree, D. C., Funt, R. C. and Madden, L. V. (1998). Effects of an apple scab-resistant cultivar on use patterns of inorganic and organic fungicides and economics of disease control. *Plant Dis.* 82: 428–33.

Evenden, M. (2016). Mating disruption of moth pests in integrated pest management: A mechanistic approach. In: J. D. Allison and R. T. Cardé (Eds), *Pheromone Communication in Moths: Evolution, Behavior, and Application* (pp. 365–93). Oakland, CA: University of California Press. ISBN: 9780520278561.

Hoffmann, M. P., Petzoldt, C. H. and Frodsham, A. C. (1996). Integrated pest management for onions. *New York State IPM Publication No. 119*, 78pp.

Howell, J. F., Knight, A. L., Unruh, T. R., Brown, D. F., Krysan, J. L., Sell, C. R. and Kirsch, P. A. (1992). Control of codling moth in apple and pear with sex pheromone-mediated mating disruption, *J. Econ. Entomol.* 85(3): 918–25. doi:10.1093/jee/85.3.918.

Hoyt, S. C. and Burts, E. C. (1974). Integrated control of fruit pests, *Annu. Rev. Entomol.* 19: 231–52. doi:10.1146/annurev.en.19.010174.001311.

Hoyt, S. C. and Tanigoshi, L. K. (1983). Economic injury levels for apple insect pests. In B. A. Croft and S. C. Hoyt (Eds), *Integrated Management of Insect Pests of Pome and Stone Fruits* (pp. 203–18). New York, NY: Wiley. ISBN: 0471053341.

Hoyt, S. C., Leeper, J. R., Brown, G. C. and Croft, B. A. (1983). Basic biology and management components for insect IPM. In B. A. Croft and S. C. Hoyt (Eds), *Integrated Management of Insect Pests of Pome and Stone Fruits* (pp. 93–151). New York, NY: Wiley. ISBN: 0471053341.

Hull, L. A., Myers, C., Ellis, N. and Krawczyk, G. (2003). Management of the internal lepidopteran complex in Pennsylvania. *Compact Fruit Tree* 36(1): 21–5.

Knight, A. L. (2008). Codling moth areawide integrated pest management. *Publications from USDA-ARS/UNL Faculty No. 651*. http://digitalcommons.unl.edu/usdaarsfacpub/651

Knight, A. L., Thomson, D. R. and Cockfield, S. D. (1998). Developing mating disruption of obliquebanded leafroller (Lepidoptera: Tortricidae) in Washington State. *Environ. Entomol.* 27(5): 1080–8. doi:10.1093/ee/27.5.1080.

Kovach, J., Petzoldt, C., Degni, J. and Tette, J. (1992). A method to measure the environmental impact of pesticides. *NY Food and Life Science Bulletin No. 139*. Geneva, 8pp. http://hdl.handle.net/1813/55750

Kovanci, O. B. and Walgenbach, J. F. (2005). Monitoring the oriental fruit moth with pheromone and bait traps in apple orchards under different management regimes. *Int. J. Pest Manag.* 51(4): 273–9. doi:10.1080/09670870500270661.

Lawson, D. S., Reissig, W. H., Agnello, A. M., Nyrop, J. P. and Roelofs, W. L. (1996). Interference with the mate-finding communication system of the obliquebanded leafroller (Lepidoptera: Tortricidae) using synthetic sex pheromones. *Environ. Entomol.* 25(5): 895–905. doi:10.1093/ee/25.5.895.

Leskey, T. C. and Wright, S. E. (2004). Monitoring plum curculio, *Conotrachelus nenuphar* (Coleoptera: Curculionidae), populations in apple and peach orchards in the Mid-Atlantic. *J. Econ. Entomol.* 97(1): 79–88. doi:10.1603/0022-0493-97.1.79.

Leskey, T. C., Bergh, J. C., Walgenbach, J. F. and Zhang, A. (2009). Evaluation of pheromone-based management strategies for dogwood borer (Lepidoptera: Sesiidae) in commercial apple orchards. *J. Econ. Entomol.* 102(3): 1085–93. doi:10.1603/029.102.0329.

Madsen, H. and Madsen, B. (1980). Response of four leafroller species (Lepidoptera: Tortricidae) to sex attractants in British Columbia orchards. *Can. Entomol.* 112(4): 427–30. doi:10.4039/Ent112427-4.

Madsen, H. F. and Vakenti, J. M. (1973). Codling moth: Use of Codlemone®-baited traps and visual detection of entries to determine need of sprays. *Environ. Entomol.* 2(4): 677–80. doi:10.1093/ee/2.4.677.

Mague, D. L. and Reissig, W. H. (1983). Phenology of the San Jose scale (Homoptera: Diaspididae) in New York State apple orchards. *Can. Entomol.* 115: 717–22. doi:10.4039/Ent115717-7.

Miller, J. R. and Gut, L. J. (2015). Mating disruption for the 21st century: Matching technology with mechanism. *Environ. Entomol.* 44: 427–53. doi:10.1093/ee/nvv052.

Nyrop, J. P., Agnello, A. M., Kovach, J. and Reissig, W. H. (1989). Binomial sequential classification sampling plans for European red mite (Acari: Tetranychidae) with special reference to performance criteria, *J. Econ. Entomol.* 82(2): 482–90. doi:10.1093/jee/82.2.482.

Nyrop, J. P., Loeb, G. M. and Roda, A. (1998). Conservation biological control of spider mites in perennial cropping systems. In P. Barbosa (Ed.), *Perspectives on the Conservation of Natural Enemies of Pest Species* (pp. 307–33). Cambridge, MA: Academic Press. ISBN: 9780120781478.

Olatinwo, R. and Hoogenboom, G. (2013). Weather-based pest forecasting for efficient crop protection. In D. P. Abrol (Ed.), *Integrated Pest Management, Current Concepts and Ecological Perspective* (pp. 59–76). Waltham, MA: Elsevier. doi:10.1016/B978-0-12-398529-3.00005-1.

Onstad, D. W., Reissig, W. H. and Shoemaker, C. A. (1985). Phenology and management of the obliquebanded leafroller (Lepidoptera: Tortricidae) in apple orchards. *J. Econ. Entomol.* 78: 1455–62. doi:10.1093/jee/78.6.1455.

Peck, G. and Merwin, I. (2010). A grower's guide to organic apples. *NYS IPM Publication No. 223*. Ithaca, NY: Cornell University, 64pp. https://ecommons.cornell.edu/handle/1813/42886.

Pfeiffer, D. G., Killian, J. C., Rajotte, E. G., Hull, L. A. and Shows, W. J. (1991). Mating disruption for reduction of damage by lesser peachtree borer (Lepidoptera: Sesiidae) in Virginia and Pennsylvania peach orchards. *J. Econ. Entomol.* 84(1): 218–23. doi:10.1093/jee/84.1.218.

Phillips, M. (2005). *The Apple Grower*, revised edition. White River Junction, VT: Chelsea Green, 242pp.

Phillips, M. (2011). *The Holistic Orchard: Tree Fruits and Berries the Biological Way*. White River Junction, VT: Chelsea Green, 414pp.

Rabb, R. L. and Guthrie, F. E. (1970). *Concepts of Pest Management: Proceedings of a Conference*. Raleigh, NC: North Carolina State University, 242pp. SERBIULA (sistema Librum 2.0).

Reissig, W. H., Barnard, J., Weires, R. W., Glass, E. H. and Dean, R. W. (1979). Prediction of apple maggot fly emergence from thermal unit accumulation. *Environ. Entomol.* 8: 51–4. doi:10.1093/ee/8.1.51.

Reissig, W. H., Nyrop, J. P. and Straub, R. (1998). Oviposition model for timing insecticide sprays against plum curculio (Coleoptera: Curculionidae) in New York State. *Environ. Entomol.* 27: 1053–61. doi:10.1093/ee/27.5.1053.

Riedl, H. (1980). The importance of pheromone trap density and trap maintenance for the development of standardized monitoring procedures for the codling moth (Lepidoptera: Tortricidae). *Can. Entomol.* 112(7): 655–63. doi:10.4039/Ent112655-7

Riedl, H., Croft, B. A. and Howitt, A. J. (1976). Forecasting codling moth physiology based on pheromone trap catches and physiological-time models. *Can. Entomol.* 108: 449–60. https://doi.org/10.4039/Ent108449-5.

Roelofs, W. L. (2016). Reminiscence of the early days. In J. D. Allison and R. T. Cardé (Eds), *Pheromone Communication in Moths: Evolution, Behavior, and Application* (pp. 3–9). Oakland, CA: University of California Press. ISBN: 9780520278561.

Schmaedick, M. A. and Nyrop, J. P. (1995). Method for sampling arthropod pests with uncertain phenology with application to spotted tentiform leafminer (Lepidoptera: Gracillariidae). *J. Econ. Entomol.* 88: 875–89. doi:10.1093/jee/88.4.875.

Sevacherian, V., Stern, V. M. and Mueller, A. J. (1977). Heat accumulation for timing *Lygus* control measures in a safflower-cotton complex. *J. Econ. Entomol.* 70(4): 399–402. doi:10.1093/jee/70.4.399.

Trimble, R. M. and Appleby, M. E. (2004). Comparison of efficacy of programs using insecticide and insecticide plus mating disruption for controlling the obliquebanded leafroller in apple (Lepidoptera: Tortricidae). *J. Econ. Entomol.* 97(2): 518–24. doi:10.1093/jee/97.2.518.

Vakenti, J. and Madsen, H. (1976). Codling moth (Lepidoptera: Olethreutidae): Monitoring populations in apple orchards with sex pheromone traps. *Can. Entomol.* 108(4): 433–8. doi:10.4039/Ent108433-4.

Vickers, R., Rothschild, G. and Jones, E. (1985). Control of the oriental fruit moth, *Cydia molesta* (Busck) (Lepidoptera: Tortricidae), at a district level by mating disruption with synthetic female pheromone. *Bull. Entomol. Res. 75*(4): 625–34. doi:10.1017/S0007485300015881.

Welch, S. M. and Croft, B. A. (1983). Models of direct fruit pests of apple. In B. A. Croft, and S. C. Hoyt (Eds), *Integrated Management of Insect Pests of Pome and Stone Fruits* (pp. 343–68). New York, NY: Wiley. ISBN:0471053341.

Welter, S. C., Pickel, C., Millar, J., Cave, F., Van Steenwyk, R. A. and Dunley, J. (2005). Pheromone mating disruption offers selective management options for key pests. *Cal. Agric.* 59(1): 16–22. doi:10.3733/ca.v059n01p16.

Wise, J. (2017). Rainfast characteristics of insecticides. *Michigan State University Extension.* http://msue.anr.msu.edu/news/rainfast_characteristics_of_insecticides_on_fruit.

Chapter 11

Optimizing disease management in fruit cultivation

Sara M. Villani, North Carolina State University, USA; Kerik D. Cox, Cornell University, USA; and George W. Sundin, Michigan State University, USA

1 Introduction

1.1 Cultivation and production of pome fruit

The pome fruit, members of the *Rosaceae* family, are produced in most temperate regions throughout the world. With an annual fresh market production of 76.2 million metric tons (Anon., 2017a) apples (*Malus domestica*) are the most widely produced and recognizable pome fruit, and will thus be the primary focus throughout the chapter. However, it is important to note other important pome fruit including common pear (*Pyrus communis*), Asian pear (*Pyrus pyrifolia*) and quince (*Cydonia oblonga*). China is currently the world leader in apple production, producing 58% of the global apple crop in 2017. A distant second to China, the United States produced approximately 6% of the world's apples in 2017 with a utilized production value of US$3.4 billion (Anon., 2017a). According to the USDA Non-Citrus Fruit and Nuts 2016 Summary (Anon., 2017b), approximately 130 300 ha of apples are in production throughout the United States with the majority of hectarage located in Washington (66 773 ha), New York (16 187 ha), Michigan (13 354 ha), Pennsylvania (8 094 ha) and California (5 059 ha). Despite advancements of apple breeding programmes for consumer preferences and disease resistance, traditional apple varieties including cvs. 'Gala', 'Red Delicious', 'Granny Smith', 'Fuji' and 'Golden Delicious'

http://dx.doi.org/10.19103/AS.2018.0040.12

continue to be the most popular apples for fresh market consumption in the United States (Anon., 2018c).

Apple growers worldwide are increasingly adopting modern production systems consisting of high-density trellised plantings (Fig. 1). It is through the utilization of dwarfing rootstocks, most commonly those in the Malling, Budagovsky and Geneva series, that modern planting spaces (~1 m × 3.5 m) can be achieved. Accelerated profitability, improved fruit quality, improved land efficiency, reduced pesticide input and compatibility with new orchard technologies are among several economic and horticultural benefits of modern orchard systems (Fazio et al., 2015; Norelli et al., 2003; Robinson et al., 2003). In addition, dwarfing rootstocks, particularly within the Geneva series, are tolerant or resistant to diseases including fire blight ('rootstock blight'), Phytophthora root rot and apple replant disease (Fazio et al., 2015; Robinson et al., 2003). Despite resistance imparted in the rootstock, horticultural practices aimed at rapidly increasing tree vigour, reducing leaf-to-fruit ratio and reducing tree width to accommodate tighter tree spacing (Westwood, 1993) can often exacerbate disease development in modern planting systems by promoting a greater period of susceptibility to foliar pathogens (Jurick and Cox, 2017) and decreasing systemic pathogen travel distance to the scion.

1.2 Cultivation and production of stone fruit

In the United States, stone fruit have an annual utilized production value of US$1.8 billion. Sweet cherries (*Prunus avium*) are the most economically important stone fruit crop, representing 43% of the U.S. stone fruit industry value in 2016. Besides sweet cherries, economically important stone fruit crops

Figure 1 A modern high-density apple planting using a trellis system. Trees are grafted to dwarfing rootstocks.

in the United States include peaches (*Prunus persica*), nectarines (*Prunus persica* var. *nucipersica*), plums/prunes (*Prunus salicina/Prunus domestica*), tart cherries (*Prunus cerasus*) and apricots (*Prunus armeniaca*). Globally, 3.1 million tons of sweet cherries were produced in 2017, with the European Union (19%), Turkey (16%) and the United States (16%) leading global production (Anon., 2017c). According to the USDA Non-Citrus Fruit and Nuts 2016 Summary (Anon., 2017b), approximately 31 100 ha of sweet cherries are in production throughout the United States with the majority of production located in Washington (15 700 ha), California (13 350 ha), Oregon (5 260 ha) and Michigan (2 710 ha). As of 2017, China continued to lead peach and nectarine production, growing ~66% of the market's share. Unlike dwarfing rootstocks utilized for apple production, with the exception of cherry production, peach and plum/prune rootstocks are still standard/seedling in size. In regard to disease resistance, breeding efforts of stone fruit rootstocks have focused on nematode, virus, oak root rot (Armillaria root rot), Phytophthora root rot and peach tree short life.

1.3 Managing diseases of pome and stone fruit: current challenges

The development of national and international government agricultural regulations, the increasing prevalence of multiple fungicide resistance, and short- and long-term changes in climate are currently among the greatest challenges to the sustainable management of pre-harvest diseases of pome and stone fruit (Campoy et al., 2011; Chapman et al., 2011; Cosmulesco et al., 2010; Fernández-Ortuño et al., 2014; Garrett et al., 2006; Romanazzi et al., 2016; Rupp et al., 2016). For more than 60 years, applications of protective and therapeutic fungicides and bactericides have widely been considered the cornerstone for tree fruit disease management. The implementation of more stringent legislation reducing maximum residue levels (MRLs) and pesticide deregistration in current and potential U.S. export markets (Anon., 2018a,b), as well as ongoing legislation aimed at promoting environmental, pollinator and other non-target safety, has encouraged stakeholders to further incorporate non-chemical, often less effective, control methods in disease management programmes. Along with the necessity to adjust chemical management programmes to satisfy federal and international regulations, the loss of safer single-site fungicides and agricultural antibiotics due to resistance (Frederick et al., 2014; Luo et al., 2008; McGhee et al. 2011; Tancos et al., 2016; Villani et al., 2015) further creates disease management challenges for tree fruit growers.

The incorporation of cultural methods to mitigate diseases of tree fruit has been less frequently relied on as a management paradigm for conventional producers in the United States compared to those in the European Union. In the United States, cultural management predominantly occurs during tree dormancy and focuses on the reduction or removal of pathogen inoculum from the orchard. Often this is accomplished by shredding leaves using specialized equipment, the application of 5% urea on fallen leaves to stimulate microbial breakdown, and pruning to remove cankers from trees (Holb, 2005, 2007; Sutton et al., 2000). While the slow adoption of cultural methods for disease management during the growing season may be reflective of more relaxed fungicide regulations compared to other nations, labour availability in the United States is an increasingly limiting factor in achieving the benefits of labour-intensive in-season cultural management practices (Villani, unpublished).

In addition, the preferential selection of horticultural attributes desirable to consumers over other traits has caused setbacks in the development of disease-resistant pome fruit and stone fruit cultivars (Brown and Maloney, 2008). For example, as regards apple scab, several cultivars with monogenic resistance have been developed, but have not been widely planted by growers due to marginal consumer acceptance (Gessler and Pertot, 2012). In South America where Glomerella leaf spot is a limiting factor for apple production, breeding efforts have largely focused on developing quantitative resistance to the disease where popular, but susceptible, cultivars like 'Gala' are crossed with more resistant cultivars. Despite the potential for monogenic resistance to specific diseases, fungicide use would likely not be minimized due to the necessity to manage other fungal diseases. Adaptation to erratic short- and long-term climate change is a current and certainly a future obstacle for tree fruit producers globally. While orchard management paradigms can be adjusted to satisfy challenges imposed by agricultural regulations and economic constraints, in established orchards there are few protective or remedial measures to alleviate losses associated with erratic climate conditions. For example, direct damage to maturing fruit by hail leads to a greater threat of fungal and bacterial colonization. Water stress, particularly in high-density systems where competition for water and nutrients is common, has become an increasing challenge for growers, particularly in Eastern and Western U.S. apple production regions. In recent years, tree decline resulting from opportunistic wood rotting pathogens including *Armillaria* spp., *Botryosphaeria dothidea* and *Phomopsis mali* has been observed in young, high-density plantings in which irrigation is absent or not used frequently and/or that have been subjected to cold injury due to rapid temperature fluctuations in fall, winter or spring.

2 Pre-harvest fungal diseases of tree fruit

2.1 Fungal diseases of pome fruit

There are several pre-harvest fungal diseases of pome fruit that pose a perennial challenge to growers throughout the world (Fig. 2). For pome fruit, the root system, woody tissue, foliage and fruit are all vulnerable to fungal pathogen colonization. The most common pathogens of apples in major growing regions

Figure 2 Apple foliar and fruit fungal diseases that occur in the United States. (a) Apple scab on leaves (*Venturia inaequalis*), (b) cedar apple rust on leaves (*Gymnosporangium juniperi-virginianae*), (c) primary powdery mildew on leaves (*Podosphaera leucotricha*), (d) flyspeck (*Schizathyrium pomi*) and sooty blotch (species complex) on fruit, (e) bitter rot on fruit (*Colletotrichum* spp.) and (f) Glomerella leaf spot on chlorotic leaf (*Colletotrichum gloeosporioides* species complex).

of the United States include *Alternaria mali* (*Alternaria* blotch), *Botryosphaeria dothidea* (white rot), *Botryosphaeria obtusa* (frogeye leaf spot and black rot), *Colletotrichum* spp. (bitter rot and Glomerella leaf spot), *Gymnosporangium* spp. (apple rust diseases), *Podosphaera leucotricha* (powdery mildew), *Schizothyrium pomi* (flyspeck), sooty blotch species complex (sooty blotch) and *Venturia inaequalis* (apple scab) (Sutton et al., 2014). Although not a fungal pathogen, the oomycete pathogen, *Phytophthora* spp., the causal agent of root, collar and crown rot in apple, can also reduce the productivity and lifespan of apple trees (Wilcox, 1993) and should be prominently mentioned in discussions concerning predominant diseases of apple in the United States. Many of the same fungal pathogens infecting apples also cause disease on common pear including *Colletotrichum* spp. (bitter rot) and *Botryosphaeria dothidea* (white rot). In addition, *Diplocarpon mespili* (Fabraea leaf spot) and *Venturia pirina* (pear scab) are two common pathogens that pear growers must consider in disease management programmes (Sutton et al., 2014).

Several of the aforementioned pathogens of pome fruit have the potential to cause losses in most apple growing areas; however, regional climate affects the severity of each disease and the resulting economic consequences. In cool, humid growing regions such as the Northeastern United States, apple scab caused by the fungal pathogen *Venturia inaequalis* is a perennial challenge for growers (MacHardy, 1996). Following leaf drop during the autumn months, the pathogen undergoes sexual reproduction, resulting in pseudothecial development on overwintering leaf lesions. Epidemics begin when ascospores are released during spring rain events and primary infection is initiated when certain wetting hour and temperature requirements are met. While ascospores can infect either leaf or fruit tissue, usually only cluster leaves have emerged during the primary infection period. Slightly raised, olive-coloured lesions develop from infection locations after a minimum of 9 days and produce asexual spores, conidia, which serve as a source of secondary inoculum for terminal leaf and fruit infection (MacHardy and Gadoury, 1989; Stensvand et al., 1997).

In warm, humid growing regions such as the Southeastern United States and Brazil, bitter rot and Glomerella leaf spot, caused by *Colletotrichum* spp., are considered the most significant threat to the apple industry. Glomerella leaf spot is primarily a concern on apple cultivars with 'Golden Delicious' parentage (Mathias et al., 2016), whereas the fruit of most popular commercial cultivars are susceptible to bitter rot (Biggs and Miller, 2001). Bitter rot and Glomerella leaf spot epidemics begin when conidia of *Colletotrichum* spp. that are overwintering in mummified fruit, crevices in bark, flower buds and cankers are splash dispersed during rainfall and land on susceptible tissue. The sexual stage of the pathogen, *Glomerella* spp., is believed to predominantly overwinter as perithecia on leaf litter along the orchard floor. After a latent period, leaf infection by *Colletotrichum* spp. (teleomorph *Glomerella* spp.) is initially observed as small purple flecks that develop into asymmetrical light tan, concentric lesions (Fig. 3a). As the disease progresses, infected leaves become chlorotic and under severe disease pressure, prematurely defoliate (Fig. 3b). Symptoms of bitter rot on fruit are first observed as sunken small, brown to black spots (Fig. 3c) that progress into large concentric rots as the fruit matures. Acervuli or masses of salmon to orange coloured conidia commonly, but not always, accompany rot lesions (Sutton et al., 2014).

Due to the potential of *Venturia inaequalis* and *Colletotrichum* spp. to infect both leaves and fruit, management of both pathogens is essential. Failure to manage either disease can lead to a complete loss of fresh market fruit in a given season. In some situations, although fruit are infected, symptoms do not develop until after they are harvested or in storage as a result of the pathogen's incubation period or the fungistatic mode of action of certain single-site

Figure 3 Symptoms of Glomerella leaf spot and bitter rot on 'Gala' apple: (a) red-purple flecks and irregular necrotic lesions on a leaf, (b) premature defoliation resulting from Glomerella leaf spot in mid-July and (c) spots on fruit caused by *Colletotrichum* spp.

fungicides. Severe leaf symptoms may lead to premature defoliation , which can negatively impact the crop during the current and following season. For instance, the impacts of premature defoliation may include premature fruit ripening, reduced winter hardiness and poor return bud set the following season (MacHardy et al., 2001).

2.2 Pre-harvest fungal diseases/pathogens of stone fruit

There are numerous pre-harvest fungal diseases that impact sustainable stone fruit production worldwide (Fig. 4) (Ogawa et al., 1995; Ogawa and English, 1991). The relative importance of stone fruit diseases is highly dependent on the stone fruit crop being grown and the region in which the crop is produced. The most common stone fruit diseases and their respective pathogens include brown rot (*Monilinia* sp.), cherry leaf spot (*Blumeriella jaapii),* peach leaf curl (*Taphrina deformans*), powdery mildew (*Sphaerotheca pannosa*), peach scab (*Cladosporium carpophilum*), perennial canker (*Leucostoma cincta*) and black knot (*Apiosporina morbosa*) (Fig. 4a–f). Of these diseases, brown rot is arguably the most economically devastating and the biggest threat to sustainable stone fruit production, particularly in China, Europe and the United States (Ogawa et al., 1995; Ogawa and English, 1991). Brown rot management often drives stone fruit disease management programmes. Despite efforts to understand and develop management practices for brown rot and other fungal diseases of stone fruit, considerable fungicide applications are required to avoid fresh fruit losses and ensure stone fruit profitability. Diseases, such as cherry leaf spot, can be quite devastating to tart cherries (*Prunus cerasus*), and drive stone fruit management programmes in regions where susceptible varieties (e.g. cv. 'Montmorency') are grown (McManus et al., 2007). In capacity, a greater discussion of diseases like brown rot and cherry leaf spot are also warranted.

Figure 4 Common pre-harvest fungal diseases of stone fruit: (a) cherry leaf spot on tart cherry (*Blumeriella jaapii*), (b) brown rot on peach (*Monilinia fructicola*), (c) peach scab (*Cladosporium carpophilum*), (d) powdery mildew on peach (*Sphaerotheca pannosa*), (e) perennial canker on peach (*Leucostoma cincta*) and (f) black knot on sweet cherry (*Apiosporina morbosa*).

2.2.1 Brown rot

Brown rot is caused by several species in the genus *Monilinia* with the most common species being *M. fructicola*, *M. laxa* and *M. fructigena* (Cox and Villani, 2010; Fulton et al., 1999; Holb, 2004; Jones and Aldwinckle, 1990; Ogawa and English, 1991). Additional species, including *M. mumecola, M. yunnanensis* and *M. polystroma,* are responsible for causing brown rot epidemics in Central Europe and Asia, but are less prevalent than the aforementioned species (Hu et al., 2011; van Leeuwen et al., 2002). Brown rot is most problematic in temperate production regions, where spring and summer rainfall is a common occurrence (Ogawa et al., 1995; Ogawa and English, 1991). At bloom, rainfall leads to sporulation of overwintering cankers and mummies and subsequent infection events, where *Monilinia* sp. causes blossom and shoot blight in addition to the economically devastating pre- and post-harvest soft fruit rots (Fig. 4b). If left unmanaged, brown rot can lead to a complete crop loss. Blossom infections can lead to reduced crop loads and sporulate causing secondary infections of young shoots, senescing bloom and young fruit. Similarly, young twigs or shoots can become blighted, sporulate and infect developing fruit. Green fruit can be infected prior to pit hardening with infections remaining latent until fruit mature, develop colour and convert starch to sugars. As fruit approach harvest maturity, they become increasingly susceptible to infection and numerous secondary infection cycles may occur. After harvest, infected fruit and blighted shoots remaining on trees may develop into fruit mummies and overwintering cankers, respectively.

There are no known sources of host resistance to this necrotrophic pathogen. However, stone fruit species, such as peach (*Prunus persica* L.), and varieties with more abundant pubescence on developing fruit will have more physical resistance to brown rot infections (Fernández et al., 2011). Aside from this physical resistance, sanitation through the removal of mummies has been shown to be effective. However, in larger operations the cost-effectiveness of inoculum removal is questionable at best. Sustainable management is reliant on the application of modern single-site fungicides in conventional production and the continual application of sulphur in organic production. These applications begin from bud break to harvest with the peak of susceptibility occurring at 'shuck-split' or 'shuck-off', where the remnants of the floral tissues give way to developing fruit. Conventional chemical success can be compromised by the development of quantitative resistance to the highly effective demethylation inhibitor (DMI) and quinone outside inhibitor (QoI) fungicides, which has been reported for populations of *M. fructicola* (Burnett et al., 2010; Villani and Cox, 2011). Increased labelled rates of DMI fungicides and fungicide class rotation have ensured that DMIs and QoIs remain in pre-harvest chemical management programmes in the Eastern United States. Sustainable management of brown rot in organic production operations is compromised by the lack of efficacy of biopesticide products and the potential phytotoxicity of sulphur to certain stone fruit species, such as sweet cherries. In this regard, organic stone fruit production is quite difficult except in arid production regions where disease pressure is typically low.

2.2.2 Cherry leaf spot

Cherry leaf spot is caused by the fungus *Blumeriella jaapii* and is the most economically devastating disease of tart cherries (*Prunus cerasus*), which are highly susceptible to infection in temperate production regions (McManus et al., 2007). Unlike the aforementioned brown fruit rot, cherry leaf spot infects the leaves leading to the development of numerous tiny (<2 mm) violet leaf spots with chlorotic halos (Fig. 4a), and subsequent premature leaf drop, which can greatly impact crop maturity, compromise vigour, reduce winter hardiness and potentially lead to tree mortality. The disease is initiated by ascospores ejected from overwintering leaves on the orchard floor during spring rains beginning at bloom. Ascospores will continue to eject during the 6–7 weeks following petal fall. Lesions from primary infections will sporulate on the underside of leaves and contribute to numerous secondary infection cycles that occur during rain events through summer and fall until trees are completely defoliated. As with brown rot, sanitation isn't always cost-effective for large plantings of susceptible tart cherry varieties, and sustainable management is highly dependent on modern single-site fungicides. Similar to the situation with brown rot, fungicide resistance concerns also drive cherry leaf spot chemical management programmes. Quantitative resistance to the DMI fungicides can be prevalent at moderate to high levels in regions where susceptible tart cherries are grown (Proffer et al., 2006). There is a considerable risk of QoI resistance, but these fungicides are still used in moderation in many regions. Resistance to the succinate dehydrogenase inhibitor (SDHI) fungicide boscalid has been detected in some regions in Michigan (United States), where tart cherries are grown. However, the newer SDHI fungicides such as fluxapyroxad and fluopyram still seem to be effective and often serve as the cornerstone of pre-harvest cover applications to manage devastating secondary infections.

2.2.3 Additional diseases

Additional diseases, including peach leaf curl, powdery mildew, black knot and perennial canker, occur sporadically in stone fruit depending on the production region, seasonal climate, varieties planted and the level of care afforded to the operation (Ogawa et al., 1995; Ogawa and English, 1991). In many instances, these diseases are managed by the programmes for brown rot or cherry leaf spot. In the case of peach leaf curl, the risk of disease can virtually be eliminated by an application of copper between bud swell and bud break in the spring. Copper applications will also help reduce overwintering inoculum in sporulating cankers, such as those produced by *Monilinia* sp., *Cladosporium carpophilum* and *Leucostoma cincta*, and protect against ascosporic inoculum released from *Monilinia* sp. and *Blumeriella jaapii*. Along these lines, the early fungicide programmes for brown rot and cherry leaf spot are also designed to manage several of the other diseases at the same time. For example, chlorothalonil, commonly used from bud break to shuck split for brown rot and cherry leaf spot, is highly effective against black knot. However, including a DMI or QoI at first cover (10–14 days after shuck split) is recommended to manage powdery mildew on susceptible varieties. As with other diseases, sanitation can be helpful, but its cost-effectiveness will depend on the labour needed to achieve sanitation. For example, it may not be cost-effective to remove the overwintering leaf litter in large (>25 Ha) cherry orchards to reduce cherry leaf spot inoculum, but it could be effective to prune out isolated cankers to slow the progression of perennial canker in small operations.

3 Pre-harvest bacterial diseases of tree fruit

Fire blight, caused by *Erwinia amylovora*, is the most important bacterial disease of pome fruit, and devastating epidemics can result in the death of large percentages of trees within orchards on a regional basis (Longstroth, 2000). The significance of fire blight includes both current season yield loss due to flower infections and systemic infections leading to rootstock blight and tree death. *E. amylovora* is capable of infecting flowers, fruits, vegetative shoots, woody tissues and rootstock crowns, leading to blossom, shoot and rootstock blight symptoms (Fig. 5). Young orchards of susceptible apple cultivars are particularly vulnerable to tree losses due to fire blight. Disease epidemics are stimulated by weather conditions that are conducive for infections which include warm and wet weather during bloom and extreme weather events such as hailstorms that can injure trees leaving a point of entry for infection by fire blight bacteria (i.e. trauma blight).

On apple, blossom blight disease symptoms (flower death) of fire blight are typically initiated in the spring on flowers as overwintering inoculum from cankers is disseminated to the surface of stigmas by rain and insects (Norelli et al., 2003). The surface of the tips of stigmas particularly favours exponential growth of *E. amylovora* in the intercellular spaces between papillae cells. Under conducive weather conditions, populations can double rapidly, and populations of 10^6–10^7 *E. amylovora* cells per flower are common (Taylor et al., 2003). The availability of free moisture through rain or heavy dew is required

Figure 5 Different stages of fire blight on apple: (a) fire blight canker in the scion, (b) blossom blight with ooze production (denoted by red arrows), (c) shoot blight and (d) bacterial ooze on shoot.

to enable *E. amylovora* cells to migrate down the style where these cells infect flowers through natural openings present in the nectaries.

Blossom blight infection initiates the start of systemic spread of *E. amylovora* within trees; pathogen cells can also emerge in ooze droplets, which consist of *E. amylovora* cells within an exopolysaccharide matrix (Slack et al., 2017). Cells in ooze can be disseminated by wind, rain and insects to susceptible shoot tips, where shoot blight infections are initiated. Shoot blight infections are initiated through infection of actively growing leaves at the shoot tip, and continue as the pathogen forms biofilms within leaves to build up populations (Koczan et al., 2009). Further systemic spread of the pathogen and downward migration through stems, often accompanied by ooze emergence, continues the cycle of disease spread throughout the summer. Cankers are also formed, typically on branches older than 1 year, and serve as an overwintering survival structure for *E. amylovora*. In the subsequent spring, as temperatures warm, ooze emerges from cankers, and insects such as flies are attracted to this ooze and spread the pathogen to newly opening flowers.

Fire blight disease is exacerbated because most of the commercially desired cultivars are highly susceptible to this disease. Breeding for fire blight resistance has been accomplished in certain rootstocks, but has been slowed in scion varieties because the sources of resistance are from *Malus* spp. with poor fruit quality. Nevertheless, advances in resistance breeding have occurred recently with identification of at least two cases of gene-for-gene interactions (Vogt et al., 2013). However, the focus of fire blight management efforts has been on strategies targeting blossom blight that are aimed at reducing *E. amylovora* populations on flower stigmas (discussed below).

Bacterial canker is the most significant disease of stone fruit and is caused by two pathovars of *Pseudomonas syringae* pv. syringae (PSS) and pv. morsprunorum. This disease is predominantly characterized by cankers with associated gummosis; cankers can girdle and kill large scaffold limbs and entire trees (Kennelly et al., 2007). Bacterial canker is a significant problem on sweet cherry, apricot, and plum and more of a minor problem on tart cherry. Additional symptoms of the disease include blossom blast, which can be facilitated by the ice nucleation activity of PSS, and leaf spots (Kennelly et al., 2007). The organism overwinters in cankers or in asymptomatic dormant buds; as flowers develop from buds in the spring, they are readily colonized by epiphytic strains of PSS. Flower infection by PSS usually requires a predisposing environmental condition such as a frost or an extended period of cool, wet weather. Epiphytic populations of PSS survive on symptomless leaves during the growing season, and bacterial colonization of dormant buds occurs through leaf scars in autumn (Crosse, 1957). Management of bacterial canker on sweet cherry is extremely difficult because of the extensive amount of pathogen inoculum at many points of the season, the lack of registered bactericides for this disease, and commercially desired highly susceptible cultivars. Copper sprays have been applied, but the sweet cherry host is highly susceptible to copper phytotoxicity, and copper resistance has developed in PSS in many regions of the world (Renick et al., 2008).

4 Viruses and virus-like agents of pome fruit and stone fruit

Viruses and virus-like agents have the potential to reduce fruit quality and the longevity of production in pome and stone fruit orchards. The majority of tree fruit viruses are transmitted during the grafting process, primarily through infected budwood, which often originates from sources that are either asymptomatic or in which symptoms are inconspicuous and consequently not identified.

Apple chlorotic leaf spot virus (ACLSV), apple stem grooving virus (ASGV) and apple stem pitting virus (ASPV), transmitted during vegetative propagation, are generally latent viruses of apple, meaning that they are asymptomatic, or fail to produce acute symptoms in the majority of commercially available apple cultivar and rootstock combinations. While each of these viruses can occur alone, often they are detected in combination with at least one of the other aforementioned viruses.

Of the three major latent viruses, ACLSV is most common in the United States and is frequently associated with trees that have been top-worked with an infected scion. Approximately 1–2 years following budding with a scion that is latently infected with ACLSV, tree decline may be observed due to virus translocation to a susceptible rootstock compromising water and nutrient allocation to aboveground portions of the tree. In infected susceptible cultivars, foliar and woody tissue symptoms include chlorotic leaf spots and line patterns, premature defoliation, terminal shoot dieback and inner bark necrosis. Economic losses due to ACLSV are primarily due to reductions in yield, which have been reported as high as 30% when the virus is present in combination with other latent viruses (Cieniewicz and Fuchs, 2016a).

Similar to ACLSV, ASPV also results in appreciable tree decline and yield reduction worldwide, especially when the host is being co-infected with an additional latent virus. In susceptible cultivar/rootstock combinations, the most common symptom of APSV is pitting (depressions) that occurs on the woody cylinder. When pitting is severe, impairment of vascular tissue can occur, reducing nutrient and water translocation ultimately leading to tree decline and death. In pear infected with APSV, highly susceptible cultivars such as 'Jules d'Arrioles' and 'Beurre Hardy' may exhibit yellowing of leaf veins, red mottling, necrotic spotting and deformed fruit (Cieniewicz and Fuchs, 2016b). ASGV, also transmitted through grafting, budding and top-working, can also cause appreciable reduction in tree size, and consequently delay tree production and reduce yield. While generally asymptomatic on most commercially available cultivars, symptoms on susceptible cultivar/rootstock combinations vary in severity and may include a swollen graft union with a necrotic line, leaf chlorosis, premature defoliation and growth stunting.

Plum pox virus (PPV, genus Potyvirus), the causal agent of plum pox disease (also referred to as sharka disease), can infect all stone fruit species including plum, prune, peach, nectarine, cherry and apricot. Globally, six strains of PPV have been currently identified. Mechanisms of PPV transmission include aphid vectors and through infected material during propagation (e.g. budding). The type and severity of symptoms resulting from PPV infection are highly dependent on the affected stone fruit species and cultivar. Fruit symptoms can range from the presence of light-coloured rings on the fruit surface (peach) to flesh discolouration, deformed and dark spots or rings (plum). On leaves, symptoms may include yellowing of veins and the presence of chlorotic rings. PPV remains latent until at least 3 years following infection (Barba et al., 2015).

Tomato ringspot virus (ToRSV), transmitted by the dagger nematode (*Xiphinema* spp.) and less commonly through weed seeds and infected nursery stock, is the causal agent of *Prunus* stem pitting disease in stone fruit and apple union necrosis in apple. In both stone fruit and apple, symptoms and progression of ToRSV appear similar, often mimicking foliar symptoms resulting from other root diseases such as *Phytophthora* spp., or abiotic disorders or injury. Careful inspection of *Prunus* spp. infected with ToRSV will reveal trunk/graft union girdling, thickened sapwood and elongated indentations or pits in the region extending a few inches above or below the soil line (Ogawa et al., 1995). In apple, ToRSV has been most commonly associated with 'Red Delicious' grafted to MM.106 rootstock. The most distinctive symptom of apple union necrosis is a necrotic line accompanied by pitting in the woody cylinder at the graft union (Rosenberger et al., 1983). In both infected apple and *Prunus* spp., weakening of the graft union may cause trees to break under mild physical stress.

Sustainable management of viruses in tree fruit is dependent on its means of transmission. The production of virus-tested propagation material through the development and utilization of clean plant programmes is imperative to reducing the incidence of diseases caused by viruses and virus-like agents in nurseries and newly established orchards (Gergerich et al., 2015). Management of weeds that serve as alternative hosts of tree fruit viruses or as hosts of insect or nematode vectors is also necessary for reducing virus spread once it is introduced.

5 Disease management of tree fruit

5.1 Disease forecasting systems

Computer-based disease forecasting systems have been used to assist with the management of tree fruit diseases since the advent of the Maryblyt programme (Steiner, 1988, 1990a), which has been used to advise pome fruit growers on fire blight management since its availability. Numerous models have been proposed to describe the epidemiology of tree fruit diseases. However, many of these models are part of a larger study, which aims to achieve a better understanding of the detailed biology and epidemiology of the host and tree fruit pathogens more so than develop forecasting tools for growers. Studies illustrating simple relationships between environmental variables and the biology needed for the infection and disease development easily lend themselves to the development of disease forecasting tools. Such studies are primarily concerned with identifying the temperature, surface wetness and the appropriate biofixes. Biofixes in disease forecasting models are typically defined as host growth stages or features of host developmental biology (often defined by degree days or chilling hours) whereby one would initiate the forecasting model. Of the numerous diseases that affect tree fruit crops, there are only a handful of diseases that are sufficiently prevalent and devastating as to warrant disease forecasting systems. These include apple scab, fire blight, flyspeck/sooty blotch, cherry powdery mildew and cherry leaf spot (Brown and Sutton, 1995; Eisensmith and Jones, 1981; Gadoury and MacHardy, 1982; Steiner, 1988, 1990a; Turechek and Biggs, 2015). Brown rot is so aggressive in its ability to initiate infection continuously despite current environmental conditions, that risk would best be defined by the presence of fungicide residues more so than forecast weather. Hence, the utility of disease forecasting systems for brown rot risk is limited. At the time of this publication, there are numerous disease forecasting systems for assessing infection risks for the aforementioned tree fruit diseases. Some of the more advanced disease forecasting systems available to tree fruit producers include Michigan State University's Enviroweather (https://enviroweather.msu.edu/), New York State IPMs' Network for Environment and Weather Applications (http://newa.cornell.edu/), the Washington State University's Decision Aid System (https://www.decisionaid.systems/), Penn State University's Seasonal Monitoring Data and Models (http://agsci.psu.edu/frec/growing-season-information) and RIMpro Cloud Service's interactive Decision Support System (http://www.rimpro.eu/).

5.1.1 Apple scab

Apple scab disease forecasting systems are based on a series of published studies that attempted to define both the maturity of ascospores, which initiate apple scab epidemics, and the environmental conditions under which primary infection of leaf tissues occur (Gadoury and MacHardy, 1982; MacHardy and Gadoury, 1989; Stensvand et al., 1997; Stensvand et al., 2005). Interestingly, there is little published literature on the conditions that define the ejection of ascospores, and many systems often use local expert rules that are only in extension literature or conference proceedings and not present in the larger body of peer-reviewed scientific literature.

5.1.2 Fire blight

Fire blight forecasting systems are typically based on either the Cougar Blight Model (Smith, 1996), the Maryblyt disease forecasting model (Biggs and Turechek, 2010; Smith, 1996; Steiner, 1988, 1990a,b; Turechek and Biggs, 2015) or both. Some of the available systems mentioned above use local rule sets that have not been validated or published in peer-reviewed scientific literature. Regardless, all fire blight risk prediction models try to predict the risk of blossom infection on the basis of accumulated heat units using a bloom biofix and the availability of free moisture or high relative humidity needed for infection. These models are fairly reliable in their predictions and are widely implemented by producers to time applications of streptomycin. Some models, such as the Maryblyt disease forecasting model, also predict the emergence of overwintering cankers and development of shoot blight symptoms occurring from trauma events.

5.1.3 Flyspeck and sooty blotch

Flyspeck and sooty blotch disease forecasting systems are based on several studies conducted in North Carolina (Brown and Sutton, 1986, 1995; Williamson and Sutton, 2000). These studies define the biofix as 2 weeks post-petal fall for initial ascospore release and describe the environmental conditions that define incubation periods of the flyspeck and sooty blotch fungi on developing apple fruit (Brown and Sutton, 1986, 1995; Williamson and Sutton, 2000). These models typically predict the risk of primary infection on the basis of accumulated surface wetness hours after the biofix, and the risk of secondary infections following accumulation of more than 2 inches of rain. This is meant to provide an indication of depletion of protective fungicide residues. As reviewed by Cooley et al. (2011) and Duttweiler et al. (2008), there are newer studies where predictions may better define the behaviour of regional species and use of relative humidity, instead of leaf wetness, to determine primary infection. To the best of our knowledge, this newer information has not been implemented in tree fruit disease forecast systems.

5.1.4 Cherry leaf spot

Cherry leaf spot disease forecasting systems use a model developed by Eisensmith and Jones (1981), whereby the risk of infection is defined by the duration of leaf wetness and temperature during the period of leaf wetness. Much like the models for apple scab, the cherry leaf spot model implements rules for interrupted leaf wetness and combined wetting events in cases where drying is less than eight consecutive hours.

Overall, disease forecasting systems are a useful tool for predicting risk of infection for many key diseases in tree fruit crops. In many cases, grower adoption and utility of forecasting models is highly dependent on the accessibility of the forecasting system in terms of regional specificity, cost, reliability and ease of use and interpretation by the end user. Systems that are overly cumbersome or detailed may give the impression that they may be more accurate in prediction, but the actual chances of type I and type II errors by any forecasting system is truly dependent on the quality of weather forecast data and appropriateness of model logic.

5.2 Fungicides

Applications of broad-spectrum multi-site protectant fungicides including captan (pome fruit and stone fruit), mancozeb (pome fruit) and chlorothalonil (stone fruit) have served as the foundation for pome and stone fruit disease management programmes since the 1960s. The longevity of protectant fungicides is based on their low risk for developing resistance and the fact that most of them control multiple pathogens. Unlike modern fungicides, protectants target several enzymes in order to prevent spore germination. While this mode of action is beneficial as regards consistency, their potential to harm non-targets such as beneficial microbes and insects has long been a topic of research and heightened regulation, particularly in recent years (Kovach et al., 1992; McHart et al., 2017). Also a limiting characteristic of protectant fungicides is their lack of post-infection activity.

The development of single-site fungicides with curative or post-infection activity including dodine (1957), methyl benzimidazole carbamates in the 1960s, sterol DMIs in the 1970s and 1980s, anilinopyrimidines (APs) in the 1990s, QoIs in the 1990s and next-generation SDHI fungicides during the twenty-first century have enabled growers to manage key fungal diseases prior to and following infection, while minimizing harmful effects to non-target organisms and the environment (Köller, 1999, 2003; Köller et al., 1997, 2004; Morton and Staub, 2008; Schnabel et al., 2004; Sierotzki and Scalliet, 2013). The curative activity of single-site fungicides is reflective of their locally systemic movement in host tissue. Unfortunately, because these fungicides are single-site inhibitors (i.e. they block a single biochemical pathway in fungi), repetitive use selects for the emergence of reduced sensitive and resistant populations of fungal pathogens. Provided is a brief overview of the three most important groups of single-site fungicides currently used in pome fruit and stone fruit production in the United States.

5.2.1 Demethylation inhibitors

The DMI fungicides, also referred to as the Class I sterol biosynthesis inhibitors, are a class of broad-spectrum, locally systemic compounds that are highly effective against several plant pathogenic fungi, including yeasts, and fungi that cause scab, rusts, powdery mildews and some fruit rots (Schnabel and Jones, 2001). DMI fungicides are members of FRAC group 3 and are considered to be at medium risk for fungicide resistance development. Examples of FRAC 3-active ingredients commonly used for management in pome fruit and or stone fruit include difenoconazole, fenbuconazole, flutriafol, myclobutanil, propiconazole, tebuconazole and triflumizole. DMI fungicides target the C14-demethylase enzyme encoded by the fungal *CYP51* gene. Specifically, applications of DMIs disrupt ergosterol biosynthesis by inhibiting demethylation at the 14-α carbon of lanosterol, a sterol intermediate of ergosterol, which is essential for hyphal growth (Köller and Scheinpflug, 1987). Although DMI fungicides have activity as protective fungicides, historically they are most efficacious when used following an infection event (Köller and Wilcox, 1999; O'Leary et al., 1987; Schwabe et al., 1984) due their fungistatic physical mode of action.

5.2.2 Quinone Outside inhibitors

The QoI fungicides are also broad-spectrum, locally systemic compounds that are highly efficacious against scabs, powdery mildew and fruit rots – particularly those caused by members of the *Colletotrichum* and *Botryosphaeria* genera. The QoIs exhibit varying degrees of translaminar activity, meaning that they can redistribute from the upper sprayed surface of the leaf to the unsprayed bottom surface. Members of FRAC group 11, QoI fungicides are considered to be at high risk for resistance development. Examples of FRAC 11-active ingredients commonly used for management in pome fruit or stone fruit include azoxystrobin, kresoxim-methyl, pyraclostrobin and trifloxystrobin. QoI fungicides inhibit mitochondrial respiration by binding to the outer quinone oxidizing pocket of the cytochrome bc_1 enzyme complex III. Specifically, these fungicides target the Q_o site within the cytochrome b protein that is encoded by the mitochondrial *cyt b* gene (Zheng and Köller, 1997). This site is vital to electron transport within the mitochondrion and the production of ATP. The binding of a QoI fungicide inhibits electron transfer in the respiratory pathway, which terminates cellular respiration and thus the production of ATP (Bartlett et al., 2002; Gisi et al., 2002; Köller, 1999). Although QoI fungicides are registered for both protective and curative control of fungal pathogens, their efficacy is greatest when applied prior to an infection event because of their ability to prevent conidial germination.

5.2.3 Succinate Dehydrogenase inhibitors

The SDHI fungicides represent a diverse class of compounds belonging to eight distinct chemical groups within FRAC group 7. The efficacy of specific SDHI fungicides is highly dependent on the target pathogen, but in general SDHIs have shown moderate to high levels of activity against plant pathogenic fungi causing scabs, powdery mildew and some leaf spots (e.g. *Blumeriella jaapi*). Examples of FRAC 7-active ingredients commonly used for management in pome fruit or stone fruit include benzovindiflupyr, boscalid, fluopyram, fluxapyroxad, isofetamid and penthiopyrad. SDHIs are considered to be at a medium to high risk for resistance development. Similar to QoI fungicides, SDHIs also inhibit mitochondrial respiration by shutting down the succinate dehydrogenase reductase enzyme, a component of complex II in the mitochondrial electron transport chain. The target enzyme contains four nuclear-encoded subunits: SDHA, SDHB, SDHC and SDHD, encoded by the nuclear *sdhA*, *sdhB*, *sdhC* and *sdhD* genes, respectively. Currently, all SDHIs labelled for crop protection target the ubiquinone-binding pocket, defined by the interface between the SDHB, SDHC and SDHD subunits (Cecchini, 2003). Although SDHI fungicides are recommended for both protective and curative control of fungal pathogens, their efficacy is greatest when applied prior to an infection event, because of their ability to prevent conidial germination (Villani et al., 2016).

5.3 Fungicide resistance

Fungicide resistance management has been a primary focus for the tree fruit community because of the importance of chemical control in disease management of these crops. Frequent reliance on single-site fungicides combined with the formerly recommended practices of low rate applications, applying on extended and post-infection schedules, and applying to alternate row middles have supported the development of resistant populations (Beckerman et al., 2015). Resistance of an isolate or of a fungal population to single-site fungicides is commonly characterized as being quantitative or qualitative. An isolate exhibiting qualitative resistance to a particular fungicide has complete resistance to that fungicide and will not respond to higher application rates. In the case of qualitative resistance within a population, a bimodal distribution of isolate sensitivity exists where some isolates are fully sensitive and others are fully resistant (Fernández-Ortuño et al., 2008; Villani and Cox, 2014). Point mutations that alter the coding sequence of the fungicide target gene are resistance mechanisms that are most commonly associated with qualitative resistant phenotypes. Alternatively, an isolate exhibiting quantitative resistance to a particular fungicide exhibits partial, or reduced, sensitivity to that fungicide. Within a population, there is a normal distribution or skewed distribution in which isolates exhibit a range in sensitivity phenotypes. Often, mechanisms of quantitative resistance are associated with overexpression of energy-dependent ABC and MSF transporters encoding drug efflux pumps or overexpression of the fungicide target gene.

Prolonging the efficacy of fungicides is imperative for achieving sustainable management of pome fruit and stone fruit. To combat fungicide resistance, the Fungicide Resistance Action Committee (FRAC) was established in 1981 to provide fungicide resistance management guidelines, to prolong the efficacy of at-risk fungicides and to limit crop losses if resistance does occur. A cornerstone of resistance management is minimizing field populations of plant pathogens. Within a given fungal population, while the ratio of resistant to sensitive members remains relatively similar regardless of population size, the absolute number of resistant individuals will be greater in larger populations. As a result, in the absence of fitness costs, the rate of resistance development is accelerated as fungal population size increases. In addition to cultural practices aimed at reducing overwintering and in-season inoculum (discussed below), fungicide application practices are important for resistance management (Beckerman et al., 2015). Applying fungicides prior to an infection event with the aid of predictive models, applying the highest recommended rates and achieving proper coverage by calibrating spray equipment can reduce both population size and selection pressure for resistance. The practice of tank mixing fungicide products, in which a single-site fungicide is applied with either a protectant fungicide or a single-site fungicide with a different mode of action, is another tactic for managing resistance. Lately there has been a trend followed by the U.S. agrichemical industry to release an increasing number of products as premixes that combine two or more active ingredients to promote mandatory compliance with resistance management. Lastly, rigorous fungicide resistance monitoring efforts by industry stakeholders and university specialists to identify shifts in practical resistance in the field or shifts in fungicide sensitivity through classical and molecular biology techniques are needed to mitigate the development and spread of fungicide-resistant populations.

5.4 Bactericides and biological controls for fire blight

The most effective method for blossom blight management is through the use of antibiotics such as streptomycin, oxytetracycline or kasugamycin (McGhee and Sundin, 2011; McManus et al., 2002). Antibiotics have been in use for fire blight since the 1950s, and resistance to streptomycin has impacted fire blight management in several U.S. states (Coyier and Covey, 1975; Forster et al., 2015; McGhee et al. 2011; Tancos et al., 2016). In addition, because antibiotic use is banned in Europe and in organic production in the United States, many other management efforts have been tested. For example, a variety of biological control strategies have been developed and evaluated. These strategies started with the use of bacterial antagonists such as the epiphytes *Pantoea agglomerans* and *Pseudomonas fluorescens* which were first used as competitive antagonists (Johnson and Stockwell, 1998). The next wave of antagonists has included the yeast *Aureobasidium pullulans*, which appears to be more efficacious than the bacterial antagonists (Johnson and Temple, 2013). It should also be noted that biological control agents containing living organism antagonists are differentially effective in regions differing in climate. For example, in the humid climates of the Midwestern and Eastern United States, many biological controls are much more limited in efficacy compared to in the dry climates of the Western United States (Sundin et al., 2009). There has also been some effort in using bacteriophages as biocontrol agents; however, although effective phage have been isolated (Gill et al., 2003), more work is needed on deployment strategies such as increasing protection of the phage from environmental degradation. Another strategy for blossom blight management has been the use of inducers of systemic acquired resistance such as acibenzolar-*S*-methyl to activate a defence response in the apple host (Maxson-Stein et al., 2002).

Management of the shoot blight phase of fire blight also became possible thanks to a fortuitous observation that the compound prohexadione-calcium (ProCa), an inhibitor of gibberellin biosynthesis originally used to inhibit shoot growth to lower pruning costs, could also reduce the incidence of shoot blight infection (Yoder et al., 1999). As stated above, *E. amylovora* infects shoots that are actively growing; the inhibition of shoot growth was shown to result in increased cell wall widths in parenchymal cells of leaves at shoot tips (McGrath et al., 2009). It was hypothesized that *E. amylovora* cells were unable to infect the plant parenchymal cells with wider cell wall widths, thus resulting in shoot blight control (McGrath et al., 2009).

5.5 Cultural management

Cultural management for both bacterial and fungal diseases of pome fruit and stone fruit production in the United States begins with the reduction of overwintering inoculum from late fall until early spring. To manage foliar diseases of apple such as apple scab and more recently Glomerella leaf spot, shredding leaf litter using a flail mower has been utilized in an attempt to increase the rate of microbial leaf degradation and to redirect the ascospore release angle. The success of this method has been variable, however, with reduction in ascospore inoculum ranging from 40% to 95% (Sutton et al., 2000). Alternatively, a single application of 5% urea to the orchard floor has also been used to stimulate microbial degradation of leaves. The removal of cankers and diseased woody tissue during dormancy can also mitigate overwintering inoculum pressure. Cankers caused by bacterial pathogens *E. amylovora* (fire blight), *Pseudomonas syringae* (bacterial canker) and *Xanthomonas arboricola* (bacterial spot) should be pruned beyond the advancing margin. Although infections by *P. syringae* and *X. arboricola* tend to be more localized, woody tissue infected by *E. amylovora* can extend well beyond the canker zone. Thus, pruning of fire blight infected shoots should be at least 12 inches beyond the advancing infection margin. The removal of mummified fruitlets during dormancy is also advised, as they can harbour several pathogens including *Monilinia fructicola* (American brown rot), *Botryosphaeria* spp. (black and white fruit rot) and *Colletotrichum* spp. (bitter rot). The removal of infected fruit and woody tissue can also be accomplished during the season but special care should be taken to limit the spread of inoculum through an orchard. If pruning for bacterial canker during the season, a protective copper application should be made to prevent additional pathogen colonization.

5.6 Host resistance

Although U.S. apple breeding programmes have historically prioritized the development of apple cultivars with superior horticultural traits over those with qualitative disease resistance, the development of apple and stone fruit rootstocks with resistance to some apple and stone fruit diseases has been a focus of breeding programmes in the United States. Dwarfing rootstocks in the Geneva series, such as G.41 and G.935 are considered very resistant to fire blight and tolerant to *Phytophthora* crown and root rots. For stone fruit, Mariana 2624 is a semi-dwarfing rootstock for plum and has resistance to Armillaria root rot, crown rot, crown gall and root rot nematodes. For peaches, the rootstock Guardian is resistant to peach tree short life. In addition to stakeholder acceptance and regulatory concerns (Woo et al., 2015) the utilization of genetic transformation technology for disease resistance in commercial fruit trees has been challenged by a paucity of cloned resistance genes in *Malus* and *Prunus*, efficiency of plant regeneration and resistance durability (Cusin et al., 2017; Gessler and Pertot, 2012). In perennial orchard systems, deployment of cultivars with single R-gene resistance has been met with the development of new virulent fungal pathogen races (Cusin et al., 2017). The introduction of foreign antifungal genes into apple has also been used to more broadly enhance disease resistance to several pathogen races or species. For example, the antifungal wheat protein, pinB, has been expressed in apple to enhance resistance to several races of *Venturia inaequalis* (Faize et al., 2004). Most recently, targeted genome editing utilizing the CRISPR/Cas9 endonuclease system in tree fruit offers an efficient and promising new tool for the development of disease-resistant tree fruit without the introduction of foreign genes or antibiotic-resistant markers.

6 Conclusion and future trends

Tree fruit production worldwide will likely increase with consumer demand and increasing global populations. This trend will result in an intensification of tree fruit production, where producers will need to plant higher densities of trees that are highly productive on the limited land available. Such increases in production intensity may require increased labour to manage production operations effectively and to reduce the threat of diseases and pests. However, labour availability for agricultural production has only become scarcer as producers must limit labour costs to remain competitive and the sociopolitical views are limiting access to inexpensive migrant labour. Efforts to achieve mechanized production may help solve labour issues, but non-optimized procedures could lead to wounding in these highly vigorous plantings and increase the occurrence of canker diseases, fire blight and fruit rots.

This paradigm of dense vigorous production will serve as the perfect environment for disease, and for selection for single-site fungicide resistance when more applications are needed to deal with higher inoculum pressure and an increasing number of secondary cycles as growing seasons become longer due to climate change. The development of fungicide resistance in conventional plantings will only continue to exacerbate the social impetus for growers to subscribe to organic certification programmes. This social pressure is founded on the popular assumption that organic food is healthier for the consumer and that modern single-target fungicides with ephemeral persistence are more harmful to the consumer and the environment than naturally derived copper and sulphur fungicides used in organic production.

This impact on producers is already being felt as commercial fruit processors are establishing more restrictive MRLs for modern single-site fungicides. In many instances, the MRLs seem arbitrary, are set at a level of zero detection and are not grounded in peer-reviewed scientific literature. Achieving MRLs of zero will be complicated by the fact that the limits of detection by modern analytical techniques become lower with every improvement in detection technology.

As we begin to forsake modern chemical disease management tools to meet social perceptions, new disease problems, previously held in check by modern chemical management programmes, will present unforeseen problems to tree fruit growers. Moreover, erratic climatic conditions resulting from global climate change will also present new and challenging disease problems to tree fruit producers. The perennial tree fruit varieties currently in production have not been bred to tolerate erratic climatic conditions. In the last 5 years, daily temperature changes in the order of 20°C, droughts of 30 or more days and daily precipitation events in the order of 10–15 cm are commonplace. These erratic conditions expose tree fruit plantings to considerable stress, such as low-temperature injury, flooding and drought stress. This appears to be allowing new previously described opportunist pathogens and novel pathogens to become endemic problems. To ensure sustainable disease management and tree fruit production, future pathologists will need to better understand climate change, epidemiology and biology of tree pathogens, host genomics and plant breeding, and, most importantly, social science and marketing.

7 Where to look for further information

Most of the key references for managing tree fruit diseases are available online and can be located with a routine search of the internet. Land-grant universities, such as the ones to which the authors belong: Michigan State University, Cornell University and North Carolina State University all have high-quality and publicly available fact sheets and resources on the management of tree diseases. In particular, production guidelines produced by land-grant universities are of particular value to regional production as they contain highly specific information for managing local diseases and pest concerns. For a more in-depth coverage of tree fruit diseases, the American Phytopathological Society (APS) webpage (APSnet.org) can be an excellent place to start for the most up-to-date information. Finally, both the *Compendium of Apple and Pear Diseases and Pests*, second edition (APS Press), and the *Compendium of Stone Fruit Diseases* (APS Press) are some of the best resources available.

8 References

Anon. 2017a. Fresh deciduous fruit: World markets and trade (apples, grapes, & pears). Foreign Agricultural Service/United States Department of Agriculture. Available at: https://apps.fas.usda.gov/psdonline/circulars/fruit.pdf (accessed 25 January 2018).

Anon. 2017b. Noncitrus fruits and nuts 2016 summary. United States Department of Agriculture, National Agricultural Statistics Service. Available at: http://usda.mannlib.cornell.edu/usda/current/NoncFruiNu/NoncFruiNu-06-27-2017.pdf (accessed 25 January 2018).

Anon. 2017c. Fresh peaches and cherries: World markets and trade. Foreign Agricultural Service/United States Department of Agriculture. Available at: https://apps.fas.usda.gov/psdonline/circulars/StoneFruit.pdf (accessed 29 January 2018).

Anon. 2018a. EU legislation on MRL. European Commission. Available at: https://ec.europa.eu/food/plant/pesticides/max_residue_levels/eu_rules_en (accessed 28 January 2018).

Anon. 2018b. Maximum residue limits (MRL) database. Foreign Agricultural Service/United States Department of Agriculture. Available at: http:www.fas.usda.gov/maximum-residue-limits-mrl-database (accessed 24 January 2018).

Anon. 2018c. US Apple Association popular varieties. Available at: http:// http://usapple.org/the-industry/apple-varieties/ (accessed 28 February 2019).

Barba, M., Ilardi, V. and Pasquini, G. 2015. Control of pome and stone fruit virus diseases. *Adv. Virus Res.* 91, 47–83. doi:10.1016/bs.aivir.2014.11.001.

Bartlett, D. W., Clough, J. M., Godwin, J. R., Hall, A. A., Hamer, M. and Parr-Dobrzanski, B. 2002. The strobilurin fungicides. *Pest Manag. Sci.* 58(7), 649–62. doi:10.1002/ps.520.

Beckerman, J. L., Sundin, G. W. and Rosenberger, D. A. 2015. Do some IPM concepts contribute to the development of fungicide resistance? Lessons learned from the apple scab pathosystem in the United States. *Pest. Manag. Sci.* 71(3), 331–42. doi:10.1002/ps.3715.

Biggs, A. R. and Miller, S. S. 2001. Relative susceptibility of selected apple cultivars to *Colletotrichum acutatum*. *Plant Dis.* 85(6), 657–60. doi:10.1094/PDIS.2001.85.6.657.

Biggs, A. R. and Turechek, W. W. 2010. Epidemiological concepts comprising the Maryblyt forecasting program. *Plant Health Prog.* (online). doi:10.1094/PHP-2010-0315-01-RS.

Brown, S. and Maloney, K. 2008. Scab-resistant cultivars (varieties). *N. Y. Fruit Q.* 16, 3–6.

Brown, E. M. and Sutton, T. B. 1986. Control of sooty blotch and flyspeck of apple with captan, mancozeb, and mancozeb combined with dinocap in dilute and concentrate applications. *Plant Dis.* 70(4), 281–4. doi:10.1094/PD-70-281.

Brown, E. M. and Sutton, T. B. 1995. An empirical model for predicting the first symptoms of sooty blotch and flyspeck of apples. *Plant Dis.* 79(11), 1165–8. doi:10.1094/PD-79-1165.

Burnett, A., Lalancette, N. and McFarland, K. 2010. First report of the peach brown rot fungus *Monilinia fructicola* resistant to demethylation inhibitor fungicides in New Jersey. *Plant Dis.* 94(1), 126. doi:10.1094/PDIS-94-1-0126A.

Campoy, J. A., Ruiz, D. and Egea, J. 2011. Dormancy in temperate fruit trees in a global warming context: A review. *Sci. Hortic.* 130(2), 357–72. doi:10.1016/j.scienta.2011.07.011.

Cecchini, G. 2003. Function and structure of complex II of the respiratory chain. *Annu. Rev. Biochem.* 72, 77–109. doi:10.1146/annurev.biochem.72.121801.161700.

Chapman, K. S., Sundin, G. W. and Beckerman, J. L. 2011. Identification of resistance to multiple fungicides in field populations of *Venturia inaequalis*. *Plant Dis.* 95(8), 921–6. doi:10.1094/PDIS-12-10-0899.

Cieniewicz, E. and Fuchs, M. 2016a. Apple chlorotic leaf spot virus. New York State IPM Program.

Cieniewicz, E. and Fuchs, M. 2016b. Apple stem pitting virus. New York State IPM Program.

Cooley, D. R., Rosenberger, D. A., Gleason, M. L., Koehler, G., Cox, K. D., Clements, J. M., Sutton, T. B., Madeiras, A. and Hartman, J. R. 2011. Variability among forecast models for the apple sooty blotch/flyspeck disease complex. *Plant Dis.* 95(1), 1179–86.

Cosmulesco, S., Baciu, A., Cichi, M. and Gruia, M. 2010. The effect of climate changes on phenological phases in plum tree (*Prunus domestica*) in south-western Romania. *S.-W. J. Hortic. Biol. Environ.* 1, 9–20.

Cox, K. D. and Villani, S. M. 2010. Confirmation of European brown rot caused by *Monilinia laxa* on tart cherry, *Prunus cerasus*, in Western New York. *Plant Dis.* 94(6), 783. doi:10.1094/PDIS-94-6-0783B.

Coyier, D. L. and Covey, R. P. 1975. Tolerance of *Erwinia amylovora* to streptomycin sulfate in Oregon and Washington. *Plant Dis. Rep.* 59, 849–52.

Crosse, J. E. 1957. Bacterial canker of stone-fruits III. Inoculum concentration and time of inoculation in relation to leaf-scar infection of cherry. *Ann. Appl. Biol.* 45(1), 19–35. doi:10.1111/j.1744-7348.1957.tb00440.x.

Cusin, R., Revers, L. F. and dos Santos Maraschin, F. 2017. New biotechnological tools to accelerate scab-resistance trait transfer to apple. *Genet. Mol. Biol.* 40(1 suppl 1), 305–11. doi:10.1590/1678-4685-GMB-2016-0043.

Duttweiler, K. B., Gleason, M. L., Dixon, P. M., Sutton, T. B., McManus, P. S. and Monteiro, J. E. B. A. 2008. Adaptation of an apple sooty blotch and flyspeck warning system for the Upper Midwest United States. *Plant Dis.* 92(8), 1215–22. doi:10.1094/PDIS-92-8-1215.

Eisensmith, S. P. and Jones, A. L. 1981. A model for detecting infection periods of *Coccomyces hiemalis* on sour cherry. *Phytopathology* 71(7), 728–32. doi:10.1094/Phyto-71-728.

Faize, M., Sourice, S., Dupuis, F., Parisi, L., Gautier, M. F. and Chevrea, E. 2004. Expression of wheat puroindoline-b reduces scab susceptibility in transgenic apple (*Malus* × *domestica* Borkh.). *Plant Sci.* 167(2), 347–54. doi:10.1016/j.plantsci.2004.04.003.

Fazio, G., Robinson, T. L. and Aldwinckle, H. S. 2015. The Geneva apple rootstock breeding program. In: Janick, J. (Ed.), *Plant Breeding Reviews: Volume 39*. Hoboken, NJ: John Wiley & Sons, Inc.

Fernández, V., Khayet, M., Montero-Prado, P., Heredia-Guerrero, J. A., Liakopoulos, G., Karabourniotis, G., del Río, V., Domínguez, E., Tacchini, I., Nerín, C., et al. 2011. New insights into the properties of pubescent surfaces: Peach fruit as a model. *Plant Physiol.* 156(4), 2098–108. doi:10.1104/pp.111.176305.

Fernández-Ortuño, D., Torés, J. A., de Vicente, A. and Pérez-García, A. 2008. Mechanisms of resistance to QoI fungicides in phytopathogenic fungi. *Int. Microbiol.* 11(1), 1–9.

Fernández-Ortuño, D., Grabke, A., Bryson, P. K., Amiri, A., Peres, N. A. and Schnabel, G. 2014. Fungicide resistance profiles in *Botrytis cinerea* from strawberry fields of seven Southern U.S. states. *Plant Dis.* 98(6), 825–33. doi:10.1094/PDIS-09-13-0970-RE.

Forster, H., McGhee, G. C., Sundin, G. W. and Adaskaveg, J. E. 2015. Characterization of streptomycin resistance in isolates of *Erwinia amylovora* in California. *Phytopathology* 105(10), 1302–10. doi:10.1094/PHYTO-03-15-0078-R.

Frederick, Z. A., Villani, S. M., Cooley, D. R., Biggs, A. R., Raes, J. J. and Cox, K. D. 2014. Prevalence of quinone outside inhibitor resistance and the stability of qualitative QoI resistance in *Venturia inaequalis*. *Plant Dis.* 98(8), 1122–30. doi:10.1094/PDIS-10-13-1042-RE.

Fulton, C. E., van Leeuwen, G. C. M. and Brown, A. E. 1999. Genetic variation among and within *Monilinia* species causing brown rot of stone and pome fruits. *Eur. J. Plant Pathol.* 105(5), 495–500. doi:10.1023/A:1008711107347.

Gadoury, D. M. and MacHardy, W. E. 1982. A model to estimate the maturity of ascospores of *Venturia inaequalis*. *Phytopathology* 72(7), 901–4. doi:10.1094/Phyto-72-901.

Garrett, K. A., Dendy, S. P., Frank, E. E., Rouse, M. N. and Travers, S. E. 2006. Climate change effects on plant disease: Genomes to ecosystems. *Annu. Rev. Phytopathol.* 44, 489–509. doi:10.1146/annurev.phyto.44.070505.143420.

Gessler, C. and Pertot, I. 2012. Vf scab resistance of *Malus*. *Trees* 26(1), 95–108. doi:10.1007/s00468-011-0618-y.

Gill, J. J., Svircev, A. M., Smith, R. and Castle, A. J. 2003. Bacteriophages of *Erwinia amylovora*. *Appl. Environ. Microbiol.* 69(4), 2133–8. doi:10.1128/AEM.69.4.2133-2138.2003.

Gisi, U., Sierotzki, H., Cook, A. and McCaffery, A. 2002. Mechanisms influencing the evolution of resistance to Qo inhibitor fungicides. *Pest Manag. Sci.* 58(9), 859–67. doi:10.1002/ps.565.

Holb, I. J. 2004. Yield loss and disease development of *Monilinia fructigena* (Aderh. & Ruhl.) Honey in an organic apple orchard. *J. Agric. Sci.* 15, 519–28.

Holb, I. J. 2005. Effect of pruning on apple scab in organic apple production. *Plant Dis.* 89(6), 611–8. doi:10.1094/PD-89-0611.

Holb, I. J. 2007. Effect of four non-chemical sanitation treatments on leaf-infection by *Venturia inaequalis* in organic apple orchards. *Eur. J. Hort. Sci.* 72, 60–5.

Hu, M. J., Cox, K. D., Schnabel, G. and Luo, C. X. 2011. *Monilinia* species causing brown rot of peach in China. *PLoS ONE* 6(9), e24990 (doi:10.1371/journal.pone.0024990). doi:10.1371/journal.pone.0024990.

Jones, A. L. and Aldwinckle, H. S. 1990. *Compendium of Apple and Pear Diseases*. St. Paul, MN: APS Press, 100pp.

Johnson, K. B. and Stockwell, V. O. 1998. Management of fire blight: A case study in microbial ecology. *Annu. Rev. Phytopathol.* 36, 227–48. doi:10.1146/annurev.phyto.36.1.227.

Johnson, K. B. and Temple, T. N. 2013. Evaluation of strategies for fire blight control in organic pome fruit without antibiotics. *Plant Dis.* 97(3), 402–9. doi:10.1094/PDIS-07-12-0638-RE.

Jurick, W. M. and Cox, K. D. 2017. Pre- and postharvest fungal apple diseases. In: Evans, K. (Ed.), *Achieving Sustainable Cultivation of Apples*. Philadelphia, PA: Burleigh Dodds Science Publishing Limited, pp. 371–82.

Kennelly, M. M., Cazorla, F. M., de Vicente, A., Ramos, C. and Sundin, G. W. 2007. *Pseudomonas syringae* diseases of fruit trees – Progress towards understanding and control. *Plant Dis.* 91(1), 4–17. doi:10.1094/PD-91-0004.

Koczan, J. M., McGrath, M. J., Zhao, Y. and Sundin, G. W. 2009. Contribution of *Erwinia amylovora* exopolysaccharides amylovoran and levan to biofilm formation: Implications in pathogenicity. *Phytopathology* 99(11), 1237–44. doi:10.1094/PHYTO-99-11-1237.

Köller, W. 1999. Chemical approaches to managing plant pathogens. In: Ruberson, J. R. (Ed.), *Handbook of Pest Management*. New York, NY: Marcel Dekker, pp. 337–76.

Köller, W. 2003. Fungicides, sterol biosynthesis inhibitors. In: Plimmer, J. R. (Ed.), *The Encyclopedia of Agrochemicals*. New York, NY: John Wiley.

Köller, W. and Scheinpflug, H. 1987. Fungal resistance to sterol biosynthesis inhibitors: A new challenge. *Plant Dis.* 71(12), 1066–74. doi:10.1094/PD-71-1066.

Köller, W. and Wilcox, W. F. 1999. Evaluation of tactics for managing resistance of *Venturia inaequalis* to sterol demethylation inhibitors. *Plant Dis.* 83(9), 857–63. doi:10.1094/PDIS.1999.83.9.857.

Köller, W., Wilcox, W. F., Barnard, J., Jones, A. L. and Braun, P. G. 1997. Detection and quantification of resistance of *Venturia inaequalis* populations to sterol demethylation inhibitors. *Phytopathology* 87(2), 184-90. doi:10.1094/PHYTO.1997.87.2.184.

Köller, W., Parker, D. M., Turechek, W. W., Avila-Adame, C. and Cronshaw, K. 2004. A two-phase resistance response of *Venturia inaequalis* populations to the QoI fungicides kresoxim-methyl and trifloxystrobin. *Plant Dis.* 88(5), 537–44. doi:10.1094/PDIS.2004.88.5.537.

Kovach, J., Petzoldt, C., Degnil, J. and Tette, J. 1992. A method to measure the environmental impact of pesticides. *N. Y. Food Life* 139, 1–8.

Longstroth, M. 2000. The 2000 fireblight epidemic in southwest Michigan. Van Buren County: Michigan State University – extension. Available at: http://www.msue.msu.edu/vanburen/fb2000.htm.

Luo, C.-X., Cox, K. D., Amiri, A. and Schnabel, G. 2008. Occurrence and detection of the DMI resistance-associated genetic element 'Mona' in *Monilinia fructicola. Plant Dis.* 92(7), 1099–103. doi:10.1094/PDIS-92-7-1099.

MacHardy, W. E. 1996. *Apple Scab: Biology, Epidemiology, and Management*. St. Paul, MN: The American Phytopathological Society.

MacHardy, W. E. and Gadoury, D. M. 1989. 'A revision of Mills' criteria for predicting apple scab infection periods. *Phytopathology* 79(3), 304–10. doi:10.1094/Phyto-79-304.

MacHardy, W. E., Gadoury, D. M. and Gessler, C. 2001. Parasitic and biological fitness of *Venturia inaequalis*: Relationship to disease management strategies. *Plant Dis.* 85(10), 1036–51. doi:10.1094/PDIS.2001.85.10.1036.

Mathias, M. F., Velho, A. C., Gonçalves, A. E., Mondino, P. E., Alaniz, S. M. and Stadnik, M. J. 2016. Genetic structure of *Colletotrichum fructicola* associated to apple bitter rot and Glomerella leaf spot in Southern Brazil and Uruguay. *Phytopathology* 106(7), 774–81. doi:10.1094/PHYTO-09-15-0222-R.

Maxson-Stein, K. L., He, S.-Y., Hammerschmidt, R. and Jones, A. L. 2002. Effect of treating apple trees with acibenzolar-s-methyl on fire blight and expression of pathogenesis-related protein genes. *Plant Dis.* 86(7), 785–90. doi:10.1094/PDIS.2002.86.7.785.

McGhee, G. C. and Sundin, G. W. 2011. Evaluation of kasugamycin for fire blight management, effect on nontarget bacteria, and assessment of kasugamycin resistance potential in *Erwinia amylovora*. *Phytopathology* 101(2), 192–204. doi:10.1094/PHYTO-04-10-0128.

McGhee, G. C., Guasco, J., Bellomo, L. S., Blumer-Schuette, S. E., Shane, W. W., Irish-Brown, A. and Sundin, G. W. 2011. Genetic analysis of streptomycin-resistant (Sm^R) strains of *Erwinia amylovora* suggests that dissemination of two genotypes is responsible for the current distribution of Sm^R E. amylovora in Michigan. *Phytopathology* 101, 182–91.

McGrath, M. J., Koczan, J. M., Kennelly, M. M. and Sundin, G. W. 2009. Evidence that prohexadione-calcium induces structural resistance to fire blight infection. *Phytopathology* 99(5), 591–6. doi:10.1094/PHYTO-99-5-0591.

McHart, S. H., Urbanowicz, G., McCoshum, S., Irwin, R. E. and Adler, L. S. 2017. Landscape predictors of pathogen prevalence and range contractions in US bumblebees. *Proc. R. Soc. B* 284, 20172181. doi:10.1098/rspb.2017.2181.

McManus, P. S., Stockwell, V. O., Sundin, G. W. and Jones, A. L. 2002. Antibiotic use in plant agriculture. *Annu. Rev. Phytopathol.* 40, 443–65. doi:10.1146/annurev.phyto.40.120301.093927.

McManus, P. S., Proffer, T. J., Berardi, R., Gruber, B. R., Nugent, J. E., Ehret, G. R., Ma, Z. and Sundin, G. W. 2007. Integration of copper-based and reduced-risk fungicides for control of *Blumeriella jaapii* on sour cherry. *Plant Dis.* 91(3), 294–300. doi:10.1094/PDIS-91-3-0294.

Morton, V. and Staub, T. 2008. A short history of fungicides. *APSnet Features* (online). doi:10.1094/APSnetFeature-2008-0308.

Norelli, J. L., Jones, A. L. and Aldwinckle, H. S. 2003. Fire blight management in the twenty-first century: Using new technologies that enhance host resistance in apple. *Plant Dis.* 87(7), 756–65. doi:10.1094/PDIS.2003.87.7.756.

Ogawa, J. M. and English, H. 1991. Diseases of temperate zone tree fruit and nut crops. Publication 3345, Oakland, CA: University of California, Division of Agricultural and Natural Resources, 461pp.

Ogawa, J. M., Zehr, E. I., Bird, G. W., Ritchie, D. F., Uriu, K. and Uyemoto, J. K. 1995. *Compendium of Stone Fruit Diseases*. St. Paul, MN: APS Press, 98pp.

O'Leary, A. L., Jones, A. L. and Ehret, G. R. 1987. Application rates and spray intervals for apple scab control with flusilazol and pyrifenox. *Plant Dis.* 71(7), 623–6. doi:10.1094/PD-71-0623.

Proffer, T. J., Berardi, R., Ma, Z., Nugent, J. E., Ehret, G. R., McManus, P. S., Jones, A. L. and Sundin, G. W. 2006. Occurrence, distribution, and polymerase chain reaction-based detection of resistance to sterol demethylation inhibitor fungicides in populations of *Blumeriella jaapii* in Michigan. *Phytopathology* 96(7), 709–17. doi:10.1094/PHYTO-96-0709.

Renick, L. J., Cogal, A. G. and Sundin, G. W. 2008. Phenotypic and genetic analysis of epiphytic *Pseudomonas syringae* populations from sweet cherry in Michigan. *Plant Dis.* 92(3), 372–8. doi:10.1094/PDIS-92-3-0372.

Robinson, T., Aldwinckle, H., Fazio, G. and Holleran, T. 2003. The Geneva series of apple rootstocks from Cornell: Performance, disease resistance, and commercialization. *Acta Hortic.* 622, 513–20. doi:10.17660/ActaHortic.2003.622.56.

Romanazzi, G., Smilanick, J. L., Feliziani, E. and Droby, S. 2016. Integrated management of postharvest gray mold on fruit crops. *Postharvest Biol. Technol.* 113, 69–76. doi:10.1016/j.postharvbio.2015.11.003.

Rosenberger, D. A., Harrison, M. B. and Gonsalves, D. 1983. Incidence of apple union necrosis and decline, tomato ringspot virus, and *Xiphinema* vector species in Hudson Valley orchards. *Plant Dis.* 67(4), 356–60. doi:10.1094/PD-67-356.

Rupp, S., Weber, R. W. S., Rieger, D., Detzel, P. and Hahn, M. 2016. Spread of *Botrytis cinerea* strains with multiple fungicide resistance in German horticulture. *Front. Microbiol.* 7, 2075. doi:10.3389/fmicb.2016.02075.

Schnabel, G. and Jones, A. L. 2001. The 14α-demethylase (*CYP51A1*) gene is overexpressed in *Venturia inaequalis* strains resistant to myclobutanil. *Phytopathology* 91(1), 102–10. doi:10.1094/PHYTO.2001.91.1.102.

Schnabel, G., Bryson, P. K., Bridges, W. C. and Brannen, P. M. 2004. Reduced sensitivity in *Monilinia fructicola* to propiconazole in Georgia and implications for disease management. *Plant Dis.* 88(9), 1000–4. doi:10.1094/PDIS.2004.88.9.1000.

Schwabe, W. F. S., Jones, A. L. and Jonker, J. P. 1984. Greenhouse evaluation of the curative and protective action of sterol-inhibiting fungicides against apple scab. *Phytopathology* 74(2), 249–52. doi:10.1094/Phyto-74-249.

Sierotzki, H. and Scalliet, G. 2013. A review of current knowledge of resistance aspects for the next-generation succinate dehydrogenase inhibitor fungicides. *Phytopathology* 103(9), 880–7. doi:10.1094/PHYTO-01-13-0009-RVW.

Slack, S. M., Zeng, Q., Outwater, C. A. and Sundin, G. W. 2017. Microbiological examination of *Erwinia amylovora* exopolysaccharide ooze. *Phytopathology* 107(4), 403–11. doi:10.1094/PHYTO-09-16-0352-R.

Smith, T. J. 1996. A risk assessment model for fire blight of apple and pear. *Acta Hortic.* 411(411), 97–104. doi:10.17660/ActaHortic.1996.411.22.

Steiner, P. W. 1988. Maryblyt: A predictive model for eastern apple fire blight. *Phytopathology* 78, 1571.

Steiner, P. W. 1990a. Predicting apple blossom infections by *Erwinia amylovora* using the MARYBLYT model. *Acta Hortic.* 273(273), 139–48. doi:10.17660/ActaHortic.1990.273.18.

Steiner, P. W. 1990b. Predicting canker, shoot and trauma blight phases of apple fire blight epidemics using the MARYBLYT model. *Acta Hortic.* 273, 149–58.

Stensvand, A., Gadoury, D. M., Amundsen, T., Semb, L. and Seem, R. C. 1997. Ascospore release and infection of apple leaves by conidia of *Venturia inaequalis* at low temperatures. *Phytopathology* 87(10), 1046–53. doi:10.1094/PHYTO.1997.87.10.1046.

Stensvand, A., Eikemo, H., Gadoury, D. M. and Seem, R. C. 2005. Use of rainfall frequency threshold to adjust a degree-day model of ascospore maturity of *Venturia inaequalis. Plant Dis.* 89(2), 198–202. doi:10.1094/PD-89-0198.

Sundin, G. W., Werner, N. A., Yoder, K. S. and Aldwinckle, H. S. 2009. Field evaluation of biological control of fire blight in the eastern United States. *Plant Dis.* 93(4), 386–94. doi:10.1094/PDIS-93-4-0386.

Sutton, D. K., MacHardy, W. E. and Lord, W. G. 2000. Effects of shredding or treating apple leaf litter with urea on ascospore dose of *Venturia inaequalis* and disease buildup. *Plant Dis.* 84(12), 1319–26. doi:10.1094/PDIS.2000.84.12.1319.

Sutton, T. B., Aldwinckle, H. S., Agnello, A. M. and Walgenbach, J. F. 2014. *Compendium of Apple and Pear Diseases and Pests*. St. Paul, MN: APS Press.

Tancos, K. A., Villani, S., Kuehne, S., Borejsza-Wysocka, E., Breth, D., Carol, J., Aldwinckle, H. S. and Cox, K. D. 2016. Prevalence of streptomycin-resistant *Erwinia amylovora* in New York apple orchards. *Plant Dis.* 100(4), 802–9. doi:10.1094/PDIS-09-15-0960-RE.

Taylor, R. K., Hale, C. N., Henshall, W. R., Armstrong, J. L. and Marshall, J. W. 2003. Effect of inoculum dose on infection of apple (*Malus domestica*) flowers by *Erwinia amylovora. N. Z. J. Crop Hortic. Sci.* 31(4), 325–33. doi:10.1080/01140671.2003.9514268.

Turechek, W. W. and Biggs, A. R. 2015. Maryblyt v.7.1 for Windows: An improved fire blight forecasting program for apples and pears. *Plant Health Prog.* 16(1), 16–22. doi:10.1094/PHP-RS-14-0046.

van Leeuwen, G. C. M., Baayen, R. P., Holb, I. J. and Jeger, M. J. 2002. Distinction of the Asiatic brown rot fungus *Monilia polystroma* sp. nov. from *M. fructigena. Mycolog. Res.* 106(4), 444–51. doi:10.1017/S0953756202005695.

Villani, S. M. and Cox, K. D. 2011. Characterizing fenbuconazole and propiconazole sensitivity and the prevalence of 'Mona' in isolates of *Monilinia fructicola* from New York. *Plant Dis.* 95(7), 828–34. doi:10.1094/PDIS-12-10-0943.

Villani, S. M. and Cox, K. D. 2014. Heteroplasmy of the cytochrome b gene in *Venturia inaequalis* and its involvement in quantitative resistance and practical resistance to trifloxystrobin. *Phytopathology* 104(9), 945–53. doi:10.1094/PHYTO-06-13-0158-R.

Villani, S. M., Biggs, A. R., Cooley, D. R., Raes, J. J. and Cox, K. D. 2015. Prevalence of myclobutanil resistance and difenoconazole insensitivity in populations of *Venturia inaequalis. Plant Dis.* 99(11), 1526–36. doi:10.1094/PDIS-01-15-0002-RE.

Villani, S. M., Ayer, K. and Cox, K. D. 2016. Molecular characterization of the *sdhB* gene and baseline sensitivity to penthiopyrad, fluopyram, and benzovindiflupyr in *Venturia inaequalis. Plant Dis.* 100, 1709–16.

Vogt, I., Wohner, T., Richter, K., Flachowsky, H., Sundin, G. W., Wensing, A., Savory, E. A., Geider, K., Day, B., Hanke, M.-V., et al. 2013. Gene-for-gene relationship in the host-pathogen system *Malus × robusta* 5 – *Erwinia amylovora. New Phytol.* 197, 1262–75.

Westwood, M. N. 1993. *Temperate-Zone Pomology: Physiology and Culture* (3rd edn.). Timber Press, Inc.

Wilcox, W. F. 1993. Incidence and severity of crown and root rots on four apple rootstocks following exposure to *Phytophthora* species and waterlogging. *J. Am. Soc. Hort. Sci.* 118, 63–7.

Williamson, S. M. and Sutton, T. B. 2000. Sooty blotch and flyspeck of apple: Etiology, biology, and control. *Plant Dis.* 84(7), 714–24. doi:10.1094/PDIS.2000.84.7.714.

Woo, J. W., Kim, J., Kown, S. I., Corvalan, C., Cho, S. W., Kim, H., Kim, S. G., Kim, S. T., Choe, S. and Kim, J. S. 2015. DNA-free genome editing in plants with preassembled CRISP-Cas9 ribonucleoproteins. *Nat. Biotechnol.* 33(11), 1162–4. doi:10.1038/nbt.3389.

Yoder, K. S., Miller, S. S. and Byers, R. E. 1999. Suppression of fire blight in apple shoots by prohexadione-calcium following experimental and natural inoculation. *Hortic. Sci.* 34(7), 1202–4.

Zheng, D. and Köller, W. 1997. Characterization of the mitochondrial cytochrome b gene from *Venturia inaequalis*. *Curr. Genet.* 32(5), 361–6. doi:10.1007/s002940050289.

Chapter 12

Pre- and post-harvest strategies to optimize fruit quality and shelf life

Peter M. A. Toivonen, Agriculture and Agri-Food Canada, Canada

1 Introduction

Post-harvest technologies offer the promise of storage/shelf life extension, but they do not take into account the biology and variations of biological processes in individual fruits that are handled using these technologies. In reality, during commercial harvesting, and the storage and shipping of fresh fruits, it must be accepted that there is biological variation within any orchard, which will result in the loss of fruit not meeting a minimum standard. It is therefore imperative to work towards minimizing such variation in losses or modifying the application of technologies such that the variation is less determinant of consequent losses.

Harvest maturity is a very important determinant of post-storage or post-shipping quality of tree fruits and, in general, maturity indices are reliable. However, as new, non-destructive tools are developed, it is becoming clear that existing approaches for some fruit, for example apples, may not be as reliable as the new tools that are more closely based on physiological parameters and are firmly linked to the ripening process (Toivonen, 2015a). As such it is important to re-evaluate existing approaches to maturity testing.

http://dx.doi.org/10.19103/AS.2018.0040.13
Published by Burleigh Dodds Science Publishing Limited, 2019.

Non-destructive tools have been developed, and are ever-improving, to sort fruit during the packing process to enhance the uniformity of fruit that has a higher standard of internal quality. The process of sorting 'high-grade' fruit is required to ensure more uniform quality at the marketplace and hopefully fewer problems during post-harvest handling and transportation. As a consequence, there must also be some emphasis on alternative marketing strategies for sound fruit that will not tolerate longer-term storage or long-distance shipping.

Post-harvest technologies can enhance quality but when dealing with biological variance, the precise control of quality decline may sometimes be overshadowed by an increased risk of loss or injury to fruit not meeting a minimum standard. Hence, there needs to be reliance on precise atmospheric control in certain situations, evaluation of the use of other technologies when temperature control cannot be achieved, and finally, approaches to ensure better temperature control must be considered.

Achieving a sustainable storage, shipping and marketing continuum depends on evaluating and improving the uniformity of the fruit from each orchard, harvesting the fruit at the optimal harvest time, applying non-destructive approaches to select the best fruit for long-term storage or shipping, and modifying the approaches of long-term storage or shipping such that remaining differences in biological variance will be minimized and quality outcomes will be consistent at the marketplace.

2 Managing biological variance to improve the success of post-harvest technologies

A challenge to managing apple maturity is that while measures such as background peel greenness, firmness and ethylene production relate to physiological optima for harvest, the determining factor for commercial harvest is often the development of red blush on a bicolor or red apple cultivar (CFIA, 2011). While anthocyanin synthesis, which is responsible for the red blush color in apples, has been associated with ripening (Chalmers et al., 1973), it is not solely modulated by physiological maturation processes (Espley et al., 2007; Honda et al., 2014; Dallabetta et al., 2017). Another factor that is most often used to gauge physiological maturity is starch clearing, which is measured by a decline in iodine-starch staining as the apple matures (Smith et al., 1979; Blanpied and Silsby, 1992). However, it has recently become apparent that starch accumulation and clearing can be significantly influenced by orchard microclimate and/or yearly weather patterns (Alegre et al., 2006; Lötze and Bergh, 2012; Toivonen, 2015a). There have been reports in the past indicating that the position in a tree can influence the starch content of apples (Feng et al., 2014) and hence maturity, which is determined via starch clearing index values (Krishnaprakash et al., 1983; Brookfield et al., 1997). However, recent

Published by Burleigh Dodds Science Publishing Limited, 2019.

research conducted using relative peel chlorophyll content, measured as peel light absorbance differences at 670 nm and 720 nm, which is designated as the I_{AD} value (Ziosi et al., 2008) and is a measure of maturity, has indicated that both firmness and maturity are not generally affected by position on a tree (Dallabetta et al., 2017). Since harvest maturity assessments are generally focused on ensuring good firmness (Blanpied and Silsby, 1992; CFIA, 2011) and both firmness and peel chlorophyll content (measured as the I_{AD} value, Toivonen, 2015a) are very closely associated physiologically (Fig. 1, Johnston et al., 2009), the measure of peel chlorophyll probably provides the best information on which to base harvest decisions. Hence, with reference to I_{AD} values, apple physiological maturity does not vary with position on the tree; however, internal quality, including starch content, varies with position on the tree (Krishnaprakash et al., 1983; Feng et al., 2014; Dallabetta et al., 2017).

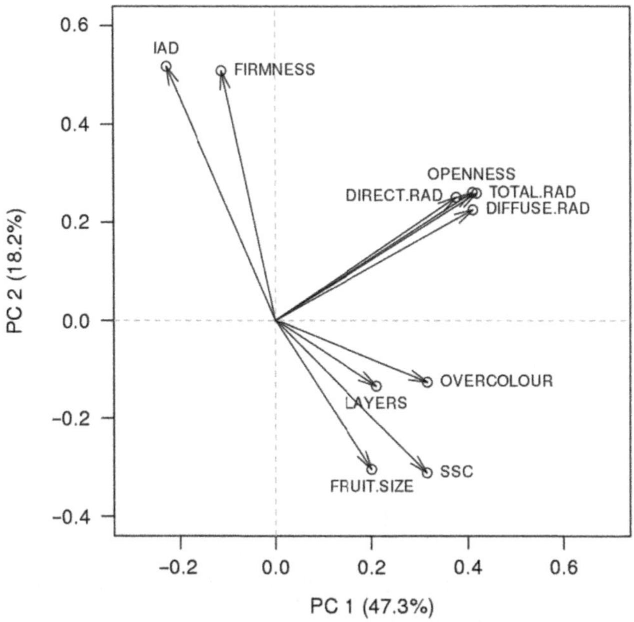

Figure 1 Principal components analysis of different quality parameters of 'Gala' apple fruit harvested from different layers (positions on the tree) and in association with measured environmental factors. From the data it is clear that firmness of the apples is very closely associated with I_{AD} values (i.e. measures of peel chlorophyll content). The fruit size, blush (overcolour) and soluble solids content (SSC) of the fruit should be noted. Soluble solids content is generally closely associated with starch clearing index, however starch clearing was not assessed in this work. These results reflect those obtained by Toivonen (2015a), showing the I_{AD} values at harvest which were the maturity measure most closely associated with firmness after storage in apples. This figure is reproduced, with permission from Dallabetta et al. (2017).

This poses a challenge to the commercial apple industry, optimizing internal quality that varies with tree position while also managing physiological harvest maturity, which is much less influenced by tree canopy position of the apple.

To complicate matters, the response of fruit to environmental conditions changes as the fruit develops physiologically and increases in size (Nordey et al., 2014), and so maturation and internal quality under varying prevailing weather conditions in a particular season can be very difficult to model accurately. More research in this area is required to improve the outcomes when applying different post-harvest technologies to tree fruit, especially as it has been observed that unpredictable variations in weather patterns within growing seasons have a far greater effect on crop growth, health, yield and quality than average annual temperature increases as predicted by global warming models (Wittwer, 1995). In the future, developments to better model growth, maturation and ripening will require complex model development and significant interaction between a multidisciplinary team comprising expertise that includes advanced mathematical modelling, meteorology, crop production and post-harvest physiology.

There have been reports of research to elucidate biological variance in the orchard that is focused on the determination of internal quality (Nilsson and Gustavsson, 2007; Feng et al., 2014) within the tree canopy or internal and external fruit quality within the orchard (Tijskens et al., 2009; Gwanpua et al., 2013). Some work has also focused on understanding the variance of internal apple quality from different growing regions in order to reconcile variations in market quality based on the source of the fruit (Bobelyn et al., 2010). However, in these studies, the determination of maturity was based on commercially determined maturity, which was most probably based on grower or packer experience and was often not clearly defined by uniform standards. Therefore, these studies reflect the expectations under current commercial practices and may not represent what may be possible with improved approaches to predicting optimum harvest maturity.

The management of harvest for non-climacteric fruits, such as sweet cherries or blueberries, is more straightforward than for climacteric fruit. Physiological maturity in reference to ethylene-associated ripening is not a factor since this phenomenon does not occur as these fruits come to harvest maturity (Toivonen and Beveridge, 2005; Toivonen, 2015a). The size and internal quality of these types of fruit is the prime determinant for harvest (Toivonen and Beveridge, 2005). For example, for sweet cherries, harvest is most often determined by the shade of red colouration of the peel and/or flesh and soluble solids content (Chockchaisawasdee et al., 2016). Since internal quality is the primary determinant for harvest, the maturity determined by these criteria can exhibit wide ranges in colour development and/or soluble solids concentration due to location within the tree canopy, which is dependent on the tree architecture and whether it is on a dwarfing rootstalk (Lauri and Claverie, 2008).

3 Measuring harvest maturity

The management of harvest maturity can significantly impact upon post-storage or post-shipping quality. The key to maximizing quality in the marketplace is to harvest fruit at an advanced stage of maturity as it is possible then to ensure good sensory quality, while also harvesting sufficiently early so that the fruit will retain acceptable textural attributes at market. In the case of climacteric fruit, the harvest decision is often based on some time point before the onset of the climacteric ripening pattern, whereas in non-climacteric fruit the harvest decision attempts to approximate harvesting at a point prior to the cessation of vacuolated growth, when tissues are near their largest size while also being turgid and tender (Toivonen and Beveridge, 2005). Both climacteric and non-climacteric fruit continue to grow in size and accumulate carbohydrates and pigment molecules in their vacuoles until they are harvested (Reid, 2002; Toivonen and Beveridge, 2005; Hu et al., 2016; Pervaiz et al., 2017). While pigment accumulation can be used as a clue for harvest maturity in non-climacteric fruit, it is not often used as an indicator for harvest maturity in climacteric fruit, except in the case of tomatoes (Gross et al., 2016).

Non-climacteric fruit are often harvested on the basis of colour (Gross et al., 2016) and in the case of sweet cherries and strawberries, a colour chip series is used to classify relative maturity (CTIFL, 2017). However, the use of colour chips to gauge maturity in the orchard is often made very difficult due to variations in lighting and so these chips are most reliably used in a laboratory situation where lighting is uniform and stable (Toivonen, 2018, personal observation). Recently, there have been advances made in the prediction of relative anthocyanin pigment content in sweet cherries using one of two specialized light absorption-based instruments (Costa et al., 2017; Overbeck et al., 2017). The advantage of using non-destructive instruments to measure colour is that they will measure peel and the flesh beneath the peel, which is important since flesh colour change is also associated with fruit maturation (Long et al., 2007). There are positive indications from current research that these non-destructive tools will be useful and practical for measuring sweet cherry maturity in the orchard (Costa et al., 2017; Overbeck et al., 2017).

Dry matter (i.e. specific gravity) and/or soluble solids content are very useful to estimate quality and maturity in sweet cherries (Reid, 2002). Soluble solids are typically measured by destructive sampling and the use of a handheld refractometer (Reid, 2002); however, destructive measuring destroys fruit. There have been significant advances in developing handheld spectrometer (non-destructive) techniques to measure these components (Escribano et al., 2017; Toivonen et al., 2017) and relate the values at harvest to sensory quality of the fruit (Escribano et al., 2017). In the future, as awareness of the value of non-destructive assessment of internal quality and maturity improves, the benefits of using such instruments will be identified. It is also highly likely that use of these measures will increase due to the demand by buyers to verify fruit quality prior to shipping. Over time and with varying ranges of eating experience, buyers and consumers are becoming more discerning in their understanding of eating quality, especially for high-value fruit such as sweet cherries. In addition, new miniaturized, pocket-sized technology is emerging and will probably be much less expensive than existing handheld spectrometers (Kaur et al., 2017); hence, it is expected that international fruit buyers will have the tools to be able to verify their arrivals for dry matter or soluble solids content which they will then equate to eating quality (Escribano et al., 2017).

The current and past determination of maturity are destructive in nature, which leads to significant quantities of wasted fruit and also limits the number of samples a grower or packer is willing to measure. For example, often for apples in British Columbia, a five-fruit sample is destroyed to determine the maturity level of one orchard. A small number of samples, considering variability in these destructive measures within a tree (Krishnaprakash et al., 1983; Costes et al., 1999; Feng et al., 2014), puts into question how well such measures estimate the overall orchard maturity. Non-destructive assessment is relatively rapid and the number of samples measured can be much larger, depending on the time that the grower or field manager has to assess orchard maturity. The other aspect of non-destructive measuring is that maturity patterns in an orchard can be mapped (Subedi et al., 2013). While in some fruit, dry matter accumulation can be used to assess harvest maturity (Reid, 2002), in apples, dry matter accumulation ceases well before the onset of harvest maturity (Schechter et al., 1993), which has been observed in our research with the Ambrosia apple (Toivonen, 2016, 2017, personal observations). Therefore, dry matter prediction using non-destructive tools would not be useful to assess harvest maturity in apples. However, there have been numerous reports indicating that peel chlorophyll content is probably the most useful indicator of apple maturity (Toivonen, 2015a). This is probably due to the fact that loss of peel chlorophyll is most closely linked to firmness change in apples in comparison to other indicators that have been used in the past to assess maturity (Johnston et al., 2009). However, instrumental measures of chlorophyll to assess apple maturity must be performed using a standard protocol in order to avoid light and heat environmental artefacts on localized chlorophyll content over the apple surface that might mask the true maturity (Toivonen et al., 2016; Toivonen, 2018). The chlorophyll content of peel in stone fruit (Ziosi et al., 2008; Costa et al., 2017; Zhang et al., 2017)

Published by Burleigh Dodds Science Publishing Limited, 2019.

and also for kiwi fruit (Costa et al., 2017) has also been shown to be useful to assess harvest maturity.

4 Improved sorting operations

Over the past several decades there have been great advances in the development of non-destructive tools that enable improved sorting of tree fruit via their internal and external quality attributes. Machine vision technologies are most often used to determine the external colour, size, shape and presence or absence of defects on tree fruits (Gowen et al., 2007; Zou and Zhao, 2015). Near-infrared reflectance spectroscopy technologies have been integrated into sorting line operations to assess internal quality (total soluble solids, dry matter, acidity etc.) and internal defects such as bruising (Nicolaï et al., 2006). Recently, techniques of multi-spectral and hyperspectral imaging have emerged as powerful tools to evaluate variations in quality or defect attributes in two-dimensional images of fruit (Nicolaï et al., 2006; Gowen et al., 2007). The major difference between the two approaches is that multi-spectral imaging analyses 3–10 very wide bands of the electromagnetic spectrum in each of the pixels within a captured image, whereas hyperspectral imaging analyses hundreds of much narrower bands (10–20 nm in width) in each pixel. This allows a spatial and qualitative analysis of each piece of fruit. In addition, since the range of wave bands measured extends beyond those measured in machine vision, internal or invisible defects can be identified on the packing line using these newer techniques (Nicolaï et al., 2006; Gowen et al., 2007). These new techniques, once imbedded into packing lines already fitted with other machine vision capabilities, will provide enhanced capability to remove fruit unacceptable for fresh consumption prior to marketing and thus will maintain confidence of consumers for the quality expectations of that fruit. The use of hyperspectral imaging will probably see considerable growth for fruit sorting in the coming decades, as equipment and research data develop further.

One aspect of sorting that is often overlooked is the common practice to sort at the pack house, which for some tree fruit crops in certain growing areas is not the optimal approach to minimize losses. In very large industries, such as the tree fruit industry of the Pacific Northwest of North America, tree fruit production practices are such that very low cull rates occur in apples, for example (Mizushima and Lu, 2011). However, in other areas and with other apple cultivars, the cull rate after sorting can be high (Mizushima and Lu, 2011). In these larger industries, a commit-to-pack strategy is sometimes used, where the fruit coming in from the orchard is pre-sized and packed in boxes before storage (http://treefruit.wsu.edu/web-article/apple-pre-sizing-and-packing/). However, in smaller industries, this approach is not economically feasible and fruit are generally placed into storage in the same bins in which they were harvested (Mizushima and Lu, 2011). In this latter scenario, cull apples left in the bin during storage can be a 'ground zero' site for the development of decay and loss of fruit. Recently, a very low cost, but efficient, in-field pre-sorting line has been developed that is fitted onto a harvesting platform (Mizushima and Lu, 2011). The implementation of such a system in smaller industries that have apple cultivars with variable on-tree quality could provide a realistic approach to removing culls at harvest to ensure both good utilization of those culls (i.e. before they decay) and also upgrade their pack of superior quality fruit for premium markets with greater returns (Mizushima and Lu, 2011).

5 Post-harvest technologies and temperature control

Temperature management is the most important factor for controlling quality change during storage or transport, and maintaining fruit at the lowest possible temperature is generally recommended as the best practice since the lower the temperature the lower the metabolic rate, which ensures that quality changes are minimized (Morris and Brady, 2005). However, with controlled atmosphere, dynamic atmosphere and ultra-low oxygen storage technologies, further modulation of metabolic activity can be achieved to improve quality retention over longer durations of storage or transportation (Yahia, 2009). Similarly, modified atmosphere packaging or pallet covers can also enhance quality retention during shipping and distribution (Yahia, 2009).

Despite the benefits of controlled or modified atmosphere technologies, temperature continues to be an important determinant of overall quality retention in fruit (Yahia, 2009; Toivonen, 2014). For example, when sweet cherries are handled in modified atmosphere box liners, shipping temperature can have a significant influence on the quality and firmness of sweet cherries for up to three or more weeks, even when there are small differences in holding temperatures (Fig. 2; Toivonen, 2014). These types of results indicate that handling and holding temperatures continue to be the largest influence on fruit quality.

However, it must be kept in mind that not all fruit are amenable to low temperatures due to susceptibility to chilling-temperature-induced injury (Wang, 1990). While the application of controlled or low oxygen atmosphere technologies can effectively minimize the development of some disorders such as superficial scald, which is a chilling injury (Zanella, 2003), they cannot control the development of other chilling-related injuries such as soft scald (Meheriuk et al., 1994). In such cases, a compromise of employing slightly higher storage temperatures with the application of controlled atmosphere storage can often ensure good firmness retention while minimizing the incidence and severity of soft scald (Fig. 3; Meheriuk et al., 1994).

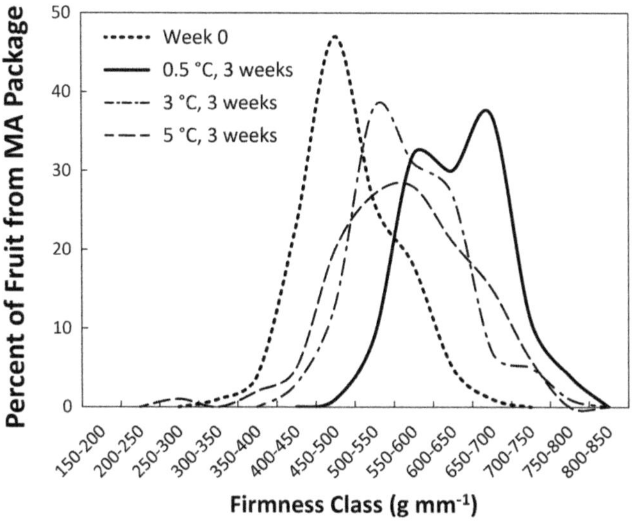

Figure 2 Distribution of firmness classes for 'Sweetheart' sweet cherries at harvest and after 3 weeks of storage in modified atmosphere box liners (Lifespan®, Amcor Limited, Southbank, Victoria, Australia) at three different temperatures. For each temperature, 125 fruit were randomly sampled from a box of commercially grown fruit and individually measured using a FirmTech firmness tester (BioWorks, Inc., Wamego, Kansas, USA) at the time of harvest (week 0) and after 3 weeks storage at 0.5°C, 3°C and 5°C. Note that the distribution shifts to firmer classifications with time in storage and that the most dramatic shift is with cherries stored at 0.5°C. Unpublished data from Toivonen (2012).

6 Case study: post-harvest strategies in cherry cultivation in British Columbia

The British Columbia sweet cherry industry relies heavily on international export markets that are willing to pay premium prices for their fruit and this requires that the industry deliver high-quality large fruit, with green stems, on a consistent basis. In addition, transport cost is a significant issue to profitability for export marketing and so the industry has to be able to deliver the premium quality fruit using long-distance transport in refrigerated sea containers to reduce the cost of transport. To achieve this goal, research and industry needed to cooperate to identify factors and processes that would ensure good quality retention during the 3–5 weeks of sea container shipment.

The first issue to be addressed was the optimal harvest maturity index for successful long-distance shipping. In this work, 'Sweetheart' cherries harvested at three progressively darker colour stages showed a progressive increase in soluble solids and titratable acidity concentration (Toivonen, 2015b). When considering quality change during storage/simulated shipping, soluble solids did not change during cold storage but the titratable acidity declined significantly, with the greatest declines occurring in the least mature fruit

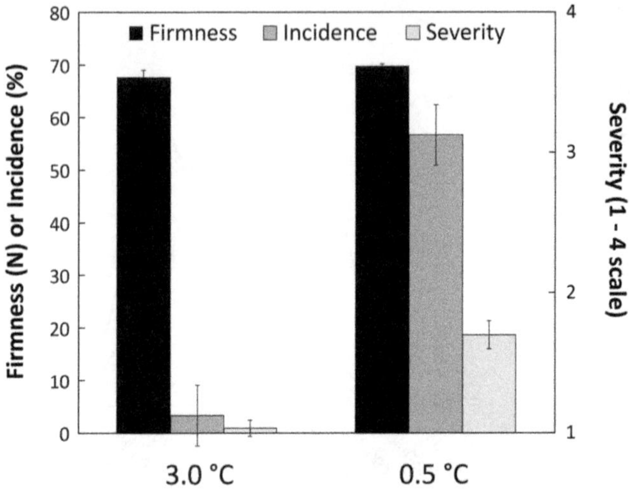

Figure 3 Firmness, incidence and severity of soft scald for Ambrosia™ apples after 5 months of storage in controlled atmosphere (1.2% O_2 + 1% CO_2) at 0.5°C or 3°C. Each histogram represents the mean of four replicates containing 15 apples and the error bars represent the standard deviations of those means. Note that firmness values for apples from the two temperatures are not significantly different, but apples stored in 3°C are almost free of soft scald, whereas those stored at 0.5°C have over 50% of the fruit affected by soft scald and the symptoms are clearly visible as indicated by a severity rating level of approximately 0.75. Unpublished data from Toivonen (2016).

(Toivonen, 2015b). The firmness of the cherries at all three harvest maturities was considered acceptable for market and the firmness increased with time in storage/simulated shipping (Toivonen, 2015b). Later harvested cherries had a slightly higher stem browning at 6 weeks in storage/simulated shipping (Toivonen, 2015b). Considering all of the factors assessed in this part of the study the target for premium quality export of 'Sweetheart' sweet cherries was set at an advanced harvest maturity of between colour stages 5 and 6 (CTIFL, 2017).

Preventing the effects of sweet cherry exposure to both sun and low humidity was demonstrated to be important in controlling browning and shrivelling of stems (Schick and Toivonen, 2002). Protecting the fruit with special reflective covers was critical even though stem dehydration (shrivelling) could be reversed with hydro-cooling. Despite the potential for rehydration, the injury (solarization) leading to browning was irreversible and so browning proceeded similarly whether the exposed fruit was hydro-cooled or not.

The impact of harvest on packing house logistics was also found to be very important in determining resultant quality after storage/simulated shipping. Delays in sorting, cooling and packing were associated with increases in stem browning incidence and severity, stem yellowing and greater loss of titratable acidity during storage/simulated shipping (Toivonen, 2012, 2015b). Delays in

Published by Burleigh Dodds Science Publishing Limited, 2019.

cooling, sorting and packing clearly result in accelerated degradation of flavour quality of the fruit.

Temperature management during the shipping of packed cherries was found to be extremely important to quality retention and avoidance of decay after storage/simulated shipping (Toivonen, 2014). The packing house hydro-cooler operation results in a minimum core temperature for cherries of 2°C and in some cases, depending on the volume of cherries being packed, the temperature in the box was found to be as high as 6°C (Toivonen, 2015b). When evaluating the response of sweet cherries to different cooling temperatures, it was determined that as the core temperature of the cherry increased from 0.5°C to 5°C, significant differences in firmness resulted after 3 weeks in cold storage (Toivonen, 2015b). The only way to reduce core temperatures of the cherries in the box to 0.5°C, or lower, is to apply forced-air cooling for at least 14 h after packing (Toivonen, 2015b). These results highlight the importance of forced-air cooling of assembled pallets after the hydro-cooling and packing of cherries if they are to be shipped by container.

Respiratory heat is an important factor in determining cherry temperature in palletized loads and different cultivars have different respiration rates (Kappel et al., 2002). 'Sweetheart' has the lowest respiration rate pattern as temperatures increase from 0°C to 10°C and hence would be expected to exhibit the least rise in respiratory heat when palletized. This would mean that this cultivar has the best performance during long-distance shipping, which is supported by the real-world British Columbia export industry. More recently, newly released Staccato™ and Suite Note™ have also been found to have similar respiration profiles and hence similar potential shipping days as 'Sweetheart'.

In this case study, there were many factors that could have been investigated for the improvement of the containerized shipping of sweet cherries but it was important to study the question in the context of current commercial practices. This meant conducting the research in an industry setting as done for the British Columbia sweet cherries. The important findings of our work, conducted over the past two decades, has led to a list of factors that are most important to quality retention in sweet cherries that are shipped over long distance by ocean containers. The application of modified atmosphere packaging technologies provide control of water loss and decay via high carbon dioxide partial pressures, however when different technologies have been assessed, there has been no evidence of any advantage of one technology over another (Toivonen et al., 2008).

Temperature management is probably the most important concern for long-distance sweet cherry shipping. However, there continues to be a lack of understanding related to the fact that packing line technology alone cannot provide the terminal low temperatures required for successful long-distance refrigerated container shipping. From our analysis, even the best packing line operation cannot reduce the fruit core temperature to near 0°C before packing the fruit. This is a reality that must be understood and consequent post-packing cooling (forced-air) of assembled pallets must be practised in order to ensure acceptable temperatures of container-shipped sweet cherries.

Sweet cherries, like other fruit, are living entities and respire even in the package and while low temperatures will reduce the respiration rate, the consequence of respiration is heat generation, which can lead to heat accumulation in palletized fruit (Waelti, 1986). If cherries are shipped for short distances via truck or long distances via air, the time in pallet format is in the range of days to a week. However, containerized shipment takes at least 25–30 days and therefore the issue of heat accumulation in palletized cherries becomes very important. We found that different cultivars have different respiration rates at lower temperatures and hence different propensities for heat accumulation. As a consequence, we are now beginning to classify new cultivars for shipping potential, based on respiration rate measurements. The future of containerized shipping of sweet cherries should focus on selecting cultivars with lower respiration rates at higher temperatures.

7 Conclusion

Ensuring a sustainable tree fruit industry, with regard to post-harvest quality management, involves engaging approaches and technologies that encompass both pre- and post-harvest issues. The term often used for this outlook is precision agriculture (Zude-Sasse et al., 2016). From the above discussion it is clear that fruit on the tree is often anything but uniform in quality and maturity. The adoption of new, non-destructive technologies is important in order to quantify and evaluate the challenges to the management of fruit quality in post-harvest systems. Non-destructive tools will also provide rapid assessment methods to evaluate the effect of pre-harvest management practices and microclimate on the post-harvest performance of fruit from differing locations or in-field management practices. The rapid, real-time prediction of post-harvest performance should accelerate the accumulation of knowledge needed to improve quality at the marketplace, and hopefully reduce losses due to culling.

While it is necessary to modernize the approach to assessing harvest maturity and quality using new, non-destructive tools, there continues to be a need for post-harvest tools to cull out fruit not meeting quality standards and allow the industry to have a premium quality classification for specialized markets. The post-harvest sorting/packing process is further complicated by the fact that there are two contrasting scales of industry – a very large industry capable of implementing the latest tools with little regard for cost and small-scale industry where cost and equipment scale are of concern. The larger-scale industries generally have large volumes of fruit and have access to large-scale sorting and packing operations that allow them to use various strategies to manage culls and grades of fruit. Smaller industries have fewer resources and would be better off if they could cull and grade fruit in the orchard and so in-field technologies would certainly be of benefit to those industries and could significantly improve their sustainability.

The challenge today is the fact that there are many post-harvest technologies that promise great improvements in quality retention, however, the basic factor of temperature is often minimized in the evaluation of making changes. Most often, good temperature management has the greatest impact on quality preservation, but for some situations where chilling injury is a factor, the reliance on technologies such as controlled, ultra-low oxygen or dynamic controlled atmosphere storage with elevated, non-chilling temperatures, could be essential to sustainability. This last statement is strongly supported by the example of the case study discussed above – all factors for successful storage or shipping of tree fruit must be evaluated and optimized in order to have a sustainable and reliable industry. Therefore, research must be performed in an integrative manner to achieve a complete, multifaceted and functional understanding of how to improve post-harvest performance under real-world conditions.

There are two aspects of rapidly evolving research that could improve the evaluation of maturity and internal quality of fruit. The first aspect revolves around the development of non-destructive tools based on light absorbance or reflectance properties of tree fruit (Zude-Sasse et al., 2016). As these tools are developed and robust models are built (Nicolaï et al., 2007), a second aspect emerges and involves spatial analysis of on-tree location and differences in quality and maturity in the orchard (Costes et al., 1999; Subedi et al., 2013; Feng et al., 2014; Costa et al., 2017). Along with the implementation of these instruments, there are emerging modelling platforms that will enable a significant shift in the capability of mapping the dynamics of quality and maturity changes in a very reliable manner (Costes et al., 2008; Lescourret et al., 2011), which will lead to better targeted production research to reduce quality variation within a tree and within the orchard (Dallabetta et al., 2017). The advantage of using some non-destructive measures such as dry matter content is that they are considered to be predictive of the post-storage/shipping consumer quality (Palmer et al., 2010; Escribano et al., 2017), without the need to implement extensive and expensive post-harvest analyses of production practice effects on market outcomes.

Cultivar improvement is an ongoing requirement for the sustainability of smaller-scale industries in particular, providing tree fruit with premium quality and superior post-harvest handling characteristics. Smaller-scale industries will only be sustainable if they can provide premium quality fruit in the market at much higher prices to offset the efficiency-of-scale disadvantage that they experience (Mizushima and Lu, 2011). Export market success, as found in the British Columbia sweet cherry industry, is largely due to the small-scale industry approach to delivering exceptional quality to compete with the slightly lower quality offered by larger-scale competitors (Toivonen, 2012, 2015b).

Moreover, further improvements in sorting and culling capabilities will enhance the consistency of quality at the marketplace and enable the efficient utilization of fruit that does not meet the fresh market standards (Mizushima and Lu, 2011). In small-scale industries, moving towards lower in-field pre-sorting systems may provide the advantage needed to enhance market quality and minimize losses (Mizushima and Lu, 2011). In larger-scale industries, improved sorting operations employing tools such as hyperspectral imaging will assist in predicting which fruit will be culled and thus diverting the fruit to alternative uses to avoid losses from time in storage (Nicolaï et al., 2006; Gowen et al., 2007).

Finally, since cooling and temperature management are central to successful long-term storage and long-distance shipping, future research should focus on understanding the realities of temperature abuse within a shipping and distribution path (Gwanpua et al., 2013). This understanding should identify factors to improve logistics and/or design protocol or technologies that will compensate for potential abuse events.

8 Where to look for further information

The most recent comprehensive reference on the subject is E. M. Yahia, *Modified and Controlled Atmospheres for the Storage, Transportation, and Packaging of Horticultural Commodities* (see Yahia, 2009). There are several sources of standard recommendations for harvesting and post-harvest handling and storage of all fruits:

- USDA Agricultural Research Service Handbook # 66 (https://www.ars.usda.gov/ARSUserFiles/oc/np/CommercialStorage/CommercialStorage.pdf)
- University of California Post-harvest Center website (http://postharvest.ucdavis.edu/) and
- le Centre Technique Interprofessionnel des Fruits et Légumes (CTIFL) website (http://www.ctifl.fr/)

There is also information available on the internet regarding commercially available technologies for post-harvest. Two websites feature up-to-date information:

- http://www.poscosecha.com
- http://www.postharvest.biz

Information on the optimal protocols for use of non-destructive spectrometers for predicting internal quality in fruit can be found in a series of publications

Published by Burleigh Dodds Science Publishing Limited, 2019.

which are listed in the References and include Peirs et al. (2000), Nicolaï et al. (2007), Gwanpua et al. (2013), Subedi et al. (2013), Toivonen et al. (2016), Kaur et al. (2017), Costa et al. (2017), Escribano et al. (2017) and Toivonen (2018).

Almost every jurisdiction in the developed world has local or regional information resources to support the implementation and development of improved harvest and post-harvest handling procedures. Some of these are based at university departments active in the area, some are global in nature such as the post-harvest information site of the Food and Agriculture Organization of the United Nations, found at http://www.fao.org/inpho/en/, or the CGIAR repository of global post-harvest research and development, found at https://cgspace.cgiar.org/discover?scope=%2F&query=postharvest&submit=Go.

9 References

Alegre, S., Molina, D.P., Recasens, I., Casals, M., Bonany, J., Carbó, J., Casero, T. and Iglesias, I. (2006), 'Seasonal trends in harvest indices for 'Golden Smoothee'® apples in Spain', *J. Fruit Ornam. Plant Res.* 14(Suppl. 2), 65–75.

Blanpied, G. D. and Silsby, K. J. (1992), 'Predicting harvest date windows for apples', Cornell Cooperative Extension Information Bulletin 221, 12p. http://www.uvm.edu/~fruit/treefruit/tf_horticulture/AppleHortBasics/Readings/Predicting+Harvest+Date+Window+for+Apples.pdf (accessed 8 January 2018).

Bobelyn, E., Serban, A.-S., Nicu, M., Lammertyn, J., Nicolai, B. M. and Saeys, W. (2010), 'Postharvest quality of apple predicted by NIR-spectroscopy: Study of the effect of biological variability on spectra and model performance', *Postharvest Biol. Technol.* 55, 133–43. doi:10.1016/j.postharvbio.2009.09.006.

Brookfield, P., Murphy, P., Harker, R. and Macrae, E. (1997), 'Starch degradation and starch pattern indices: Interpretation and relationship to maturity', *Postharvest Biol. Technol.* 11, 23–30. doi:10.1016/S0925-5214(97)01416-6.

CFIA (Canadian Food Inspection Agency) (2011), '"Apples", fruit inspection manual', last modified on 24 April 2011. http://www.inspection.gc.ca/food/fresh-fruits-and-vegetables/quality-inspection/fruit-inspection-manuals/apples/eng/1303668473869/1303672406197 (accessed 16 November 2017).

Chalmers, D. J., Faragher, J. D. and Raff, J. W. (1973), 'Changes in anthocyanin synthesis as an index of maturity in red apple varieties', *J. Hortic. Sci.* 48, 387–92. doi:10.1080/00221589.1973.11514541.

Chockchaisawasdee, S., Golding, J. B., Vuong, Q. V., Papoutsis, K. and Stathopoulos, C. E. (2016), 'Sweet cherry: Composition, postharvest preservation, processing and trends for its future use', *Trends Food Sci. Technol.* 55, 72–83.

Costa, G., Rocchi, L., Farneti, B., Busatto, N., Spinelli, F. and Vidoni, S. (2017), 'Use of non-destructive devices to support pre- and postharvest fruit management', *Horticulturae* 3, 12. doi:10.13390/horticulturae3010012.

Costes, E., Sinoquet, H., Godin, C. and Kelner, J. J. (1999), '3D digitizing based on tree topology: Application to study the variability of apple quality with the canopy', *Acta Hortic.* 499, 271–80.

Costes, E., Smith, C., Renton, M., Guédon, Y., Prusinkiewicz, P. and Godin, C. (2008), 'MAppleT: Simulation of apple tree development using mixed stochastic and biomechanical models', *Funct. Plant Biol.* 35, 936–50.

CTIFL (le Centre Technique Interprofessionnel des Fruits et Légumes) (2017), 'Color codes'. http://www.ctifl.fr/Pages/Kiosque/ListeOuvrage.aspx?idType=3 (accessed on 11 December 2017).

Dallabetta, N., Costa, F., Guerra, A., Pasqualini, J., Giordan, M., Zorer, R., Wehrens, R., George, E., Magnanini, E. and Costa, G. (2017), 'Effects of light availability and training systems on apple fruit quality', *Acta Hortic.* 1160, 237–43. doi:10.17660/ActaHortic.2017.1160.35.

Escribano, S., Biasi, V. W., Lerud, R., Slaughter, D. C. and Mitcham, E. J. (2017), 'Non-destructive prediction of soluble solids and dry matter content using NIR spectroscopy and its relationship with sensory quality in sweet cherries', *Postharvest Biol. Technol.* 128, 112–20. doi:10.1016/j.postharvbio.2017.01.016.

Espley, R. V., Hellens, R. P., Putterill, J., Stevenson, D. E., Kutty-Amma, S. and Allan, A. C. (2007), 'Red colouration in apple fruit is due to the activity of the MYB transcription factor, MdMYB10', *Plant J.* 49, 414–27. doi:10.1111/j.1365-313X.2006.02964.x.

Feng, F., Li, M., Ma, F. and Cheng, L. (2014), 'Effects of location within the tree canopy on carbohydrates, organic acids, amino acids and phenolic compounds in the fruit peel and flesh from three apple (*Malus × domestica*) cultivars', *Hortic. Res.* 1, 14019. doi:10.1038/hortres.2014.19.

Gowen, A. A., O'Donnell, C. P., Cullen, P. J., Downey, G. and Frias, J. M. (2007), 'Hyperspectral imaging – an emerging process analytical tool for food quality and safety control', *Trends Food Sci. Technol.* 18, 590–8.

Gross, K. C., Wang, C. Y. and Saltveit, M. (Eds). (2016), 'The commercial storage of fruits, vegetables, and florist and nursery stocks', USDA Agricultural Research Service Handbook 66.

Gwanpua, S. G., Verlinden, B. E., Hertog, M. L. A. T. M., Van Impec, J., Nicolaï, B. M. and Geeraerd, A. H. (2013), 'Towards flexible management of postharvest variation in fruit firmness of three apple cultivars', *Postharvest Biol. Technol.* 85, 18–29.

Honda, C., Bessho, H., Murai, M., Iwanami, H., Moriya, S., Abe, K., Wada, M., Moriya-Tanaka, Y., Hayama, H. and Tatsuki, M. (2014), 'Effect of temperature on anthocyanin synthesis and ethylene production in the fruit of early- and medium-maturing apple cultivars during ripening stages', *HortScience* 49, 1510–17.

Hu, D.-G., Sun, C.-H., Ma, Q.-J., You, C.-X., Cheng, L. and Hao, Y.-J. (2016). 'MdMYB1 regulates anthocyanin and malate accumulation by directly facilitating their transport into vacuoles in apples', *Plant Physiol.* 170(3), 1315–30. doi:10.1104/pp.15.01333.

Johnston, J. W., Gunaseelan, K., Pidakala, P., Wang, M. and Schaffer, R. J. (2009), 'Co-ordination of early and late ripening events in apples is regulated through differential sensitivities to ethylene', *J. Exp. Bot.* 60, 2689–99.

Kappel, F., Toivonen, P., McKenzie, D.-L. and Stan, S. (2002), 'Storage characteristics of new sweet cherry cultivars', *HortScience* 37, 139–43.

Kaur, H., Künnemeyer, R. and McGlone, A. (2017), 'Comparison of hand-held near infrared spectrophotometers for fruit dry matter assessment', *JNIRS* 25(4), 267-77. doi:10.1177/0967033517725530.

Krishnaprakash, M. S., Aravindaprasad, B., Krishnaprasad, C. A., Narasimham, P., Ananthakrishna, S. M., Dhanaraj, S. and Govindarajan, V. S. (1983), 'Effect of apple position on the tree on maturity and quality', *J. Hortic. Sci.* 58, 31–6. doi:10.1080/00221589.1983.11515087.

Lauri, P. E. and Claverie, J. (2008), 'Sweet cherry tree architecture, physiology and management: Towards an integrated view', *Acta Hortic.* 795, 605–14.

Lescourret, F., Moitrier, N., Valsesia, P. and Génard, M. (2011). 'QualiTree, a virtual fruit tree to study the management of fruit quality. I. Model development', *Trees* 25, 519–30. doi:10.1007/s00468-010-0531-9.

Long, L. E., Whiting, M. and Nuñez-Elisea, R. (2007), 'Sweet cherry cultivars for the fresh market', Pacific Northwest Extension Publication 604, Oregon State University in cooperation with Washington State and Idaho Extension services. http://horticulture.oregonstate.edu/system/files/u1473/pnw604.pdf (accessed on 6 January 2018).

Lötze, E. and Bergh, O. (2012), 'Evaluating the Streif index against commercial subjective predictions to determine the harvest date of apples in South Africa', *S. Afr. J. Plant Soil* 29(2), 53–6.

Meheriuk, M., Prange, R. K., Lidster, P. D. and Porritt, S. W. (1994), 'Postharvest disorders of apples and pears', Agriculture Canada Publication 1737/E.

Mizushima, A. and Lu, R. (2011), 'Cost benefits analysis of in-field presorting for the apple industry', *Appl. Eng. Agric.* 27, 33–40.

Morris, J. R. and Brady, P. L. (2005), 'Temperature effects on produce degradation', in Lamikanra, O., Imam, S. and Ukuku, D. (Eds), *Produce Degradation, Pathways and Prevention*, CRC Press, Boca Raton, FL, pp. 599–647.

Nicolaï, B. M., Lötze, E., Peirs, A., Scheerlinck, N. and Theron, K. I. (2006), 'Non-destructive measurement of bitter pit in apple fruit using NIR hyperspectral imaging', *Postharvest Biol. Technol.* 40, 1–6.

Nicolaï, B. M., Beullens, K., Bobelyn, E., Peirs, A., Saeys, W., Theron, K. I. and Lammertyn, J. (2007), 'Nondestructive measurement of fruit and vegetable quality by means of NIR spectroscopy: A review', *Postharvest Biol. Technol.* 46, 99–118.

Nilsson, T. and Gustavsson, K.-E. (2007), 'Postharvest physiology of "Aroma" apples in relation to position on the tree', *Postharvest Biol. Technol.* 43, 36–46. doi:10.1016/j.postharvbio.2006.07.011.

Nordey, T., Léchaudel, M., Saudreau, M., Joas, J. and Génard, M. (2014), 'Model-assisted analysis of spatial and temporal variations in fruit temperature and transpiration highlighting the role of fruit development', *PLoS ONE* 9(3), e92532, 12p. doi:10.1371/journal.pone.0092532.

Overbeck, V., Schmitz, M. and Blanke, M. (2017), 'Non-destructive sensor-based prediction of maturity and optimum harvest date of sweet cherry fruit', *Sensors* 17, 277. doi:10.3390/s17020277.

Palmer, J. W., Harker, F. R., Tustin, D. S. and Johnston, J. (2010), 'Fruit dry matter concentration: A new quality metric for apples', *J. Sci. Food Agric.* 90, 2586–94. doi:10.1002/jsfa.4125.

Peirs, A., Lammertyn, J., Ooms, K. and Nicolaï, B. M. (2000), 'Prediction of the optimal picking date of different apple cultivars by means of VIS/NIR-spectroscopy', *Postharvest Biol. Technol.* 21, 189–99. doi:10.1016/S0925-5214(00)00145-9.

Pervaiz, T., Songtao, J., Faghihi, F., Salman Haider, M. and Fang, J. (2017), 'Naturally occurring anthocyanin, structure, functions and biosynthetic pathway in fruit plants', *J. Plant Biochem. Physiol.* 5, 2. doi:10.4172/2329-9029.1000187.

Reid, M. S. (2002), 'Maturation and maturity indices', in A. A. Kader (Ed.), *Postharvest Technology of Horticultural Crops*. University of California Agriculture and Natural Resources Publication 3311, pp. 55–65. ISBN 1-879906-51-1.

Schechter, I., Proctor, J. T. A. and Elfving, D. C. (1993), 'Reappraisal of seasonal apple fruit growth', *Can. J. Plant Sci.* 73, 549–56.

Schick, J. and Toivonen, P. M. A. (2002), 'Reflective tarps at harvest reduce stem browning and improve fruit quality of cherries during subsequent storage', *Postharvest Biol. Technol.* 25, 117–21.

Smith, R. B., Lougheed, E. C., Franklin, E. W. and McMillan, I. (1979), 'The starch iodine test for determining stage of maturation in apples', *Can. J. Plant Sci.* 59, 725–35.

Subedi, P., Walsh, K. and Purdy, P. (2013), 'Determination of optimum maturity stages of mangoes using fruit spectral signatures', *Acta Hortic.* 992, 521–7. doi:10.17660/ActaHortic.2013.992.64.

Tijskens, L. M. M., Unuk, T., Tojnko, S., Hribarm J. and Simčič, M. (2009), 'Biological variation in the colour development of Golden Delicious apples in the orchard' *J. Sci. Food Agric.* 89, 2045–51. doi:10.1002/jsfa.3689.

Toivonen, P. M. A. (2012), 'Optimizing handling for superior fresh market quality of sweet cherries and potential value-added product', in D. Martinez-Romero, D. Valero, M. Serrano, S. Castillo, F. Guillén and P. J. Zapata (Eds), *Ciruela y Cereza: del Campo al Consumidor (Plum and Sweet Cherry: from Field to Consumer)*, Proceedings of the International Plum and Sweet Cherry Symposium, 24–25 October 2012, pp. 159–64. ISBN 978-84-939989-3-6.

Toivonen, P. M. A. (2014), 'Relationship of typical core temperatures after hydrocooling on retention of different quality components in sweet cherry', *HortTechnology* 24, 457–62.

Toivonen, P. M. A. (2015a), 'Comparison of I_{AD} and starch-iodine indices at harvest and how they relate to post-storage firmness retention in Ambrosia™ apples over three growing seasons', *Can. J. Plant Sci.* 95, 1177–80. doi:10.4141/cjps-2015-185.

Toivonen, P. M. A. (2015b), 'Integrated analysis for improving export of sweet cherries and how a small industry can compete by focusing on premium quality', *Acta Hortic.* 1079, 71–82.

Toivonen, P. M. A. (2016), 'Postharvest physiology of fruits and vegetables', in S. Pareek (Ed.), *Postharvest Ripening Physiology of Crops. Innovations in Postharvest Technology Series*, CRC Press, Boca Raton, FL, pp. 49–79. ISBN: 9781498703802.

Toivonen, P. M. A. (2018), 'Chapter 1.3. A case study on the impact of maturity stage on CA/MAP requirements', in M. I. Gil and R. M. Beaudry (Eds), *Controlled and Modified Atmosphere for Fresh and Fresh-Cut Produce*, Academic Press, Cambridge, Massachusetts (in press). ISBN: 9780128045992.

Toivonen, P. M. A. and Beveridge, T. H. J. (2005), 'Maturity, ripening and quality relationship', in O. Lamikanra, S. H. Imam and D. O. Ukuku (Eds), *Produce Degradation: Reaction Pathways and their Prevention*, CRC Press, Boca Raton, FL, pp. 55–77.

Toivonen, P. M. A., Kappel, F., Lannard, B. and MacKenzie, D.-L. (2008), 'Comparison of two commercial modified atmosphere box-liners for sweet cherries', www.ultraperf.com/pdf/Cherry_Liner_Report.pdf (accessed 8 January 2018).

Toivonen, P. M. A., Lu, C. and Lannard, B. (2016), 'Visible spectroscopy I_{AD} measures in apple: Impact of stress and shading on maturity indexing', *Acta Hortic.* 1119, 243–9. doi:10.17660/ActaHortic.2016.1119.34.

Toivonen, P., Batista, A. and Lannard, B. (2017), 'Development of a predictive model for "Lapins" sweet cherry dry matter content using a visible/near infrared spectrometer and its potential application to other cultivars', *Can. J. Plant Sci.* 97, 1030–5. doi:10.1139/cjps-2017-0013.

Waelti, H. (1986), 'Forced air cooling of cherries', *PostHarvest Pomol. Newsl.* 4, 7. Accessible at: http://postharvest.tfrec.wsu.edu/pages/N4I1A.

Wang, C. Y. (1990), *Chilling Injury of Horticultural Crops*, CRC Press, Boca Raton, FL. ISBN: 9780849357367.

Wittwer, S. H. (1995), *Food, Climate, and Carbon Dioxide: The Global Environment and World Food Production*, CRC (Lewis Publishers), Boca Raton, FL.

Yahia, E. M. (2009), *Modified and Controlled Atmospheres for the Storage, Transportation, and Packaging of Horticultural Commodities*, CRC Press, Boca Raton, FL.

Zanella, A. (2003), 'Control of apple superficial scald and ripening - a comparison between 1-methylcyclopropene and diphenylamine postharvest treatments, initial low oxygen stress and ultra low oxygen storage', *Postharvest Biol. Technol.* 27, 69–78. doi:10.1016/S0925-5214(02)00187-4.

Zhang, B., Peng, B., Zhang, C., Song, Z. and Ma, R. (2017), 'Determination of fruit maturity and its prediction model based on the pericarp index of absorbance difference (I_{AD}) for peaches', *PLoS ONE* 12(5), e0177511. doi:10.1371/journal.pone.0177511.

Ziosi, V., Noferini, M., Fiori, G., Tadiello, A., Trainotti, L., Casadoro, G. and Costa, G. (2008), 'A new index based on vis spectroscopy to characterize the progression of ripening in peach fruit', *Postharvest Biol. Technol.* 49, 319–29. doi:10.1016/j.postharvbio.2008.01.017.

Zou, X. and Zhao, J. (2015), *Non-destructive Measurement in Food and Agro-products*, Springer Netherlands, Dordrecht. doi:10.1007/978-94-017-9676-7_2.

Zude-Sasse, M., Fountas, S., Gemtos, T. and Abu-Khalaf, N. (2016), 'Applications of precision agriculture in horticultural crops', *Eur. J. Hortic. Sci.* 81, 78–90. doi:10.17660/eJHS.2016/81.2.2.

Chapter 13

Bioactive/nutraceutical compounds in fruit that optimize human health benefits

Federica Blando and Miriana Durante, Institute of Sciences of Food Production (ISPA), Italy; and B. Dave Oomah, formerly Pacific Agri-Food Research Centre, Canada

1 Introduction

In the past decade, numerous reports have demonstrated that high consumption of fruits and vegetables is beneficial for health, being associated with a reduced risk of degenerative diseases (Miller et al., 2017). Fruits are an important part of the human diet and a major source of bioactive compounds (BCs) with human health benefits. Unfortunately, this 'important part' of the human diet has been a decreasing proportion of the Western diet for the last 50 years. The 'Five a Day' program was introduced by national campaigns in several Western countries according to World Health Organization (WHO) and United Nations Food and Agriculture Organization (FAO) recommendations in their joint initiative (launched in 2003) to promote fruits and vegetables for health worldwide. This program has been implemented to better communicate updated dietary guidelines, which recommend more than five servings of fruits and vegetables. A recent English study found that eating seven or more portions of fruits and vegetables reduces the specific risks of death by cancer and heart disease (25% and 31%, respectively) (Oyebode et al., 2013).

This chapter provides a brief description of the chemistry of BCs and their presence in (temperate) fruits, and discusses recent advances in strategies towards improving sustainable crop production for nutraceuticals.

http://dx.doi.org/10.19103/AS.2018.0040.14

Among different BCs present in fruits are polyphenols, carotenoids and some vitamins (C), which are all correlated with antioxidant potential of fruits. The content of BCs in fruits can vary depending on different factors, primarily genetics and the environmental conditions where fruit trees are cultivated. In the last 30 years, the genetic improvement of fruit trees has mainly been driven by productivity issues and consumer preferences; fruit quality traits have only recently been recognized as important additions to production traits (size, high productivity etc.). The nutraceutical characterization of fruits has become an additional factor to consider for genetic improvement programmes. Unfortunately, the selective pressure for commercial traits has resulted in a reduction of BC content in most modern cultivars, relative to older varieties. For this reason, the preservation and valorization of fruit germplasm recently became a key issue, with the re-discovery of ancient varieties strongly linked with their territory of origin. This has become a leading trend in modern fruit breeding and cultivation, at least in Europe.

2 Polyphenols

2.1 Polyphenols in general

Phenolics, one of the most diverse plant food antioxidants, are characteristic features of all plant tissues, forming an extensive group of compounds with a common hydroxy-substituted benzene ring structure. Chemically, phenols are reactive substances, very susceptible to oxidation (leading to quinones during storage or processing of plant foods), but they are also very good antioxidants, thanks to hydroxyl groups, especially at the *ortho*-position (Harborne, 1980; Parr and Bolwell, 2000).

Phenolics are mostly synthesized from phenylalanine through phenylalanine ammonia-lyase (PAL) and further metabolism of the simple C6-C3 phenylpropanoids.

Most often, phenolics are grouped into non-flavonoid (hydroxybenzoic acid and hydroxycinnamic acid derivatives) and flavonoid classes (Fig. 1). The phenolics hydroxycinnamic and chlorogenic acids predominate in the non-flavonoid phenolic class, whereas anthocyanins represent the most common sub-class of the flavonoid group. Other investigators divide phenolic compounds into sixteen classes based on their molecular structure, with flavonoids and phenolic acids most prevalent in fruits (Manganaris et al., 2014).

Phenolics have been associated with health benefits due to their antioxidant capacity, which varies with the chemical structure. A broad definition of antioxidant is 'any substance that, when present at low concentrations compared to those of an oxidizable substrate, significantly delays or prevents oxidation of that substrate' (Halliwell, 1994). Dealing with antioxidant properties

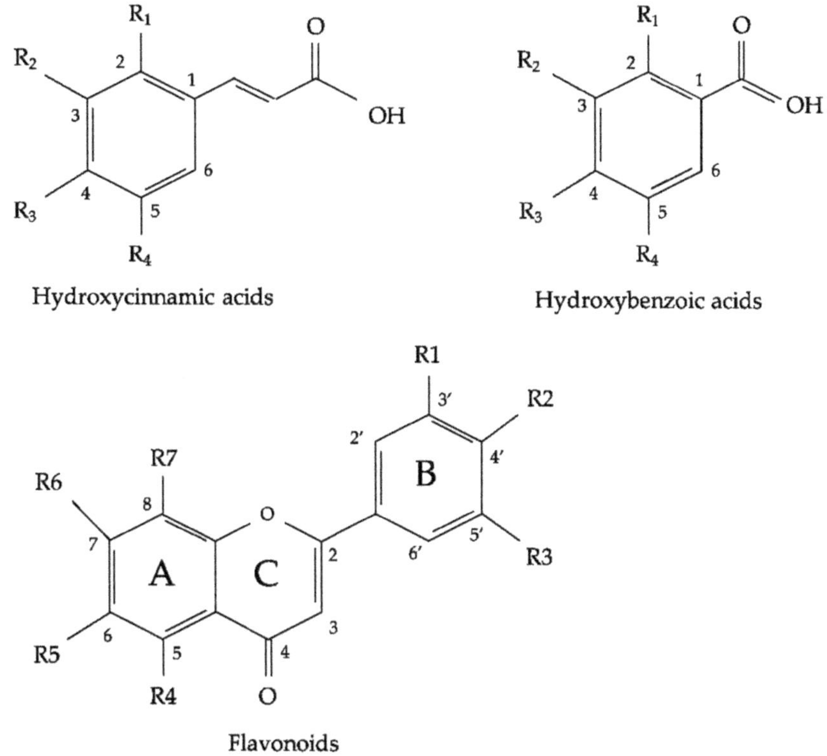

Figure 1 Basic structure of phenolic compounds.

and their relation with biological systems, it is important to underline that the antioxidant compound needs to remain stable after its reducing activities have been exerted.

Phenolics can act as antioxidants in a number of ways: chelating metals responsible for non-enzymatic free radical generation, breaking the cycle of generating new radicals or inhibiting radical generation from pro-oxidant enzymes. Phenolics are able to act not only as antioxidants, but they have shown anticarcinogenic action, in many studies, both blocking/suppressing carcinogenesis, inhibiting enzymes, and showing specific receptor interactions (Parr and Bolwell, 2000).

The antioxidant capacity of fruit extracts can be assayed *in vitro* by different tests, based on the scavenging activity of the BC against different radicals, such as the hydroxyl radical in the Trolox Equivalent Antioxidant Capacity assay (TEAC), and the peroxyl radical in the Oxygen Radical Absorbance Capacity assay (ORAC), and others. Total antioxidant capacity can be used as a screening tool for use in fruit breeding programmes designed to increase antioxidant phytochemicals available for human nutrition (Scalzo et al., 2005).

2.2 Anthocyanins

Anthocyanins are a group of naturally occurring pigments responsible for the red–blue colour of many fruits and vegetables. They belong to the polyphenolic flavonoids sub-class. Over 700 different anthocyanin structures have been identified so far in the plant kingdom, revealing the plasticity of the anthocyanin biosynthetic pathway; 277 of these structures have been identified in the last 20 years (Andersen and Jordheim, 2006).

Anthocyanins are particularly present in fruits (Mazza and Miniati, 1993). Temperate fruits containing anthocyanins include cherries, plums, peaches, nectarines, apples, pear, grape and berries. Anthocyanins have the structure of heterosides, the aglycon (anthocyanidin) being linked to glycosidic units. Sugars (often mono-, di- or tri-glycosides) can be, in addition, acylated with different organic acids (cinnamic, caffeic, ferulic or sinapic acid).

To date, there are 17 known anthocyanidins occurring in nature, but six principal anthocyanidins are present in most fruits and vegetables: pelargonidin, cyanidin, delphinidin, peonidin, petunidin and malvidin (Fig. 2). The glycosides of the three non-methylated anthocyanidins (pelargonidin, cyanidin and delphinidin) are the most widespread structure present in nature, while cyanidin 3-glucoside is predominant in fruits.

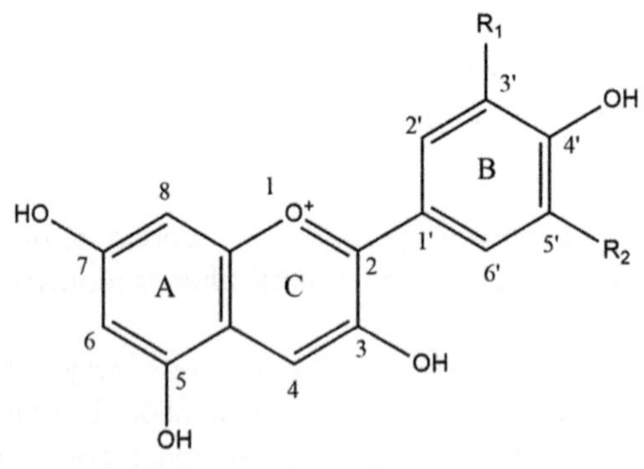

Anthocyanidins	R_1	R_2
Pelargonidin	H	H
Cyanidin	OH	H
Delphinidin	OH	OH
Peonidin	OCH_3	H
Petunidin	OH	OCH_3
Malvidin	OCH_3	OCH_3

Figure 2 Basic structure of six anthocyanidins and substitutions (R).

The hue of anthocyanins varies according to different substitution groups, and colour saturation increases with the number of hydroxyl groups and decreases with the addition of methoxyl groups. Anthocyanins are highly soluble in water, where four equilibrium forms co-exist. Under highly acidic conditions, the flavilium ion is present and relatively stable, showing a strong red colour. Under weakly acidic or neutral conditions, the red anthocyanin is converted into a highly unstable pseudo-base, the carbinol, which can be converted into the chalcone. At alkaline pH, the red flavilium cation is converted into the blue quinoidal base.

Anthocyanins can act as antioxidants thanks to their positively charged oxygen atom, which makes the structure a potent hydrogen-donating molecule, more than other flavonoids or general phenolics. Anthocyanins can act as antioxidants not only as hydrogen donors, but also as metal chelator and protein-binding compounds (Kong et al., 2003). Reduction of lipid peroxyl radicals and inhibition of lipid peroxidation have also been evaluated (Tsuda et al., 1996).

Anthocyanins have been reported to bind to DNA, forming anthocyanin-DNA complex which can protect DNA from oxidative damage after exposure to free radicals (Fig. 3) (Sarma and Sharma, 1999; Wang et al., 2011). Mas et al. (2000) suggested that anthocyanins have the ability to stabilize the DNA

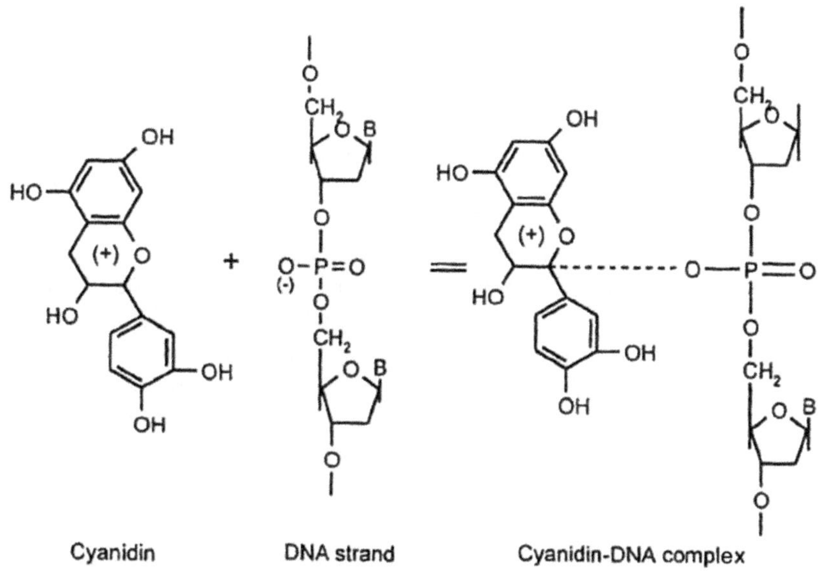

Figure 3 Proposed mechanism for cyanidin-DNA interaction which leads to the formation of cyanidin-DNA co-pigmentation complex. From Kong et al. (2003), by kind permission of Professor Jinming Kong.

triple-helical structure. This feature may have *in vivo* physiological functions, and may contain the basis of the possible defence mechanism of anthocyanins against oxidative damage.

2.3 Occurrence and role of polyphenols in fruits

Due to the increasing interest in nutraceuticals and functional foods, plant breeders have initiated selection of fruit crops with higher than normal phenolic antioxidant contents. Breeding programmes aimed at developing red-fleshed peaches [*Prunus persica* (L.) Batsch], plums (*Prunus salicina* Erhr. and hybrids) and red-peel apricots (*Prunus armeniaca* L.) with high levels of beneficial phenolic compounds for the fresh produce and processing market have been initiated in different US and EU university and research stations (Cevallos-Casals et al., 2006; Bureau et al., 2009). In pear [*Pyrus communis* L. or *Pyrus pyrifolia* (Burm.f.) Nakai], red skin colour has become a quality marker for cultivar breeding programmes, releasing new red-skinned pear varieties (Steyn et al., 2005; Feng et al., 2010 and references therein).

The different classes of phenolics which can be found mostly in fruits are discussed below.

2.3.1 Phenolic acids

Phenolic acids are the simplest phenolic compounds found in fruits. Apple (*Malus domestica* Borkh), grape (*Vitis vinifera* L.), kiwifruit (*Actinidia deliciosa* C.F. Liang & A.R. Ferguson), pears and plums are important sources of phenolic acids (hydroxycinnamic and chlorogenic acids). Red fruits are also relevant sources of gallic, *p*-hydroxybenzoic and protocatechuic acids, while caffeic acid is the predominant phenolic acid (75–100% of the total hydroxycinnamic acid) in most fruits (Nollet and Gutierrez-Uribe, 2018 and references therein). Phenolic monomers vary by species and account for 5–9% of the *Rubus* sp. phenolic composition consisting of phenolic acids (hydroxycinnamic and hydroxybenzoic acids), flavonol (catechin and epicatechin) and flavonol-glycosides (quercetin- and kaempferol-glycosides). Ellagic, chlorogenic and gallic acids are abundant phenolic acids in berries; blueberry (*Vaccinium corymbosum* L. and *Vaccinium angustifolium* Aiton) contains up to 2 g/kg (fresh weight) of chlorogenic acid, whereas ellagic acid accounts for about 50% of total phenolic compounds in cranberries (*Vaccinium macrocarpon* Ait.) and raspberries (*Rubus idaeus* L.) (Olas, 2018 and references therein). Generally speaking, the phenolic concentration depends on fruit types; for example, chokeberry (*Aronia melanocarpa* Michx.) is high in phenolics (2080 mg/100 g fruits) compared to blueberries and blackberries (*Rubus* sp.) (525 and 248 mg/100 g fruits), respectively (Olas, 2018 and references therein).

2.3.2 Flavonoids

Flavonoids (anthocyanins, flavonols, flavanols, flavones and isoflavone) are a large category of phenolic compounds, highly present in fruits. Anthocyanins are particularly abundant in red fruits: blackberry, blueberry, chokeberry, strawberry (*Fragaria × ananassa* Duch.), cherry (*Prunus avium* L. and *Prunus cerasus* L.), red grape, plum and blood orange cultivars (cvs), but are also present in some peach and nectarine cvs, some apricot cvs of recent constitution, some apple and pear cvs (Hakkinen et al., 1999; Josuttis et al., 2012; Gao and Mazza, 1995; Cantos et al., 2002; Zhu et al., 2012; Usenik et al., 2009; Lo Piero, 2015; Tomás-Barberán et al., 2001; Ruiz et al., 2005b; Mazza and Velioglu, 1992; Dussi et al., 1995). Anthocyanin accumulation during development differs among domesticated cultivars of apple, pear, peach, plum and apricot. Fruit varieties vary widely in regard to the content of phenolic (particularly anthocyanin) compounds (e.g. blond vs. blood oranges). This could be due to the genetic differences of the evaluated samples since modern fruit varieties have been obtained from breeding programmes that used parents coming from different genetic origins. As a general rule, the peel tissues contain higher amounts of phenolics, with anthocyanins and flavonols almost exclusively confined in this tissue.

2.3.3 Tannins

Ellagitannins or hydrolysable tannins are high molecular weight phenolic compounds and, as well as proanthocyanidins, are able to precipitate proteins and alkaloids. They are particularly present in red fruits such as strawberries, raspberries and blackberries. The main phenolic polymer comprises 55–71% of *Rubus* phenolic composition (Lee et al., 2012 and references therein). Proanthocyanidins or condensed tannins are oligomeric or polymeric end products of the flavonoids biosynthetic pathway and are the second most abundant natural phenolic after lignin. They are present in apples, pears, apricots, red berries, grapes, cherries and peaches (Nollet and Gutierrez-Uribe, 2018 and references therein).

Fruit phenolics are generally involved in the defence against ultraviolet radiation and pathogens (Olas, 2018). The photoprotective function of anthocyanin accumulation (induced by low temperature) in apple and pear peel has been demonstrated (Steyn et al., 2002, 2009). Resveratrol in grape is produced in response to UV radiation or microbial infection; its production decreases naturally during fruit ripening (Nollet and Gutierrez-Uribe, 2018 and references therein). This may explain the beneficial effect of using post-harvest UV-B radiation as a valid eco-friendly strategy to obtain phenolic-enriched apple fruit. UV-B treatment (219 kJ/m^2; 36 h) prior to storage (7, 14 and 21 days) increased and decreased hydroxycinnamic acids (38%) and flavonols (−45%), respectively, in apple skin. However, hydroxycinnamic acids, flavonols and anthocyanins increased in UV-B treated apples at the end of the storage period. The treatment presumably scavenged reactive oxygen species (ROS) and free radical species, thereby reducing total phenols and flavonoids (Assumpção et al., 2018). Fruit peel or skin acts as the primary defence against pathogen penetration, thereby inducing polyphenol synthesis. Thus, apple, cherry, nectarine and peach peels contain 1.6–4.5-fold more phenolic compounds than the pulp, while mango and blueberry present even higher peel/pulp ratios (32.5 and 14 folds, respectively) (Nollet and Gutierrez-Uribe, 2018 and references therein). Total phenolics and individual phenolics such as hydroxycinnamic acids (caffeic, chlorogenic and 4-*p*-coumaroylquinic acids) and flavonols (quercetin 3-glycosides) were higher in apple peel than in pulp (Jakopic et al., 2012).

Protection against (1) early feeding on fruits and (2) fungal pathogens in developing fruit tissues is provided by the presence and abundance of tannins in immature fruits. Furthermore, tannin concentration decreases during fruit ripening since the *Fahyprp* (hybrid proline-rich protein) gene related to the ripening process may be involved in polyphenols (lignin and condensed tannin) anchoring to the cell membranes (Blanco-Portales et al., 2004). Phenolic acid content has been proposed as a differentiating parameter for scab (*Venturia inaequalis*) resistance in apple cultivars due to their inherent variation observed in moderately and highly susceptible cultivars. 'Golden Delicious' (rich in phenolic acids) had less scab incidence than the highly susceptible 'Red Delicious' (90–95% disease intensity) with its low phenolic acids (tannic, caffeic, ferulic, benzoic and salicylic acids) (Singh et al., 2015). Furthermore, scab-infected 'Braeburn' apple peel contained almost 4-fold more flavan-3-ols than healthy peel (Slatnar et al., 2010).

3 Carotenoids

3.1 Carotenoid chemistry and biosynthesis

Carotenoids are the most widespread class of isoprenoid pigments synthesized by plants, algae and photosynthetic bacteria, and are responsible for the yellow, orange and red colours in various fruits and vegetables (Namitha and Negi, 2010). Carotenoids include about 600 isoprenoid compounds containing up to 15 conjugated double bonds divided into the two major groups of carotenes and xanthophylls. Carotenes, such as α-carotene, β-carotene and lycopene, are hydrocarbons that are either linear or cyclized at one or both ends of the molecule. Xanthophylls, such as β-cryptoxanthin, lutein and zeaxanthin, are the oxygenated derivatives of carotenes (Rodriguez-Amaya, 2001). The presence of a polar group in the structure (epoxy, hydroxyl and keto) alters the polarity of carotenoids and affects biological functions (Britton, 2008).

The carotenoid biosynthesis pathway is well established. The first committed step is the head-to-head condensation of two geranylgeranyl pyrophosphate (GGPP) molecules to form colourless phytoene catalysed by phytoene synthase (PSY). This is a major rate-limiting step of carotenoid biosynthesis (Cazzonelli and Pogson, 2010). The red carotenoid lycopene is synthesized by the action of phytoene desaturase (PDS), ζ-carotene isomerase (ZISO), ζ-carotene desaturase (ZDS) and carotene isomerase (CRTISO) (Naik et al., 2003). Lycopene is subsequently cyclized to synthesize α-carotene and ß-carotene, and xanthophylls are formed after a series of oxygenation reactions. After lycopene synthesis, the pathway of carotenoid biosynthesis has two main branches. In one branch, lycopene ß-cyclase (LCYb) introduces two ß-rings to lycopene to form ß-carotene. In the other branch, lycopene ε-cyclase (LCYe) and LCYb introduce one ß- and one ε-ionone ring into lycopene to form α-carotene. α-Carotene conversion to lutein occurs by sequential hydroxylation catalysed by ε-ring hydroxylase (HYe) and ß-ring hydroxylase (HYb). ß-Carotene is converted to zeaxanthin via ß-cryptoxanthin by two hydroxylation catalysed by Hyb. Zeaxanthin is converted to violaxanthin *via* antheraxanthin by zeaxanthin epoxidase (ZEP). Violaxanthin is converted into neoxanthin by the action of neoxanthin synthase (NSY).

Carotenoids are fat-soluble micronutrients that cannot be synthesized within the body and must be supplemented through daily intake to exert beneficial effects on human health and well-being. About 50 carotenoid structures are present in a typical human diet, but only 20 of them have been identified in human blood and tissues; among these, the most represented are β-carotene and α-carotene, lycopene and cryptoxanthin (Milani et al., 2017).

Carotenoids are well-known for their health benefits because of their pro-vitamin A activity. Of all the compounds, α-carotene, β-carotene and β-cryptoxanthin are the main precursors of vitamin A synthesis (Haskell, 2013). Apart from this, carotenoids are also important for antioxidant activity, intercellular communication and immune system activity (Skibsted, 2012; Stephensen, 2013). Studies indicate that consumption of carotenoid-rich fruits and vegetables may help prevent lung, head, neck and prostate cancer, cardiovascular diseases, age-related macular degeneration and cataract formation (Reboul et al., 2006; Meyers et al., 2014; Sharoni et al., 2012). Recently, lycopene- and carotene-rich oleoresins were reported to trigger significant delay of Caco-2, SAOs and NSCLC A549 malignant human cell growth (Bruno et al., 2017; Russo et al., 2017).

3.2 Occurrence and role of carotenoids in fruits

Content and types of plant carotenoids depend on several factors such as genotype, degree of maturation, cultivation method and climatic conditions.

Carotenoids contribute to fruit exocarp (peel) and mesocarp (flesh) colouration. The exocarp is generally the region of highest carotenoids concentration and this is marked in citrus and apple fruits. The qualitative and quantitative distributions of carotenoid compounds vary between peel and flesh in many fruits. Mature oranges accumulate xanthophylls, which are responsible for their orange-yellow colour, while mandarin fruit are characterized by ß-cryptoxanthin and ß-citraurin that contribute to their intense orange colour. Mature white grapefruits accumulate low amount of carotenoids, mainly phytoene and violaxanthin, while the distinctive colour of red grapefruits is due to lycopene, an unusual carotene in citrus (Alquézar et al., 2013). Furthermore, in the red grapefruits, the carotenoid amount is similar in the peel and the flesh (Rodrigo et al., 2013; Alquézar et al., 2013).

Ampomah-Dwamena et al. (2012) observed that the carotenoid content was higher in peel than in flesh of commercial apple cultivars such as Royal Gala and Granny Smith, whereas in Aotea, a non-commercial cultivar, the carotenoids content was similar in the peel and flesh of the fruit. The study showed that yellow-fleshed apple genotypes accumulated more carotenoids than the white-fleshed genotypes. Carotenoid composition also differed in yellow-, red- and green-skinned commercial apple varieties (Delgado-Pelayo et al., 2014). In that study, 'Granny Smith' and 'Fuji' from France had the highest and lowest carotenoid content in the peel (151.7 and 17.1 µg/g dry weight, DW, respectively). In most apple cultivars, a characteristic carotenoid profile was observed in the peel, where lutein was the most abundant compound followed by violaxanthin, neoxanthin and ß-carotene. Nevertheless, apple flesh was characterized by high ß-carotene followed by neoxanthin and violaxanthin values. In another study, ß-carotene was the most representative compound in apples from Portugal (Dias et al., 2009), with carotenoid levels varying with species, varieties, harvest time and production site. A wide difference was reported in ß-carotene content among apple fruits of different cultivars (Tuscan, Royal Gala and Fuji Beni Shogun) and strains of Fuji (Fuji Yanfu 3, Fuji Yanfu 6 and Fuji Beni Shogun) (Jia et al., 2011). ß-carotene and total carotenoid content of the peel was higher than that of the flesh, and Fuji Beni Shogun had the highest pigment content in both the peel and the flesh. Thus, it is possible to produce higher ß-carotene content apples by selecting genotype (Jia et al., 2011).

Apricot peel generally contains higher carotenoid content than the flesh. Fruit flesh hue angle correlated with fruit flesh colour and carotenoid content of 37 apricot varieties from different genetic origins (Ruiz et al., 2005a). In that study, the total carotenoid content varied from 13.6 µg/g fresh weight (FW) in the flesh of a white variety to 38.5 µg/g FW in the peel of an orange flesh apricot variety. Furthermore, ß-carotene was the major carotenoid in apricots, followed by ß-cryptoxanthin and γ-carotene. The ß-carotene ranged from 48%

to 88% in a white and orange flesh apricot. ß-cryptoxanthin ranged from 5% in an orange flesh apricot to 28% in a yellow flesh apricot. γ-carotene varied from 5% in an orange variety to 26% in a white flesh one. High levels of ß-carotene were reported in apricot by Sass-Kiss et al. (2005) and Leong and Oey (2012).

Sweet cherry fruit carotenoids consist of ß-carotene, ß-cryptoxanthin, lycopene, lutein, zeaxanthin and α-carotene (Leong and Oey, 2012; Demir, 2013). Di Matteo et al. (2016) evaluated bioactive compounds such as carotenoids from 25 autochthonous sweet cherry cultivars, selecting Mulignana and Pagliarella cultivars with a higher concentrations of carotenoids (374 and 400 μg/100 g FW, respectively).

The main carotenoids identified in peaches and nectarines were ß-carotene and ß-cryptoxanthin, with higher ß-carotene abundance than ß-cryptoxanthin, and higher total carotenoid compound in the peel of fruits compared to the flesh. As expected, yellow cultivars contained more carotenoids than white flesh cultivars; carotenoids content in yellow- and white-fleshed nectarines were 6–8 and 5–14 μg/100 g FW, respectively, and 7–20 and 71–210 μg/100 g FW in yellow- and white-fleshed peaches, respectively (Gill et al., 2002).

The carotenoid profiles of various fruit cultivars have generally been characterized to provide useful information in selective breeding programmes directed to improve their nutritional value, as well as to increase our knowledge of the carotenoids biosynthesis and accumulation in fruits. Generally, the total carotenoid content increases during ripening: carotenogenesis will take over while the chlorophylls commonly are degraded, thereby a greater amount of carotenoid compounds are synthesized in the chromoplasts than the chloroplast (Gross, 1991). Many studies have focused on carotenogenesis during development and/or amongst cultivars. Ampomah-Dwamena et al. (2012) observed changes of carotenoids accumulation pattern in different apple genotypes during fruit development. Metabolic and gene expression analysis of the carotenoid pathway suggested that expression of several rate-limiting carotenogenic steps regulate carotenoid accumulation during fruit ripening. These studies also indicated that transcriptional regulation of the phytoene gene appears to be the major regulatory mechanism responsible for determining the carotenoid pools in chromoplasts.

Carotenoid concentration and composition in citrus fruit vary greatly during maturation, and many studies have been carried out on the carotenoids biosynthesis pathway to understand regulation of colour and carotenoid composition (Kato et al., 2004; Kato, 2012; Rodrigo et al., 2013). ß-cryptoxanthin content is low in unripe citrus and it increases progressively during ripening (Kim et al., 2001). Similarly, the amount of total carotenoids and ß-cryptoxanthin was positively associated during ripening (Yoo and Moon, 2016), indicating that ß-cryptoxanthin was the major carotenoid. The content of ß-cryptoxanthin, zeaxanthin and ß-carotene increased with maturation in three different citrus cultivars grown in Korea (Yoo and Moon, 2016), while the amount of canthaxanthin, astaxanthin, lycopene and α-carotene decreased with maturation. During fruit development, total carotenoid increased three-fold in the flavedo and 10-fold in the juice sac.

The increase in carotenoid content correlates with changes in the expression of genes of the total carotenoid biosynthetic pathway (Kato et al., 2004; Rodrigo et al., 2004). The flavedo colour changed from green to orange during maturation. In the green stage, *LCYe* gene is highly expressed, resulting in the predominant accumulation of α-carotene and lutein (ß, ε carotenoids). During the transition of peel colour from green to orange, expression simultaneously increased in genes participating in the synthesis of ß-cryptoxanthin, zeaxanthin, all-*trans* violaxanthin and 9-*cis* violaxanthin (ß,ß-xanthophylls). In the juice sacs of the green fruit, pathway change was earlier than in the flavedo because *LCYe* transcripts were undetected. As fruit maturation progressed, a simultaneous increase in the expression of *PSY*, *PDS*, *ZDS*, *LCYb* and *ZEP* led to massive ß,ß-xanthophyll accumulation in both the flavedo and juice sacs. The results are in agreement with those reported by Fanciullino et al. (2008) for Shamouti (normal orange colour) and Sanguinelli ('blood cultivar' purple colour) cultivars, that are characterized by the highest percentages of ß,ß-xanthophylls. Thus, the carotenoid accumulation during citrus fruit maturation is due to changes in regulating carotenoid biosynthetic genes at the transcriptional level. In the flavedo of Navel oranges during fruit development and maturation, the shift from ß, ε carotenoids to the ß,ß xanthophylls was explained by the downregulation of *LCYe* and upregulation of *LCYb* and *HYb* (Rodrigo et al., 2004). However, Marty et al. (2005) suggested post-transcriptional regulation, end products feedback and ethylene regulation as carotenoid accumulation mechanisms in apricot fruit. Therefore, the identification of the rate-limiting steps of carotenogenesis has paved the way for breeding programmes aimed at increasing carotenoids concentration.

4 Vitamin C

4.1 Vitamin C chemistry and biosynthesis

Vitamin C is ubiquitous in all plants, but is not synthesized in humans due to gene mutations for gulonolactone oxidase, the terminal enzyme of the ascorbic acid synthesis pathway. Chemically, it is a reducing agent, donating electrons that account for all of its known physiological effects. Electrons from ascorbate can reduce metals (copper and iron) resulting in superoxide and hydrogen peroxide formation, subsequently generating reactive oxidant species (Padayatty and Levine, 2016). Vitamin C is chemically the simplest and, as such, the most industrially produced of all the vitamins. Despite being a simple molecule, its enediol structure provides it with a highly complex chemistry. For example, its redox chemistry involves comparatively stable radical intermediates that can be modified by the acidic properties of the molecule. L-ascorbic acid can act as both a redox and a complexing agent. It is the major precursor of tartaric acid in grapes as well as oxalic acid in rhubarb (*Rheum rhabarbarum*) (Davies et al., 1991).

Vitamin C is an essential, water-soluble micronutrient that exists predominantly as the ascorbate anion under physiological pH conditions. The average ascorbate concentration in the plasma of healthy humans is ~50 μM and concentrations below 11.4 μM are considered deficient and are associated with a risk of developing scurvy. The current recommended dietary allowances are 90 mg and 75 mg ascorbate for adult males and females, respectively; the daily adequate intake for infants is ~40–50 mg. Systemic ascorbate deficiency is caused predominantly by inadequate dietary intake. Over 7% of the US population is estimated to be deficient (plasma concentration < 11.4 μM) in ascorbate, according to a 2003–2004 survey (Young et al., 2015 and references therein). Vitamin C, similar to vitamin A, is considered to be a shortfall nutrient, that is defined as being under-consumed relative to the estimated average requirement or adequate intake through biochemical nutritional status indicators. These nutrient inadequacies may be associated directly with a specific health condition. In fact, ascorbate, the dominant form of vitamin C, regulates epigenomic processes demethylating the genome, presumably by mediating the interface between the genome and environment, highlighting its potential roles in various diseases (Young et al., 2015). Therefore, consumption of vitamin A- and C-rich foods, particularly fruits and vegetables, is highly recommended.

Vitamin C is almost exclusively (>90%) supplied by fruits and vegetables in the human diet. Ascorbic acid (AsA) is the predominant biologically active form of, and the main contributor to, the total vitamin C content (>85%), except for nopal (*Opuntia ficus-indica*) and avocado (*Persea americana*) (26% and 58%, respectively) (Corral-Aguayo et al., 2008). The oxidized L-dehydroascorbic acid (DHA) form contributes less than 10% of the total vitamin C (AsA + DHA) in most horticultural crops and generally increases during storage.

In higher plants, ascorbate biosynthesis and metabolism is complex and affected by several factors; changes in ascorbate status alter the signalling pathway and plant response to various stresses. AsA accumulation is not dependent on a single master regulator, which increases the challenge in vitamin C bio-fortification. AsA accumulates in fruits from glucose, predominantly through the L-galactose pathway mediated by two genes, *GDP-D-mannose pyrophosphorylase* (*VTC1* or *GMP*) and *GDP-D-mannose-3,5-epimerase* (*GME*) (Mellidou and Kanellis, 2017). *GMP* (coding for a limiting enzyme in the Smirnoff-Wheeler's pathway for AsA biosynthesis) expression is correlated with AsA concentrations in acerola (*Malpighia emarginata*) (one of the highest vitamin C containing fruit), but not in kiwifruit (*Actinidia deliciosa*) or blueberry. Apparently, other genes involved in AsA biosynthesis (*L-galactose-1-phosphate phosphatase*, *L-galactono-1,4-lactone dehydrogenase* and *D-galacturonic reductase*) as well as high *GMP* expression may be responsible for the elevated AsA in acerola (Badejo et al., 2008). However, reduction in vitamin C content of acerola due to environmental (location) effect and/or maturity stage has been ascribed to ascorbic acid oxidase enzyme, since its activity is higher in mature than immature fruits. Similarly, repressed ascorbate oxidase activity has been associated with elevated AsA content in mature carambola (*Averrhoa carambola*) fruit (Zainudin et al., 2014). Interestingly, lower AsA acerola cultivars 'Florida Sweet' and 'Will#2' have been developed for the fresh market and juice processing since the wild varieties are considerably too tart due to high vitamin C levels (Delva, 2012). Although *GME* is not the rate-limiting step in AsA biosynthesis, its transcripts are correlated with AsA in apple and blueberry (Mellidou and Kanellis, 2017 and references therein). *GDP-L-galactose phosphorylase* (*GGP*) also regulates AsA biosynthesis (Smirnoff-Wheeler pathway) in citrus, blueberry, strawberry and kiwifruit, where *GGP* expression correlated significantly with changes in AsA levels between high- and low-AsA species. *GGP* acts synergistically with *GME*, and both genes are regulated diurnally by light (at least in kiwifruit).

In apple, grape, orange and strawberry, AsA can also accumulate through the alternate D-galacturonate reductase (GalUR) pathway (Mellidou and Kanellis, 2017 and references therein). *GalUR* expression correlates positively with AsA content in strawberry (Agius et al., 2003).

Dehydroascorbate reductase (*DHAR*) regulates AsA redox state, and its expression along with *monodehydroascorbate reductase* (*MDHAR*) correlate with AsA accumulation in blueberry. *MDHAR* silencing can increase AsA accumulation in fruits since its overexpression and silencing reduced and significantly increased AsA accumulation, respectively, in tomato (*Solanum lycopersicum*) (Gest et al., 2013). The genes/factors modulating the higher AsA levels of wild accessions, compared to modern cultivars (up to 10-fold in kiwifruit), have not been completely elucidated to date and need to be investigated. The most successful attempt to engineer elevated AsA has involved silencing of *malate dehydrogenase* (*MDH*) or overexpression of *DHAR* that elevated AsA content ~6-fold in tomato and corn (*Zea mays*), respectively (Goggin et al., 2010 and references therein).

4.2 Occurrence and role of vitamin C in fruits

Several factors influence vitamin C content in fruits including genotype, environment, cultural practices, maturity and harvesting methods, and post-harvest procedures, particularly storage conditions (Lee and Kader, 2000). AsA concentration is highly dependent on fruit species with persimmon (*Diospyros kaki*), strawberry, kiwifruit and citrus fruit considered as excellent sources. However, variations are wide within cultivars of the same species. For example, AsA content varies 29–80 mg and 14–103 mg/100 g FW in kiwifruit and berry fruits, respectively (Kanellis and Manganaris, 2014). Vitamin C contents of pulp extracts of 21 citrus fruits (orange, satsuma, clementine, mandarins, tangor, bergamot, lemon, tangelos, kumquat, calamondin and pamplemousse) commonly grown in Mauritius ranged from 166 to 677 μg/mL (Ramful et al., 2011).

Genotypic variation in vitamin C content is well documented in several fruits. Wide vitamin C range (2.5–21.4 mg/100 g FW) occurs in sour cherry that on average contains 12 mg of vitamin C/100 g FW. However, vitamin C variation was low (3–9 mg/100 g FW) in fruits of 11 Hungarian sour cherry genotypes and in three dominant, commercially grown, well-fertilized Polish sour cherry genotypes hand-picked at maturity (17.7–21.1 mg/100 g FW). Furthermore, fruits of the most and least productive Polish cultivars displayed the highest and least vitamin C contents, respectively (Borowy et al., 2018 and references therein). In kiwifruit, vitamin C content varies between and within species: 50–450 mg and 20–250 mg/100 g for *Actinidia chinensis* and *A. deliciosa*, respectively (Huang et al., 2004). However, vitamin C contents of 'SunGold' (*A. chinensis* var. *chinensis* 'Zesy003') and 'Sweet Green' (*A. chinensis* var. *chinensis* × *A. chinensis* var. *deliciosa* 'Zesh004') (161 and 150 mg/100 g edible flesh, respectively) were higher than those of the common green variety 'Hayward' (*A. deliciosa*) (85 mg/100 g) (Sivakumaran et al., 2018). Russian sea buckthorn cultivars (*Hipophea rhamnoides* subsp. *mongolica*) grown in Poland and high in AsA content (53–131 mg/100 g FW) may range from 0.5 to 3.3 g/kg, differing from those of subsp. *turkistanica* (2.5–4.2 g/kg) (Teleszko et al., 2015). Total vitamin C (AsA + DHA) content of 31 apple cultivars varied 3.6-fold (401–1448 nmol/g FW) grown at the same location in Belgium; this indicates the limited genetic potential for breeding for increased fruit L-AA contents (Davey and Keulemans, 2004). Furthermore, a genotype-environment interaction was also important in vitamin C concentration (320 vs. 75 μg/g) indicating that breeding lines should be evaluated in multiple locations and seasons (Patil et al., 2014).

AsA content varied (19.3–71.5 mg/100 g FW) in six strawberry cultivars from four locations (Lee and Kader, 2000) and was not affected by altitude when grown in Italy under the same cultural and environmental practices (Doumett et al., 2011). The climate can considerably affect AsA content of fruits. For example, the subtropical climate, along with soil composition, accounted for the higher (2- to 7-fold for strawberry and blueberries, respectively) AsA content of berries grown in Brazil compared to temperate production zones. Furthermore, blackberry (*Rubus* spp.) plants produce large quantities of fruit in subtropical areas, with some varieties producing higher amounts of vitamin C compared to those in temperate zones (De Souza et al., 2014). Similarly, native wild Chinese sea buckthorn berries (*Hipophea rhamnoides* subsp. *sinensis*) contain 5–10 times more vitamin C in the juice fraction than those of European and Russian subspecies (*rhamnoides* and *mongolica*, respectively) (Kallio et al., 2002).

Growing season also can affect AsA content. For example, Argentinian-grown oranges harvested in 2011 had significantly higher AsA content (55 vs. 49 mg/100 mL) than those harvested in the preceding year (2010) notable for its high precipitation (Stinco et al., 2015). However, ascorbate concentration in six highbush blueberry cultivars differed less than 15% between two growing seasons in Poland, except for 'Duke' with ~85% drop between seasons (Łata and Wińska-Krysiak, 2010). Pulp vitamin C content did not differ significantly in similar citrus varieties harvested at different periods except for tangor, tangelo, kumquat and pamplemousse (Ramful et al., 2011).

Developmental stages influence AsA content such that it increases from the early stages of development to full maturity in a cultivar-dependent manner in strawberries grown in the same environment in Brazil. However, the final total AsA content (40–85 mg/100 g) at maturity was not related to pattern of accumulation in the tissues during fruit development (Cordenunsi et al., 2002). Furthermore, AsA and anthocyanin contents in strawberry are inversely associated such that these components are presumed to be complementary, probably as antioxidant agents protecting cells against oxidative damage (Cordenunsi et al., 2002). In pomegranate (*Punica granatum*), AsA decreases with fruit maturation and ripening (Erkan and Kader, 2011).

AsA is one of the major non-enzymatic antioxidants in plants, involved in both enzymatic and non-enzymatic quenching of ROS, particularly in the scavenging of hydrogen peroxide (Suza et al., 2010). Most fruits with high AsA content (strawberries, bananas [*Musa* spp.], lemons and oranges) also exhibited high antioxidant capacity (Kevers et al., 2007). It correlates highly with free radical scavenging activity of commercial orange juices and antioxidant activity (TEAC, $r = 0.73$) of fresh orange juice. Vitamin C in hydrophilic extracts of tropical fruits (avocado, black sapote, guava, mango, papaya, prickly pear and strawberry) correlated highly ($r \geq 0.88$) with antioxidant activity evaluated using the DPPH, DMPD, FRAP and TEAC methods (Corral-Aguayo et al., 2008). Vitamin C contributes 65–100% of antioxidant activity of beverages derived from citrus fruits (Ramful et al., 2011 and references therein) and only 0.35–8.6% for eleven fruits (cranberry, apple, red grape, strawberry, pineapple, banana, peach, lemon, orange, pear and grapefruit) (Sun et al., 2002). AsA is poorly or not correlated with antioxidant activity of many fruits (nectarines, peaches, plums) and exotic tropical fruits; it is inversely ($r = -0.80$) associated with antioxidant activity for strawberries, raspberries and blueberries (Ramful et al., 2011 and references therein).

5 Production practices that influence bioactive compound synthesis

Genetic characteristics of cultivars strongly determine their BC contents. Landraces and wild unexplored genotypes often accumulate high BCs that may be valuable for breeding purposes. However, fruits with high secondary metabolites (e.g. total phenolics) may confer bitter and astringent tastes (Manganaris et al., 2018 and references therein). Total polyphenolic content of wild apple germplasm varies considerably and is generally higher than that of commercial cultivars, similar to the trend for AsA. Old apple cultivars also contain higher polyphenolic and vitamin C contents of flesh and antioxidant capacity relative to new cultivars (Kschonsek et al., 2018). The lower phenolic and flavonol contents of new apple cultivars are presumed to result from breeding for their consumer-preferred sweet flavour and low enzymatic browning. The higher average content of total polyphenols, particularly flavonols, of old apple cultivars may also be due to species diversity (Kschonsek et al., 2018).

Breeding strategies to enhance polyphenolic content can also improve quality traits (colour, aroma, disease resistance and browning disorders) related to the phenylpropanoid pathway (Manganaris et al., 2018). Breeding strategies should focus on the greater within-family, rather than among family, variation in phenolics, including assessments of genotypic performance in several locations and years to differentiate environmental effects. The phenolic compositions among apple cultivars are compounded by differences in growing regions, highlighting that certain combinations of cultivar and growing region can result in significantly higher levels of phenolic compounds (Brovelli, 2006 and references therein). In strawberry, the role of genetic background (different cultivars) has been demonstrated in relation to their antioxidant activities (Scalzo et al., 2005).

Classical genetic breeding is time-consuming and the release of new cultivars with desired characteristic takes many years. The genetic background of the plant is the main factor determining the quality and content of phenolic compounds in plant tissues. Even if the genetic approach relies on the variability existing in the genetic pool, and therefore is more potent to create high BC content cultivars, the development of agronomic strategies to enhance the production of useful metabolites provides an opportunity to optimize the currently available cultivars. The agronomic approach can be valuable to achieve a moderate (~two-fold) increase by changing only the crop growing conditions (Poiroux-Gonord et al., 2010).

5.1 Polyphenols: environmental variation and agronomic conditions

Over the past few years, many of the genes coding for enzymes involved in the phenylpropanoid pathway have been identified, cloned and characterized. Scientists have been able to dissect the regulatory mechanisms of the phenylpropanoid pathway, increasing our understanding of the abiotic and biotic factors governing polyphenols accumulation. The information can now be applied to enhance or optimize the production of health benefitting polyphenols in plants. Apart from the transgenic approach (which is still problematic for consumer acceptance), the phenylpropanoid pathway can also be manipulated through agronomic and physiological factors involving exposure to specific light regimes, nutrient supply, microbe-host plant interaction, elicitor applications, cultural management and water stress imposition (Dorais and Ehret, 2008; Jaganath and Zainal, 2016). In the following paragraphs, we highlight the environmental factors that can contribute to the modulation of polyphenols, particularly anthocyanins, in fruits.

5.1.1 Effect of light and temperature on the polyphenols (anthocyanin) content

One of the factors that dramatically influence polyphenols (particularly anthocyanins) content is light exposure, both photoperiodic as well as quality and intensity. Fruit bagging and shading experiments have demonstrated the importance of light conditions on the biosynthesis and accumulation of anthocyanins (Zoratti et al., 2014a). The anthocyanin function in vegetative (photosynthetic) tissues has been proposed as being photoprotective, since it has been shown that anthocyanins reduce photoinhibition and photobleaching of chlorophyll under high light stress conditions (Steyn et al., 2002). Similarly, anthocyanins afford photoprotection to fruit peel (apple and pear) during low-temperature-induced light stress and the protection is not a fortuitous side effect of light absorption by anthocyanin (Steyn et al., 2009).

The regulation of flavonoid biosynthesis in fruit crops has been studied extensively and reviewed during the last 10 years, leading to knowledge of the key transcription factors that coordinate and regulate flavonoid structural genes (Jaakola, 2013). One of these transcription factors (R2R3 MYB) has been found to be inducible with specific light wavelengths, thus leading to distinct flavonoid biosynthesis in apple, pear, nectarine, Chinese bayberry (*Myrica rubra*), strawberry, litchi (*Litchi chinensis*) and grapevine (Kim, 2003; Zoratti et al., 2014a). In grape berries, the anthocyanin content was significantly enhanced by blue light filter film and suppressed by red, orange and green films compared to white film (control) (Cheng et al., 2015). High light intensity upregulates the expression of an array of both early and late flavonoid biosynthesis genes in the grape berry skin, leading to an increased content of anthocyanins and general flavonoids (Zoratti et al., 2014a). Moreover, following the illumination with blue, red, far-red or white light during the berry ripening process in bilberry, it has been shown that all the three monochromatic lights influenced the accumulation of total anthocyanins in the mature fruits (Zoratti et al., 2014b). Therefore, in controlled production conditions (e.g. greenhouses, plastic covers, high tunnels), it is crucial to use the correct spectrum of light if an increase of anthocyanin is desired (see Fig. 4).

Flavonoid biosynthesis is influenced by light in a species- and cultivar-specific manner. It is possible that white-fleshed apples have lost the ability to produce pigmentation throughout the fruit due to a mutation in one or more biosynthetic or regulatory gene(s), as has been observed for white grapes (Boss et al., 1996). Not all species require strong light exposure to accumulate anthocyanins. Bilberry (*Vaccinium myrtillus*), one of the best sources of anthocyanins, grows and produces better in shaded light conditions. Light is recognized as an important regulation factor in numerous crops belonging to the *Rosaceae* family, including strawberry, peach, apple and pear (Esplay et al., 2007; Feng et al, 2010; Zoratti et al., 2014a).

Another environmental factor that can significantly influence anthocyanin accumulation in fruit crops is temperature. Low night temperatures induce anthocyanin synthesis in grape berries, with higher *UFGT* gene expression at low temperature (Mori et al., 2005), while higher temperatures inhibit anthocyanin

Figure 4 'Robada' apricot grown under a 'Luminence THB' polyethylene-covered high tunnel (left) and in the open orchard (right) in Michigan, USA. Photos by G. Lang.

accumulation (inhibiting PAL activity) or promote their degradation (De Pascual-Teresa and Sanchez-Ballesta, 2008). Studies in apple and pear support the view that anthocyanins are adaptable light screens deployed to modulate light absorption in sensitive tissues such as fruit peel in response to environmental triggers such as cold fronts (Steyn et al., 2002). Blood orange varieties (Tarocco, Moro and Sanguinello) require a wide day–night thermal range to induce the biosynthesis of anthocyanins and maximize colour formation. Production of red oranges characterized by high anthocyanin levels is limited to areas with characteristic climate conditions, such as the area around Mount Etna in Sicily (Italy), where high colour intensity and quality blood orange fruits are produced (Lo Piero, 2015). Cooling by micro-sprinkler irrigation has been found to increase red colour and anthocyanin content in apple (Iglesias et al., 2005).

5.1.2 Effect of cultural practices on the polyphenols content

Many cultural practices influence fruit polyphenolic content. Moderate pruning, cluster thinning or regulated irrigation significantly increased phenolic content of grapes (Ruiz-García and Gómez-Plaza, 2013). The pruning presumably imparts shock/stress to the plants, thereby elevating phenolic synthesis. Furthermore, pruning can promote other desirable characteristics, such as high fruit firmness and reduced disease incidence (Asrey et al., 2013). It is well-known that environmental stress, such as drought, diverts considerable amounts of substrates from primary to secondary metabolism, leading to a reduced growth rate and competitive ability of the plant, while accumulating various phenolic compounds (Dixon and Paiva, 1995). Application of water stress during different phases of crop development must be managed in such a way as not to hinder plant development. Mild stress treatments seem to be promising for increasing BC content, without negatively affecting yield (Poiroux-Gonord et al., 2010).

Generally, phenolic compounds and anthocyanins increase with reduced water supply, enabling the use of deficit irrigation strategies to improve fruit quality and water use efficiency. In grape, water stress applied during the pre-veraison period increased the concentration of phenolic compounds in wines (Intrigliolo and Castel, 2010). Water management/stress (15% of water potential) can elevate grape skin tannins and anthocyanins, and also modify/shift grape anthocyanins composition and profiles (Tavarini et al., 2011 and references therein). Other investigations on peaches and apricots have shown that deficit irrigation during fruit growth can have a positive effect on fruit quality (Crisosto et al., 1994; Torrecillas et al., 2000; Pérez-Pastor et al., 2007). In pomegranate, deficit irrigation at specific times (fruit set or growth) enhanced fruit quality attributes, such as soluble solids and anthocyanin content. Depending on the phenological period of application of water shortage, deficit irrigation can be used as a field practice to control ripening attributes (Laribi et al., 2013). However, high phenolic levels due to water stress can result in internal browning of stone fruits. Moreover, the drought-induced variations in gene expressions for polyphenols production were highly cultivar-specific. Water stress increased total hydroxycinnamic acid of peach depending on cultivar and rootstock (Tavarini et al., 2011).

It is known that physiological responses of fruit trees are influenced by rootstock, because trees on different size-controlling rootstocks result in different water relations, gas exchange and vegetative growth. Even if rootstock has a much greater effect on tree growth and yield, it can also influence fruit quality (Castle, 1995), particularly phytochemicals such as phenolics in apricots (Bartolini et al., 2014) and antioxidant activity in peach (Scalzo et al., 2005). In cherries, the regulation of fruit quality is more dependent on the cultivar (and the crop load), although the effects of rootstock genotype and cultivar × rootstock interaction were also statistically significant (Gonçalves et al., 2006). In peach, total antioxidant capacity and levels of some phytochemicals (phenols, ascorbic acid and β-carotene) are significantly influenced by rootstock, even if it is not possible to define a common behaviour in terms of rootstock vigour. Indeed, rootstocks of similar vigour produced fruits with very different nutritional characteristics, indicating that the rootstock effect is more complex than just vigour (Remorini et al., 2008).

5.1.3 Effect of nutrient management on the polyphenols content

Nutrient management is an important feature of healthy plant production for food purposes. Optimal mineral nutrition is a prerequisite for providing co-factors for many enzymes of the flavonoid pathway. Carbohydrate availability is another prerequisite for phenylpropanoid accumulation. Reduction or accumulation of various phenolic compounds in plant tissues has been reported in relation to specific nutrients. For example, nitrogen (N) supplementation was associated, in apple, with reduction of anthocyanin and flavan-3-ol levels (Awad and Jager, 2002). Conversely, excess N may inhibit anthocyanin production in green apple cultivars such as 'Granny Smith' where red colour is undesirable. Phosphorus supplementation increased the percentage of red skin (i.e. anthocyanin content) at harvest for different apple cultivars (Paliyath et al., 2002).

Arbuscular mycorrhizal fungi (AMF) refers to symbiosis between plants and soil-borne fungi. Several examples demonstrate enhanced synthesis of phenolic compounds in mycorrhiza-inoculated plants. Higher anthocyanin content was observed in fruits from strawberry plants inoculated with *Glomus intraradices* (Castellanos-Morales et al., 2010). AMF can therefore be used as a strategy not only to improve the growth of crops, but also to enhance the production of specific phenolic compounds.

Organic fertilization can directly result in lower N content and pesticide residues, and a corresponding increase in content of phenolic defence compounds (Veberič, 2016). This can increase plant resistance to pests and diseases, although at the cost of a lower growth rate and yield (Brandt and Mølgaard, 2001). Organic fertilization increases phenolic content in apple, marionberry (blackberry), strawberry and white seedless grape compared to conventional fertilization. White seedless grapes managed under organic fertilization exhibited the highest total phenolic, catechin, quercetin-3-galactoside, total flavonoids and antioxidant (DPPH) activity compared to those with conventional or integrated fertilization. However, physical fruit characteristics of white seedless grapes under conventional fertilization were considerably higher than those with organic or integrated fertilization (Koureh et al., 2018).

Apples in organic production contained higher concentrations of hydroxycinnamic acids, flavan-3-ols, flavonols and total phenolics, particularly in the peel, than those in integrated or improved integrated pest management (IPM) production systems (Jakopic et al., 2012). Organically grown strawberries ('Candonga' and 'Ventana' genotypes) exhibited significantly higher antioxidant activity than IPM strawberries (Blando et al., 2012). The higher phenols concentration in organically grown fruits can be attributed to severe stress (biotic or abiotic) that induces phenolic biosynthesis. However, other factors, such as ultraviolet radiation, low temperature, nutrient deficiencies and pest or pathogen attacks, may contribute to differences in concentration of fruit phenolic compounds independent of production system/management.

5.1.4 Effect of treatments with elicitors, stimulating agents and plant activators on polyphenols content

Physical elicitors (e.g. high/low temperatures, light (UV, gamma radiation)) can trigger plant defence mechanisms that may concomitantly increase polyphenolic contents of fruit, such as grapes and blueberries (Ruiz-García and Gómez-Plaza, 2013). Phenolic acid contents of cherries generally decrease during storage at 1–2°C and increase at 15 ± 5°C, while anthocyanin levels increase at both temperature ranges without affecting flavonols and flavan-3-ol contents (Valero, 2015). Total phenolics and phenolic compounds (neo-chlorogenic, hydroxycinnamic and 3-*p*-coumaroylquinic acids) increase with cherry maturity and during storage.

Chemical elicitors (chitosan, benzothiadazole [BTH], harpin and 1-methylcyclopropane) that stimulate salicylic acid (SA) and jasmonic acid (JA) signalling can activate plant defence responses and PAL, thereby elevating phenolic content in various fruits [apples, bananas, grapes, mangoes, peaches and berries (strawberries, raspberries, blackberries, bayberries)]. SA activates the phenylpropanoid pathway, while JA induces de novo synthesis of phenolic compounds; both confer resistance to plant diseases (Ruiz-García and Gómez-Plaza, 2013). Other elicitors include oxalic acid (OA), calcium chloride (enhances the defence-related enzymes-β-1-3-glucanase and PAL) and potassium silicate. SA (or its analog acetylsalicylic acid) increases total phenolic acid content in fruits including sweet cherry, pomegranate and plum (Valero, 2015).

Pre-harvest applications (foliar sprays) of SA (0.5 mM) or acetyl salicylic acid (1 mM) to sweet cherry trees elevated the total phenolics, total anthocyanins and hydrophilic total antioxidant activity of fruit at harvest (commercial ripening stage) and during post-harvest storage (Giménez et al., 2017). Furthermore, the treatments also enhanced the activity of antioxidant enzymes (catalase, peroxidase, superoxide dismutase and ascorbate peroxidase), thereby delaying the post-harvest ripening process. OA increases defence-related secondary metabolites, such as phenolics, and effectively delays fruit ripening in plums and peaches by inhibiting ethylene biosynthesis. Pre-harvest OA treatment (2 mM) of cherry trees increased total phenolics and anthocyanin concentrations considerably during the last 2 weeks of on-tree fruit ripening. Similar increases in total phenolics have been reported as a result of post-harvest treatment during cold storage of sweet cherry. Post-harvest treatment (1 mM SA, ASA or OA) before cherry fruit storage can delay the post-harvest ripening process and, additionally, maintain higher levels of antioxidants (phenolics and anthocyanins) (Valero, 2015). Pre-harvest BHT treatment increased anthocyanin and other flavonoid biosynthesis (resveratrol) in grape (Iriti et al., 2004).

Plant hormones, such as abscisic acid (ABA), seem to affect phenylpropanoid biosynthesis. ABA applications on grape and cherry increased the anthocyanin content of the skin (Kondo and Inoue, 1997). The role of ethylene in anthocyanin biosynthesis has not been fully elucidated: an ethylene-related increase of PAL activity was described in the first studies around 1970–80; recently ethylene-independent anthocyanin biosynthesis has been proposed in apples and cherries (Treutter, 2010 and references therein).

Substances generally known as 'biostimulants' have recently been considered to activate secondary metabolite biosynthesis in plants. A biostimulant is any substance and/or microorganism that influences plant physiology and provides a health benefit for growth, development, response to abiotic stress and improvement of quality (Calvo et al., 2014; Du Jardin, 2015). For example, a commercial new chitosan formulation not only protected grapevines from powdery mildew (*Uncinula necator* Schwein.), but also improved total phenols and antioxidant activity in both fruits and wine (Iriti et al., 2011). A commercial oak aqueous extract has been used on grapes at veraison, affecting fruit composition and resulting in higher polyphenols content of the wine, which was less alcoholic and acidic with high colour stability (Pardo-García et al., 2014). Oak aqueous extracts foliar application to the grapevine

(Martínez-Gil et al., 2011) and more recently vine-shoot extracts (Sánchez-Gómez et al., 2017) have been used as 'biostimulants' to improve grape yield and wine quality, particularly in aromatic cultivars.

5.2 Carotenoids: effects of light, sugars and plant probiotic bacteria on carotenoid accumulation

Light and sugars are involved in fruit ripening, and can affect carotenoid accumulation in fruits such as citrus, strawberry and grape. In citrus species, light exposure during fruit ripening generally enhances carotenoid accumulation and peel colour (Cronje et al., 2011, 2013). In contrast, Lado et al. (2015) observed that red grapefruit developed more intense red colouration when grown under low light conditions. Zhang et al. (2012) showed that blue, but not red, light induced total carotenoid accumulation in Satsuma mandarin, Valencia orange and Lisbon lemon. Furthermore, *PSY* expression was upregulated by blue light treatment. The results showed that *PSY* and *LCYb* gene expression was increased, respectively, by sucrose and mannitol treatment, contributing to increased carotenoid content. Carotenoid metabolism during bilberry fruit development under different light conditions showed a metabolic flux towards formation of β-branch of carotenoid compounds, along with increased enzymatic carotenoid cleavage and degradation (Karppinen et al., 2016).

Recently, Rahman et al. (2018) evaluated the effects of plant probiotic bacterial strains BChi1 and BRRh-4 on various types of bioactive compounds such as carotenoids in fresh strawberry fruits. Application of plant probiotic bacteria increased total carotenoid content compared to the non-treated control. The highest carotenoid content, expressed as mg of lutein equivalent, was recorded in plants treated with BRRh-4 (7.71 mg lutein/g FW), followed by treatment with BChi1 (6.46 mg lutein/g FW) and the control plants (2.82 mg lutein/g FW).

5.3 Vitamin C: environmental variation and agronomic conditions

Plants can be considered as 'living bioreactors' for vitamin C production. Moreover, the baseline production can be increased as a response to stress conditions (Locato et al., 2013). During the growing season, light intensity quantitatively influences AsA accumulation; low light intensity exposure results in low-AsA content in plant tissues (Lee and Kader, 2000). Therefore, fruit on the same plant exposed to maximum sunlight contains higher vitamin C levels than those in shade or the interior of the canopy. This may account for the beneficial effects of tree pruning and thinning on vitamin C content of fruits. Other agronomic practices that enhance light interception, such as highly reflective mulches, significantly increase total vitamin C levels of strawberries (Atkinson et al., 2006); for example, white (Extenday) reflective mulch increased fruit AsA concentration in 'Elsanta' by 11% and 14% over hessian and black polythene, respectively. Branch girdling, another extensively used practice to promote flowering and improve fruit set in citrus production, significantly increased AsA content in Satsuma mandarin fruit peel or pulp (Yang et al., 2013). This elevated AsA accumulation is presumably due to sugar or phytohormone modulation that regulates AsA biosynthesis-related gene expression in fruit tissues.

Temperature also influences AsA content: mandarins grown under cool temperatures (20–22°C day, 11–13°C night) contain higher vitamin C than at hot temperatures (30–35°C day, 20–25°C night). Similarly, grapefruits grown in coastal areas of California generally contain more vitamin C than those grown in the desert (Lee and Kader, 2000). AsA content generally decreases gradually as storage temperature or duration increases; in apples, AsA content can drop to less than 50% of the original amount during cold storage after 6 months (Lee and Kader, 2000 and references therein). In most fruits, AsA content is relatively stable but decreases rapidly during the first day of storage, for example apricot, banana, melon, cherry and citrus fruits (Kevers et al., 2007). Cold storage (6°C, 65 days) positively influenced vitamin C content in blond orange varieties, but reduced its content in blood oranges (Rapisarda et al., 2008). Longer cold storage (1°C, 3 months) reduced total AsA content (35%) of apples, but it increased slightly in some cultivars, suggesting an acclimation response to low temperature (and other stress) (Davey and Keulemans, 2004). Strawberries lost 20–30% of total AsA during storage (1°C, 8 days) (Lee and Kader, 2000 and references therein). AsA loss occurs during cold storage in chilling-sensitive fruits such as pomegranate; however, AsA increases significantly (11.5%) following OA (6 mM) application to cold stored (2°C, 84 days) pomegranates (Sayyari et al., 2010).

Environmentally sustainable agricultural practices, such as organic agro-ecosystem production, elevates vitamin C (AsA + DHA) and AsA content (9.7%; 0.621 vs. 0.566 mg/g FW; $p = 0.009$), and AsA/DHA ratio (>3-fold) compared to conventionally grown strawberries in California (Reganold et al., 2010). This favourable antioxidant status of the organically grown strawberries increases its anti-proliferative effects on breast (MCF-7) and colon (HT-29) cancer cells compared to conventionally grown berries, with AsA concentrations inversely correlated with cancer cell proliferation (Olsson et al., 2006). Other pre-harvest practices aimed at delaying fruit ripening, such as foliar K treatments, also increase AsA content (21%) in muskmelons grown in high K soils (Jifon and Lester, 2009). The positive response of AsA to foliar K applications is presumably due to improved metabolic processes such as leaf photosynthesis, sugar production, photoassimilate transport from leaves to fruits and substrate availability for AsA biosynthesis. Calcium treatments (2% and 4% $CaCl_2$) increased AsA content (up to 29% and 68%, respectively) in apples as well as increased fruit firmness (Lee and Kader, 2000 and references therein) and in pomegranate (Erkan and Kader, 2011). Post-harvest application of SA to delay ripening and maintain fruit quality attributes generally increases antioxidant enzymes during storage. The changes in AsA content in response to SA treatment differ depending on the fruit; its level is maintained in pomegranate exposed to SA (Sayyari et al., 2010). Similarly, JA application elevated AsA levels (140%) in Japanese plum (Khan and Singh, 2007).

6 Future trends and conclusion

It is important to underline the economic impact of managing/elevating the content and profile of BCs in fruit crops and their related health benefits. Consumers' choices and market trends can follow different paths. Fruit colour often is a consumer selection criterion, for example selection of red apple strains in spite of a higher price, even if red colour is not always associated with high fruit quality (Treutter, 2010).

Research progress, through genetics and agronomic practices, can influence and enhance the content of BCs in fruits. 'Classical' genetic breeding is time-consuming and the release of new cultivars with desired BC characteristics takes years. Genetic engineering and modification of genetic traits (GMO) has not been successful, at least in Europe. Someday, 'genome editing' and subsequent '*cis*-genesis' might be more readily accepted, both by consumers and governmental agencies.

Plant physiological research can improve agronomic strategies to increase BC contents. Studies on the adaptation and response of physiological processes are fundamental in designing crop production techniques that can yield reliable and realistic increases in BC contents in fruits. From basic research, new mathematical models fitting the interaction between environment × physiological processes for a specific organ (fruit) and genome (cv) need to be developed. New production technologies, such as biostimulants tailored to a specific target, are expected to emerge, as well as new post-harvest and storage solutions which enable preservation and/or improvement of fruit quality in terms of BC contents.

From a grower's point of view, it is important to assess the feasibility of the agronomic approach (at both the technical and economic levels). In fact, the increase of BC contents, for example by means of reduced N application or moderate drought stress, can incur yield reductions. The balance between the higher price paid for a higher quality fruit and a lower yield needs to be always positive.

7 Where to look for further information

In this section readers can find references helping them to explore further on bioactive compounds in fruits and their sustainable elicitation.

Books:

- Mazza, G. and Miniati, E. (Eds) (1993), *Anthocyanins in Fruits, Vegetables and Grains*, Boca Raton, FL, USA: CRC Press.
- Yahia, E. M. (Ed.) (2017), *Fruit and Vegetable Phytochemicals: Chemistry and Human Health*, 2nd Edition, John Wiley & Sons, Ltd. doi:10.1002/9781119158042.

Reviews:

- Gonzalez-Aguilar, G., Robles-Sánchez, R., Martínez-Téllez, M., Olivas, G., Alvarez-Parrilla, E. and De La Rosa, L. (2008), 'Bioactive compounds in fruits: Health benefits and effect of storage conditions', *Stewart Postharvest Review*, 4(3), 1–10. doi:10.2212/spr.2008.3.8.
- Gniech Karasawa, M. M. and Mohan, C. (2018), 'Fruits as prospective reserves of bioactive compounds: A review', *Natural Products and Bioprospecting*, 8, 335–46. doi:10.1007/s13659-018-0186-6.

General information on bioactive compounds and functional foods can be found in different useful websites, referring to research organizations:

- http://www.choosecherries.com/cherry-marketing-institute/
- https://www.functionalfoodscenter.net/
- http://www.inaf.ulaval.ca/

Below are some references to universities or research centres where international research projects on bioactive compounds are carried out:

- Michigan State University, Department of Horticulture, East Lansing, MI, USA
- Université Laval, Quebec, Canada; University of British Columbia, Faculty of Land and Food Systems, Vancouver, Canada
- University of Helsinki, Department of Food and Nutrition, Helsinki, Finland; Natural Resources Institute Finland (LUKE), Helsinki, Finland

These are only few examples; it must be emphasized that many universities, all over the world, deal with programmes on nutraceuticals.

8 References

Agius, F., González-Lamothe, R., Caballero, J. L., Muñoz-Blanco, J., Botella, M. A. and Valpuesta, V. (2003), 'Engineering increased vitamin C levels in plants by overexpression of a D-galacturonic acid reductase', *Nat. Biotechnol.*, 21, 177–81. doi:10.1038/nbt777.

Alquézar, B., Rodrigo, M. J., Lado, J. and Zacarías, L. (2013), 'A comparative physiological and transcriptional study of carotenoid biosynthesis in white and red grapefruit (*Citrus paradisi* Macf.)', *Tree Genet. Genom.*, 9, 1257–69.

Ampomah-Dwamena, C., Dejnoprat, S., Lewis, D., Sutherland, P., Volz, R. K. and Allan, A. C. (2012), 'Metabolic and gene expression analysis of apple (*Malus × domestica*) carotenogenesis', *J. Exp. Bot.*, 3, 4497–511. doi:10.1093/jxb/ers134.

Andersen, Ø. M. and Jordheim, M. (2006), 'The anthocyanins', *in* Andersen, Ø. M. and Markham, K. R. (Eds), *Flavonoids: Chemistry, Biochemistry and Applications*, Boca Raton, FL: CRC Press, pp. 471–551.

Asrey, R., Patel, V. B., Barman, K. and Pal, R. K. (2013), 'Pruning affects fruit yield and postharvest quality in mango (*Mangifera indica* L.) cv. Amrapali', *Fruits*, 68, 367–80.

Assumpção, C. F., Hermes, V. S., Pagno, C., Castagna, A., Mannucci, A., Sgherri, C., Pinzino, C., Ranieri, A., Flôres, S. H. and De Oliveira Rios, A. (2018), 'Phenolic enrichment in apple skin following post-harvest fruit UV-B treatment', *Postharvest Biol. Technol.*, 138, 37–45.

Atkinson, C. J., Dodds, P. A. A., Ford, Y. Y., Le Mière, J., Taylor, J. M., Blake, P. S. and Paul, N. (2006), 'Effects of cultivar, fruit number and reflected photosynthetically active radiation on *Fragaria × ananassa* productivity and fruit ellagic acid and ascorbic acid concentrations', *Ann. Bot.*, 97, 429–41.

Awad, M. and Jager, A. (2002), 'Relationships between fruit nutrients and concentrations of flavonoids and chlorogenic acid in "Elstar" apple skin', *Sci. Hort.*, 92, 265–76.

Badejo, A. A., Tanaka, N. and Esaka, M. (2008), 'Analysis of GDP-D-mannose pyrophosphorylase gene promoter from acerola (*Malpighia glabra*) and increase in ascorbate content of transgenic tobacco expressing the acerola gene', *Plant Cell Physiol.*, 49, 126–32.

Bartolini, S., Leccese, A., Iacona, C., Andreini, L. and Viti, R. (2014), 'Influence of rootstock on fruit entity, quality and antioxidant properties of fresh apricots (cv Pisana)', *New Zeal. J. Crop Hort. Sci.*, 42(4), 265–74. doi:10.1080/01140671.2014.894919.

Blanco-Portales, R., López-Raéz, J. A., Bellido, M. L., Moyano, E., Dorado, G., González-Reyes, J. A., Caballero, J. L. and Muñoz-Blanco, J. (2004), 'A strawberry fruit-specific and ripening-related gene codes for a HyPRP protein involved in polyphenol anchoring', *Plant Mol. Biol.*, 55, 763–80.

Blando, F., Spirito, R., Gerardi, C., Durante, M. and Nicoletti, I. (2012), 'Nutraceutical properties in organic strawberries from South Italy', *Acta Hort.*, 926, 683-90.

Borowy, A., Chrzanowska, E. and Kapłan, M. (2018), 'Comparison of three sour cherry cultivars grown in central-eastern Poland', *Acta Sci. Pol. Hortoru.*, 17(1), 63–73.

Boss, P. K., Davies, C. and Robinson, S. P. (1996), 'Expression of anthocyanin biosynthesis pathway genes in red and white grapes', *Plant Mol. Biol.*, 32(3), 565–9.

Brandt, K. and Mølgaard, J. P. (2001), 'Organic agriculture: Does it enhance or reduce the nutritional value of plant foods?' *J. Sci. Food Agric.*, 81, 924–31.

Britton, G. (2008), 'Functions of intact carotenoids', *in* Britton, G., Liaaen-Jensen, S. and Pfander, H. (Eds), *Carotenoids*, Basel: Birkhäuser, pp. 189–212.

Brovelli, E. A. (2006), 'Pre- and postharvest factors affecting nutraceutical properties of horticultural products', *Stewart Postharvest Rev.*, 2(2), 1–6.

Bruno, A., Durante, M., Marrese, P. P., Migoni, D., Laus, M., Pace, E., Pastore, D., Mita, G., Piro, G. and Lenucci, M. S. (2017), 'Shades of red: Comparative study on supercritical CO_2 extraction of lycopene-rich oleoresins from gac, tomato and watermelon fruits and effect of the α-cyclodextrin clathrated extracts on cultured lung adenocarcinoma cells' viability', *J. Food Comp. Anal.*, 65, 25–32.

Bureau, S., Renard, C. M. G. C., Reich, M., Ginies, C. and Audergon, J.-M. (2009), 'Change in anthocyanin concentrations in red apricot fruits during ripening', *LWT*, 42(1), 372–7.

Calvo, P., Nelson, L. and Kloepper, J. W. (2014), 'Agricultural uses of plant biostimulants', *Plant Soil*, 383, 3–41. doi:10.1007/s11104-014-2131-8.

Cantos, E., Espín, J. C. and Tomás-Barberán, F. A. (2002), 'Varietal differences among the polyphenol profiles of seven table grape cultivars studied by LC-DAD-MS-MS', *J. Agric. Food Chem.*, 50, 5691–6.

Castellanos-Morales, V., Villegas, J., Wendelin, S., Vierheilig, H., Eder, R. and Cárdenas-Navarro, R. (2010), 'Root colonisation by the arbuscular mycorrhizal fungus *Glomus intraradices* alters the quality of strawberry fruits (*Fragaria* × *ananassa* Duch.) at different nitrogen levels', *J. Sci. Food Agric.*, 90, 1774–82.

Castle, W. S. (1995), 'Rootstock as a fruit quality factor in citrus and deciduous tree crops', *New Zeal. J. Crop Hort. Sci.*, 23(4), 383–94. doi:10.1080/01140671.1995.9513914.

Cazzonelli, C. I. and Pogson, B. J. (2010), 'Source to sink: Regulation of carotenoid biosynthesis in plants', *Trends Plant Sci.*, 15, 266–74.

Cevallos-Casals, B. A., Byrne, D., Okie, W. R. and Cisneros-Zevallos, L. (2006). 'Selecting new peach and plum genotypes rich in phenolic compounds and enhanced functional properties', *Food Chem.*, 96, 273–328. doi:10.1016/j.foodchem.2005.02.032.

Cheng, J. H., Wei, L. Z. and Wu, J. (2015), 'Effect of light quality selective plastic films on anthocyanin biosynthesis in *Vitis vinifera* L. cv. Yatomi Rosa', *J. Agr. Sci. Tech.*, 17, 157–66.

Cordenunsi, B. R., Do Nascimento, J. R. O., Genovese, M. I. and Lajolo, F. M. (2002), 'Influence of cultivar on quality parameters and chemical composition of strawberry fruits grown in Brazil', *J. Agric. Food Chem.*, 50, 2581–6.

Corral-Aguayo, R. D., Yahia, E. M., Carrillo-Lopez, A. and González-Aguilar, G. (2008), 'Correlation between some nutritional components and the total antioxidant capacity measured with six different assays in eight horticultural crops', *J. Agric. Food Chem.*, 56(22), 10498–504.

Crisosto, C. H., Johnson, R. S., Luza, J. G. and Crisosto, G. M. (1994), 'Irrigation regimes affect fruit soluble solids concentration and rate of water loss of "O'Henry" peaches', *HortSci.*, 29, 1169–71.

Cronje, P. J. R., Barry, G. H. and Huysamer, M. (2011), 'Postharvest rind breakdown of "Nules Clementine" mandarin is influenced by ethylene application, storage temperature and storage duration', *Postharvest Biol. Technol.*, 60, 192–201.

Cronje, P. J. R., Barry, G. H. and Huysamer, M. (2013), 'Canopy position affects pigment expression and accumulation of flavedo carbohydrates of "Nules Clementine" mandarin fruit, thereby affecting rind condition', *J. Am. Soc. Hortic. Sci.*, 138, 217–24.

Davey, M. W. and Keulemans, J. (2004), 'Determining the potential to breed for enhanced antioxidant status in *Malus*: Mean inter- and intravarietal fruit vitamin C and glutathione contents at harvest and their evolution during storage', *J. Agric. Food Chem.*, 52(26), 8031–8.

Davies, M. B., Austin, J. and Partridge, D. A. (1991), *Vitamin C: Its Chemistry and Biochemistry*, Royal Society of Chemistry, Cambridge, UK, p. 150.

De Pascual-Teresa, S. and Sanchez-Ballesta, M. T. (2008), 'Anthocyanins: From plant to health', *Phytochem. Rev.*, *7*, 281-99.

De Souza, V. R., Pereira, P. A. P., Silva, T. L. T., Lima, L. C. O. L., Pio, R. and Queiroz, F. (2014), 'Determination of the bioactive compounds, antioxidant activity and chemical composition of Brazilian blackberry, red raspberry, strawberry, blueberry and sweet cherry fruits', *Food Chem.*, 156, 362-8.

Delgado-Pelayo, R., Gallardo-Guerrero, L. and Hornero-Méndez, D. (2014), 'Chlorophyll and carotenoid pigments in the peel and flesh of commercial apple fruits varieties', *Food Res. Int.*, 65, 272-81.

Delva, L. (2012), 'Acerola (*Malphighia emarginata* DC): Phenolic profiling, antioxidant capacity, antimicrobial property, toxicological screening, and color stability', *PhD Thesis*, University of Florida, 174p.

Demir, T. (2013), 'Determination of carotenoid, organic acid and sugar content in some sweet cherry cultivars grown in Sakarya, Turkey', *J. Food Agric. Environ.*, 11, 73-5.

Di Matteo, A. D., Russo, R., Graziani, G., Ritieni, A. and Di Vaio, C. (2016), 'Characterization of autochthonous sweet cherry cultivars (*Prunus avium* L.) of southern Italy for fruit quality, bioactive compounds and antioxidant activity', *J. Sci. Food Agric.*, 97, 2782-94.

Dias, M. G., Camões, M. F. G. F. C. and Oliveira, L. (2009), 'Carotenoids in traditional Portuguese fruits and vegetables', *Food Chem.*, 113, 808-15.

Dixon, R. and Paiva, N. (1995), 'Stress-induced phenylpropanoid metabolism', *Plant Cell*, 7(7), 1085-97. doi:10.1105/tpc.7.7.1085.

Dorais, M. and Ehret, D. L. (2008), 'Agronomy and the nutritional quality of fruit', *in* Tomás-Barbéran, F. and Gil, I. M. (Eds), *Improving the Health- Promoting Properties of Fruit and Vegetable Products*, Cambridge, UK: Woodhead Publishing Limited, pp. 346-91.

Doumett, S., Fibbi, D., Cincinelli, A., Giordani, E., Nin, S. and Del Bubba, M. (2011), 'Comparison of nutritional and nutraceutical properties in cultivated fruits of *Fragaria vesca* L. produced in Italy', *Food Res. Intern.*, 44, 1209-16.

Du Jardin, P. (2015), 'Plant biostimulants: Definition, concept, main categories and regulation', *Sci. Hort.*, 196, 3-14. doi:10.1016/j.scienta.2015.09.021.

Dussi, M. C., Sugar, D. and Worlstad, R. E. (1995), 'Characterizing and quantifying anthocyanins in red pears and the effect of light quality on fruit color', *J. Am. Soc. Hort. Sci.*, 120, 785-9.

Erkan, M. and Kader, A. A. (2011), 'Pomegranate (*Punica granatum* L.)', *in* Yahia, E. M. (Ed.), *Postharvest Biology and Technology of Tropical and Subtropical Fruits: Mangosteen to White Sapote*, Cambridge, UK: Woodhead Publishing, pp. 287-311.

Esplay, R. V., Hellens, R. P., Putterill, J., Stevenson, D. E., Kutty-Amma, S. and Allan, A. C. (2007), 'Red colouration in apple fruit is due to the activity of the *MYB* transcription factor, *MdMYB10*', *Plant J.*, 49, 414-27. doi:10.1111/j.1365-313X.2006.02964.x.

Fanciullino, A. L., Cercós, M., Dhique-Mayer, C., Froelicher, Y., Talón, M., Ollitrault, P. and Morillon R. (2008), 'Changes in carotenoid content and biosynthetic gene expression in juice sacs of four orange varieties (*Citrus sinensis*) differing in flesh fruit color', *J. Agric. Food Chem.*, 56, 3628-38.

Feng, S., Wang, Y., Yang, S., Xu, Y. and Chen, X. (2010), 'Anthocyanin biosynthesis in pears is regulated by a *R2R3-MYB* transcription factor *PyMYB10*', *Planta*, 232, 245-55.

Gao, L. and Mazza, G. (1995), 'Characterization, quantitation, and distribution of anthocyanins and colorless phenolics in sweet cherries', *J. Agric. Food Chem.*, 43, 343–6.

Gest, N., Garchery, C., Gautier, H. Jiménez, A. and Stevens, R. (2013), 'Light-dependent regulation of ascorbate in tomato by a monodehydroascorbate reductase localized in peroxisomes and the cytosol', *Plant Biotechnol. J.*, 11, 344–54.

Gill, M. I., Tomás-Barberán, F. A., Hess-Pierce, B. and Kader A. A. (2002), 'Antioxidant capacities, phenolic compounds, carotenoids, and vitamin C contents of nectarine, peach, and plum cultivars from California', *J. Agric. Food Chem.*, 50, 4976–82.

Giménez, M. J., Serrano, M., Valverde, J. M., Martínez-Romero, D., Castillo, S., Valero, D. and Guillén, F. (2017), 'Preharvest salicylic acid and acetylsalicylic acid treatments preserve quality and enhance antioxidant systems during postharvest storage of sweet cherry cultivars', *J. Sci. Food Agric.*, 97, 1220–8.

Goggin, F. L., Avila, C. A. and Lorence, A. (2010), 'Vitamin C content in plants is modified by insects and influences susceptibility to herbivory', *Bioessays*, 32(9), 777–90.

Gonçalves, B., Moutinho-Pereira, J., Santos, A., Silva, A. P., Bacelar, E., Correia, C. and Rosa, E. (2006), 'Scion-rootstock interaction affects the physiology and fruit quality of sweet cherry', *Tree Physiol.*, 26(1), 93–104.

Gross, J. (Ed.) (1991), *Pigments in Vegetables: Chlorophylls and Carotenoids*, New York, NY: Van Nostrand, pp. 3–74

Hakkinen, S., Heinonen, M., Karenlampi, S., Mykkanen, H., Ruuskanen, J. and Torronen, R. (1999), 'Screening of selected flavonoids and phenolic acids in 19 berries', *Food Res. Int.*, 32, 345–53.

Halliwell, B. (1994), 'Free radicals, antioxidants, and human disease: Curiosity, cause, or consequence?' *The Lancet*, 344, 721–4.

Harborne, J. B. (1980), 'Plant phenolics', *in* Bell, E. A. and Charlwood, B. V. (Eds), *Encyclopedia of Plant Physiology, volume 8, Secondary Plant Products*, Berlin, Germany: Springer, pp. 329–402.

Haskell, M. J. (2013), 'Provitamin A carotenoids as a dietary source of vitamin A', *in* Tanumihardjo, S. A. (Ed.), *Carotenoids and Human Health. Nutrition and Health*, Totowa, NJ: Humana Press, pp. 249–60.

Huang, H. W., Wang, Y., Zhang, Z. H., Jiang, Z. W. and Wang, S. M. (2004), '*Actinidia* germplasm resources and kiwifruit industry in China', *HortSci.*, 39(6), 1165–72.

Iglesias, I., Salvia, J., Torguet, L. and Montserrat, R. (2005), 'The evaporative cooling effects of overtree microsprinkler irrigation on "Mondial Gala" apples', *Sci. Hort.*, 103, 267–87.

Intrigliolo, D. S. and Castel, J. R. (2010), 'Response of grapevine cv. "Tempranillo" to timing and amount of irrigation: Water relations, vine growth, yield and berry and wine composition', *Irrig. Sci.*, 28, 113–25. doi:10.1007/s00271-009-0164-1.

Iriti, M., Rossoni, M., Borgo, M. and Faoro, F. (2004), 'Benzothiadiazole enhances resveratrol and anthocyanin biosynthesis in grapevine, meanwhile improving resistance to *Botrytis cinerea*', *J. Agric. Food Chem.*, 52(14), 4406–13.

Iriti, M., Vitalini, S., Di Tommaso, G., D'Amico, S., Borgo, M. and Faoro, F. (2011), 'New chitosan formulation prevents grapevine powdery mildew infection and improves polyphenol content and free radical scavenging activity of grape and wine', *Aust. J. Grape Wine R.*, 17(2), 263–9. doi:10.1111/j.1755-0238.2011.00149.x.

Jaakola, L. (2013), 'New insight into the regulation of anthocyanin biosynthesis in fruits', *Trends Plant Sci.*, 18(9), 477–83.

Jaganath, I. B. and Zainal, A. (2016), 'Controlled environment for enhanced and consistent production of (poly)phenols in specialty crops', *in* Asaduzzaman, Md. (Ed.), *Controlled Environment Agriculture - Production of Specialty Crops Providing Human Health Benefits through Hydroponics*, New York, NY: Nova Science Publishers, pp. 1–32.

Jakopic, J., Slatnar, A., Štampar, F., Veberič R. and Simoncic, A. (2012), 'Analysis of selected primary metabolites and phenolic profile of "Golden Delicious" apples from four production systems', *Fruits*, 67 (5), 377–86.

Jia, D., Fan, L., Liu, G., Shen, J., Liu, C. and Yuan, T. (2011), 'Effects of genotypes and bagging practice on content of β-carotene in apple fruits', *J. Agric. Sci.*, 3, 196–202.

Jifon, J. L. and Lester, G. E. (2009), 'Foliar potassium fertilization improves fruit quality of field-grown muskmelon on calcareous soils in south Texas', *J. Sci. Food Agric.*, 89, 2452–60.

Josuttis, M., Carlen, C., Crespo, P., Nestby, R., Toldam-Andersen, T. B., Dietrich, H. and Krüger, E. (2012), 'A comparison of bioactive compounds of strawberry fruit from Europe affected by genotype and latitude', *J. Berry Res.*, 2, 73–95. doi:10.3233/JBR-2012-029.

Kallio, H., Yang, B. and Peippo, P. (2002), 'Effects of different origins and harvesting time on vitamin C, tocopherols, and tocotrienols in sea buckthorn (*Hippophaë rhamnoides*) berries', *J. Agric. Food Chem.*, 50, 6136–42.

Kanellis, A. and Manganaris, G. A. (2014), 'Antioxidants and bioactive compounds in fruits', *in* Nath, P., Bouzayen, M., Matoo, A. K. and Pech J. C. (Eds), *Fruit Ripening: Physiology, Signalling and Genomics*, Wallingford, UK: CAB International, pp. 99–126.

Kato, M. (2012), 'Mechanism of carotenoid accumulation in citrus fruit', *J. Jpn. Soc. Hort. Sci.*, 81, 219–33.

Kato, M., Ikoma, Y., Matsumoto, H., Sugiura, M., Hyodo, H. and Yano, M. (2004), 'Accumulation of carotenoids and expression of carotenoid biosynthetic genes during maturation in citrus fruit', *Plant Physiol.*, 134, 824–37.

Kevers, C., Falkowski, M., Tabart, J., Defraigne, J. O., Dommes, J. and Pincemail, J. (2007), 'Evolution of antioxidant capacity during storage of selected fruits and vegetables', *J. Agric. Food Chem.*, 55(21), 8596–603.

Khan, A. S. and Singh, Z. (2007), 'Methyl jasmonate promotes fruit ripening and improves fruit quality in Japanese plum', *J. Hort. Sci. Biotechnol.*, 82, 695–706.

Karppinen, K., Zoratti, L., Sarala, M., Carvalho, E., Hirsimäki, J, Mentula, H., Martens, S., Häggman, H. and Jaakola, L. (2016), 'Carotenoid metabolism during bilberry (*Vaccinium myrtillus* L.) fruit development under different light conditions is regulated by biosynthesis and degradation', *BMC Plant Biol.*, 16, 95. doi:10.1186/s12870-016-0785-5.

Kim, S.-H. (2003), 'Molecular cloning and analysis of anthocyanin biosynthesis genes preferentially expressed in apple skin', *Plant Sci.*, 165, 403–13.

Kim, I. J., Ko, K. C., Kim, C. S. and Chung, W. I. (2001), 'Isolation and characterization of cDNAs encoding β-carotene hydroxylase in *Citrus*', *Plant Sci.*, 161, 1005–10.

Kondo, S. and Inoue, K. (1997), 'Abscisic acid (ABA) and 1-aminocyclopropane-l-carboxylic acid (ACC) content during growth of "Satohnishiki" cherry fruit, and the effect of ABA and ethephon application on fruit quality', *J. Hortic. Sci.*, 72(2), 221–7. doi:10.1080/14620316.1997.11515509.

Kong, J.-M., Chia, L.-S., Goh, N.-K., Chia, T.-F. and Brouillard, R. (2003), 'Analysis and biological activities of anthocyanins', *Phytochem.*, 64(5), 923–33. doi:10.1016/S0031-9422(03)00438-2.

Koureh, O. K., Bakhshi, D., Pourghayoumi, M. and Majidian, M. (2018), 'Comparison of yield, fruit quality, antioxidant activity, and some phenolic compounds of white seedless grape obtained from organic, conventional, and integrated fertilization', *Int. J. Fruit Sci.*, 18, 1–12.

Kschonsek, J., Wolfram, T., Stöckl, A. and Böhm, V. (2018), 'Polyphenolic compounds analysis of old and new apple cultivars and contribution of polyphenolic profile to the *in vitro* antioxidant capacity', *Antioxidants*, 7, 20. doi:10.3390/antiox7010020.

Łata, B. and Wińska-Krysiak, M. (2010), 'Cultivar and seasonal variation in bioactive compounds of highbush blueberry fruits (*Vaccinium corymbosum* L.)', *Folia Horticul.*, 22(1), 31–5.

Lado, J., Zacarías, L., Gurrea, A., Page, A., Stead, A. and Rodrigo, M. J. (2015), 'Exploring the diversity in *Citrus* fruit coloration to decipher the relationship between plastid ultrastructure and carotenoid composition', *Planta*, 242, 645–61.

Laribi, A. I., Palou, L., Intrigliolo, D. S., Nortes, P. A., Rojas-Argudo, C., Taberner, V., Bartual, J. and Pérez-Gago, M. B. (2013), 'Effect of sustained and regulated deficit irrigation on fruit quality of pomegranate cv. "Mollar de Elche" at harvest and during cold storage', *Agric. Water Manag.*, 125, 61–70.

Lee, S. K. and Kader, A. A. (2000), 'Preharvest and postharvest factors influencing vitamin C content of horticultural crops', *Postharvest Biol. Technol.*, 20(3), 207–20.

Lee, J., Dossett, M. and Finn, C. E. (2012), '*Rubus* fruit phenolic research: The good, the bad, and the confusing', *Food Chem.*, 130(4), 785–96.

Leong, S. Y. and Oey, I. (2012), 'Effects of processing on anthocyanins, carotenoids and vitamin C in summer fruits and vegetables', *Food Chem.*, 53, 1577–8.

Lo Piero, A. R. (2015), 'The state of the art in biosynthesis of anthocyanin and its regulation in pigmented sweet oranges (*Citrus sinensis* L. Osbeck), *J. Agric. Food Chem.*, 63, 4031–41.

Locato, V., Cimini, S. and De Gara, L. (2013), 'Strategies to increase vitamin C in plants: From plant defense perspective to food biofortification', *Front. Plant Sci*, 4, 152.

Manganaris, G. A., Goulas, V., Vicente, A. R. and Terry, L. A. (2014), 'Berry antioxidants: Small fruits providing large benefits', *J. Sci. Food Agric.*, 94(5), 825–33.

Manganaris, G. A., Goulas, V., Mellidou, I. and Drogoudi, P. (2018), 'Antioxidant phytochemicals in fresh produce: Exploitation of genotype variation and advancements in analytical protocols', *Front. Chem.*, 5, 95. doi:10.3389/fchem.2017.00095.

Martínez-Gil, A. M., Garde-Cerdán, T., Martínez, L., Alonso, G. L. and Salinas, M. R. (2011), 'Effect of oak extract application to Verdejo grapevines on grape and wine aroma', *J. Agric. Food Chem.*, 59, 3253–63.

Marty, I., Bureau, S., Sarkissian, G., Gouble, B., Audergon, J. M. and Albagnac, G. (2005), 'Ethylene regulation of carotenoid accumulation and carotenogenic gene expression in colour-contrasted apricot varieties (*Prunus armeniaca*)', *J. Exp. Bot.*, 56, 1877–86.

Mas, T., Susperregui, J., Berké, B, Chèze, C., Moreau, S., Nurich, A. and Vercauteren, J. (2000), 'DNA triplex stabilization property of natural anthocyanins', *Phytochem.*, 53, 679–87.

Mazza, G. and Miniati, E. (1993), *Anthocyanins in Fruits, Vegetables and Grains*, Boca Raton, FL, USA: CRC Press.

Mazza, G. and Velioglu, Y. S. (1992), 'Anthocyanins and other phenolic compounds in fruits of red-flesh apples', *Food Chem.*, 43(2), 113–17. doi:10.1016/0308-8146(92)90223-O.

Mellidou, I. and Kanellis, A. K. (2017), 'Genetic control of ascorbic acid biosynthesis and recycling in horticultural crops', *Front. Chem.*, 5, 50. doi:10.3389/fchem.2017.00050.

Meyers, K. J., Mares, J. A., Igo, R. P., Truitt, B., Liu, Z., Millen, A. E., Klein, M., Johnson, E. J., Engelman, C. D., Karki, C. K., Blodi, B., Gehrs, K., Tinker, L., Wallace, R., Robinson, J., LeBlanc, E. S., Sarto, G., Bernstein, P. S., SanGiovanni, J. P. and Iyengar, S. K. (2014), 'Genetic evidence for role of carotenoids in age-related macular degeneration in the carotenoids in age-related eye disease study (CAREDS)', *Invest. Ophthalmol. Vis. Sci.*, 55(1), 587–99.

Milani, A., Basirnejad, M., Shahbazi, S. and Bolhassani, A. (2017), 'Carotenoids: Biochemistry, pharmacology and treatment', *Brit. J. Pharmacol.*, 174(11), 1290–324.

Miller, V., Mente, A., Dehghan, M., Rangarajan, S., Zhang, X., Swaminathan, S., Dagenais, G., Gupta, R., Mohan, V., Lear, S., Bangdiwala, S. I., Schutte, A. E., Wentzel-Viljoen, E., Avezum, A., Altuntas, Y., Yusoff, K., Ismail, N., Peer, N., Chifamba, J., Diaz, R., Rahman, O., Mohammadifard, N., Lana, F., Zatonska, K., Wielgosz, A., Yusufali, A., Iqbal, R., Lopez-Jaramillo, P., Khatib, R., Rosengren, A., Kutty, V. R., Li, W., Liu, J., Liu, X., Yin, L., Teo, K., Anand, S., Yusuf, S., Prospective Urban Rural Epidemiology (PURE) study investigators (2017), 'Fruit, vegetable, and legume intake, and cardiovascular disease and deaths in 18 countries (PURE): A prospective cohort study', *The Lancet*, 390(10107), 2037–49. doi:10.1016/S0140-6736(17)32253-5.

Mori, K., Sugaya, S. and Gemma, H. (2005), 'Decreased anthocyanin biosynthesis in grape berries grown under elevated night temperature condition', *Sci. Hort.*, 105, 319–30. doi:10.1016/j.scienta.2005.01.032.

Naik, P. S., Chanemougasoundharam, A., Paul Khurana, S. M. and Kalloo, G. (2003), 'Genetic manipulation of carotenoid pathway in higher plants', *Curr. Sci.*, 85, 1423–30.

Namitha, K. K. and Negi, P. S. (2010), 'Chemistry and biotechnology of carotenoids', *Crit. Rev. Food Sci.*, 50, 728–60.

Nollet, L. M. L. and Gutierrez-Uribe, J. A. (2018), *Phenolic Compounds in Food Characterization and Analysis*, Boca Raton, FL: CRC Press, p. 460.

Olas, B. (2018), 'Berry phenolic antioxidants – implications for human health?' *Front. Pharmacol.*, 9, 78.

Olsson, M. E., Andersson, C. S., Oredsson, S., Berglund, R. H. and Gustavsson, K-E. (2006), 'Antioxidant levels and inhibition of cancer cell proliferation in vitro by extracts from organically and conventionally cultivated strawberries', *J. Agric. Food Chem.*, 54(4), 1248–55.

Oyebode, O., Gordon-Dseagu, V., Walker, A. and Mindell, J. S. (2013), 'Fruit and vegetable consumption and all-cause, cancer and CVD mortality: Analysis of Health Survey for England data', *J. Epidemiol. Commun. Health*, 68, 856–62. doi:10.1136/jech-2013-203500.

Padayatty, S. J. and Levine, M. (2016), 'Vitamin C physiology: The known and the unknown and Goldilocks', *Oral Dis.* 22(6), 463–93.

Paliyath, G., Schofield, A., Oke, M. and Taehyun, A. (2002), 'Phosphorus fertilization and biosynthesis of functional food ingredients', *in* T. W. Bruulsema (Ed.), *Fertilizing Crops for Functional Foods*. Symposium Proceedings, 11 November 2002, Indianapolis, Indiana, USA. Potash and Phosphate Institute/Potash and Phosphate Institute of Canada, Section 5, pp. 5–6.

Pardo-García, A. I., Martínez-Gil, A. M., Cadahía, E., Pardo, F., Alonso, G. L. and Salinas, M. R. (2014), 'Oak extract application to grapevine as a plant biostimulant to increase wine polyphenols', *Food Res. Int.*, 55, 150–60. doi:10.1016/j.foodres.2013.11.004.

Parr, A. J. and Bowell, G. P. (2000), 'Phenols in the plant and in man: The potential for possible nutritional enhancement of the diet by modifying the phenols content or profile', *J. Sci. Food Agric.*, 80, 985–1012.

Patil, B. S., Crosby, K., Byrne, D. and Hirschi, K. (2014), 'The intersection of plant breeding, human health, and nutritional security: Lessons learned and future perspectives', *HortSci.*, 49(2), 116–27.

Pérez-Pastor A., Ruiz-Sànchez, M. C., Martinez, J. A., Nortes, P. A., Artes, F. and Domingo, R. (2007), 'Effect of deficit irrigation on apricot fruit quality at harvest and during storage', *J. Sci. Food Agric.*, 87, 2409–15.

Poiroux-Gonord, F., Bidel, L. P. R., Fanciullino, A.-L., Gautier, H., Lauri-Lopez, F. and Urban, L. (2010), 'Health benefit of vitamins and secondary metabolites of fruits and vegetables and prospects to increase their concentrations by agronomic approaches', *J. Agric. Food Chem.*, 58, 12065–82. doi:10.1021/jf1037745.

Rahman, M. R., Sabir, A. A., Mukta J. A., Khan M. M. A., Mohi-Ud-Din M., Miah, M. G., Rahman M. and Islam T. (2018), 'Plant probiotic bacteria *Bacillus* and *Paraburkholderia* improve growth, yield and content of antioxidants in strawberry fruit', *Sci. Rep.*, 8, 2504.

Ramful, D., Tarnus, E., Aruoma, O. I., Bourdon, E. and Bahorun, T. (2011), 'Polyphenol composition, vitamin C content and antioxidant capacity of Mauritian citrus fruit pulps', *Food Res. Intern.*, 44(7), 2088–99.

Rapisarda, P., Bianco, M. L., Pannuzzo, P. and Timpanaro, N. (2008), 'Effect of cold storage on vitamin C, phenolics and antioxidant activity of five orange genotypes [*Citrus sinensis* (L.) Osbeck]', *Postharvest Biol. Technol.*, *49*(3), 348–54.

Reboul, E., Richelle, M., Perrot, E., Desmoulins-Malezet, C., Pirisi, V. and Borel, P. (2006), 'Bioaccessibility of carotenoids and vitamin E from their main dietary sources', *J. Agric. Food Chem.*, 54, 8749–55.

Reganold, J. P., Andrews, P. K., Reeve, J. R., Carpenter-Boggs, L., Schadt, C. W., Alldredge, J. R., Ross, C. F., Davies, N. M. and Zhou, J. (2010), 'Fruit and soil quality of organic and conventional strawberry agroecosystems', *PLoS ONE*, 5(9), e12346. doi:10.1371/journal.pone.0012346.

Remorini, D., Tavarini, S., Degl'Innocenti, E., Loreti, F., Massai, R. and Guidi, L. (2008), 'Effect of rootstocks and harvesting time on the nutritional quality of peel and flesh of peach fruits', *Food Chem.*, 110, 361–7. doi:10.1016/j.foodchem.2008.02.011.

Rodrigo, M. J., Marcos, J. F. and Zacarías, L. (2004), Biochemical and molecular analysis of carotenoid biosynthesis in flavedo of orange (*Citrus sinensis* L.) during fruit development and maturation', *J. Agric. Food Chem.*, 52, 6724–31.

Rodrigo, M. J., Alquézar, B., Alós E., Medina V., Carmona L., Bruno M., Al-Babili S. and Zacarías, L. (2013), 'A novel carotenoid cleavage activity involved in the biosynthesis of *Citrus* fruit-specific apocarotenoid pigments', *J. Exp. Bot.*, 64(14), 4461–78.

Rodriguez-Amaya, D. B. (2001), *A Guide to Carotenoid Analysis in Foods*, Washington DC, USA: ILSI Press, pp. 41–3.

Ruiz, D., Egea, J., Tomás-Barberán, F. A. and Gil, M. I. (2005a), 'Carotenoids from new apricot (*Prunus armeniaca* L.) varieties and their relationship with flesh and skin colour', *J. Agric. Food Chem.*, 53, 6368–74.

Ruiz, D., Egea, J., Gil, M. I. and Tomás-Barberán, F. A. (2005b), 'Characterization and quantitation of phenolic compounds in new apricot (*Prunus armeniaca* L.) varieties', *J. Agric. Food Chem.*, 53(24), 9544–52. doi:10.1021/jf051539p.

Ruiz-García, Y. and Gómez-Plaza, E. (2013), 'Elicitors: A tool for improving fruit phenolic content', *Agriculture*, 3(1), 33–52.

Russo, M., Moccia, S., Bilotto, S., Spagnuolo, C., Durante, M., Lenucci, M. S., Mita, G., Volpe, M. G., Aquino, R. P. and Russo, G. L. (2017), 'A carotenoid extract from a Southern Italian cultivar of pumpkin triggers nonprotective autophagy in malignant cell', *Oxid. Med. Cell. Longev.* doi:10.1155/2017/7468538.

Sánchez-Gómez, R., Zalacain, A., Pardo, F., Alonso G. L. and Salinas, M. R. (2017), 'Moscatel vine-shoot extracts as grapevine biostimulant to enhance wine quality', *Food Res. Int.*, 98, 40–9. doi:10.1016/j.foodres.2017.01.004.

Sarma, A. D. and Sharma, R. (1999), 'Anthocyanin-DNA copigmentation complex: Mutual protection against oxidative damage', *Phytochem.*, 52(7), 1313–18.

Sass-Kiss, A., Kiss, J., Milotay, P., Kerek, M. M. and Toth-Markus, M. (2005), 'Differences in anthocyanin and carotenoid content of fruits and vegetables', *Food Res. Int.*, 38, 1023–9.

Sayyari, M., Valero, D., Babalar, M., Kalantari, S., Zapata, P. J. and Serrano, M. (2010), 'Pre-storage oxalic acid treatment maintained visual quality, bioactive compounds, and antioxidant potential of pomegranate after long-term storage at 2°C', *J. Agric. Food Chem.*, 58(11), 6804–8.

Scalzo, J., Politi, A., Pellegrini, N., Mezzetti, B. and Battino, M. (2005), 'Plant genotype affects total antioxidant capacity and phenolic contents in fruit', *Nutrition*, 21, 207–13. doi:10.1016/j.nut.2004.03.025.

Sharoni, Y., Linnewiel-Hermoni, K., Khanin, M., Salman, H., Veprik, A., Danilenko, M. and Levy, J. (2012), 'Carotenoids and apocarotenoids in cellular signaling related to cancer: A review', *Mol. Nutr. Food Res.*, 56, 259–69.

Singh, K, P., Singh, A. and Singh, U. P. (2015), 'Phenolic acid content of some apple cultivars with varying degrees of resistance to apple scab', *Int. J. Fruit Sci.*, 15(3), 267–80.

Sivakumaran, S., Huffman, L., Sivakumaran, S. and Drummond, L. (2018), 'The nutritional composition of Zespri® SunGold kiwifruit and Zespri® Sweet Green kiwifruit', *Food Chem.*, 238, 195–202.

Skibsted, L. H. (2012), 'Carotenoids in antioxidant networks colorants or radical scavengers', *J. Agric. Food Chem.*, 60, 2409–17.

Slatnar, A., Petkovsek, M. M., Halbwirth, H., Štampar, F., Stich, K. and Veberič, R. (2010), 'Response of the phenylpropanoid pathway to *Venturia inaequalis* infection in maturing fruit of "Braeburn" apple', *J. Hort. Sci. Biotechnol.*, 85, 465–72.

Stinco, C. M., Baroni, M. V., Naranjo, R. D. D. P., Wunderlin, D. A., Heredia, F. J., Meléndez-Martínez, A. J. and Vicario, I. M. (2015), 'Hydrophilic antioxidant compounds in orange juice from different fruit cultivars: Composition and antioxidant activity evaluated by chemical and cellular based (*Saccharomyces cerevisiae*) assays', *J. Food Comp. Anal.*, 37, 1–10.

Stephensen, C. B. (2013), 'Provitamin: A carotenoids and immune function', *in* S. A. Tanumihardjo (Ed.), *Carotenoids and Human Health, Nutrition and Health*. Humana Press, Totowa, NJ, pp. 261–70.

Steyn, W. J., Wand, S. J. E., Holcroft, D. M. and Jacobs, G. (2002), 'Anthocyanins in vegetative tissues: A proposed unified function in photoprotective', *New Phytol.*, 155, 349–61.

Steyn, W. J., Wand, S. J. E., Holcroft, D. M. and Jacobs, G. (2005), 'Red colour development and loss in pears', *Acta Hort.*, 671, 79–85.

Steyn, W., Wand, S. J. E., Jacobs, G., Rosecrance, R. C. and Roberts, S. C. (2009), 'Evidence for a photoprotective function of low-temperature-induced anthocyanin accumulation in apple and pear peel', *Physiol. Plant.*, 136, 461–72. doi:10.1111/j.1399-3054.2009.01246.x.

Sun, J., Chu, Y. F., Wu, X. and Liu, R. H. (2002), 'Antioxidant and antiproliferative activities of common fruits', *J. Agric. Food Chem.*, 50(25), 7449–54.

Suza, W. P., Avila, C. A., Carruthers, K., Kulkarni, S., Goggin, F. L. and Lorence, A. (2010), 'Exploring the impact of wounding and jasmonate on ascorbate metabolism', *Plant Physiol. Biochem.*, 48(5), 337–50.

Tavarini, S., Gil, M. I., Tomás-Barberán, F. A., Buendia, B., Remorini, D., Massai, R., Degl'Innocenti, E. and Guidi, L. (2011), 'Effects of water stress and rootstocks on fruit phenolic composition and physical/chemical quality in Suncrest peach', *Ann. Appl. Biol.*, 158(2), 226–33.

Teleszko, M., Wojdyło, A., Rudzińska, M., Oszmiański, J. and Golis, T. (2015), 'Analysis of lipophilic and hydrophilic bioactive compounds content in sea buckthorn (*Hippophae rhamnoides* L.) berries', *J. Agric. Food Chem.*, 63(16), 4120–9.

Tomás-Barberán F. A., Gil, M. I., Cremin, P., Waterhouse, A. L., Hess-Pierce, B. and Kader, A. A. (2001), 'HPLC-DAD-ESIMS analysis of phenolic compounds in nectarines, peaches and plums', *J. Agric. Food Chem.*, 49, 4748–60.

Torrecillas A, Domingo R., Galego R. and Ruiz-Sánchez M. C. (2000), 'Apricot tree response to withholding irrigation at different phenological periods', *Sci. Hortic.*, 85, 201–15.

Treutter, D. (2010), 'Managing phenol contents in crop plants by phytochemical farming and breeding – visions and constraints', *Int. J. Mol. Sci.*, 11, 807–57.

Tsuda, T., Shiga, K., Ohshima, K., Kawakishi, S. and Osawa, T. (1996), Inhibition of lipid peroxidation and the active oxygen radical scavenging effect of anthocyanin pigments isolated from *Phaseolus vulgaris* L., *Biochem. Pharmacol.*, 52, 1033–9.

Usenik, V., Štampar, F. and Veberič, R. (2009), 'Anthocyanins and fruit colour in plums (*Prunus domestica* L.) during ripening', *Food Chem.*, 114(2), 529-34. doi:10.1016/j.foodchem.2008.09.083.

Valero, D. (2015), 'Recent developments to maintain overall sweet cherry quality during postharvest storage', *Acta Hort.*, 1079, 83–94.

Veberič, R. (2016), 'The impact of production technology on plant phenolics', *Horticulturae*, 2, 8. doi:10.3390/horticulturae2030008.

Wang, D., Wang, X., Zhang, C., Ma, Y. and Zhao, X. (2011), 'Calf thymus DNA-binding ability study of anthocyanins from purple sweet potatoes (*Ipomea batatas* L.)', *J. Agric. Food Chem.* 50(13), 7405–9.

Yang, X-Y., Wang, F-F., Teixeira da Silva, J. A., Zhong, J., Liu, Y.-Z. and Peng, S.-A. (2013), Branch girdling at fruit green mature stage affects fruit ascorbic acid contents and expression of genes involved in L-galactose pathway in citrus. *New Zeal. J. Crop Hort. Sci.*, 41(1), 23-31.

Yoo, K. M. and Moon, B. K. (2016), 'Comparative carotenoid compositions during maturation and their antioxidative capacities of three citrus varieties', *Food Chem.*, 196, 544–9.

Young, J. I., Züchner, S. and Wang, G. (2015), 'Regulation of the epigenome by vitamin C', *Annu. Rev. Nutr.*, 35(1), 545–64.

Zainudin, M. A. M., Hamid, A. A., Anwar, F., Osman, A. and Saari, N. (2014), 'Variation of bioactive compounds and antioxidant activity of carambola (*Averrhoa carambola* L.) fruit at different ripening stages', *Sci. Hort.*, 172, 325–31.

Zhang, L., Ma, G., Kato, M., Yamawaki, K., Takagi, T., Kiriiwa, Y., Ikoma, Y., Matsumoto, H., Yoshioka, T. and Nesumi, H. (2012), 'Regulation of carotenoid accumulation and the expression of carotenoid metabolic genes in citrus juice sacs *in vitro*', *J. Exp. Bot.*, 63, 871–86.

Zhu, L., Zhang Y. and Lu, J. (2012), 'Phenolic contents and compositions in skins of red wine grape cultivars among various genetic backgrounds and originations', *Int. J. Mol. Sci.*, 13, 3492–510.

Zoratti, L., Karppinen, K., Luengo Escobar, A., Häggman, H. and Kaakola, L. (2014a), 'Light-controlled flavonoid biosynthesis in fruits', *Front. Plant Sci.*, 5, 534. doi:10.3389/fpls.2014.00534.

Zoratti, L., Sarala, M., Carvalho, E., Karppinen, K., Martens, S., Giongo, L., Häggman, H. and Kaakola, L. (2014b), 'Monochromatic light increases anthocyanin content during fruit development in bilberry', *BMC Plant Biol.*, 14, 377. doi:10.1186/s12870-014-0377-1.

Index